TECHNIQUES AND MANAGEMENT OF FIELD CROP PRODUCTION

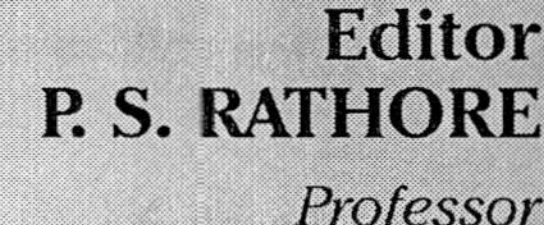

Editor

P. S. RATHORE

Professor
P. G. Department of Agronomy
College of Agriculture
Rajasthan Agricultural University
Bikaner 334 006

AGROBIOS

TITLE: Techniques and Management of Field Crop Production
EDITOR: P. S. Rathore

Published by:
AGROBIOS (INDIA)
Agro House, Behind Nasrani Cinema
Chopasani Road, Jodhpur 342 003
Phone: 91-0291-2643993
Website: www.agrobiosonline.com
E. mail: agrobiosindia@gmail.com

AGROBIOS

First Published : 2000
Reprinted:
2001-2017
Reprinted: 2025

ISBN No. (13): 978-81-7754-054-3 (HB)
ISBN No. (13): 978-81-7754-175-5 (PB)

AGROBIOS (INDIA)
Managing Director: Dr. Updesh Purohit
Founder & Editor: Dr. S. S. Purohit

Services
Book Design & Layout: Yashee Computers
Illustrator & Printers: Agrobios Digital
Cover Illustration: Reena

PRINTED IN THE INDIA

PREFACE

Agriculture is the main enterprise for more than 70 per cent of Indian population. With diverse climatic conditions, all most all crops are grown extensively throughout India. Green and yellow revolution has changed the status of Indian Agriculture from 'begging bowl' to one of the exporter. But to feed growing population (1.8 per cent) and to enhance crop productivity we have to achieve an annual growth rate of 4.5 per cent, inspite of declining use of inputs and rapidly deteriorating natural resources. Hi-tech agriculture is the only way to meet out these challenges.

Growth in Agriculture in terms of increased sustainable productivity has to come from location specific, low input, eco-friendly and demand oriented techniques. Lot of research work has been carried out in different projects in different parts of India, but there is lack of up-to-date and compiled information to be useful for under graduate and postgraduate students, research workers and for progressive farmers.

This publication has been brought out with an objective of collecting information regarding location specific research and recommended technologies for harvesting high productivity on sustainable basis. To make this publication more informative and relevant, chapters on different crops have been invited from scientist working in the field or have a vast experience. Efforts have been made to present each chapter in an useful way and readable manner for quick understanding.

I sincerely express my gratitude to all authors and co-authors of chapters for contributing up-to-date, informative and comprehensive script without which it would be very difficult to complete this task.

I take this opportunity to express gratitude to Dr. S. S. Manohar, Dean, College of Agriculture, Bikaner for constant encouragement and guidance to get this task completed. Sh. S. K. Sharma, Assistant Professor, Agronomy, College of Agriculture, Bikaner deserve special thanks for giving useful suggestions in improving quality of this publication.

Thanks are also extended to Dr. S. S. Purohit, Head, Department of Botany, Dungar College, Bikaner and Dr. Rajesh Sharma, Asstt. Prof., Department of Agriculture Economics, College of Agriculture, Bikaner for exerting encouraging attitude and for giving constructive suggestions. Mr. Updesh Purohit M/s Agrobios, Jodhpur deserves special mention for necessary assistance and publishing this manuscript with in short span.

My father Th. Anoop Singh Rathore, mother Smt. Sada Kanwar, wife Smt. Kamlesh and son Jitendra, daughter Sonu, who provided required environment at home and constant encouragement to complete this work, deserves very special appreciation.

P. S. Rathore
EDITOR

CONTENTS

Introduction **1**

CEREAL AND MILLET CROPS

1. Rice **9**
K. P. Jha, B. T. S. Moorthy and K. S. Rao

2. Maize **48**
Dayanand

3. Sorghum **71**
P. S. Rathore

4. Pearlmillet **85**
R. C. Gautam

5. Finger Millet **99**
P. S. Rathore

6. Kodo Millet **103**
P. S. Rathore

7. Wheat **106**
S. N. Sharma

8. Barley **132**
B. N. Mishra and B. G. Shivakumar

9. Oats **156**
N. P. Shukla, J. B. Singh and Shiva Dhar

OIL SEED CROPS

10. Groundnut **169**
P. K. Ghosh, Rinku Dey, K.K. Pal and R. K. Mathur

11. Sesame **196**
T. N. Balasubramanian and SP Palaniappan

12. Castor **217**
D. D. Malaviya, V. K. Poshiya and H. L. Dhaduk

13. Sunflower **236**
Gajendra Giri

14. Rapeseed and Mustard 254
P. S. Rathore
15. Linseed 269
P. S. Rathore
16. Safflower 276
P. S. Rathore

PULSE CROPS

17. Soybean 285
P. S. Rathore
18. Pigeonpea 294
D. S. Yadav
19. Greengram 311
S. K. Sharma
20. Black Gram 323
P. S. Rathore
21. Kidney Bean 330
P. S. Rathore and R. S. Rathore
22. Cowpea 336
S. K. Sharma
23. Bengal Gram 345
I. P. S. Ahlawat
24. Lentil 365
P. S. Rathore

FIBRE CROPS

25. Cotton 373
M. V. Venugopalan
26. Jute 397
V. N. Saraswat

SPICES AND MEDICINAL CROPS

27. Fennel 417
P. S. Rathore and D. S. Bhati
28. Fenugreek 422
D. S. Bhati and P. S. Rathore
29. Cumin 429
G. R. Chaudhary
30. Coriander 437
D. S. Bhati and P. S. Rathore

31. Dill Seed **444**
Avtar Singh and R. K. Mahey

32. Isabgol **450**
G. S. Randhawa and B. S. Gill

NARCOTIC CROP

33. Tobacco **459**
Aditya Kumar Singh

FODDER CROPS

34. Napier **479**
Virendra Singh

35. Anjan Grass **484**
A. S. Gill, R. L. Arya and A. Singh

36. Cluster Bean **492**
N. D. Yadava

37. Lucerne **503**
S. K. Patil

38. Berseem **510**
S. K. Sharma

SUGAR CROPS

39. Sugarcane **523**
B. Sundara

40. Sugar Beet **547**
P. S. Rathore

TUBER CROP

41. Potato **557**
S. S. Lal and K. C. Sud

APPENDIX

42. Cultivated Field Crops **581**

43. Important Herbicides **584**

44. Availability of Biofiterlizer in India **585**

45. Common Weeds **586**

LIST OF CONTRIBUTORS

1. **Ahlawat I. P. S.**
   Sr. Scientist
   Division of Agronomy
   IARI, New Dehli-110 012
2. **Balasubramanian T. N.**
   Prof. of Agronomy (Agric. Meteorology)
   Tamil Nadu Agric. Univ
   Coimbatore - 641 003.
3. **Bhati D. S.**
   Agronomist
   ARS, Mandore 342 002
4. **Choudhary G. R.**
   Agronomist.
   AICRP on Spices, SKN College of Agric.
   Jobner 303 329
5. **Dayanand**
   Pr. Investigator
   Directorate of Maize Research
   Cumming's Laboratory
   IARI, New Dehli-110 012
6. **Dey, Rinku**
   Scientist (Microbiology)
   NRC for Groundnut, Junagarh
7. **Dhaduk H. L.**
   Agric. Officer
   Deptt. of Botany, College of Agriculture
   GAU, Junagdh - 362 001.
8. **Dhar, Shiva**
   IGFRI, Jhansi
9. **Gautam R. C.**
   Professor & Head, Deptt. of Agronomy
   IARI, New Delhi 110 012
10. **Ghosh P. K.**
    Sr. Scientist (Agronomy)
    NRC for Groundnut
    Jungadh 362 001
11. **Gill A. S.**
    Pr. Scientist (Agronomy)
    IGFRI, Jhansi 284 003
12. **Gill B. S.**
    Deptt. of Agronomy
    Punjab Agricultural University
    Ludhiana - 141 004
13. **Giri Gajendra**
    Sr. Scientist
    Division of Agronomy
    IARI, New Dehli-110012
14. **Jha K. P.**
    Pr. Scientist & Head.
    CRRI, Cuttak 753006
15. **Lal S. S.**
    Pr. Scientist (Agronomy)
    CPRI, Shimla-171007
16. **Mahey R. K.**
    Professor of Agronomy
    Punjab Agric. University
    Ludhiana 141 004
17. **Malvia D. D.**
    Principal
    College of Agriculture
    Jhunagadh 362 001
18. **Mathur R. K.**
    Scientist & Head of Plant Breeding
    NRC for Groundnut
    Junagadh 362 001
19. **Mishra B. N.**
    Sr. Scientist
    Division of Agronomy
    IARI, New Dehli-110 012

20. Moorthy B. T. S.
Division of Agronomy
CRRI, Cuttack - 753 006.

21. Pal, Kamal, Krishna
Scientist & Head (Microbiology)
NRC for Groundnut, Junagarh

22. Paliniappan S. P.
Head Crop Production
Nagarjuna Agriculture Research & Development Institute
Vikarampuri-
Secundearabad-500 009

23. Patil S. K.
Assoc. Prof. Agronomy
NARP (Agroforestry) Scarcity Zone
Deptt. of Botany
Mahatma Phule Krishi Vidyapeeth
Rahuri - 413 722

24. Poshiya V. K.
Professor & Head
Deptt. of Botany, College of Agric.
GAU, Junagadh - 362 001.

25. R. L. Arya
IGFRI, Jhansi 284 003

26. Randhawa G. S.
Sr. Agronomist
Deptt. of Agronomy, Punjab Agric. Univ.
Ludhiyana 141 004

27. Rao K. S.
Division of Agronomy
CRRI, Cuttack - 753 006

28. Rathore P. S.
Assoc. Prof. & Head
Deptt. of Agronomy
College of Agriculture,
Bikaner 334 006

29. Rathore Rajendra Singh
Plant Physiologist
SKN College of Agriculture
Jobner 303 329

30. Sarswat V. N.
Principle Scientist (Agronomy)
Central Research Institute of Jute & Allied Fibers
Barrackpur 743 101

31. Sharma S. K.
Asstt. Prof. Agronomy
College of Agriculture,
Bikaner 334 006

32. Sharma S. N.
Sr. Scientist, Division of Agronomy
IARI, New Delhi 110 012

33. Shivakumar B. G.
Scientist
Division of Agronomy
IARI, New Delhi 110 012

34. Shukla N. P.
Sr. Scientist Crop Production Division.
Jhansi (Uttar Pardesh)-284003

35. Singh A. K.
Asstt. Professor Agronomy
Rajasthan College of Agriculture
Udaipur

36. Singh A.
IGFRI, Jhansi 284 003

37. Singh Virendra
Professor of Agronomy & Joint Director Research, College of Agric.
G.B. Pant Univ. of Agric. & Tech.
Pantnagar 263 145.

38. Singh, A.
Agronomist, Regional Research Station
Kheri (Sangrur)

39. Singh, J. B.
IGFRI, Jhansi

40. Sud K. C.
Sr. Scientist of Soil Science
CPRI, Shimla

41. Sundara B.
Sr. Scientist (Agronomy)
Sugarcane Breeding Institute
Combatore 641 007

42. Venugopalan, M. V.
Scientist Crop Production Division
CICR, Nagpur – 440010

43. Yadav D. S.
Professor of Agronomy
N.D. Univ. of Agric & Tech
Narendra Nagar,
Faizabad - 224 229

44. Yadav N. D.
Sr. Scientist
CAZRI, Regional Station
Bikaner 334 006

INTRODUCTION

Agriculture may be merely means of earning livelihood elsewhere but in India it has been a way of life. For centuries our culture, festivals, traditions and even trade & business were linked to agriculture. Even today it is common saying that the real India lives in villages. In ancient time, agriculture was considered as a pride, notable and honorable occupation. Agriculture is the mainstay of Indian economy not only in terms of contribution to gross domestic product but also the number of people dependent upon it.

Role of Agriculture sector in industrialization can not be over looked. Important segments of industries based on Agriculture produce are sugar industry, textile industry, small/village industries, oil mills, dal mills, flourmills etc. About 30 per cent of country's GDP is contributed by Agriculture. In addition to its contribution in national economy, approximately 20 per cent of exporting earning of country is also accounted by agricultural commodities. Two third of labour force and 70 per cent population are still dependent on Agriculture.

1. AGRICULTURAL SCENARIO

Agriculture in India has received appreciable attention. History of Indian Agriculture turned to grand slab during period from 1951-1961 and 1966. In the first period main attempt were made for institutional changes and land reforms and to expand irrigation infrastructure. During this period agriculture output grew at an average compound rate of 3.1 per cent per annum with 58 and 42 per cent increase in total area and production. In the second spell after 1960-61, the main focus was on increasing yield with the use of modern inputs particularly quality seed, fertilisers and improved technology. In mid sixties green revolution has completely changed the agricultural scenario, particularly of wheat and rice. Wheat production which was 6.46 mt during 1950-51 increased to 68 mt in 1997-98 registering ten times increase Whereas, rice production touched 81 mt mark. Oilseed production increased from 11 mt to 25 mt due to yellow revolution, but increase in pulse production was low 12.42 mt to 14.12 mt. and still we have to depend on import of pulses. During 1994-95 about 555 thousand tonnes of pulses of Rs 5735 millions have been imported.

Total food grain production increased by about four times during last 50 years. With the concentrated efforts of farmers, agricultural scientists and the Government, India has largely solved the problem of domestic food deficit, but still we have future challenges ahead for ever-growing population and to compete globally. The gain in production in different crops is given in table 1 and India's position in World and Asia is given in Table 2.

TABLE 1.1: Gain in production of crops in India from 1950-51 to 1996-97

Crop	Production (mt)	
	1950-51	1996-97
Rice	20.60	80.53
Wheat	6.46	68.00*
Maize	1.70	10.13
Sorghum	5.50	10.53
Pearl millet	2.60	7.39
Cotton	3.04	14.16
Pulses	8.41	14.02
Oilseeds	5.16	24.41
Sugarcane	57.05	67.48
Food grain	50.90	192.12
Productivity	**522 kg/ha**	**1499 kg/ha**

* Million bales of 170 kg each.

TABLE 1.2: Estimated food grain requirement (mt) in India

Commodity	Year		
	2000	2010	2020
Rice	81.5	98.6	116.7
Wheat	62.5	77.2	92.7
Pulses	13.5	18.4	24.2
Food grains	**180.2**	**219.5**	**261.8**

2. FUTURE CHALLENGES

Agriculture being the most vital sector of the economy, the strategy for agricultural development should not be only in achieving self-sufficiency in food, but also in generating surplus for export. From *begging bowl* now India attained a status of exporting country in some of food items, particularly wheat and rice. Inspite of this fact we have to produce more for increasing population, growing @ 1.8% per annum. The estimated food grain requirement for year 2000, 2010 and 2020 are given table 3. Indian export also increased by about 21 per cent over 1986-87.

Diminishing per capita availability of agricultural land, scarcity of water, nutrient depletion, ecological imbalance and environmental degradation are major constraints in increasing production. Rather agriculture development around 2000 AD would be guided not only by compulsion of improving, food and nutritional security but also by the concern for environmental protection, profitability and sustainability. Commercialization of agriculture globally will induce shift in cropping pattern & resources. Labour scarcity may emerge largely because of rapid migration and occupational changes. Unemployment or under employment are going to be major problem.

TABLE 1.3: India's position in agricultural production as compared to Asia and wcrld (1996)

Crop	Asia		World	
	Area	Production	Area	Production
Wheat	2	2	3	3
Paddy	1	2	1	2
Maize	2	3	4	10
Sorghum	1	1	1	2
Millets	1	1	1	1
Chick pea	1	1	1	1
Lentil	1	1	1	1
Soybean	2	2	5	5
Ground nut	1	2	1	2
Castor	1	1	1	1
Sun flower	1	1	2	6
Sesame	1	1	1	1
Rapeseed and mustard	2	2	2	2
Linseed	1	2	1	3
Safflower	1	1	1	1
Seed cotton	1	2	1	3

The required productivity (kg/ha) to meet the future demand is given in Table 4.

TABLE 1.4: Required crop productivity (kg/ha) for year 2000, 2010 and 2020

Crop	Required productivity (kg/ha)		
	2000	2010	2020
Rice	2024	2456	2895
Wheat	2644	3270	3918
Coarse grain	974	1137	1333
Total cereals	1864	2263	2682
Pulses	742	987	1282
Food grains	**1662**	**2034**	**2447**

To achieve these targets, we have to raise production levels as given in Table 5.

TABLE 1.5: Required increase in productivity level to meet out the future demand

Crop	Required increase (%)		
	2000	2010	2020
Rice	9.4	32.7	56.4
Wheat	9.3	35.1	61.9
Coarse grain	0.0	16.1	36.1
Total cereals	10.4	34.1	58.9
Pulses	25.1	66.4	116.2
Food grains	**11.2**	**36.6**	**64.3**

3. STRATEGY TO MEET CHALLENGES

To achieve these targets land is a vital natural resource. India has 2 per cent of total world geographical area but sustain about 18 per cent world's livestock. This resource is limiting, because of its demand for industrial sector and urban expansion. Further, due to immense pressure land is degrading fast. Current data show that 60 per cent of land suffer from soil erosion, water logging and salinity problem including 16 per cent area as ravines.

To sustain availability of water is another important factor. In India 70 per cent land area is under dry farming, where crop production gamble with monsoon. Efforts are also required to achieve water needs of 109 million hectare meter by 2025 with 40 per cent irrigation efficiency and 80 million hectare meter with 60 per cent efficiency.

Thus, looking to these facts enhancement in crop productivity, using generated technology and their adoption by farmer, is the only way to achieve these targets. The productivity of majority of crops is very low in India as compared to other countries in world. In India more than 2/3 crop area is having the average yield below the national average. Thus, there is vast untapped potential existing in India.

To get the required increase, the following approaches may be considered.

1. Producing crops with available and generated location specific techniques. The concept of planning for agriculture in terms of homogeneous agro-climatic areas needs to be intensified.
2. *Greening of gray area:* Over 67 per cent cropped area are rainfed. Even if the irrigation potential of the country is achieved, approximately 55 per cent of cultivable area may remain rainfed. Emphasis should be given to greening this area for food production. On farm moisture conservation, improved dry land technology and scientifically planned sustainable contingent cropping system should be considered.
3. To enhance and sustain productivity of irrigated agriculture by using generated technology and new research.
4. Developing genetically superior high yielding and hybrid varieties. The genetic manipulation of plant may raise genetic yield ceiling, provide resistance against insect pest and diseases and abiotic stresses. Modifying plant architect also help in better suitability of plants to given conditions.
5. Use of modern improved farm machinery for insuring timely field operations and effective application of inputs.
6. Providing infra-structural supports in rural areas for transfer of modern technology. The extension strategy for transfer of technology is concentrated on demonstrations. We have to reorient our extension methods and research so that Indian farming advances from its improved stage to scientific stage and highly productive on a sustainable basis.
7. Strengthening the agricultural research information system (ARIS).
8. Women should be used as producers.
9. Weeds pose major problem in crop production. The problem of changing weed flora should be tackled in integrated manner.

10. Since there is very meager scope to bring additional area under cultivation, intercropping, double cropping and multiple cropping are the only ways to enhance crop productivity per unit of area.
11. Use of good quality improved seeds and assured supply at village level.
12. Eradicate soil problems i.e. salinity, alkalinity, water logging and other problems, wherever feasible.
13. Efficient water management with adequate provision for drainage. To maximise water productivity and profitability, there is an urgent need to diversify crop and cropping pattern based on the availability of water and rainfall.
14. Strengthening agriculture research. Widening the base of research to cover large number of potential crops suitable for diverse agro-climatic conditions.
15. Development of integrated pest management (IPM) for efficient management of insect pest and disease problem with concern on environmental protection.
16. Judicious use of organic manure, fertilisers, micronutrients and bio fertilisers. Efforts should be made on integrated nutrient management for sustainable crop productivity.
17. To promote agricultural exports for acquiring foreign currency and to sustain the tempo of development of farming sector. It is also necessary to explore possibilities and to make our agriculture to be globally competitive. Processing, product development, packaging, storage and marketing may be given due emphasis.

4. REFERENCES

Paroda, R.S. (1997), Options and priority areas, *The Hindu survey of Indian Agriculture*, 1997.

Kumar, P. (1997), Food security: Supply and demand perspectives, *Indian Farming*, 47(9): 4-10.

Paroda, R.S. (1999), Towards sustainable agriculture export- New paradigms, National Academy of Agriculture Science.

CEREAL AND MILLET CROPS

1. Rice

K. P. Jha, B. T. S. Moorthy and K. S. Rao

Rice is an important food crop of India it has the world's largest area (28%) covering 42.3 million hectares with a total production of 80 million tonnes annually. Its area is 37% of the total area under food grains as against 20% under wheat, another important food crop. But globally it stands next to wheat in harvested area. It is grown in diversified soil, topographical and hydrological situations ranging from sloppy uplands to deep water areas of above 1 metre depth. It is adopted to cold temperatures that exist in hilly areas and also in adverse soil conditions such as salinity, alkalinity and acidity.

It is a staple food crop of 60 per cent of world's population. Mostly it is cooked with water. Other edible uses includes rice flakes and puffed rice, rice wafers and canned rice. It is also used in starch and brewing industries. The byproduct of rice milling i.e., rice husk and bran are used as a cattle and poultry feed. Hard board and paper industries also uses these byproducts. Rice bran oil is used in cooking after refinement. It is also used in making soap. Rice straw is good cattle feed as well as it is also used in making hats, mats and ropes.

1. ORIGIN

Rice is regarded as a first cultivated crop of Asia. Preserved rice grains were found in China around 3000 B. C. Paddy grains found during excavation at Hastinapur (India) around 1000-750 B.C. considered as a oldest sample in the world. Vavilov (1926) concluded that south-west Himalayas has various types and varieties and indicated probable centre of origin. de Condolle (1886) mentioned south India as its centre of origin. Thus based on all this Archaeological evidences India is considered centre of origin of Rice.

2. DISTRIBUTION

Rice cultivation is limited to as far as north as 49° in Czechoslovakia and as far as south as 35° in new South Wales. But most of the world's rice is grown in tropics which includes countries of south and south east Asia, west Africa and central and south America. India stands first in area and second with regards to production, China is first. Area and production of important rice growing countries is given below:

TABLE 1.1: Area and production of rice in world (1997)

Countries	Area (000 ha)	Production (000 t)	Productivity (kg/ha)
India	42880	121812	2846
China	31107	197074	6206
Indonesia	11332	51102	4509
Bangladesh	9966	26531	2662
Thailand	10137	22016	2172

Rice is grown through out India in all the states. It is a cosmopolitan crop grown in high altitude hills and also in regions below mean sea level (Kuttanad area of Kerala).

The major rice growing states in India are West Bengal, Bihar, Madhya Pradesh, Orissa, Andhra Pradesh and Uttar Pradesh where 69.8 % of area exists. Rice area under assured irrigation is around 48.3 % of the total area. Highest percentage area under irrigated condition exists in the states of Tamil Nadu, Punjab, Haryana, Andhra Pradesh. In eastern India rainfed rice predominates over irrigated rice (Table 1.2).

TABLE 1.2: Area and production of rice in India (1996-97)

State	Area (000ha)	Average yield (t/ha)	% area under irrigation	% area under HYV	Fertiliser consumption (kg/ha)
Assam	2503.4	1.11	33.8	49.9	12.9
Bihar	5031.1	1.37	37.5	59.6	73.9
Orissa	4529.0	1.38	35.5	67.4	24.4
West Bengal	5953.4	2.00	24.6	64.5	97.7
Punjab	2161	3.13	99.2	94.4	65.8
Uttar Pradesh	5575.7	1.87	58.0	87.4	102.0
Andhra Pradesh	3678.8	2.50	95.0	94.4	138.0
Tamil Nadu	2228.5	3.39	92.7	85.9	105.5
Madhya Pradesh	5168.2	1.10	22.9	66.7	33.3
All India	42910.1	1.85		72.4	74.4

The national average yield level of rice in India is around 1.9 t/ha which is much less compared to that of the countries like Japan, Spain, Italy where more than 5 t/ha yields are achieved. Also among the different states of India there are regional imbalances with regard to productivity levels. In north-west India and southern India (Tamil Nadu and Andhra Pradesh, the yield levels are fairly higher than the national average because of assured irrigation and better technology adoption. In general the yield levels in eastern states of India (excepting West Bengal), are not only low but stagnating since several years. The reasons are quite obvious. As already mentioned the rainfed areas account for around 70% of the total area under rice in these states. The area covered by modern varieties and fertiliser consumed in these states is also lower (Table 1.2). Drought, floods, insect and disease problems besides socio-economic constraints are responsible for limiting yields in these states. Hence these factors reflect in the overall production from the se states. But with the work on development of modern varieties which are high productive and technologies that are appropriate to diversified situations carried out at the Central Rice Research Institute, Cuttack and various State Agricultural Universities in different states in the country, there is an ample scope to improve the yield levels of rice with effective technology transfer.

3. CLASSIFICATION

Rice belongs to family graminae. There are 24 of species of genus *Oryza* but only two species are under cultivation *viz., Oryza sativa* and *Oryza glaberrima. Oryza sativa* is most popular and grown in the most of the rice growing areas of the world, where as *Oryza glaberrima* is mainly grown in south Africa.

Oryza sativa is further grouped into three i.e., *indica, japonica* and *javonica* rice. The characteristics of each type are given below:

Characteristics	*indica*	*japonica*	*javonica*
Area of Cultivation	India	Japan	Indonesia
Plant Height	Tall	Dwarf	Tall
Leaves	Broad, light green, slightly pubescent	Narrow, dark green	Broad, hard, light green
Tillering	Very high	Medium	Low
Grain	Thin, flattened, elongated	Small and rounded	Broad and thick
Shattering	High	Low	Low

4. CLIMATE

Rice is heat and water loving plant, requiring high temperature and adequate moisture supply. For germination a minimum of 10 °C and for flowering a temperature of 22-25°C is considered as an optimum. Grain formation takes place at 20-21°C. Low temperature particularly during 10-11 days before heading causes considerable loss in grain yield because of sterile spiklets. At ripening temperature should be in between 20-25 °C. Rice is a short day, thermosensitive plant but now many thermoinsensitive verities have been developed. It is mostly grown in areas receiving 1000-1500 mm rain. Crop needs heavy rains (125 mm) during vegetative phase and 200 mm monthly rain for low land and 100 mm for upland rice. In India it is grown from 8° to 35° north latitude from sea level to mountaneous regions up to 3000 meters. Heavy winds specially after heavy rainfall or heavy irrigation results in severe lodging.

5. SOILS AND SOIL PREPARATION

There is a great diversity of edaphic situations where rice is cultivated in different states of India. The important soil groups are alluvial, red, red and yellow, lateritic, black, mixed red and black, greyish brown, brown hill soils, submountaneous, saline alkaline, peaty and saline -peaty soils.

Rice crop is generally grown not only on flat lands but also on a wide range of slopes. In humid Bengal, Assam belt and eastern humid region and Bay islands, soils are mostly alluvial. In West Bengal, old alluviums, red acidic soils deficient in P are present. In Assam valley acidic soils high in available P and K exist. In Subsumed Suttees Gangetic Alluvial Plains Region in UP allluvial soils of varying texture exist. In Punjab, the rice soils are light in texture and alkaline in reaction. Organic matter and N are low.

In sub-humid to humid eastern and southeastern Uplands, red and yellow soils in MP are sandy in texture with pH ranging from 5.5-8.5. In Orissa ultisols and oxisols deficient in P are found. In coastal belt rice is grown on alluvial soils and saline soils in this state. In Andhra Pradesh, the allluvial soils of delta are deep and well drained with clay loam to clay texture. Red soils occupy larger areas. In the states of Haryana, Rajasthan and Gujarat alluvial red, yellow and moderately deep black soils constitute major groups of the arid western plains. Maharashtra has lateritic soils. In west Madhya Pradesh, soils are mostly alluvial.

In coastal areas of Malnad, the soils are ultisols and oxisols. Acid ultisols are found in Malabar coast in Kerala. In Thanjavur district of Tamil Nadu, the soils are poor in N, P and organic matter. Deep black soils are found in all the districts except Nilgiri.

6. GROWING SEASONS

Rice is grown in different seasons in India. The temperature is the deciding factor for such adaptability of rice crop. In southern states like Tamil Nadu rice is grown throughout the year while in northern Indian states like Punjab and Haryana only during wet season rice is cultivated. The different rice growing seasons in various states of India are as follows:

State	Common Name	Seasons
Assam	Ahu	March-June
	Sali	June-October
	Boro	November-May
Orissa	Beali	June to October
	Sarad	June to December
	Dalwa	January-May
West Bengal	Early *Aus*	April-September
	Amman	June-December
	Boro	December-June
Bihar	Early *kharif* (summer crop)	March-April - July-August
	Kharif (Autumn crop)	June - October
	Winter crop	June/July -November/December
Meghalaya	*Kharif* (autumn)	Mid February-April-August
	Kharif (winter)	March/July-October/November
Nagaland	*Kharif* (autumn)	March/May-July/August
	Kharif (winter)	May/July-October/November
Tamil Nadu	*Kuruvai* (early *kharif*)	April/July, May/September
	Samba (*kharif*)	July to December
	Thaladi	September to January
	Summer (*Navarai*)	December-April
Maharastra	Kharif	June/July-November/December
Karnataka	*Mungar* (*kharif*, early summer)	June-December
	Kar (rabi)	January-May; February-June
Punjab	Kharif	June-October

State	Common Name	Seasons
Haryana	Kharif	May/June-September/October
Madhya Pradesh	*Kharif* (early) *Kharif* (late)	June – September June -November
Uttar Pradesh	Kharif	May- October

7. RICE UNDER DIFFERENT ECOSYSTEMS

Rice crop in India is irrigated to an extent of 48.3 % while the rest is grown under rainfed situations. Rainfed rice constitutes both upland rice which represents deficit moisture situation but rainfed lowland rice represents both deficit and excess water situations. The cultivation practices for different types of rice grown under different ecosystems and special types are mentioned below:

8. RAINFED UPLAND

Rainfed uplands are those where aerobic soil environment exists for major period of crop growth and development. There is no water stagnation on the soil surface 48 hours after cessation of rainfall. These uplands are either bunded or unbunded, the later type exists in sloppy terrains.

India has about 7.1 million ha of upland rice of which 85% is located in the states of Assam, Bihar, West Bengal, Orissa and eastern parts of Madhya Pradesh and Uttar Pradesh. The rainfall in this zone is in the range of 1000 to 2000 mm or more and temperature ranges from 25 to 41°C in July and from 6 to 25°C in January. Red, laterite and lateritic soil such as mixed red and yellow, red sandy, red loam, lateritic and mixed red and brown hill soils account for about 55% of the total rice area in the east zone. Next in the order of occurrence is alluvial soil which occupies about 27% of the total rice area.

Rice is grown under rainfed condition in these uplands in monsoon season. The average productivity of this crop grown under moisture stress condition is distressingly low (0.65 t/ha) due to the following constraints as reported by Pande and Reddy (1988), and Sharma *et. al.* (1995).

1. Soil moisture stress due to erratic distribution of rainfall which causes drought.
2. Heavy infestation of weeds (C_4 types) which are highly competitive with the crop.
3. Leaching loss of N on account of light textured soils and soil erosion in unbunded uplands.
4. Lack of organic matter, low N content and poor availability of phosphorus, zinc and iron especially in high pH calcareous soils.
5. Stresses like soil acidity, P deficiency, high P fixation, Al toxicity and Mn toxicity in acid soils and Fe deficiency in alkaline and calcareous soils.
6. Biotic stresses caused by insect pests like termites, gundhi bugs, mealy bugs, stem borer, and nematodes etc.
7. Inadequate crop stand caused by traditional method of broadcast seeding of low N responsive and lodging prone tall varieties.

8. Difficulties faced at the time of harvesting, threshing and drying in the event of heavy rains in September and October.
9. Lack of seed dormancy in the varieties being presently used by the farmers.
10. Soil erosion in sloppy & terrace lands.
11. Stray cattle menace during early growth of upland rice.
12. Lack of investment ability and poor resource base of farmers.

To meet the growing food demand of the country, there is a need to step up production of rice from this fragile ecosystem. Concerted efforts have been made in this direction to develop production technologies based on different problem situations in the country. By large scale adoption of these technologies it is possible to increase the average productivity upto 3.0 t/ha.

8.1. Variety

Farmers traditionally grow tall, low productive and lodging-prone rice varieties in uplands. The ideal characters for upland rice varieties are early seedling vigour, intermediate height, moderate tillering, early maturity (75-105 days), deep roots, drought tolerance, early maturity, heavy panicles, tolerance to major diseases like blast, brown spot and insect pests (stem borer, termites and gundhi bug), adequate seed dormancy, acceptable grain quality, response to moderate to high management (Sinha *et. al.*, 1996). So far about 92 varieties have been released for upland situation in the country.

In rainfed unbunded upland situation, where there is a limitation of soil moisture, fast maturing varieties of 70-90 days are needed. Such varieties include Vandana, Kalinga-III, Kalyani-II, Heera, Vanaprabha (Jha and Manna, 1989; Behera and Jha 1998) For bunded uplands (especially in alluvial belt) varieties having comparatively longer duration (100-110 days) like Annada, Annapurna and Pathara yield around 4.0 t/ha. The varieties like Sneha and Heera which mature in around 75 days are suitable for cultivation in rice-based sequence cropping. Subhadra, Luit Kubilee and Kalyani II are released for cultivation for uplands of Assam. Birsa Gora and Prabhat for Bihar and Bhupen for West Bengal. A list of promising varieties for different situations is given in Table 1.3.

TABLE 1.3: Promising rice varieties for upland situation

Variety	Maturity duration	Special features	Regions of adoption	Yield potential (t/ha)
Vandana	90	Tolerant to drought	Bihar, Orissa, Madhya Pradesh	3.5
Sneha	75	Tolerant to gundhi bug	Bihar, Orissa	3.0
Sattari	75	Early maturing mutant	Assam, eastern Madhya Pradesh, Orissa, West Bengal	3.0
KalingaIII	85	Cold tolerant, intermediate tall	Assam, Bihar, eastern Madhya Pradesh, West Bengal, eastern Uttar Pradesh, Orissa	3.5

Variety	Maturity duration	Special features	Regions of adoption	Yield potential (t/ha)
Birsadhan I	85	Suitable with intercropping with pigeonpea	Assam, Bihar, eastern Uttar Pradesh	3.0
Vanaprabha	90	Tolerant to tungro, suitable laterite soil, intermediate tall	Orissa, Assam, eastern Madhya Pradesh, Bihar, West Bengal	3.0
Poorva	90	Fine grain	Assam, eastern Madhya Pradesh	3.0
Subhadra	90	Drought tolerant	Orissa, Assam	3.5
Pathara	90	Tolerant to stem borer and drought	Orissa	3.0
Birsadhan 2	95	Tolerant to blast	Bihar, eastern Uttar Pradesh	3.5
Tripti	95	Suitable for light textured soils	Eastern Madhya Pradesh, Bihar, Orissa	3.5
Keshari	95		Orissa	3.5
Parijat	95	Moderately resistant	Orissa	3.5
Tara	100	Resistant to gall midge, green leaf hopper, tungro virus and white backed plant hopper	Orissa, West Bengal, Assam, eastern Madhya Pradesh	4.0
Annapurna	100	Red kernel, slightly drought tolerant, moderately resistant to blast, bacterial leaf blight, stem borer and resistant to brown spot	Orissa and West Bengal	3.0
Annada	105	High yielding, suitable for alluvial belt of eastern Madhya Pradesh	Assam, Bihar, Orissa, West Bengal, eastern Uttar Pradesh	5.0
Suphala	105	Early seedling vigour, tolerant to bacterial leaf blight and brown spot	Orissa	4.0
Narendral 1	105		Eastern Uttar Pradesh, Bihar	3.5
Govind	110	Tolerant to bacterial leaf blight	Eastern Uttar Pradesh	4.0
Kiran	110		Bihar	3.5
Kalyani II	62	Intermediate tall	Assam, Orissa, eastern Madhya Pradesh	2.0
Shankar	80	Resistant to gall midge and BPH	Orissa	3.0
Rasi	110	Resistant to blast; Drought tolerant	Assam, Bihar, Tamil Nadu, Madhya Pradesh, Orissa	3.5
Neela	90	Resistant to gall midge and BPH	Orissa	3.5
Rudra	90		Orissa	3.5
Prasanna	90		Tamil Nadu, Madhya Pradesh, Assam	3.5

Variety	Maturity duration	Special features	Regions of adoption	Yield potential (t/ha)
Anupama	110	Drought tolerant; Resistant to blast	Madhya Pradesh	
Abha	95	Drought tolerant	Madhya Pradesh	3.5
Heera	68	Resistant to gall midge, tungro and blast; seed dormant for 12 days	Assam, Bihar, Madhya Pradesh, Orissa, West Bengal, eastern Uttar Pradesh	3.5

8.2. Crop Management Practices

8.2.1. Land Preparation

Proper land preparation and leveling should be followed for ensuring adequate crop stand and finally achieving higher yield. The land has to be opened after the harvest of the previous crop preferably with a mould board plough. This prevents the land from getting hard on drying and conserves moisture at lower depths of soil. Pre-monsoon showers in April and May help to break the clods. Summer ploughing help in controlling weeds like *Cyperus rotundus* effectively. Deep ploughing and sub-soiling across the slopes help to conserve moisture in the wet season. It enables enhanced root growth and extraction of soil moisture from deeper layers. Besides it facilitates row seeding.

8.2.2. Stand Establishment

Optimum time: Optimum time of sowing is one of the important non-monetary input for raising crop production. It is decided by the onset of monsoon rains. Work carried out under FSR/E project at CRRI, Cuttack indicated that 15-23 June is the optimum time for realizing higher yields from rainfed uplands of eastern India (Jha *et. al.*, 1989).

Seeding densities and method of sowing: Plant population is the basic prerequisite and key factor in determining the quantum of yield of any crop. Deficiencies in crop stand could lead to inefficient utilization of costly inputs. Farmer's traditional practice of broadcasting method of sowing results in uneven stand. Dibble seeding behind the plough with an inter-row spacing of 20 cm and intra-row spacing of 10-15 cm proved superior in establishing optimum crop stand Jha *et. al.*, 1989; Singh *et. al.*, 1995). It may be mentioned that under dibble seeding, the seeds could be placed at proper depth (5 cm) which saves the rainfed crop to a great extent from drought. Besides, mechanical interculture or weeding and fertiliser incorporation are feasible in this system. Row seeding can be done by making furrows with a country plough and seeding in those furrows. Alternatively seed drills can be used. Manual seed drills of 2-3 rows can cover 0.2 ha/day. Bullock-drawn seed drills of 3 rows and tractor-drawn seed drills of 9 rows can cover 0.4 and 2 ha/day, respectively.

Seed rates of 70-90 kg/ha for drilling and 100-120 kg/ha for broadcasting was recommended. Moorthy and Mittra (1991) reported that a seed rate of 75 kg/ha was ideal for establishing adequate crop stand and realizing higher yields. Line sowing is a better method than broadcasting as it ensures better crop stand and facilitates easy mechanical weed control.

Fertiliser management: Most of the uplands are poor in fertility status. Crop responses to application of nutrients is universal. Due to higher elevation and light textured soils, percolation losses may result in low availability of N due to excessive leaching. Generally the recovery of N is low (20-30%) because of moisture stress and weed growth (Mishra *et. al.,* 1995). The optimum N requirement of upland rice in different locations are presented in Table 1.4.

TABLE 1.4: Optimum nitrogen requirement for major upland growing regions of India

Zone	State	Location	Optimum N requirement (kg/ha)
East	Bihar	Hazaribagh	40-63
		Ranchi	40-60
	West Bengal	Kharagpur	60-80
		Bankura	60
	Orissa	Cuttack	60
	Uttar Pradesh	Faizabad	40
		Varanasi	40
North-east	Manipur	Imphal	90
	Meghalaya	Shillong	57-63
West	Maharashtra	Rahuri	80
North	Uttar Pradesh	Pantnagar	60-90
South	Kerala	Pattambi	60
		Trivandrum	50

Deficiency of phosphorus is common in red and lateritic soils. Hence application of P_2O_5 to an extent of 30 kg/ha is considered essential. Similarly application of K_2O to an extent of 30 kg/ha helps in improving the yield. The crop under red and lateritic soils which are acidic responds to lime application. A dose of 1.5 t/ha is necessary to improve crop performance (Singh *et. al.,* 1995).

8.2.3. Weed Control

In upland rice weed competition is maximum. Weeds emerge along with rice crop and compete with it for light, nutrients moisture and space. The extent of losses in yield is influenced by competitive efficiency of weeds, weed density, duration of crop-weed competition, cultivar, fertility level, row spacing etc. Generally the competition from weeds is severe in upland direct-seeded rice, where weed dry matter remained higher than that of the crop up to 60 days (Bhan, 1983). The first 30-40 days period is considered critical for crop-weed competition (Sharma *et. al.*, 1977; Tewary and Singh, 1991). Weeds compete severely for nutrients and depending on intensity of weed growth, the depletion was estimated to be 19.3-26.6 kg N, 1.6-11.6 kg P and 19.4-68.1 kg K (Tosh *et. al.*, 1981; Moorthy and Mittra, 1990).

Crop-weed competition is severe in short duration varieties of rice (generally grown in monsoon season on uplands) because weeds mature rapidly and hence short duration varieties have weed competition for proportionately longer time than long duration varieties. Among the short duration cultivars, weed competition is more in semi-dwarf than semi-tall varieties (Moody, 1992).

Weeds are effectively controlled by mechanical and chemical methods. Several weeders are recently developed and can be used to control weeds effectively. Moorthy and Das (1992) observed that 2 rounds of working with finger weeder or wheel hoes at 15 and 30 days after sowing supplemented with hand weeding gave effective and economic weed control. Singh and Singh (1989) reported that better weed control was achieved with blade hoe at 20 and 40 days after sowing in upland rice.

Chemical weed control offers a better alternative cost effective technology to hand weeding which is highly expensive and labour intensive. Often labour availability for weeding in upland rice is a problem as it coincides with transplanting of lowland rice. Efficiency of post-emergence herbicides propanil @ 2-3 kg/ha in controlling graminaceous weeds when applied at 12-15 days after sowing at 2-3 leaf stage of weeds was reported (Dixit and Singh, 1981). Pre-emergence herbicides butachlor, thiobencarb, pendimethalin @1.5-3.0 kg/ha and anilofos at 0.4-0.5 kg/ha proved to be effective in controlling weeds (Moorthy and Mittra, 1991; Govindra Singh *et. al.*, 1988; Shivamadiah *et. al.*, 1987; Patro *et. al.*, 1989). One supplementary hand weeding after herbicide application was found necessary after herbicide use to control the second flush of weeds (Moorthy and Manna, 1989; Bhagat *et. al.*, 1991).

9. RAINFDED LOWLAND RICE

The rainfed lowland rice areas include all the areas where the water stagnates upto a depth of 100 cm or more excluding manageable irrigated lowlands. The rice crop in this situation occasionally experiences moisture deficit in the early or terminal stages of growth but mostly excess water stress at any stage which sometimes submerges the crop as well. This area mostly located in eastern India has the total annual rainfall ranging from 1200 to 3100 mm and marked with saucer shaped physiography with drainage congestion causing accumulation of water to 100 cm or more which is further aggravated by floods from the heavily silted rivers.

9.1. Classification of Rainfed Lowlands Based on Water Depth

Water starts accumulating as the monsoon becomes active and reaches its peak in August-September and thereafter comes down gradually towards crop maturity in November- December. Frequent and heavy rains bring in flash-floods because of over flowing of rivers in the catchment areas and drainage congestion in the coastal areas. Based on the depth of water stagnation in the fields, the rainfed lowland rice areas in India are classified into i) shallow lowlands (0-30 cm), ii) Intermediate lowlands (0-50 cm), iii) semi-deep lands (0-100 cm) and iv) deep water lands (more than 100 cm). In India, out of the 39 per cent rice area under rainfed lowlands 17, 9, 7 and 6 % is under shallow Intermediate, Semi-deep and deep water conditions respectively. In eastern India, the corresponding area under such lands is 21, 13, 10 and 7% (Sharma *et. al.,* 1995)

1. ***Rainfed shallow lowlands (0-30 cm):*** In these areas the water stagnation in the field is temporary and is confined mostly to peak period of rainfall during monsoon season. The crop is direct-seeded with the help of pre-monsoon showers and sometimes it suffers due to drought at seedling stage. The transplanted crop which generally covers large area, under the said situation, faces the problem of drought at early stage or at maturity, if the

dry spell continues for a longer period. On the other hand, heavy rains during August-September months bring flash floods inundating the crop a few days to a week.

2. ***Intermediate lowlands (0-50 cm):*** The water depth in the field ranges from 0-50 cm or more at different growth stages. Here the crop is mostly direct-seeded and partially transplanted. The drought at seedling stage is likely to be a problem in these areas due to scanty rains leading to poor stand. Further heavy rains in catchment areas of river basins result in water logging /flooding of the rice fields at later growth stages which hinders tillering resulting in low panicle numbers. The flash floods caused by overflow of rivers also sometimes submerge the crop partially or completely for a few days to a week or more causing mortality of plants. The crop damage under such situations depends on flood character like rise in water level, its current, duration, depth and turbidity of water. In some areas, after the flood water recedes, the water depth in the field is more than 30-40 cm which hampers tillering and growth of the crop.
3. ***Semi-deep lowlands (0-100 cm):*** In these areas water stagnates for 5-6 months during wet season. Locally these lands are called as '*Chaur*' in Bizarre, '*Beel*' in West Bengal, 'Asro' in Assam, '*Talo*', '*Jarhania*' in Uttar Pradesh and '*Pata*' in Orissa. The depth of water in these areas varies between 0-100 cm. Direct seeding of rice is done in April-May or even earlier to ensure the plants of sufficient size and vigour in order to sustain water stagnation to varying depths for longer duration. Synchronizing with the accumulation of water, plants start elongation with the rise of water by extensive growth of leaf sheath & internodes. Yields under this situation are generally low.
4. ***Deep water lands (more than 100 cm):*** In these areas generally the water level is more than one meter and it may reach 3-5 meters. These areas are locally known as '*Jheels/Chaurs/Mon/Vasi/ Bheels*' and the rice crop grown in such areas is called as floating rice. In these areas, the water rise is generally slow, but at certain locations the increase in water is as much as 60 cm per day during first few days.

9.2. Production Constraints

The following are considered to be the important constraints limiting the productivity of lowland rice as reported by Sharma *et. al.*, 1995. Rice productivity in rainfed lowland ecologies is constrained due to various types of abiotic and biotic stresses, soil related factors, socio-economic environment and lack of appropriate technologies. The following are considered to be important constraints in limiting the productivity of rainfed lowland rice:

1. Inadequate crop stand due to early drought
2. Suppressed tillering and mortality of plants due to deeper submergence
3. Low use of inputs particularly fertilisers
4. Fe toxicity in acid soils
5. Premature lodging of tall varieties grown predominantly
6. Delayed planting due to inadequate draught power and labour constraints
7. Incidence of pests such as gall midge and stem borer and diseases such as bacterial leaf blight, sheath blight and brown spot

8. Coastal salinity due to high water table containing saline water (5-8 m mhos/ cm)

9.3. Production Technology

Considering the risky nature of the rainfed lowland production system, the technology should aim at risk avoidance and produce stable yields. Maintenance of adequate crop stand, fertiliser management and plant protection are important for realizing higher yields. Further diversification and integration of different cropping enterprises is essential for improving income of farmers.

9.4. Varieties

Most traditional varieties of tall *indica* rice grown in these areas differ from one another in stress tolerance, growth habit and yield potential. Therefore, development of varieties that have a broader spectrum of tolerance to stress and yield potential are desirable.

In areas of unreliable climatic and hydrological conditions, rice cultivars possessing one or more of the following traits may be needed to provide security against risk (Siddiq and Kundu 1992).

1. Seedling ability to stay in the seed-bed 2-3 weeks longer than the normal period as a precaution against droughts or floods that might prevent transplanting in time
2. Seedling capacity to elongate rapidly or to tolerate submergence or both as safeguards against uncontrollable flooding depth at transplanting time or immediately afterwards
3. Plasticity to tiller over a relatively longer period as a means of recovery from early drought or flood damage
4. Capacity to tiller rapidly for a crop that must fit a growing season that is short, either naturally or because of delayed planting
5. Tolerance of drought, acidity, salinity, cold or combinations of such constraints both at early and late growth stages
6. Quick maturity and lack of photoperiod sensitivity when a crop has to fit in a short and abbreviated growing season, and
7. Photosensitivity when a crop has to fit a short growing period at the end of the growing season, especially in latitudes where day length varies significantly. Under delayed transplanted conditions, development of quick-growing, day-neutral cultivars or weakly photosensitive cultivars combining tolerance to low temperature and blast disease may reduce the risk and increase the productivity (Siddiq, 1992). Rice varieties for different sub-ecosystems of rainfed lowlands are very specific (Table 1.5).

TABLE 1.5: Promising high yielding rice varieties for different rainfed ecosystems in eastern states

State	Rainfed lowland			
	Shallow	Intermediate	Semi-deep	Deepwater
Bihar	Radha Sughandha Jayashree	Rajshree Sudha Vaidehi	Varidhi Janaki	
West Bengal	Neerja Jogen Mahsuri Pankaj Swarna	Mandira Biraj Amulya Dinesh	Matangini Nalini Suresh	Jaladhi-1 Jaladhi-2 Sabita
Orissa	Savitri Ramakrishna Kanchan Mahalaxmi Moti Padmini Samlei Aditya CR 1014 Dhritri	Kalashree Panidhan Utkalprabha Gayatri Tulasi CR1014 Kalashrree	Rambha -	
Assam	Bhogali Rangili Kushal Maniram Piolee Ranjeet	Manoharsali		
Uttar Pradesh	Lal Lahare Mahsuri	Madhukar	Jalpriya	Jalmagna Jalnidhi
Madhya Pradesh	Madhuri Mahsuri Swarna			

9.5. Cultural Management

9.5.1. Stand Establishment Practices

In the rainfed lowland areas, the crops are established either through direct seeding in relatively dry soil or are transplanted after accumulation of water in the field. Direct-seeded crops may experience drought at early growth stages and in severe cases crop stand become sub-optimal. Transplanting in excess water results in a shock or disturbance to the seedlings and greater part of the plant remains in water, resulting in poor stand establishment. Direct-seeded crop does not experience any such shock and the crop already established acclimatizes faster to the greater depths of water stagnation and maintain most of the plant parts above the water, thereby resulting in greater survival. Therefore, direct seeding is better than transplanting under rainfed flood-prone condition. At Cuttack, direct seeding before the onset of monsoon by May end increased the grain yield more than two times over that of transplanting after onset of monsoon in July, in higher depths of water (25 cm) (Reddy *et. al.*, 1989). Light showers received during May should be used to prepare the land well. It would be more appropriate to give first ploughing after harvest of the previous *kharif* crop in December/January itself to control difficult weeds and this also facilitates also better pulverization of the lands in May. Such timely sowing helps in establishing better crop stand and crop can tolerate onslaught of floods better. Bullock drawn/manual/tractor-drawn seed drills or even country plough should be utilized for the purpose row seeding. Hence dibble seeding adopting a spacing of 20 □ 15 cm with a seed rate of 75-100 kg/ha is considered more useful than broadcast method of seeding which results in uneven crop stand.

Although direct seeding has been found to be superior to transplanting in rainfed lowland areas prone excessive water-logging, there are still large areas where the crop is established through transplanting. This is because of the shorter time period available for sowing, flooding of low-lying fields due to heavy rains early in the season and accumulation of water to greater depths during the initial seedling and tillering stages. In the locations having higher population of wild rice, transplanting is still better than direct seeding as far as the yield stability is concerned. Transplanting also is risky and the crop performance depends mainly on the water depth at or immediately after planting and vigour of seedlings used. Excess water stress in the early stages results in greater plant mortality and reduced tillering, leading to decreased crop stand. The extent of damage varies considerably from year to year and between locations.

Nursery management helps in improving crop performance under transplanted condition. The seedlings raised conventionally in nursery seed-beds do not establish properly when planted in water which is too deep. In such situations nursery manuring helps in producing better and vigorous seedlings. Nitrogen has to be applied at level of 100 kg/ha to the nursery which amounts to about 10 kg/ha for main field. Therefore, planting with the seedlings from fertilized nursery results in their better establishment in the initial stages due to their greater portion remaining above the water surface, thus causing less mortality and ensuring good crop stand. Further, the use of older seedlings of up to 60 days made no change the performance of the crop planted with a long-duration, photosensitive variety due to their better tolerance to flooding (Sharma 1992).

In some of the lowland areas the rice fields get inundated early in the season due to heavy rains or overflow of river. Water stagnates to a depth of 40-70 cm and such situation continues during the later period of growth. The rice crop transplanted in the said situation, particularly with nursery seedlings gets severely damaged. In these areas having 40-50 cm of standing water a technique known as 'double transplanting or clonal propagation' has been found to improve yield of transplanted crop considerably. The clonal tillers are taller with more dry weight than the conventional nursery seedlings of equal age. Therefore, the clonally-propagated crop establishes well and acclimatizes faster in the similar flooded environment compared with the nursery seedlings which did not establish properly as they have been previously growing in an aerobic upland environment. Hence, the grain yield of the crop transplanted with clonal tillers is significantly higher than that planted with conventional nursery seedlings.

9.6. Specific Cultural Practices

9.6.1. Beaushening a Low Cost Weed Control Practice

The traditional practice of *Beaushaning, is* widely practiced for the control of weeds in direct seeded shallow lowland rice in varied forms in different parts of India by the farmers, particularly in Orissa, Madhya Pradesh and Bihar, and to a lesser extent in West Bengal, Assam and Uttar Pradesh for optimization of crop stand. This system involves direct seeding of rice during the onset of first pre-monsoon showers, followed by cross-ploughing with a country plough, planking, redistribution of seedlings (gap filling called as *khelua*) and hand weeding. The latter operations are normally taken 30-40 days after sowing when the water depth in the field is around 10-20 cm. These operations are unique in the sense that they help overcome the problems associated with direct seeding such as severe weed infestation, high plant population and high percolation loss of water. Further, *beushaning* stimulates plant growth by loosening/churning the soil, causing better root pruning and soil aeration.

Improved *beushaning* involves sowing of rice in lines and interculturing with a country plough in between the inter-row spaces. Interculturing with cono-weeder also results in similar yield advantage as with *beushaning* (Moorthy and Rao 1991). It is also feasible to grow rice and dhaincha simultaneously and incorporating the latter at an appropriate stage after accumulation of water in the field during the *beushaning* operation. This way the N requirement of rice is also met partially, equivalent to 30-40 kg N/ha of urea fertiliser (Manna *et. al.* 1988). However, this practice has to be followed with caution in excess water situations, because *beushaning* reduces plant population which may not be compensated through increased tillering.

9.6.2. Mixed Cropping

Lodging is the common problem in the traditionally grown tall *indica* varieties in rainfed lowlands which results in reduced yields. To reduce this mixed cropping of early and late maturing rice varieties may resist lodging and produce more yield due to the penetration of more sunlight into the canopy. This system may give stability in production by providing two crop harvests at different times from the same land, benefiting poor farmers with limited resources. Such mixed cropping is practiced in some deepwater areas of north-eastern India, Kerala and Tamil Nadu.

The possibility of growing two rice varieties together differing in plant height and growth duration has been explored under intermediate lowlands at CRRI, Cuttack. A semi-tall (130 cm) upland variety - Vanaprabha (95 days) or Kalinga-III (85 days) was sown along with a conventional semi-dwarf (100 cm) late maturing (170 days) photosensitive variety - Gayatri, Seema or CR 260-77 in alternate lines 15 cm wide. The grain yield in such cropping system was up to 19% more than the conventional system of growing a sole crop of the late variety alone.

9.6.3. Mixed Cropping of Rice with Non-Rice Crops

Cultivation of deepwater rice alone is mostly not profitable. Therefore, it is mixed-sown with various non-rice crops including cereals, pulses and oilseeds in some parts of eastern India. In the deepwater areas of Bihar and eastern Uttar Pradesh, previously only one long-duration rice used to be grown in a year. Normally in these areas, the land is tillable early in February and the water level starts rising with the onset of rains in late June. To increase the productivity in such areas, some farmers have started mixed/intercropping short-duration, non-rice crops like greengram, sesame, blackgram, jute, maize, forage sorghum etc. with deepwater rice. Both the crops are sown in March-April and the intercrop is harvested before the advent of flooding in June. In case of greengram, only the pods are picked and the remaining portion provides nutrition to rice plant after decomposition.

9.7. Manures and Fertilisers

Most rainfed lowland farmers do not apply fertiliser due to the risk involved in successful crop production. Traditional varieties have generally been considered unresponsive to fertilisers, particularly N, which tends to increase the dry matter production without improving grain production. However, many experiments with improved varieties have shown response to increased nutrient supply, ensuring higher plant survival under adverse conditions. Modern varieties developed for lowland situations respond to the application of N in terms of improved crop vigour and better ability to withstand greater depths of water stagnation.

The magnitude of response to nitrogen under flood-prone lowlands is lower due to greater nutrient losses, general low productivity and high native soil fertility. Application of 40-60 kg N/ha is adequate for rice under rainfed lowland and flood-prone conditions. Work conducted at CRRI, Cuttack indicated that In the shallow lowlands where semi-dwarf varieties like Savitri, Gayatri, Dharitri etc. are grown, a dose of 60 kg N/ha has to be applied. In higher water depth situations, where semi-tall varieties like Panidhan, Tulasi and CR 301-3066 are grown, a lower dose of 40 kg N/ha is enough.

Application of P is essential to enable the rice crop to withstand overhead flooding better during the early vegetative stage. Rice plants fertilized with both N and P tolerated flooding and produced significantly higher grain yields than the unfertilized or those fertilized with N or P alone (Reddy *et. al.*, 1985). Therefore, a moderate application rate of 20-40 kg P_2O_5/ha combined with 60 kg N/ha is recommended at sowing time to improve the population density, growth and yield of rice under submergence conditions.

In case of direct seeded rice, the entire quantity of N, P and K need to be placed in the subsoil in plough furrow along with the seeds, while in case of transplanted crop it should be applied at the time of planting.

Several studies have been conducted with coated urea fertilisers in lowland rice. Drilling sulphur coated urea behind plough at sowing has been found to be promising in prolonging N availability to rice plants under excess water conditions (Sharma 1994). However, results with coating of urea with other materials, such as, neem cake, lac, coaltar, wax, rockphosphate etc. were not very encouraging, mostly on par with broadcast incorporation of prilled urea. In transplanted rice, the grain yield was highest with urea super granule placement at 30 days after transplanting followed by sulphur coated urea broadcast incorporation at planting. However, in view of the high cost of sulphur, and non-availability of urea super granules and difficulty in applying it in standing water, prilled urea drilled behind plough in direct-sown rice and broadcast incorporation in transplanted rice are recommended for improving nitrogen use efficiency in excess water conditions.

Top dressing of N fertiliser is not feasible under excess water conditions and therefore, the crop may suffer due to N deficiency at the later stages of growth. The effect of organic sources like farmyard manure and *in situ* green manuring with dhaincha was studied at CRRI, Cuttack for ensuring a prolonged supply of N in addition to saving on chemical fertiliser costs. Application of 10 t FYM/ha about a week before sowing yielded up to 1 t/ha extra yield over the unmanured crop, and was comparable with 40 kg N/ha through urea fertiliser (Sharma 1994). However, basal N fertilization at 20 kg/ha has to be supplemented with FYM application for achieving the potential yield. The practice of green manuring is primarily confined to irrigated conditions where rice is grown by transplanting after puddling. There are possibilities of green manuring in direct sown lowland rice also with dhaincha (*Sesbania aculeata)* by growing two crops together simultaneously and incorporating the latter at an appropriate stage after water accumulation in the field. Broadcasting dhaincha seeds along with those of rice and ploughing in dhaincha plants after 40 days during *beushaning* has been found to increase the grain yield of shallow lowland rice significantly by providing 30-40 kg N/ha (Panda *et. al.,* 1988, Manna *et. al.,* 1988).

9.8. Weed Control

Irrespective of rainfed or irrigated conditions, under shallow depth of submergence, normally, the weed flora is constituted of the important sedges like *Cyperus difformis*/iria, *Fimbristylis miliacea*, *Scirpus* spp., *Eriocaulon* Spp., etc while *Sphenoclea zeylanica*, *Ludwigia* sp., *Hydrolea zeylanica*, etc among dicots. Under intermediate and semideep rainfed lowlands the weed flora remains highly divergent and complex, since it is often composed of typical aquatic species belonging to submerged, emerged and floating ferns of green algae and ferns, in addition to semi-aquatic grasses, including wild rice. Irrespective of transplanted and direct seeded rice, increasing menace caused by few typical aquatic weeds like *Chara zeylanica*, wild *Nitella* sp., *Eichhornia crassipes* (Mart) in the eastern India., *Salvinia molesta* in southern India., *Potamogeton distinctus* in Kashmir valley and wild rice all most all over the traditional rice belt is being viewed as problem of first order. Reduction in yield caused by *Chara zeylanica* and *Nitella* sp. has been reported to be as 40 per cent. Such problem is, however, specific to lowlands which remains waterlogged in contrast to that of lowlands having a continuous flow of water. Further the severe incidence of a typical weed like *Coix lachryma* has been observed in the semi-deep lowland with flowing water system. The species of *Chara*, an algal weed, is a major problem in deep water areas. The water hyacinth *Eichhornia crassipes* become the major problem in flood prone deep water areas. It migrates with flood water and effects plant survival wherever it stays and multiplies. *Hydrilla* sp., *Ipomea aquatica* and *Scirpus articulatus* are the other major weeds

The most widely adopted low cost cultural operation for traditional control of weeds is beasushening which was described earlier for control of all pre-submergence weeds. However, this operation has least impact on the incidence of typically submerged aquatic weeds as mentioned earlier, since they normally appear after this operation and coincide normally with peak period of rainfall and maximum depth of submergence around mid-August.

Irrespective of land situations, the practice of hot weather cultivation involves opening the land with a soil-inverting plough during summer after the harvest of the previous crop or after getting premonsoon showers helps to receive, retain and conserve rainwater in the field. besides ensuring timely sowing, good initial crop establishment and early control of weeds, particularly rhizomes and tubers of sedges. A deep-summer ploughing in March-April preferably with a mould-board plough after the harvest of previous crop incorporates the stubbles, inverts the soil and exposes it to the hot sun. Results of studies under intermediate lowlands at Cuttack have indicated that off-season tillage particularly in summer along with pre-emergence application of butachlor followed by hand weeding at about 40 days of growth (before accumulation of water) proved effective in controlling weeds and increasing the productivity of direct seeded rice significantly (Sharma and Das, 1993). An additional yield of about 1 t/ha could be achieved by summer ploughing alone compared with the conventional practice of ploughing fields just before sowing. Therefore, integrated weed management strategy for weed control in lowland rice includes summer ploughing, early application of pre-emergence herbicides like butachlor or thiobencarb @ 2.0 kg/ha within a week of germination in moist soil and one hand weeding after about a month, if necessary.

Wild rice is a strong competitor in direct-sown lowland rice in specific situations and populations of only a few plants/m^2 can markedly reduce the grain yield and can even completely eliminate the cultivated plants. Shattering and grain losses in deepwater areas are to a great extent due to infestation of fields with wild rice, such as, *Oryza rufipogon* and *O. nivara* and their subsequent crossing with cultivated varieties of *O. sativa*. Controlling wild rice is difficult because its growth habit is similar to but more vigorous than cultivated rice. The possible measures for controlling wild rice as recommended by Wirjahardja *et. al.* (1983) and Richharia (1964) are as follows:

1. By direct seeding of sprouted seed on fields that have been flooded
2. By transplanting rice in lines after puddling the field
3. By puddling the soil before broadcasting rice seed
4. By row seeding where wild rice growing between rows can be easily recognised and weeded out
5. By growing short duration rice cultivars that are harvested long before wild rice shatters its grain
6. By deep ploughing to bury wild rice seeds and prevent them from germinating. Ducks may also be used to graze the seed and seedlings of wild rice
7. By using crop rotation of rice and secondary crop to reduce the infestation of new wild rice

After roguing the wild rice plants, they should be burnt rather than keeping them on bunds which is practiced by the farmers traditionally. Growing of high yielding purple stem varieties also solves the problem to some extent. A variety Kalashree developed at CRRI, Cuttack is well suited for wild rice infested situations. However since wild and domesticated rice can cross pollinate, it is possible that wild rice may also have purple leaf or purple stem.

9.9. Rice-Fish Culture in Rainfed Lowlands

Monocropping of rice in rainfed lowlands is not always profitable. Hence there is a need to diversify farm enterprises like growing fish along with rice for improvement of farm incomes. Rice is a semi-aquatic plant, whereas fish is completely aquatic in nature. Integration of aquaculture with rice farming in flooded paddies augments farm output and income and also benefits rice by improving its environment. Besides, such a system provides insurance in the risk-prone rainfed lowland environments. This system has been found to be economically beneficial, besides reducing stem borer infestation and working as predator of several pests of rice (Sinhababu and Venkateswarlu 1995). Further, rice-fish-horticulture/silviculture systems provide more returns and help in maintaining ecological balance. Many fish species like catfish, snake head, *Anabas* sp. and *Trichogaster* sp. are adapted to swamp conditions.

For successful rice-cum-fish farming, the fields are made to hold water for longer periods with strong and raised bunds to avoid flooding and a suitable fish refuge/trench system. Fish rearing requires additional inputs like organic manure and fish feed, whereas rice farming remains essentially the same as in monocropping system except avoidance of toxic pesticides.

Raising of marketable fish and prawn in rainfed lowlands including coastal saline areas (brackish water paddies) along with rice has been accomplished successfully in West Bengal, Orissa and coastal Andhra Pradesh. In shallow lowlands, 500-700 kg fish or prawn/ ha and 5-6 t/ha of rice are being obtained by farmers under such mixed farming systems. In deepwater situations where water remains in the field for longer periods, the yield of fish/prawn may go up to 2000 kg/ha, although the yields of rice are lower under such conditions. Cultivation of vegetables and horticultural crops on bunds raised around the field with the soil excavated from the trenches further increase the productivity to a great extent.

10. IRRIGATED ECOSYSTEM

As already mentioned only 48.3 % of the total rice area of 42.8 million hectares is irrigated. While the entire *boro* rice area 3 million hectares is irrigated *kharif* rice is partially irrigated. Even though the yield levels in irrigated rice are higher than those of rainfed rice, currently plateauing trends are appearing in some states like Punjab. In other states the potential yields in irrigated conditions are yet to be realized evidently due to some constraints. The following are the important constraints for limiting yields in irrigated rice growing areas.

1. Erratic monsoon delaying the canal opening and consequent delay in planting.
2. Problem of water-logging in *kharif* season
3. Degradation of soil due to rising water table and salinity levels in states like Punjab and Haryana.
4. N-driven depletion of soil fertility and related problems due to excessive soil mining.

5. Changes in physico-chemical properties of the soil and the reduced nutrient capacity of the soil.
6. New and emerging problems of multi-nutrient deficiencies.
7. Severe incidence of pest and disease problems.

A brief account of the varietal suitability and production technology that is well suited for realizing higher yields for irrigated lands has been enumerated as follows:

10.1. Varieties

With the introduction of the semi-dwarf varieties Taichung Native 1 and IR8 to India in the 1960s signaled a new era in India's rice breeding programs. Afterwards the national rice improvement efforts using those dwarfing genes were made at CRRI, Cuttack and AICRIP, Hyderabad which resulted in the development of a number of semi-dwarf varieties with high yielding potential, pest and diseases tolerance and quality grain. Most of these varieties were well suited to the favourable situations with assured irrigation yielding upto 7 t/ha. Jaya was the first variety which made a significant impact throughout the country and it was a cross between TN1 and T141.

Out of 427 varieties so far developed in the country, 227 are for irrigated conditions. The most important varieties recommended for different regions of the country are given in Table 1.6.

TABLE 1.6: Rice varieties (mid-early/medium duration) suited for irrigated situation.

State	Varieties
Andhra Pradesh	Tella Hamsa, Pushkali, Satya, Rudramma, Varshi, Dhanya, Lakshmi, IR64, Vikas, Rasi, Vamshi, Divya, Vikramarya, Sonasali, Prabhat, Suraksha, Jaya, Lakshmi, Madhu, Pratibha, Vajram, Chaitanya, Pinakini, Phalguna, Kakatiya, Surekha, Pothana, Vibhava, Abhaya, Krishnaveni
Assam	Lachit, Chellari, Rasi, Vikas, Sasyasree, Pusa 2-21, Pusa 33, Sonasali, Vikramarya, Govind, Saket 4
Bihar	Vishnu, Sita, Sonasali, Vikramarya, Sujata, Radha, Prahlad, Narsing, Rajendradhan 202, Kamini, NDR 359
Gujarat	Gaur4, Gaur10, Gaur11, Gaur100, Gr101, Ambica
Karnataka	Mangala, Pushpa, Mahaver, Pragati, prakash, Mandyavani, Avinash, Jayanti, Sabari, Bhadra, Sonasali, Vikram, Mahaveer, Karna.
Kerala	Jyotih, Pavizham, Onam, Kerali, Aathina, Aruna,
Madhya Pradesh	Madhuri, Kranti, Garima, Jagriti, Sonasali, Vikramarya, Ruchi, Safri 17, Asha, Usha, Samridhi, Abhaya.
Maharashtra	Parbhavati, Ratnagiri, Kayat1, ACK5, Ambica, Khudhika, terna, Syeer1, saholi 6, Indryani, SKL 47-8, Palghar-1, Sindewali, Kundalika
Orissa	Ratna, IR36, Daya, Seema, Neela, Tara, Annada, Sravani, Lalat, Jajati, Udaya, Gauri, Khera, Bhuvan, NDR359, Punithavathi, Ajaya, IR36, Saamalei, sarasa, Shakti.
Pondicherry	Bharathidarshan, Jawahar, Punithavathi, Ajaya
Punjab	PR103, PR106, PR108, PR109, PR110, PR 4141Basmati 385, Pusa 44-33, Ajaya, Jaya
Tamil Nadu	TKM9, CO 41, Adt 36, ADU-1, IR 50, ASD 16, ADT 37, AD 39, ASD 17, ASD 18, MDU 4, IR 64, JJ 92, CO 43, CO 44, ADU 2, Thirupathisaran, TPS-2, Adt 38, MDU 3, TKM 10, Bharatidasan MDU3, Vikramarya, rajarajan, Rasi.

State	Varieties
West Bengal	Lakshmi, Khitish, Kunti, Panke, Vikramarya, Sonasali, Swarnadhan, IR36
Tripura	JRC Borodhan-1
Haryana	Vikas, Jaya, PR 106, PR 4141, PR 109
Uttar Pradesh	Saket 4, Govind, Pantdhan 4, Manhar, Pant dhan 6, Pusa 169
Rajasthan	Chambal, BK190, BK79 and RRB94

10.2. Crop Management

10.2.1. Crop Stand Establishment

There are two systems of crop stand establishment under irrigated situations- (i) transplanted system (ii) direct seeding on puddled soil. In case of the first system which is most prevalent in India, nurseries are raised in advance to produce seedlings. Later transplanting is done using the seedlings. It is highly labour intensive system involving more number of labour force for nursery raising, uprooting and transplanting. But it has got the advantage of less weed problems. In the latter system land is well puddled and sprouted seeds are either broadcast sown or dibble seeded using a spacing of 20×10 or 15×15 cm. A seed rate of 75 kg/ha is sufficient. This system saves labour as it does not involve nursery raising, uprooting of seedlings and transplanting.

Under transplanted system, for raising seedlings, normally two types of nurseries are followed by the farmers in India. Those are dry and wet bed nurseries. In case of wet bed method, pregerminated seeds are broadcast uniformly on raised beds of puddled soil. Generally wet nursery is preferred in areas where the weed jungle grass (*Echinochloa colonum*) is a problem. Wet nurseries are also practiced in dry season when temperatures are cool during the initial stages.

Dry bed method is followed in areas where the water is insufficient to follow wet bed method. Instead of wet tillage, dry tillage is followed. However, the dry nurseries are well suited for wet season and the seedlings obtained from these nurseries are hardy and can tolerate adverse conditions better. But weed problem is more than that of wet bed method.

For raising nursery, fertile well drained field is selected near source of water. To raise seedlings for one hectare 500 Sq. metre area under normal sown conditions and 750-800 Sq. meter area under late sown conditions is required. After adequate land preparation, 200-300 kg of FYM has to be applied and incorporated for every 100 m^2 area 2-3 weeks before sowing. In the absence of it, 0.5-1.0 kg N, 0.5 kg each of P_2O_5 and K_2O/ha may be applied, irrespective of dry or wet nursery. For dry nursery, sowing seeds along with P_2O_5 and K_2O in hand-drawn shallow furrows spaced at 8-10 cm and covering with soil is preferable over broadcast seeding in case of raised beds seed rate depends upon the size of grain, bold seeded varieties sown with 45-60 kg seeds/ha where as fine grain varieties are sown at 35-40 kg/ha. It is better to sow pregerminated seeds for quick growth. Seedlings are ready in 20-25 days. A seed rate of 500-600 kg/ha may be used irrespective of dry or wet nursery. The seedlings produced from the well managed nurseries are expected to give better yields.

There is another method known as *dapog* method of nursery which was developed in Laguna province of Philippines and is commonly used in that region. In this method, the land is prepared as in the wet bed method, but banana leaves or plastic sheets are used to cover the soil. The dapog bed should be about 1.5 m wide, and its length will depend upon the area to be planted. Banana bracts are planted on the seedbed with bamboo pegs to keep the leaves in position. In this *dapog* method, the seeding rate is 3000 g/m^2, which gives a rate of 100 kg/ha, about twice that of the wet bed method. *Dapog* seedlings are ready for transplanting 9-14 days after the seed is sown. The advantage of this method are labour savings, short period for raising nurseries, and relative ease of transport of seedlings. However this method has not become so popular in India because of the varied cultural and hydrological conditions.

10.2.2. Cultural Practices

Adequate crop stand establishment is quite essential in realising higher yields. Without sufficient plant population, the costly applied resources get wasted. For achieving this early planting planting in July in the wet season and the first week of January in dry season are considered essential. Progressive delay in planting results in reduction of grain yield. Similarly line planting ensures proper crop stand and facilitates working of rotary weeder for controlling weeds as well as incorporating fertiliser nitrogen during crop growth. Shallow planting (3-4 cm deep) of 25-30 days old seedlings with a spacing of 20□15 cm for long duration varieties and 20□10 cm for short duration varieties @ 2-3 seedlings per hill is desirable. If need arises, gap filling has to be done twice at 7 and 15 days after transplanting. Generally optimum age of seedlings is 25 and 30 days during *kharif* and 30-35 days during *rabi* season. However, medium and long duration rice varieties (135-150 days) exhibit a flexibility with regard to age of seedlings upto 45 days.

10.2.3. Water Management

The land areas where water management is easily achievable are the medium lands. In these land situations, rice is grown under continuously flooded condition with 2-7 cm standing water throughout the season. This practice is effective for efficient nutrient use and weed control. As the irrigation is becoming expensive day by day, the water input needs to be judiciously used.

In rice, water requirement is the total amount of water needed from sowing till ripening, to meet the losses due to transpiration through leaves, losses due to evaporation from soil and crop surface together with other economically unavoidable losses in the form of deep percolation and seepage. The first two types of losses, collectively represent the actual consumptive use of water by the crop and are therefore termed as evapotranspiration or consumptive use of water by the crop. The water requirement of rice also includes water required for raising the nursery and for special purposes, like puddling in case of transplanted crop.

The total water requirement of rice in India as reported by various workers varies widely from less than 800 mm to more than 2500 mm depending upon the duration of the variety and characteristics of the soil and climatic conditions of the locality.

Regarding irrigation/water requirement scheduling, all the research workers in the country are unanimous that continuous and deep submergence are not needed for rice. Majority of the workers advocated continuous shallow submergence for achieving higher yield of rice. Others have advocated saturation and submergence at different stages of growth for higher yield and better economy. It was also observed that continuous saturation was as good as continuous shallow submergence. Trials at CRRI, Cuttack, Faizabad and other places of north India under All India Coordinated Project on Water Management indicated that application of 5-7 cm of water 3-4 days after the disappearance of applied water does not affect the final yield adversely. Phasic submergence has been advocated by many for higher yield and water economy. It was also reported that saturation at seedling establishment stage, lag vegetative and ripening phase and submergence at active vegetative phase and reproductive phases gave the highest yield and was better than continuous submergence in case of semi-dwarf varieties. From these reports it can be inferred that the optimum water requirement for rice depends upon the soil characteristics including permeability, season of growth and associated climatic factors, depth of ground water table and variety.

10.2.4. Fertiliser Management

Irrigated rice responds well to fertiliser application if applied judiciously and with proper cultural management. Responses are generally higher in dry season as compared to wet season. This is mainly due to higher solar radiation and lack of rains and better possibility of irrigation water management. With the current prices of fertilisers and the existing socio-economic constraints, a rate of 40-50 kg N/ha with a maximum 60 kg N/ha may be considered desirable for wet season; for the dry season, a rate of 80-100 kg N/ha with a maximum of 120 kg N/ha may be considered as optimum for eastern India. However, still higher levels are used in southern and north-western states of India where intensive cultural practices are adopted and high yields are obtained.

Regarding the form of fertilisers, ammonical forms are better than nitrates. However, urea, which is the cheapest fertiliser is widely used in the country. Nitrogen recovery in lowland rice is low (30-40%). N use efficiency in irrigated rice can be increased by effective water control. The other factors which influence the N use efficiency are soil, genotypes, incidence of pests and diseases, source and method of fertiliser application and synergistic effects with the other nutrients.

A better return can be obtained from a given amount of N when it is applied in splits and the application is suitably synchronised with active growth of the crop. The quantity of each split is decided by keeping in view the crop growth stage and the soil type and especially the cation exchange capacity of the soil. In light textured soils having lower CEC, the application of higher quantity of ammonical N will also be subjected to leaching loss. Nitrogen has to be applied in reduced zone and subsequently incorporated at the time of puddling or manually placed to avoid conversion of ammonia to NO_3 and subsequent loss by leaching. Application of N in three splits - 50% at planting, 25% each at active tillering and at panicle initiation stages have been found to be appropriate for increasing N use efficiency and yield of rice.

Nutrient management in rice-rice system warrants sustainable production and maintenance of soil health. In order to achieve this goal, adoption of green manure crops, complementary use of N through organic manures and biofertilisers like *Azolla*, provision of improved compost making manures and crop residue utilization vis-à-vis its management is needed. This system of integrated nutrient management envisages the balanced use of chemical fertilisers in conjunction with organic manures, biofertiliser legumes and other locally available nutrient sources for sustaining soil productivity.

Investigations carried out at CRRI, Cuttack revealed that when chemical fertilisers along with organic manures were used, the efficiency of applied fertilisers increases significantly. Therefore to reduce the cost of cultivation one can supplement a part of fertilisers with organic manures. Conjunctive use of nitrogen at 75 kg N/ha in the form of organics (*Sesbania aculeata*, *S. rostrata*, FYM, water hyacinth compost, *Azolla* compost) and inorganics in the proportion of 50 : 50 proved to be better from the point of soil health and sustainable productivity on a long term basis.

Green manuring practices: Studies were conducted in different parts of the country to study the potential of green manuring. *In situ* incorporation of *Sesbania aculeata* adds N which is available quickly to the rice crop. This green crop can be grown in the field and incorporated after 45-50 days after sowing. Transplanting of rice can be done immediately after incorporating the green manure crop.

Biofertilisers: The water fern *Azolla* fixes atmospheric N due to the presence of symbiotic blue green algae *Anabaena azollae* in its leaves and is important biofertiliser for rice. Field experiments at CRRI and other places have shown that consistent yield increase in rice were observed due to green manuring with *Azolla* with fertiliser N. Fresh *Azolla* biofertiliser application @ 6.0 t/ha and 10-12 t/ha saves annually atleast 30-50 kg mineral N/ha respectively. The rice yield produced by *Azolla* green manuring followed by dual cropping is equivalent to that obtained from 60 kg N/ha as fertiliser. Two *Azolla* crops in addition to its green manuring produces higher yields than the application of 60 kg N/ha but two dual crops alone are not as effective as 60 kg N/ha. It was also observed that incorporation of a layer of *Azolla* adds about 25 kg N/ha.

Similarly blue green algae add about 25-30 kg N/ha. It was also reported that by algal supplementation, one-third of the recommended N fertiliser can be saved without sacrificing the yield.

Balanced use of fertilisers: Crop should be provided with nutrients other than nitrogen in a judicious way in order to improve soil fertility and improve sustainability in production. In intensive cropping systems especially cereal-cereal (rice-rice or rice-wheat), systems, the use of phosphorus, potash and some micronutrients become necessary considering long term sustainability of production. Acid lateritic soils need application of P in adequate quantities. Similarly light textures soils also respond to potash application. The ideal combination of N, P and K would be 2 : 1 : 2 ratio. Among the micronutrients, zinc has been found to be a limiting nutrient in realising normal yield of rice in about 8 million hectares of rice area in India. The practice of applying zinc sulphate (25 kg/ha) or spraying zinc based compounds in zinc deficient soils has been found profitable bringing synergistic effect with nitrogen and phosphorus. Deficiency of zinc leads to khaira disease.

10.2.5. Weed Control

In irrigated rice under both transplanted and wet sown conditions, sedges like *Cyperus difformis*, *Cyperus iria*, *Fimbristylis miliacea*, *Scirpus articulatus* and dicots, *Sphenoclea zeylanica*, *Ludwigia parviflora* predominate and compete with the crop plants. Generally flooded conditions eliminate grassy weeds to a large extent. However, in some rice growing areas, common grasses like *Echinochloa colonum* and *E. crusgalli* are also becoming problems where water submergence is not adequate. These weeds also pose problems in the rice nurseries also.

The competitive effects in transplanted rice are generally lower than puddle seeded system because generally rice seedlings of some age are planted in this system which have a head start over weeds initially. The losses in yield due to unchecked weed competition were worked out to be 13-40% with a mean of 25% in transplanted system (Manna, 1983).

There is a need to keep fields weed free for the first four weeks after transplanting. Since hand weeding is expensive, alternate methods using herbicides for controlling weeds solve the problem. Among the herbicides, the phenoxy acetic acid compounds like 2, 4-D and MCPA @ 0.5 kg/ha control weeds that belong to the groups of sedges and dicots. However, these are not effective against grassy weeds. Pre-emergence herbicides - butachlor, thiobencarb, anilofos and pretilachlor are quite effective in controlling weeds in transplanted rice. The recommended doses for different herbicides are - butachlor, thiobencarb, anilofos and pretilachlor 1.0-1.5 kg/ha and anilofos 0.4 kg/ha. These herbicides should be applied at 7 days after transplanting for better results. Post-emergence application of propanil 15 days after planting when the weeds are at 2-3 leaf stage controls majority of the weeds in transplanted rice.

In puddle-seeded rice where sprouted seeds are sown, the weed problems are similar to those of transplanted rice. But the extent of competition is higher due to the fact that young seedlings suffer more from simultaneously emerging weed seedlings. The losses in yield were estimated to range from 18-51% with a mean of 32%. (Manna, 1983). The pre-emergence herbicides butachlor, thiobencarb and pretilachlor at reduced rates of 0.75 to 1.0 kg/ha and anilofos at 0.4 kg/ha applied 7 days after sowing rice in a thin film of water are quite effective in controlling weeds.

Effective water management also helps in controlling weeds in rice culture. Maintenance of 5 cm water in the field continuously helps in reducing the weed problems besides improving the efficiency of the herbicides. Besides with appropriate and timely weed control, the efficiency of applied inputs like fertiliser also shows substantial improvement resulting in higher crop yields.

In addition to chemical weed control, use of implements such as rotary weeders and cono-weeders in between the rows help in not only controlling the weeds but also incorporating the fertilisers.

11. SCENTED RICE

Our country's share of export in the world rice trade is less than 6%. Yet in terms of value, among the exported agricultural commodities, rice accounts for 9%. Currently since we have become self sufficient in rice production, there is a lot of scope to export quality rice which include both scented and non scented rice. The scented fine quality rice popularly known as Basmati rice are traditionally grown in the north and north-western parts of Indian sub-continent for centuries. They are characterized by pleasant aroma, long slender grains, extreme grain elongation on cooking and soft texture of cooked rice. The best quality Basmati is produced in the 'Kalar' tract of Punjab of India and Pakistan. The interaction of genetic factor, soil and environment of this unique tract is supported to culminate in producing the distinctive aroma to these rice. These varieties command an excellent price in Middle East Asian and European markets and provide the much needed foreign exchange to the countries producing them. In 1993-94 India exported 67000 tonnes of basmati rice earning a foreign exchange of Rs. 0.32 billions. During the year 1997-98, a quantity of about 4.5 lakh tonnes is expected to be exported.

Several aromatic rice are grown and consumed as basmati rice. Yet, a few like Basmati 370 (Punjab basmati 1), Type 3 (Dehradun basmati), Karnal Local (Travadi basmati) and Pakistan Basmati fulfill the quality norms required for export.

In India, basmati rice are predominantly grown in the districts of Karnal, Ambala and Kurukhetra in Haryana., Kapurthala, Gurudaspur, Amritsar and Pathankot in Punjab and Dehradun, Saharanpur, Muzaffarnagar, Nainital and Pilibhit in Uttar Pradesh. The largest area under basmati rice is in Uttar Pradesh. Basmati also is grown in Delhi, Rajasthan and Jammu & Kashmir. From a total of 7 lakh ha under basmati rice in the country, nearly 6 lakh tonnes of milled rice are produced annually. These traditional tall basmati cultivars are very poor grain yielders and lodge under high doses of nitrogen fertiliser.

11.1. Quality Aspects

Basmati rice are characterized by superfine grains (length- 6.5 mm) with a distinct aroma, intensity of which is believed to be influenced by growing conditions, excellent linear elongation of kernel on cooking which is almost double its length and soft texture of cooked rice. Good quality basmati rice have intermediate amylose content, low gelatinization temperature and medium gel consistency.

11.2. Improvement of Basmati Rice

Breeding efforts were started in mid-sixties with the objective of tranferring the unique qualiyty features of traditional basmati rice into the high yielding semi-dwarf background. None of the earlier evolved varieties could make a real impact and replace the popularly grown scented varieties like Pakistan basmati, Dehradun Basmati or Karnal Local.

Breeding for high yielding scented rices was carried out at different institutes like PAU, IARI, DRR etc. Emphasis was given to 4 important quality features *viz.,* aroma, superfine grain, linear elongation on cooking and milling recovery.

An important variety, IET 8580 (Kasturi) developed at DRR which was a cross between Basmat 370/CRR 88-17-1-4, the later parent being a derivation of Ratna/Domsiah, combined semi-dwarf stature, high yield potential with several desirable quality features of Basmati rice. At IARI, a semi-dwarf scented culture IET 10364 (Pusa 615-140-10-1) was developed and released as Pusa Basmati 1. Two varieties S-370 and Punjab-1 developed by PAU are popular with the farmers. The other varieties included Haryana Basmati-1. The scented varieties developed and released by Central and state variety release committees are given in Table 1.7 and quality characters of important scented varieties are given in Table 1.8.

TABLE 1.7: Scented varieties released in India

State	Variety
Bihar	Sugandha, Kamini
Gujarat	GR 101, Kolhapur scented
Jammu & Kashmir	Ranbir Basmati
Himachal Pradesh	Himalaya-2
Karnataka	Kusuma
Madhya Pradesh	Madhuri
Maharashtra	SYE-ER-1, Indrayani, Skoli 7, Pawana, Prabhavati, SYE-2, SKL-47-8
Punjab	Punjab Basmati1, Basmati-385
Rajasthan	BK 79, Mahisugandha and Khusbhoo
Tamil Nadu	Pusa 33, ADT 32, JJ 92

TABLE 1.8: Quality characteristics of dwarf scented varieties

Character	Pusa Basmati 1	Kasturi	Haryana Basmati 1	Ranbir Basmati
Milling %	70.0	70.0	72.0	70.0
Head rice %	55.0	61.0	60.0	54.0
Kernel length (mm)	7.2	6.94	6.65	7.13
L/B ratio	3.99	3.85	3.91	3.77
Kernel length	13.9	12.41	12.4	11.9
Elongation ratio (cooked)	1.93	1.79	1.86	1.67

11.3. Production Technology of Basmati Rice

11.3.1. Time of Planting

Sowing nursery in the first week of June is ideal for western Uttar Pradesh, Delhi, Rajasthan and Jammu & Kashmir. However, in Punjab raising nursery during the last week of May ensures strong aroma and high yields. In Haryana, although late sowing has been practice in the past to ensure the quality, early June sowing is now followed to avoid off blast diseases.

In non-conventional areas of Andhra Pradesh and Orissa studies indicated that early planting in July proved ideal for higher yields while late planting proved better for obtaining better quality.

11.4. Field Preparation and Crop Management

Dry plough 3 weeks before transplanting and submerge the field in 5-10 cm of standing water. Add 6-8 tonnes/ha of green manure and incorporate it by puddling. *Sesbania aculeata* (Dhaincha) or Sunhemp can be raised prior to transplanting in the main field for 45 days. Apply basal dose of fertilisers after final puddling and leveling. Maintain 2-5 cm of water throughout the growing season. For effective water management uniform leveling is important.

11.5. Transplanting

Transplant 20-25 days old healthy seedlings. Plant 2 to 3 seedlings per hill, preferably in lines at a spacing of 20 x 15 cm for high yielding dwarf varieties and 20 x 20 cm for tall saturated traditional basmati varieties. Adopt shallow planting (2-3 cm deep) in case of delayed planting, practice closer spacing of 15 x 15 cm for both the cultivars and increase the number of seedlings from 2-3 to 5-6 per hill. Gap filling twice at 7 and 14 days after planting has to be done.

11.6. Fertiliser Schedules

For traditional tall basmati varieties: Apply basal dose of 30 kg P_2O_5, 30 kg K_2O and 25 kg zinc sulphate/ha. Apply 60 kg N/ha in 3 splits- 50% basal, 25% at tillering and 25% at panicle initiation stage.

11.7. High Yielding Dwarf Basmati Varieties

Apply basal dose 50 kg each of P_2O_5 and K_2O/ha and 25 kg $ZnSO_4$/ha. Nitrogen @ 100 kg/ha may be applied in 3 splits- 50% basal, 25% at tillering and 25% at panicle initiation stage.

The field should be drained prior to top dressing with nitrogen and irrigated after 24-36 hours. If green manure crop has been grown and incorporated in the main field before rice planting, the quantity of fertiliser nitrogen may be reduced by 25%. Good water management facilitates proper tillering, efficient use of applied nitrogen and reduction of weed infestation.

Weed control: Hand weeding twice or thrice depending on the level of weed infestation at intervals of 3 weeks starting from 20 days after planting. Herbicides like butachlor or thiobencarb @ 1.5 kg a.i./ha or anilofos @ 0.6 kg a.i./ha applied at 5-7 days after planting to control the weeds.

Insect pests and disease: The important insects pests and disease that attack scented basmati rice and their control measures are given in Table 1.11.

12. NON-BASMATI RICE

While basmati rice enjoys an assured market abroad, no such advantage is available for non-basmati rice. In fact, 90% of the rice in international trade is of non-basmati type. For those countries like that are self sufficient in rice production, there is great scope to export non-basmati rice in a large way.

Significant increases in domestic production entailed in the substantial exports of our rice surpluses since 1978. Thus, one can distinguish two trends- one pertaining to period 1970-71 to 1977-78 when rice exports were at minimum and the other to the period 1977-78 to 1988-89 as period of relatively higher level of exports. Although non basmati rice was also exported to a limited quantity in the earlier years. Starting with a meager quantity of 27000 tonnes in 1989-90 it has reached 1.3-1.4 million tonnes during 1997-98.

Most important buyers are middle east, USSR and Europe. Saudi Arabia is the most important country importing non-basmati rice and accounted for 50% of country's exports. Generally preference exists for varieties having intermediate amylose content with intermediate gelatinization temperature and soft to medium gel consistency in international markets. We have to breed such rice varieties which can compete with high quality rice of USA, Thailand and Australia.

World markets demand six basic rice types *viz.*, (i) high quality long grain rice (ii) medium quality long grain rice (iii) short grain rice (iv) parboiled rice (v) aromatic rice (vi) glutinous rice (Efferson, 1985). Exclusive breeding programme for developing export quality non basmati rice, keeping in view the varied consumer demands from perspective importing countries would alone pave the way for development and sustenance of an assured market. There is a scope for instance to develop varieties of USA and Thai rice quality (grain length 6.6-7.75 mm and 5.5-6.6 mm with L/B ratio of over 3.0 and 2.1-3.0 respectively) which is in great demand in several rice importing countries. Following the opening up of Japanese rice market, it is possible to grow *Japonica* varieties in India especially in the foot hills of Himalayas and north eastern hilly region.

At present India exports rice varieties PR 106, Pantdhan 4, Haryana Gaurav, IR 64 among the long grain and IR8 among short grain types. A variety popular among the farmers as Terrycot possibly Sona and Haryana Gaurav also have high export demand.

13. HYBRID RICE

It was the Chinese who first demonstrated the commercial viability of hybrid rice technology in a self pollinated crop like rice, almost 21 years back. Exploiting the stable and well combing WA-based cytoplasmic male sterility restorer system, they could develop and successfully spread hybrid rice technology to over 58% of rice area in China.

The long cherished goal of developing and using rice hybrids in Indian agriculture, has been accomplished with the release of 4 hybrids for commercial cultivation in 1994. Subsequently 3 more hybrids have been released by the public sector. Few hybrids from private seed sector also being successfully marketed. Many more promising hybrids are in the offing. The prospects of large scale adoption of hybrid rice technology in India appears to be bright. During the wet season 1996, it is estimated that more than 60, 000 ha area has been planted to hybrid rice in various parts of the country with maximum area in Andhra Pradesh followed by Karnataka and Tamil Nadu. It is envisaged to cover 2.0 million ha under hybrid rice in favourable irrigated ecosystem by 2000 AD. The target areas are Punjab, Haryana, Western Uttar Pradesh and irrigated areas of Andhra Pradesh, Karnataka and Tamil Nadu.

Hybrid rice has its own specific characteristics in the development of individual plants and population structure. Further, the requirement of inputs like fertiliser, water and other cultural practices are different as compared to the conventional varieties. So mastering those characteristics and adopting suitable agronomic management practices are the key components for realizing the yield potential and successful large scale cultivation of these hybrids. A comprehensive package for cultivation of hybrids in various identified target areas in different seasons is being developed, so as to exploit the full potential of these hybrids, and thereby making cultivation of hybrids a clear cut profitable enterprise for the rice farmers.

13.1. Salient Features of Hybrid Rice

- Hybrids have a convincing yield advantage of at least one tonne per hectare more than the highest yielding inbred cultivars with similar maturity duration.
- Higher productive tillers and more number of grains per panicle
- With slight modification can be successfully grown under similar management practices as for common varieties

13.2. Suitability of Hybrids

The performance of hybrid rice is reported to be highly location-specific. At present the varietal choice is limited since only a few hybrids are available for cultivation. The suitability of these hybrids for a particular location depends on their duration, plant characteristics and the land situation in which they are grown. The currently available hybrids are well suited to irrigated tracts. A list of hybrids developed in different states and their performance are presented in Tables 1.9 and 1.10.

TABLE 1.9: Important hybrids developed for different states of India and their performance.

Region	Hybrid	Duration (days)	Yield (t/ha)	Advantage over best conventional check (t/ha)
Andhra Pradesh				
Telangana	APHR 1	130-135	7.14	1.87
Telangana, Rayalaseema and uplands of coastal Andhra	DRRH 1	125-130	7.30	1.80
Rayalaseema and coastal Andhra	APHR 2	120-125	7.52	2.31
Karnataka				
Irrigated	KRH1	120-125	6.02	1.44
Tamil Nadu				
May-June September-October planting	MGR 1	110-115	6.08	0.85
West Bengal				
Boro season	CNHR-3	125-130	8.10	1.70

TABLE 1.10: Performance of other important hybrids

Region	Hybrid	Yield (t/ha)		
		Hybrid	Conventional check	Yield advantage (t/ha)
Karnataka	KMHR 2	8.3	6.9	1.4
Andhra Pradesh	MTU HR 2008	7.1	5.7	1.4
	MTU HR2003	7.1	5.8	1.3
	DRRH1	6.7	5.4	1.3
	DRRH3	9.3	8.0	1.3
Uttar Pradesh	PHR1	6.5	6.0	0.5
	NDRH1	5.8	5.4	0.4
Haryana	PMS 2A/IR 31802	5.5	5.2	0.3
Tamil Nadu	IR 58025 A/C 22	6.1	5.1	1.0
	IR 58025A/C20	5.8	5.1	0.7

14. PRODUCTION TECHNOLOGY

14.1. Nursery Raising

Hybrid rice seeds cost more and need replacement every season. It is necessary to cut down costs on seed by optimizing seed rate through appropriate adjustment of seeding density in nursery and seedling number per hill while planting. Hence, proper management of nursery raising is important for its profitable cultivation. Nursery should be sown in the first week of June for *kharif* crop and in the second week of December for *rabi* crop. A total 600 sq. m area is required for raising nursery to transplant one hectare in the main field. Prepare 1-1.5 m wide nursery beds of any convenient length. Apply 5 kg urea, 10 kg single super-phosphate with 5-10 times well rotten FYM and 5 kg muriate of potash before final leveling. Broadcast 15 kg pre-germinated seeds (soak the seeds overnight in water, incubate in warm moist condition for 36-48 hours or until germination) on this well prepared nursery bed. The optimum seed density is 1 kg seed/40 sq. m. Thick sowing of seeds would lead to week seedlings. To check the damage by the birds and for better germination, broadcast a thin layer of well decompose powdered FYM manure immediately after broadcasting the seeds.

Keep the nursery beds moist for first few days and weed free. After the seedlings are about 1 inch high, keep a shallow layer of water. Top dress the nursery beds with 5 kg urea per 600 sq.m about 10 days before transplanting. All these practices ensure healthy and robust seedlings which are pre-requisites for higher yield.

15. LAND PREPARATION AND TRANSPLANTING

The land should be well puddled and leveled in order to reduce water loss through percolation. Give a gap of at least 10 days between initial and final puddling, for better weed control and nutrient availability.

Hybrids are more sensitive to extreme temperatures at flowering, especially at lower temperatures, than the conventionally bred varieties. Therefore it is desired to plant them in time preferably in the second week of July and January for *kharif* and *rabi* crops respectively.

The ideal and optimum plant population for good crop under normal planting conditions is 3, 33, 000 plants/ha. To achieve this the crop should be planted in the main field at a row-to-row distance of 20 cm and plant-to-plant distance of 10 cm. In case of random planting ensure 33 hills per sq. m. To ensure proper crop stand it is better to transplant 2 seedlings/hill instead of one in situations where there is possibility of flooding at initial stages particularly during wet (*Kharif*) season. Otherwise one seedling per hill is sufficient as hybrid seedlings have more tillering capacity.

An experiment conducted at CRRI, Cuttack during the wet season 1996 in order to evaluate the performance of hybrid rice (VRH 2) under different methods of planting, spacing and seedling densities showed that skipping one row after transplanting four continuous rows produced comparable yield as that of conventional planting. Among the different seedling densities, in normal season where there is no stagnation of water at planting, no significant differences in grain yield among 1, 2 and 3 seedlings/hill were observed indicating that one seedling per hill was adequate enough for achieving higher yield and saving on cost of seed. The sed rate could be minimised to 15 kg/ha by this system.

16. FERTILISER MANAGEMENT

Efficient fertiliser management is the key to improve the grain yield and response to added fertilisers in any crop production system and so in hybrid rice cultivation. The dose and time of fertiliser application depends up on many factors like basic soil fertility, the varietal reaction with pests and diseases, weed incidence and cropping pattern, besides the source and method of its application and synergistic effect with other nutrients. However for normal soils following fertiliser schedule is recommended for *kharif* and *rabi* crops.

***a) Kharif* (Wet Season)**

1. Basal dose: 50 kg N, 40 kg P_2O_5 and 50 kg K_2O/ha
2. 3 weeks after transplanting: 25 kg N/ha
3. 6 weeks after transplanting: 25 kg N/ha

***b) Rabi* (Dry Season)**

- Basal dose: 60 kg N, 50 kg P_2O_5 and 60 kg K_2O/ha
- 3 weeks after transplanting: 30 kg N/ha
- 6 weeks after transplanting: 30 kg N/ha

Delayed application of N and K coinciding with flowering can help realize the full potential of hybrid rice. Preliminary observation trial conducted at CRRI during dry season 1997 revealed that delayed N and K (fraction of the total recommended N & K) application at heading increased grain yield by reducing percentage of unfilled grains.

17. WEED CONTROL

It has been established that there is considerable reduction in grain yield if weeds are not removed within three weeks of transplanting. Therefore, in order to achieve good yield target the weeds should be removed as early as possible. The use of herbicides in controlling weeds is both efficient and cost effective. Effective weed control could be achieved by preemegence application of herbicides like butachlor or thiobencarb @ 1.5 kg a.i./ha 7 days after transplanting in 4-5 cm standing water. The liquid (EC) formulations may be mixed with 150 kg of sand per hectare and broadcast uniformly. Use of liquid formulations mixed with sand or loose soil is as effective as granular application and it is far more economical in comparison to granules

18. ECONOMIC EVALUATION OF HYBRID RICE TECHNOLOGY

People outside China still have their doubts about the economic viability of hybrid rice technology. The question in their mind is whether the level of yield increase that they are getting will be sufficient to pay the extra expenditure involved. The reports from various parts of the country suggests distinct economic advantage. In hybrid rice cultivation no extra expenses are involved except the cost of seed and to some extent extra cost on fertiliser plant protection measures. A preliminary estimate indicates that even if all these costs (10-15 per cent higher than the growing of conventional varieties) are taken into account, the yield advantage of approximately 1 t/ha can well take care of these costs and one can expect a net additional profit of about Rs. 2000-2500 per hectare over the conventional varieties. The advantage to the seed producer will be still much higher.

19. HARVESTING AND THRESHING

Drain out water from the field when grains in the lowest portion of the panicle are in the dough stage (about 20 days from 50% flowering). Allow the grains to harden. Harvest 30-35 days after flowering when stalks still remain green to avoid grain shedding. Moisture content should be around 20% at the time of harvest. The crop should be harvested when the grains of upper portion of panicles are clean and firm. At this stage atleast 80% of the grains are straw coloured. In general the crop is harvested with traditional sickles requiring 250-300 man-hr/ha. At the prevailing labour wages atleast Rs.1500/ha are required to harvest 1 ha of rice. With the improved design of sickles fabricated from high carbon steel and fitted serrated blades, harvesting is done faster. The improved design requires less muscle power and results in saving of human energy/ha. There is an overall economy in harvesting of the crop as well as reduction in fatigue to the workers during harvesting with improved sickles. Harvesting of rice is also done by reaper-winnower. These machines can be fitted with 20 HP and high powered tractor and power tilers. Self propelled 2-wheeled reapers are also available. The tractor-mounted reaper can harvest 0.3 ha in an hour, while the power tiller mounted reaper can harvest 0.2 ha in an hour. The approximate cost of a reaper winnower is Rs 30, 000 and a self propelled reaper with a diesel engine is Rs.30000. The cost of harvest by a tractor winnower is approximately Rs.600 and that of self propelled one is Rs.500 (Choudhury, *et. al.*, 1996). However these mechanization options are limited at present to high productive irrigated areas and there is a need to popularise these machines. Often heavy rains at the time of maturity of the crop (often in the case of upland rice) may cause germination of grains in the panicles. After the harvest of rice under lowland situations, the bundles are kept lying in the field for sun drying and lack of transport and immediate threshing facilities. Conventional way of threshing is the bullock treading or beating on a hard surface which consumes more time. Due to bulk trading there is breakage of the grains besides the land preparation for the successive crop gets delayed. To overcome this difficulty, for small and marginal farmers two types of threshers like pedal-operated thresher and portable low cost power thresher are available. The output of these machines are 60-70 kg/hr and 350-450 kg/hr respectively. In some areas tractor treading is also a practice which gives 40 bags/day.

20. PLANT PROTECTION

20.1. Insect Pest

Termites, gundhi bugs and mealy bugs are the important pests that attack upland rice crop causing severe yield losses. Termites infest rice crop at the time of sowing and after wards and reduce crop stand considerably. Gundhi bugs affect rice crop during post-flowering period and increase grain sterility. Termites can be controlled effectively by seed treatment with chlorpyrifos @ 0.5 kg a.i./100 kg seed (Rajamani and Jha, 1994). Gundhi bugs can be controlled by spraying the crop with monocrotophos or chlorpyrifos @ 0.5 kg a.i./ha in the forenoon hours. Mealy bugs attack during vegetative growth stage in patches. These can be controlled by spraying in patches 500 ml/ha of phosphamidon ((100 EC) or malathion (50 EC) at 1 ml/1 litre.

Blast and brown spot are the major diseases of rainfed upland rice. Some other diseases like sheath rot, sheath blight and false smut are also important. For controlling blast, dry seed dressing and spray of Hinosan @ 300 ml/ha has been found effective. Brown spot usually occurs in poorly managed rice soils and under prolonged drought spells. Deficiency and imbalance of certain nutrients influences disease intensity. Seed dressing with Indofil M-45 @ 2.0 g/kg seed and its spray @ 1250 g/ha may control incidence of brown spot. Sheath rot, which is an important disease could be controlled by early sowing. Growing of insect/disease tolerant varieties (see table 1.2) help in minimising the problem considerably.

Although irrigated and rainfed lowland rice crops share many of the same insect pests, yellow stem borer, green leaf hopper, brown plant hopper, gall midge, leaf folder, case worm, cut worm, hispa and gundhi bug are of some significance under lowland conditions in different states. Besides these, the root weevil *Echinocnemus oryzae* and mealy bug *Brevinnia* rehi are also gaining importance in some parts of the country. As there is only a limited information specifically developed for rainfed lowland situations the technology available for irrigated conditions on various aspects of pest management could be utilized in tackling pest problems.

Use of insect tolerant varieties: Growing of genetically resistant varieties is an important component of integrated pest management programme. The varieties like Panidhan, Kalashree and Tulasi possess some tolerance to stem borer, WBPH and gall midge. The variety Jagannath is moderately tolerant to green leaf hopper. Swarnadhan is tolerant to gall midge. The variety Udaya is tolerant to gall midge, BPH and green leaf hopper. By growing these varieties, it is possible to reduce investment on pesticides.

Use of insecticides: Since the rainfed lowland situations, in general, are characterised by waterlogging and flash floods, the application of insecticides in the conventional ways does not give the desired degree of pest control. Therefore, depending on the prevailing situation, the formulation of insecticide and method of its application should be determined. The information generated from the All India coordinated trials and others clearly indicated that

For transplanted crop seedling root dip prior to planting in 0.02% chlorpyrifos for 12 hours or in 0.02 % chlorpyriphos + 1% urea for 3 hours will be effective against gall midge and stem borer upto 30 days.

Soil incorporation of carbofuran or phorate or quinalphos @ 1.0 kg a.i./ha gives good control of stem borer and gall midge in the vegetative stage. For protecting the crop at heading stage, two rounds of foliar spray of quinaphos or chlorpyrifos or phosphamidon at the rate of 0.5 kg a.i./ha has proved very effective.

Foliar spray with phosphamidon, monocrotophos and carbaryl @ 0.5 kg a.i./ha were quite effective against green leaf hopper. However in RTV endemic areas Sevidol may be applied @ 1.0 kg a.i./ha area in seedling stage and again before 60 DAT to control green leaf hopper.

Chlorpyrifos, quinalphos, methyl parathion and endosulfan @ 0.4 kg a.i./ha at dough stage are effective against the climbing cut worm while quinalphos, endosulfan and chlorpyriphos @ 0.5 kg a.i./ha are effective against case worm, leaf folder, hoppers and whorl maggot.

Utilization of natural enemies: In general, parasites, predators and insect pathogens are found to be more abundant in rainfed lowland rice than in irrigated rice. Therefore, every effort should me made to conserve and encourage these natural agents in keeping the pest population under check.

20.2. Disease

The lowland rice areas depend on unpredictable monsoon rain and suffer from insufficient or excess water during the crop growing season. Such conditions greatly influence the occurrence and severity of the diseases and thereby the loss caused by the diseases. Sheath blight is the major disease followed by bacterial blight, brown spot, tungro, sheath rot and blast. However, the rice tungro virus followed by bacterial blight and sheath rot are the major diseases in deep water rice areas.

In the absence of accurate knowledge of the pathogens and diseases they cause it is difficult to formulate any standard procedure for disease management and control in rainfed lowland rice culture because the culture systems are very diverse. However, efforts are to be made to try the available information that may lead to disease management in rainfed conditions.

TABLE 1.11: Important disease and their control.

Disease	Control measures
Blast	– Adopt seed dressing with carbendazim (Bavistin 50 WP @ 4 g/kg seed) to control leaf blast at pre-tillering to mid tillering (2-5% disease severity) apply ediphenphos 35 EC (1 ml/l) or Carbendazim 50 WP @ 0.1 % (l g/l). at panicle initiation to booting (5% leaf damage) spray ediphenphos 35 EC @ 0.1% (1 ml/l) or Carbendazim 50 WP @ 0.1% (1 g/l) twice at an interval of 7 days.
Sheath blight	– Foliar application of carbendazim 50 WP @ 0.1% (1 g/l) is recommended to check disease.
False smut and Grain discolouration	– Apply Mancozeb (Dithane M-45) @ 25% during evening hours. Avoid spraying during flowering period.
BLB	– Reduce nitrogen application and apply in 2 or more split doses.

Management through host resistance: The use of resistant cultivars is still the most economical and feasible method of disease control. In high yielding cultivars grown in different states of India in lowland conditions, broad-spectrum resistance against different diseases is not available. Researches are underway to utilize some of the wild rice for developing multiple resistant varieties to these diseases.

Management of virus vector: In eastern India in the rainfed lowlands, tungro is one of the most important rice diseases. Its epidemic is closely related to the availability of inoculum sources, the vector population and the dispersal of vector. The control of tungro incidence may be achieved by suppressing the vector population in seed bed by applying insecticides or by timing the planting date.

Chemical control: Though many chemicals are found effective in controlling different fungal diseases, chemical control is not yet popular in rainfed lowland culture. This might be due to lack of accurate knowledge of the diseases and yield losses caused by them, non-availability of the chemicals to the farmers and the economically non-feasibility of the chemical control

The following plant protection measures are considered necessary for realising higher yields of irrigated rice:

1. Control pests like gall midge and stem borer in the nursery by applying carbofuran or phorate granules @1.0 kg a.i./ha or spray monocrotophos or carbaryl @0.5 kg a.i/ha.
2. Adopt seedling root dip technique before planting. Dursban 20 EC may be used for this purpose.
3. Take up need based control measures in the main field against various pests
4. For controlling stem borer, when there is 5% dead hearts or one egg mass/ sq.m.), apply carbofuran (0.75 kg a.i./ha), phorate, quinalphos or sevidol granules @1.0 kg a.i./ha.
5. For controlling brown plant hopper (at 5-10 insects/hill), spray carbaryl (0.75 kg a.i./ha or apply carbofuran @ 0.75 kg a.i./ha or phorate granules @ 1.25 kg a.i./ha.

6. Growing of insect resistant varieties in endemic areas help in reducing investment on chemicals. Eg: Shakti, Pratap, and Samalei etc. for gall midge, Sonasali and Udaya etc for BPH
7. For controlling blast, adopt the practice of seed dressing with pyroquilon or carbandezim @ 4 g/kg seed.
8. For controlling tungro, apply carbofuran granules @ 1 kg a.i./ha to control vector population.
9. For minimizing bacterial leaf blight reduce nitrogen application
10. Grow resistant varieties for the diseases. For eg: Vikramarya is tolerant to RTV; Samalei and Rasi for blast; IR36, Deepti, Govind, saket 4 for BLB.

21. REFERENCES

Behera, U.K. and Jha, K.P. 1998. Coose ideal rice varieties for different ecosystems of Kalahandi, Orissa. Indian Farming 47 (12):4-9.

Bhagat, K.L., Dahama, A.K. and Singh, Harbans. 1991. Performance of herbicides in direct seeded upland rice. Annals of Agril. Res. 12 (4): 422-424.

Bhan. V.M. 1983. Effects of hydrology, soil moisture regime and fertility management on weed population and their control in rice. In: Proc. Of the conference on weed control in rice, IRRI, held during 31 August to 4 September, 1981.

Choudhury, A.K., Parida, B.C. and Patel, S.P. 1996. Mechanization of rice cultivation and post harvest processing. Indian Farming 46 (9):38-41.

De Condelle (1886) Origin of cultivated plants, Kegan Paul, Trench & Co. London

De, R.N., Reddy, J.N., Mallik, S., Ahmed, T. And Roy, A.T. 1996. Strategies for varietal improvement to increase productivity of rainfed lowland rice in India. Indian Farming 46 (9): 29-32.

Dixit, R.S. and Singh, M.M. 1981. Studies on weed control measures in direct-seeded upland rainfed paddy. In: Abstr. Of papers, Annual Conf. Indian Soc. Weed Sci.:14.

Efferson, J.N. 1985. Rice quality in world markets. In: Rice grain quality and marketing. IRRI, Los Banos, Philippines:1-14.

Govindra Singh, Deka, J. And Singh, D. 1988. Response of upland rice to seed rate and butachlor. Indian J. Weed Sci. 20 (1: 23-30.

Jha, K.P. and Manna, G.B. 1989. On farm evaluation of promising early rice (*Oryza sativa*) varieties under rainfed uplands of coastal alluvium and red lateritic soils. Indian J. Agric. Sci. 59 (8): 490-494.

Jha, K.P., Das, F.C., Manna, G.B. and Mishra A.P. 1989. Yield and economics of rice cultivation as influenced methods of stand establishment under rainfed uplands and summer mid-lands of coastal Orissa. FSR Newsletter 3 (4):10-14.

Jha, K.P., Gangadharan, C. and Manna, G.B. 1991. On-farm evaluation of promising lowland rice (*Oryza sativa*) varieties and their response to date of sowing and nitrogen level under flood-prone lowlands of Orissa. Indian J. Agric. Sci. 61 (4): 237-242.

Katoch, P.C., Sharma, S.R., Kaushik, Gupta, J.C. 1989. Rice production in Himachal Pradesh. Indian Fmg. 38 (12):4-6.

Mahadevappa, M., Ratho, S.N., Rao, R.N., Vijaya Kumar, R. and Shenoy, V.V. 1996. Hybrid rice: a novel approach to augment rice production. Indian Farming 46 (9): 11-15.

Manna, G.B. 1983. Weed control in rice under different situations. Paper presented at the Annual Workshop of the All India Coordinated Rice Improvement Project, held during 9-12 April, 1983, Hyderabad, India.

Manna, G.B., Jha, K.P. and Rao, K.S. 1988. Potentiality of green manuring practice in the current rice production strategy. In: Biological Nitrogen Fixation Associated With Rice Production. Datta, S.K. and Sloger, C. (eds). Oxford and IBH Publ. Co., New Delhi, pp. 329-335.

Mishra, G.N., Singh, C, V, and Pyare Lal (1995). Nitrogen management in rainfed upland rice-a review. *Oryza* 32 (3): 156-165.

Moody, K. 1992. The role of the rice cultivars and related cultural practices in suppressing weed growth. Paper presented at the International Rice Res. Conf. Held at Los Banos, Philippines during 2-25 April, 1992.

Moorthy, B.T.S. 1992. Effect of methods of land preparation and herbicide use on weed control and crop performance of rainfed upland rice (*Oryza sativa* L.) in coastal Orissa. Indian J. Agric. Sci. 62 (6): 382-386.

Moorthy, B.T.S. and Das, F.C. 1992. Performance evaluation of two manually operated weeders in upland rice. Orissa J.Agric. res. 5 (1-2):36-41.

Moorthy, B.T.S. and Mittra, B.N. 1990. Uptake of nutrients by upland rice and associated weeds as influenced by nitrogen application schedules and weed management practices. Crop Res. 3 (2): 144-150.

Moorthy, B.T.S. and Mittra, B.N. 1991. Influence of seeding densities and weed management practices on the performance of upland rice. Thai J. Agric. Sci. 24: 19-27.

Moorthy, B.T.S. and Rao, K.S. 1991. Influence of method of stand establishment, and cultural and weed control practices on performance of rainfed lowland rice (*Oryza sativa*) in coastal Orissa. Indian J. Agric. Sci. 61 (8): 588-590.

Moorthy, B.T.S. and Manna, G.B.1989. Evaluation of certain preemergence herbicides for their efficacy in weed control in upland rice. Sci. & Cult. 55 (5): 176-178.

Panda, D., Samantray, R.N., Mohanty, S.K. and Patnaik, S. 1988. Green manuring with Sesbania aculeata - its role in nitrogen and yield of rice. In: Biological Nitrogen Fixation Associated With Rice Production. Datta, S.K. and Sloger, C. (eds). Oxford and IBH Publ. Co., New Delhi, pp. 305-314.

Pande HK and Reddy MD. 1988. Improved management technology for augmenting production potential of rainfed marginal lands under rice based cropping system in eastern India. Indian J. Res. Mgmt. 9: 105-111.

Patnaik, S and Panda, D. 1990. technology for increasing the production of rainfed rice in eastern India. Indian Farming 40 (9): 33-39.

Patro, G.K., Prusty, J.L. and Panda, S.C. 1989. Performance of new herbicides with nitrogen levels on upland rice in Orissa. Pesticides 23 (12):236-238.

Rajamani, S. and Jha, Sunita 1994. Management of termite and gundhi bug in uplands. Paper presented in the Mini workshop of FSR/E project held on 10 October 1994 at CRRI, Cuttack.

Reddy, B.B., Ghosh, B.C. and Panda, M.M. 1985. Flood tolerance of rice at different crop growth stages as affected by fertiliser application. Plant and Soil 83: 255-263.

Reddy, M.D., Sharma, A.R. and Panda, M.M. 1991. Effect of mixed row cropping of early and late maturing rice varieties on crop productivity under intermediate deepwater conditions. Exptl. Agric. 27: 79-85.

Richharia, R.H. 1964. Wild rice menace and possible methods of its control. CRRI publication:4 p.

Sharma, A.R. 1992. Effect of nursery fertilization and age of seedlings on flood tolerance and productivity of rainfed lowland rice (*Oryza sativa*). Indian J. Agric. Sci. 62 (10):657-663.

Sharma, A.R. 1994. Fertiliser management in rainfed lowland rice under excess water conditions. Fert. News 39 (5): 35-44.

Sharma, A.R. and Das, K.C. 1993. Weed and nitrogen management in direct sown rice (*Oryza sativa*) under rainfed flooded condition. Proc. Vol. III Indian Soc. Weed Sc. International Symp. On Integrated Weed Management for Sustainable Agriculture held at Hisar:1-5.

Sharma, A.R., Moorthy, B.T.S., Rao, K.S. and Venkateswarlu, B. 1995. Improvement of rice production technology in the conventional areas of eastern India. In Improving Crop and Animal Productivity - The Challenge of the Decade. Deb, D.L. (ed.). Associated Pub. Co., New Delhi, pp. 157-178.

Sharma, H.C., Singh, H.B. and Friegen, G.H. 1977. Competition from weeds and their control in direct sown rice. Weed Res. 17 (2):103-108.

Shivamadiah, N.C., Rama Gowda and Bomme Gowda, A. 1987. Studies on integrated weed management in drill sown rice. Curr. Res. 16 (4):51-52.

Siddiq, E.A. 1992. Rice in India - potential areas for research and development. *Oryza* 29:1-6.

Siddiq, E.A. 1996. Current status and future outlook for hybrid rice in India. In: Hybrid rice technology published by DRR: Hyderabad:1-26.

Singh, M. And Singh, M. 1989. Cultural -cum-herbicidal weed management studies in upland rice. In: Proc. 12th Asian Pacific Weed Sci. Soc. No. 2: 345-351.

Singh, Y.V., Jha, K.P., Moorthy, B.T.S. and Pradhan, S.C. 1995. Appropriate technology for increasing and stabilising crop yields in rainfed risk-prone lateritic uplands of Dhenkanal and Phulbani. Indian Farming 45 (5): 21-26.

Sinha, P.K., Singh, U.V., Singh, R.K., Mishra, G.N., Maiti, D., Shukla, V.D. and Variar, M. 1996. Rainfed upland rice-future strategies. Indian Farming 46 (9):25-28.

Sinhababu, D.P. and Venkateswarlu, B. 1995. Rice-fish in rainfed lowlands. Bulletin published by CRRI, Cuttack, Orissa.

Tewary, A.N. and Singh, M. 1991. Crop-weed competition in direct seeded rainfed rice. Indian J. Weed Sci. 23 (1&2):51-52.

Thaper, B.K., Lodh, S.B., Sobha Rani, N., Sidhu, G.S., Chaudhury, D. and Jena, K.K. 1998. Breeding quality rice for domestic and export markets In: Rainfed rice for sustainable food security published by CRRI and ARRW, Cuttack: 63-89.

Tosh, G.C., Reddy, B.S. and Nanda, K.C. 1981. Integrated weed management study in upland rice. In: Proc. Weed Sci. Conf. held at Hyderabad:14.

Vavilov, N.I., (1926) Studies on the origin of cultivated plants, Bulletin of Applied Botany and plant breeding Vol., 16 (2)

Wirdjahardja, S.E., Guharrdja, J, S.E. and Wiroatmodja, J. 1983. Wild rice and its control. In: Proc. Conf. On Weed Control in Rice, 1981, IRRI, Los Banos, Philippines: 319-325.

2. Maize

Dayanand

Over 90 per cent of the maize produce in the country is used directly as food at the sight of production. Maize is also used as an important ingredient in animal feed. It is a base ingredient of swine and poultry rations. Both wet and dry milling industry separate the fraction of the grain into germ, hull and endosperm to produce various products. Beside products, such as sleep concentration, gluten feed gallium meal, maize oil and maize oil cake are by-products. Use of starch in textile industry, paper industry, engineering, oil well drilling, batteries, leather, and foods materials.

Maize is used locally as the major cereal in the traditional areas by the local urban population in the form of chapati. The use of processed foods for breakfast and snacks is very little. In addition, green ears are consumed in roasted form. Bakery productions have played a important role in the human diet in the urban areas. Use of breed buns, rusks and biscuits etc. is becoming popular. Baby invalid protein supplementing foods, Opaque-2 variety can make much cheaper formulation of baby in valid and protein supplementing foods.

All the major sweeteners are commercially made from maize starch. Maize starch is a polymer of anhydrous glucose units (dextrose) and is the raw material for sweeteners. To make them starch chains, they are depolymerised with acids or enzymes or with both into dextrose or short chained dextrose polymers. The maize syrups and HFCS are sold as liquids. Many industries in Haryana, Gujarat, Punjab, Madhya Pradesh and Uttar Pradesh. have come up to derive starch from maize.

1. **Modified starches:** Maize starch and its derivatives are essential e.g. "binder" for high precision casting moulds, and compounds oil well mud, to cool and lubricate rotary drilling bits. Starch has 280 varieties on regular sale e.g. Polyvinyl alcohol, polyvinyl accato and acrylic adhesives both organic and synthetic products in the compiled field of sizing, finishing and adhesive products.
2. **Starch derivatives:** the main starch-derivatives i.e., dextrose and glucose (in the food confectionery and pharmaceutical industries) and to a smaller extent for destine (e.g. in foundry and oil well drilling operations).

3. **Maize oil:** Maize oil is a very fine cooking medium when fully refined with nearly 90 per cent polyunsaturated fats e.g. corn oil.
4. **Alcohol:** Alcohol both as a substitute for petrol as being developed in Brazil and as industrial ethyl alcohol.
5. **Organic/synthetic co-polymers:** Bio degradable plastics made by co-polymerizing organic (starch) and synthetic polymers are already in use for agriculture and other proposes.
6. **High fructose corn syrups (HFCS):** The first product equivalent to sucrose (normal sugar) calorically (higher perceived sweetness) other than sugarcane or sugar beet is derived from maize starch.

The utilization of high cellulose wastes, enzymic action converts tonnes of grain into 1 tonne of fermentable sugar to give 1/3 tone of alcohol so also 1 tonne of maize stalks converts into 4/5 tonne of fermentable sugar in turn 2/5 tonnes of alcohol.

1. ORIGIN

Maize has been cultivated for thousand of years. During archeological excavations in New Mexico, grains of maize and parts of maize ears were found in caves and rocks shelter, which are thought to be about 4,500 years old. Of even earlier origin, pollen grains of *Zea tripsacum* and *euchlaena* are said to be found at a depth of about 50 meters under Mexico city (Weather wax and Randolph, 1955). At the time of discovery of New world, maize was already cultivated in many parts of the American continent, especially in the more highly developed regions of Mexico, Central America and South America where it formed the staple diet of the indigenous population. Christopher Columbus writes of the new Gramineace found in the inner region of Cuba described by two members of his crew. He named this new plant as maize. It is still debatable when this valuable form of grain first found its way to Europe. Several historians suggest that C. Columbus would have brought it to Europe on the occasion of his first voyage of discovery to the new world. In any case it is known that members of the second expedition return to Spain with maize samples. Although in the first instance maize was grown in the gardens as a rarity, its value as a food crop was very soon recognized. So maize spread over large area of Spain, Portugal, France, Italy, South Eastern Europe and North Africa in the years following its introduction. The Portuguese brought it to West Coast of Africa. The beginning of 16th century and some what later to India and China.

Recently, it has been postulated that maize came to India well before Columbus discovered America in 1492.

The discovery and conquest of America in which the maize pants played a very great role, continually progressed. The Spanish and English settlers took over the whole complex of maize cultivations and usage from the Red Indians. With the development of the Mid-west in the 19th century, maize extended to its present most cultivation area, namely the States of Ohio, Indiana, Illinois and Iowa. Maize corn and the Mid West became synonymous. Maize is still king in the minds of middle west farmers, just as cotton ranked as sovereign in the hot plains of the Deep South. The varied nomenclature of maize in European languages e.g. Turkish corn, Welsh corn, Corn of Asia, Indian Corn, and so on leads one to the conclusion that people in general were of the opinions that maize originated in the eastern hemisphere. De Candolle (1886), however, in decides unequivocally in favour of a new world Zea is of Greek origin and was used as a classification of a various type of cereals. The species name mays is traceable back to the Indian world mahiz or morist. In different parts of India maize is known as Makka, Makai, Chhali, Bhutta, Cholum, Mokochana, Khaukiri,.

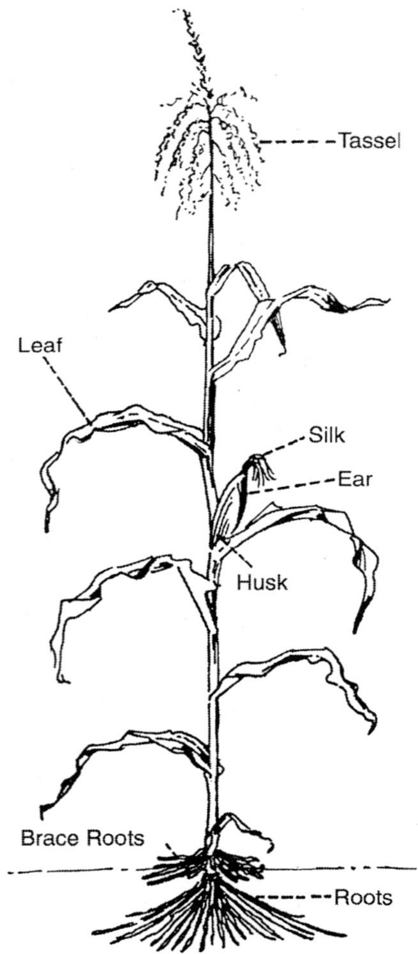

FIGURE 2.1: A maize plant

The wild forms of maize is still unknown. Various theories are put forward for its descent and region of origin.

For a long time it was thought that maize was descended from a hybrid between teosinte (Euchlaena Mexican schrad) and a known wild plant (Harshberger 1893, 1896); Collins, 1912).

The two Americans (Mangels-dorf and Reevas (1938, 1939, 1945) brought these and similar theories of origin into disrepute. After successful hybridization of maize and gama grass (*Tripsacum*) on the one hand and intensive analysis of the gene differences of maize, *Tripsacum* and *Teosinte*, they came to the conclusion that teosinte could not be the origin of maize. In their opinion the first cultivated form of maize was derived by mutation from of pod (*Tunicata* Sturt) maize indigenous to the lands of south America.

Further more these two scientists established that Euchlaena is natural hybrid of *Zea* and *Tripsacum* which accured after cultivated maize was brought to central America by man.

Weather wax (1935, 1950, 1954) believes that Zea, Tripsacum and Euchlaena arose from some common ancestor by divergent evolution. At present time generally believed that maize as we know it stems from a pod maize, the wild form of which, originating in south America, probably flourished on the eastern slopes of Andes and was also domesticated and improved their. This primitive cultivated pod maize in all probability developed many varieties, one of which mutated to the present one *Zea mays*, and in Mexico one of these also often called maize of the Andes hybridized teosinte (*Euchlaena*) and other forms of maize.

2. GROUP OF MAIZE

Various attempt have been made to classify *Zea mays* in to sub species and varieties on the basis of grain type and texture, but with little success, as most endosperm characters are now known to be simply inherited. Nevertheless, the main categories proposed by Sturtevant (1899), namely:

I	Dent corn	*(Zea mays indentata)*
II	Flint corn	*(Zea mays indurata)*
III	Sweet corn	*(Zea mays Saccharata)*
IV	Floury corn	*(Zea mays amylacea)*
V	Pop corn	*(Zea mays averta)*
VI	Pod corn	*(Zea mays tunicata)*
VII	Waxy corn	*(Zea mays ceretina)*

The genus *Zea* has only one spices *Zea mays* - is so rich in varieties that it present a very difficult task to botanist, so far as a unitary classification is concerned. The several thousand varieties are divided into various groups according to peculiarities, such as the structure of plant. Well known systems are those which has been given by Koernicke (1873), Von Sturtevant (1899) Kuleshow (1933) and Grepenscikow (1949). The principal agricultural groups with their peculiarities distribution and importance are summarized below.

2.1. *Zea mays indentata* sturt (Dent corn)

Most widely grown type of maize in the united State, Northern Mexico, Canada, Australia, South Africa, South Hungarian basin, Argentina. It probably resulted from the long continued crossing of flint x Gourd sled type. The main characteristic of the variety is a depression or dent in the crown of seed. The starch at the sides of the seeds in corneas while the soft starch extends to the apex (top). Rapid drying and shrinkage of the soft starch results in the characteristic denting.

2.2. *Zea mays indureta.* (Flint corn)

Flint corn is predominant in the agriculture of Europe (France, Italy and Balkan countries, except in the S. Hungarian basin), Asia, Central America and South America. It was grown widely in the USA in prehistoric and colonia times than at the present day.

In, general, the seeds of flint corn are hard and smooth and contain little soft starch. Flint maize is often earlier maturity, germinate better and has earlier plant vigour and move more tillers than the dent shins. The seed are smaller and therefore, it is well adopted to poultry feeding. Karnels are hard contains soft starch surrounded by hard starch.

2.3. *Zea mays,* Saccharata sturt (Sweet corn)

Sweet corn is primarily in the USA. In W. Mexico several varieties occur that may in the past have been used as source of syrups. This type of maize contains glassy, sweetish starch. It is characterized by a translucent sweet horny appearance when immature, and a wrinkled condition when dry. Sweet maize differs from dent by only one recessive gene which prevent the conversion of some of the sugar into starch. The ears are picked green for table use and canning. It contain 5% sugar as compared to 2% in dent and 2.3% in flint type.

2.4. *Zea mays amylacea* (Flour corn or softcorn)

It has been widely grown in the drier section of US and in some area in South America (Colombia, Peru and Bolivia) and in South Africa. Flour maize is some what similar to flint maize in plant and ear characteristics. It has kernels, which are composed nearly of entirely of soft starch. They especially develop no dent or only a very small one. Flour maize is one of the oldest types of maize and is frequently found in graves of the ancient Aztecst incas,. The endosperm consists of soft starch with little or no hard starch.

2.5. *Zea mays averta* sturt (Pop corn)

The cultivation is entirely restricted to the New World. Pop corn is grown in small quantities in many parts of the US especially in Iowa and Nebraska, and in C. Mexico. Pop corn does not constitute a particular type of maize, but in any kind that will pop on heating. The karnel are usually small and contain an even higher percentage of hard starch than flint corn. The corn in priority used for human consumption as freshly popped maize. The ability to pop seems to be conditioned by the hard starch when the starch grains are embedded in a tough, calcite, colloidal material which confines and resists. The steam pressure generated within the grain upon heating until it touches explosive force.

2.6. *Zea mays tunicata* sturt (Pod corn)

Pod maize is an unusual type of maize each of which is enclosed in a pod or husk. The ear is also enclosed in husks, as are the other types of maize. Pod maize belongs to one of the earliest domesticated types. It is not grown commercially, but is of considerable interest in studies of the origin of maize.

2.7. *Zea mays ceretina* kulesh (Waxy corn)

The kernels appearance is waxy. Starch molecule differs considerably from that of other types and resembles glycogen. The starch is gummy and has some of the characteristics of tapioca. Mostly found in E. Asia (N. Burma, Philippines, E. China and Manchuria). Developed in US and are now being grown commercially on a small scale to produce starch similar to tapioca starch.

3. CLIMATE

Though the crop is cultivated on a very wide range of agro-climatic conditions, yet moderate temperature with plentiful supply of water are most favourable. Temperature below 12 °C and higher that 30 C does not favour its cultivation. Wide variation in day and night temperature helps in higher production. Realization of such temperature range in Indian conditions in the traditional *kharif* season is not possible. However, this type of climatic condition can be realized during *rabi* season. Higher temperature through out the crop season curtail the total crop duration and lead to poor crop yield. Under Indian condition *kharif* crop of maize is sown at a temperature in the range of 35-40°C, which is very high. Subsequent temperature also rage in between 28-44 °C. Temperature effect on the growth of maize has been shown in the given below:

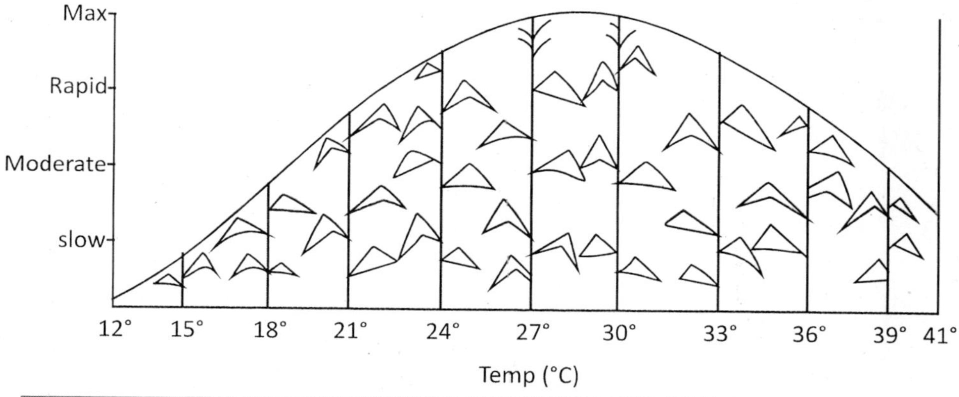

FIGURE 2.2: The relationship between temperature and growth rate.

The relationship between temperature and growth rate. face increasing difficulty in taking up water fast enough to keep plant cells turgid. When soil moisture is in short supply the temperature less than 27 °C is more suitable.

4. SOIL AND SOIL PREPARATION

The main objectives of land tilling are to prepare a good tilth in the soil for seed germination, scrap the weeds and stubble of previous crop and conserve moisture for crop. These objectives can be fulfilled by selecting suitable type of machinery for specific type of soils. By wise selection of machinery, land preparation can be completed by minimum number of operation thus cutting the cost of land preparation. Kernels need a soil that is warm, moist, well supplied with air, and find enough to give good contact between the seed and the soil.

Depending upon the agro-climatic conditions various hybrids and composites have been identified and recommended for cultivation. Undoubtedly, hybrids and composites have higher production potential than local.

5. CROPPING SYSTEM

Maize based crop rotations and intercrops in major agro-climatic zones

North-Eastern Hills Zone	
Rotations	
Maize-Wheat	Maize-Mustard
Maize-Potato	Maize-Barley
Maize-Potato-Potato	Maize-Radish-Wheat
Intercrops	
Maize+Mash	Maize+Soybean
Maize+Redgram	Maize+Cowpea
North Plains Zone	
Rotations	
Maize-Wheat	Maize-Mustard
Maize-Oats	Maize-Wheat-Moong
Maize-Potato-Moong	Maize-Wheat-Cheena
Maize-Potato-Wheat-Moong	Maize-Lahi-Wheat-Moong
Intercrops	
Maize+Blackgram	Maize+Soybean
Maize+Greengram	Maize+Cowpea
Maize+Clusterbean	
Eastern Plains Zone:	
Rotations	
Maize-Wheat	Maize-Potato
Maize-Ragi	Maize-Linseed
Maize-Sunflower	Maize-Potato-Cowpea
Maize-Potato-Wheat	Maize-Potato-Moong
Intercrops	
Maize+Blackgram	Maize+Redgram
Maize+Greengram	Maize+Soybean
Maize+Cowpea	Maize+Sesame
Southern Central Plateau:	
Rotations	
Maize-Wheat	Maize-Sunflower/Safflower
Maize-Castor	Maize-Groundnut
Maize-Linseed	Maize-Wheat-Moong
Maize-Potato-Groundnut	Groundnut-Wheat-Maize
Rice-Maize-Gram	
Intercrops	
Maize+Blackgram	Maize+Redgram
Maize+Greengram	Maize+Soybean
Maize+Cowpea	Maize+Sesame
Maize+Clusterbean	Maize+Horsegram
Maize+Sunhemp	Maize+Groundnut

Western Central Zone:	
Rotations	
Maize-Wheat	Maize-Barley
Maize-Mustard	Maize-Safflower/Sunflower/Linseed
Maize-Gram	Maize-Wheat-Moong
Maize-Mustard-Moong	Maize-Potato-Wheat-Moong
Maize-Lahi-Wheat-Moong	
Intercrops:	
Maize+Blackgram	Maize+Redgram
Maize+Greengram	Maize+Soybean
Maize+Cowpea	

6. SOWING

Sowing can be accomplished by machine or manual labour. Mechanical sowing by and large, is a better method of seed dropping in order to get desired plant population and uniform spacing. In normal case seed is dropped by hand behind the plough and also dibbled in hills. Whatever method of sowing is adopted the objective of obtaining the desired plant population should be achieved. Seed germination is not perfect when seeds are dropped in furrows behind the plough. Thus, a little higher quantity of seed is used. Time of sowing in different zones in *kharif/rabi* is given below.

TABLE 2.1: Sowing time of *kharif/rabi* maize in different zones

Zone	Kharif	Rabi
Hills of U.P., J & K, H.P., Sikkim, Assam & W.B.	1st week of April to 1st week of June	-
Delhi, Punjab & Rajasthan	Last week of June	Second week of November
U.P. & Bihar	Last week of May-15 June	First week of November
A.P., Gujarat, M.P., Karnataka, Maharashtra & T.N.	End of May to second week of June	Middle of October to first week of December
Diara area of U.P. & Bihar	25th May to third week of June	Third week of November
High altitude > 2500 ft.	Second week of April	-
Rainfed area (Rajasthan).	Third week of May	Middle of October

6.1. Time of Sowing

Time of sowing is indirectly determined by the soil and atmospheric temperature as well as by the supply of water. Thus, there is a general preference for sowing of the crop at the time of onset of monsoon. Long spell of higher temperature of summer season with low humidity desiccate both soil and air. Unless the soil and air. is brought down by rains, it is not promising to sow maize crop.

In case this rule is violated the seedlings get burnt due to high temperature of soil and air in spite of operating at high water supply which may not satisfy the soil and atmospheric need. Thus the onset of first monsoon rains or a little earlier is the optimum time of sowing the crop. Late planting in mid monsoon period creates numerous problems. Continuous and heavy rains and subsequently water stress conditions are abnormal conditions for crop growth which affect its yield adversely.

6.2. Transplanted Maize

Can be successfully cultivated by transplanting in winter season in the fields vacated by potato, toria and early harvested sugarcane. The nursery sown from 10-20 November n Punjab, Haryana and western Uttar Pradesh can be transplanted from mid-December to mid-January. For transplanting in the second fortnight of January, the nursery may be sown from 21 to 30 November. Trans-planting of nursery older than 60 days may not be done in the above parts of the country. However, only one month old seedling may be used in eastern Uttar Pradesh and Bihar where temperature is higher without frost.

To raise nursery for one hactare, sowing may be done with 25 kg seed in 1/10 ha/ by kera method, keeping a row to row distance of 20-22 cm. Well rotten farm yard manure @ 1.5 to 2.0 t/ha or 3.0 kg N, 2.5 kg P_2O_5, 1.2 kg K_2O and 1.0 kg zinc sulphate/ha may be applied to nursery area. Weed control May be carried out by pre-emergence application of weedicide as given for the directly sown crop or by hand hoeing. Uprooting of the nursery may be done under good water with *khurpa*.

The seedlings may be transplanted on southern side of east-west ridges or in flat fields. The seedlings may be placed in furrows and roots covered manually in flat fields. First irrigation should be given soon after transplanting and subsequent irrigation should be given 3-10 days after transplanting.

6.3. Method of Sowing

Three methods of sowing are commonly followed under Indian conditions.

- **Ridge Sowing:** Ridges are prepared and the sowing is done on top. This system of planting is very useful under high rainfall area of heavy soil. Excess water flows through the trenches and thus contact with seed or plant is avoided.
- **Flat Sowing:** This is a traditional method followed in most of the maize growing area. On light soils crop is sown on the smooth seedbed. If necessary crop can be earthen up subsequently to avoid plant lodging.
- **Furrow sowing:** Under low moisture condition, this is one of the most effective methods of sowing. Moisture in furrows continues to be available for a longer period. Rain water is stored in furrows. This method is practiced under dry land farming. Furrow method of irrigation can save 30-40 per cent water. This method is not only economical for water but it also produces higher grain yield, besides uniform distribution of water.

Apart from these methods of sowing, sod planting, trash planting and minimum tillage planting are practiced under different conditions.

6.4. Seed Rate

One should ensure that the seed is viable and free from external infections. Information on seed germination should be obtained for deciding the quantity of seed to be used. Seed germination standard should be 99 to 100% and the quantity of seed required for dibbling method is about 15 kg/ha. However, in planting by other mechanical methods, one may use as high as 25 kg seed/ha. The seed rate should be so adjusted as to obtain the desired plant population. The optimum plant population for achieving high yield is around 65 thousand plants/ha.

6.5. Spacing

To obtain desired plant one may sow the seeds into rows spaced from 60-90 cm apart and thereby adjusting seed to seed distance within the line. This can be achieved by maintaining 60 □ 25, 75 □ 20, 90 □16 cm between the rows and within rows. However, for practical ease a spacing of 75 □ 20 cm is more appropriate. This spacing is quite convenient for machinery drawn by tractor or bullock power. Any inter cultivation by mechanical method is less time consuming and more efficient. Thus, the plant spacing of 75 □ 20 cm is quite desirable.

6.6. Depth of Sowing

For obtaining perfect germination and uniform stand, placement of seed at a desired depth is one of the most important factors. Appropriate depth has to be decided by the user himself. It depends upon the prevailing conditions of the area, difference between soils, geographical conditions, varying temperature and the probability of soil being dry at planting time. In dry soils, the seed should be placed in moist soil, even one has to go as deep as 3 inches to 3.5 inches in clay soil, 4 inches to 4.5 inches on silts and 5 inches on sands. In cool weather at low temperature, 0.5-1 inches and at high temperature 3 inches or still lower depth can be maintained. Under optimum temperature conditions, a uniform depth of 2 inches is ideal for perfect germination. Higher temperature is responsible for high evaporation rate and thereby less moisture content in the surface soil. Therefore, seed placed in that zone will not get sufficient moisture.

7. MANNERS AND FERTILISERS

Requirement of nutrients by hybrids is higher because of its greater potentiality for grain production. Much higher responses are indicated by hybrid maize to N, P, K and sometimes Zn as compared to local varieties. Growing local material at a high nutrient level need not result in higher grain yield. But on the contrary, high levels of nutrient for hybrid seed proves beneficial.

7.1. Nitrogen

There is gradual increase in the requirement of nitrogen by the growing crop, the highest nitrogen requirement being exhibited at the flowering stage. Subsequently, the demand for nitrogen starts declining. Therefore, adequate nitrogen supply should be ensured from germination to the flowering stage. To achieve this, nitrogen is usually applied in three equal splits at sowing, knee high stage and tasseling stages. In non-sandy soils, uniform supply of nitrogen can also be accomplished by two equal application-1/2 at sowing and remaining half at knee high stage. A nitrogen level in the range of 100-120 kg/ha is applied with a view to obtain 40-50 q grain yield/ha and likewise it can be reduced or increased as per its expected yield.

7.2. Phosphorus

It is the next most important plant nutrient after nitrogen which is found deficient in in most soils. It has a beneficial effect on root growth and plant health. Maximum root development is completed until knee high stage. Thus the nutrient should be applied initially at the early stages. Because of its low solubility in water, it should be applied in moist zone to be transformed quickly for early absorption by plant. The dose of phosphorus should be balanced with the dose of nitrogen applied. Its application in single dose as a placement below seed is highly desirable.

7.3. Potassium

Although it is considered to be the third most essential fertiliser element, it is not found deficient in most of the soils. It is essential for vigorous growth of the plant and for so many other meabolic activities. The young seedling does not need much potassium, but the rate of uptake jumps upto a peak level during three weeks prior to tasseling. Deficiency symptoms, if any, usually appears during the time, the plant is one foot tall to the time of tassel emergence. Placement of 30-40 kg/ha K_2O, a little away from the seed is generally found to be quite adequate. However, this amount can be increased with increased rate of N and to balance the nutrient status of soil for better uptake of total essential nutrients.

7.4. Zinc

When we aim at higher production levels with adequate amount of fertiliser nutrients. Some of the micro-nutrients become limiting. For instance, higher uptake of other nutrients increase the demand of zinc. It is, therefore, advisable to apply zinc sulphate @20-25 kg/ha in those conditions where its deficiency is more conspicuous.

The nutrient requirement of maize is given below.

7.4.1. Fertiliser Recommendation for Different Zones of India

Zones		*Kharif* maize		*Rabi* maize	
		Doses	Method	Doses	Method
Hills of U.P., J & K, H.P., Sikkim, Assam & West Bengal	LD	90-40-30-25	M1	-	-
	MD	80-40-30	M1	-	-
	ED	60-40-20	M1	-	-
Delhi, Punjab & Rajasthan	LD	120-60-40	M1	180-60-40	M3
	MD	90-40-30	M1	120-60-40	M3
	ED	60-30-20	M1	-	
U.P. & Bihar	LD	60/120-40-30	M1	180-60-40	M3
	MD	60/90-40-30	M1	120-50-40	M3
	ED	45-30-20	M1	-	-
A.P., Gujarat, M.P., Karnataka, Maharashtra & T.N.	LD	150-75-40-10 irrigated	M2	150-60-40-10	M3
		100-50-25 rainfed	M1		

Zones		*Kharif* maize		*Rabi* maize	
		Doses	Method	Doses	Method
	MD	90-40-40	M2	100-50-40	M3
	ED	60-30-20	M2	-	-
Diara area of U.P. and Bihar	LD	120-60-40-25	M1	180-100-100	M4
	MD	-	-	120-50-40	M4
	ED	80-40-40	M1	-	-
Rainfed area of Rajasthan	LD	-	-	180-60-60	M3
	MD	90-60-30	M2	120-50-40	M3
	ED	60-40-30	M2	-	-
High altitude > 2500 ft.	LD	160-60-60	M2		

M1 = ½ at sowing + ½ at knee high stage; M2 = 1/3 at sowing + 1/3 at knee high stage + 1/3 at flowering; M3 = ¼ at sowing + ½ knee high + ¼ at flowering stage; M4 = ¾ at sowing + ¼ knee high stage.

A number of maize hybrids and composites, both yellow and white grain types, for *kharif* and *rabi* seasons have been recommended by the central and/ or state Varietal Release Committees during the past. Details of these varieties are given below:

7.4.2. Recommended Hybrids and Composites

Hybrids	Grain types	Maturity (days)	Area of adpatation
Ganga safed-2	White semi flint *Kharif*	100-110	Recommended for all states except hilly areas and Karnataka state.
Ganga-5	Orange yellow semi flint	100-110	All states except Karnataka.
Ganga-9	Yellow semi flint	100-110	Assam, Bihar, Himachal Pradesh, J & K, Manipur, Meghalaya, Nagaland, Sikkim, U.P., West Bengal, Arunachal Pradesh & Delhi.
Deccan	Yellow semi-flint	100-110	Gujarat, Andhra Pradesh, Karnataka, Madhya Pradesh and Maharashtra.
Deccan 103	Yellow semi-flint	100-110	Across the country
Him 123	Yellow semi-flint	100-110	Assam, Himachal Pradesh, J & K, Nagaland and hilly region of U.P.
Him 128	Yellow semi-flint	100-110	Himachal Pradesh, Sikkim and hilly region of U.P.
Sartaj	Yellow orange flint	95-100	Across the country.
Composites			
Vijay	Orange semi-flint	95-100	Across the country except Karnataka.
Kisan	Yellow semi-flint	95-100	Across the country except, A.P., Gujarat, H.P. Karnataka, M.P., Maharashtra, Orissa & Tamil Nadu.
Vikram	Orange yellow semi-flint	90-100	Across the country except A.P., Assam, H. P., J & K, Karnataka, Maharashtra, Sikkim, Tamil Nadu W. B. Arunachal Pradesh & Mizoram.

Hybrids	Grain types	Maturity (days)	Area of adpatation
NLD	White semi-flint	100-110	Meghalaya, Nagaland, Sikkim, West Bengal, Arunachal Pradesh and Mizoram.
COI	Yellow semi-flint	100-110	Tamil Nadu
Medium Maturity			
Ageti-76	Orange semi-glint	90-95	Across the country except Karnataka, Rajasthan, Tamil Nadu and hilly areas of U.P.
Tarun	Yellow semi-flint	85-90	Bihar, Haryana, Punjab, U.P. & Delhi.
Parbhat	Yellow semi-flint	85-95	Bihar, Gujarat, Karnataka, M.P., Maharashtra, Punjab, Tamil Nadu & U. P.
Navjot	Orange yellow semi-flint	85-95	Across of country.
Mansar	Orange yellow	85-95	Himachal Pradesh, J & K, and Hill region of U.P.
Harsha	Orange yellow semi-flint	95-100	Andhra Pradesh, Karnataka, Madhya Pradesh and Maharashtra.
Early Maturity			
Hunius	Orange semi-flint	80-85	Assam, Gujarat, Himachal Pradesh, J & K, Maharashtra, Sikkim, U.P. Delhi, Mizoram.
Diara-3	Orange yellow semi-flint	75-80	Across the country except Karnataka.
D 765	Orange semi-flint	75-80	Across the country except Karnataka.
MCU-508	Yellow semi-flint	75-80	Across the country except Haryana Karnataka, Punjab, Sikkim.
Arun	Orange yellow semi-flint	80-85	Punjab, Rajasthan, Delhi, Haryana, U. P. and Diara region of Bihar.
Renuka	Yellow semi-flint	75-80	Karnataka.
VL Makka 88	Yellow semi-flint	80-85	U.P. and H.P. hills.
Kiran	Orange flint	8-85	Across the country.
Recommended for specific states			
Varieties			
C 6	Orange yellow flint	Above 100	Jammu & Kashmir
C 15	Yellow semi-flint	Above 100	Jammu & Kashmir
VL 16	Orange yellow semi-flint	Above 80 days	Hills of Uttar Pradesh ranging from 600-2400 metres a.s.l.
VL Amber pop	Yellow pop	100-105 (Lower hills) 115-120 (Higher hills)	Hilly areas of U.P.
VL-41	yellow flint	80-85	days Hilly areas of Uttar Pradesh and high elevation of H.P.
Naveen	Yellow flint	90-95	Plains of Uttar Pradesh.
Kanchan	Orange semi flint	75-80	Uttar Pradesh
Amber	Yellow pop	80-85	Andhra Pradesh
Rohini	Orange semi flint	90-100	Andhra Pradesh
Manjari	Orange semi flint	90-100	Maharashtra
Suwan composite	Yellow semi-flint	95-100	Bihar plains

7.4.3. Recommended for Fodder

Variety	Grain type	Area of adaptability
African tall	White semi-flint	Across the country.
Rabi		
Deccan	Yellow semi-flint	Karnataka, Andhra Pradesh.
Deccan 103	Yellow semi-flint	Karnataka, Bihar, U.P., Andhra Pradesh, Gujarat, Maharashtra Tamil Nadu, M.P., West Bengal, Orissa, Punjab, Rajasthan.
Ganga-5	Yellow Orange Semi-flint	Bihar, U.P., Andhra Pradesh, Gujarat, Maharashtra, Tamil Nadu, Madhya Pradesh, West Bengal, Orissa, Rajasthan.
Ganga-9	Yellow Semi-flint	Bihar, Uttar Pradesh
Ganga-2	White Semi-flint	Rajasthan.
Hi-starch	White semi dent	Bihar, Gujarat, West Bengal, Orissa, Rajasthan, Uttar Pradesh.
COH 1	Yellow semi-flint	Tamil Nadu.
Composite		
MCU 508 (Early)	Yellow Semi-flint	Bihar, Uttar Pradesh.
Renuka (Early)	Yellow Semi-flint	Karnataka.
Hunius (Early)	Orange yellow Semi-flint	Maharashtra, (For intercropping in Sugarcane and turmeric).
CO-I	Yellow Semi-flint	Tamil Nadu.
Manjari	Orange Semi-flint	Maharashtra
Partap Composite I	Orange yellow Semi-flint	Punjab.

7.4.4. For Rabi

State	Maize hybrid/composites	
	White-grain type	Yellow-grain type
Bihar	Hi-Starch Comp. Lakshami Comp. Hemant	Deccan 103, Ganga 5, Ganga 9, Diara MFS 4
Uttar Pradesh	Hi-Starch	Deccan 103, Ganga 5, Ganga 9
Andhra Pradesh	-	Ganga 5, Deccan 101, Deccan 103
Gujarat	Hi-starch	Ganga 5, Deccan 103
Maharashtra	-	Ganga 5, Deccan 103, Manjari Composite, Hunius Composite for intercropping
Tamil Nadu	-	Ganga 5, Deccan 103, CoH 1, Co 1
Madhya Pradesh	-	Ganga 5, Deccan 103 m Deccan
West Bengal	Hi-Starch	Ganga 5, Deccan 103
Orissa	Hi-Starch	Ganga 5, Deccan 103
Punjab	-	Pratap Composite 1, Deccan 103
Rajasthan	Ganga 2 Hi-Starch	Ganga 5, Deccan 103
Orissa	Hi-starch	Ganga 5
Karnataka	-	Deccan, Deccan 103, Renuka

8. IRRIGATION

Water management of maize for high production is another important aspect so far the monetary inputs are concerned. Traditional crop of maize is greatly affected by vagaries of weather conditions. Rains in this part of the country is very uncertain and may either cause prolonged drought or water logged conditions. It is, therefore, essential to consider both the aspects.

Maize cultivation is extended to a very wide range of agro-climatic condition from drier region of Rajasthan having scanty rainfall to parts of Assam having annual rainfall of the order of 2000 mm.

A total rainfall in the rage of 500-600 mm during the crop season would be quite enough for maize, if the water losses through different sources are kept to the minimum. But in actual practice a large quantity of water is lost through runoff, leaching and evaporation and as a matter of fact very little water is left for use for metabolic activity. Monsoon rain which is generally erratic may cause either prolonged drought or water logged conditions both of which are highly detrimental to crop production.

In July precipitation generally exeeds evapo-transpiration which may result in water logged conditions. The crop at initial growth stage is very tender and is likely to suffer from water logging connotations. It would, therefore, be necessary to be prepared for such an eventuality from the beginning to minimise losses by making provision for proper drainage system. Strong bunds around the field should be provided to check inflow of water and on cessation of rain proper drainage facility should be created. If there is an adjacent sugarcane field, excess water may be transferred to that field. It is important to note that few hours water logging at germination stage may cause complete wipe out of the crop. Flooding the crop for a duration of 24 to 72 hrs. may cause 30-40% damage in crop yield. It is, therefore, essential to prevent such a high loss by proper water management. Wherever possible, maize sowing may be undertaken before onset of monsoon preferably 10 to 20 days in advance. By doing so the high rainfall period will coincide with grand growth period of the crop and may not have any adverse effect on the crop yield. Crop sowing on first monsoon shower resulted in 5 to 25 per cent increase in yield. Yield reduction due to unfavourable moisture condition can also be prevented by selection of suitable crop variety.

Coarse soils cause faster drainage of water and retain little moisture and are thus likely to create early drought condition. Retention of more water (28 per cent) by clay soil is quite helpful to the crop in drought spell but it may cause set back to the crop in a period of prolonged flooding. Thus the property of soil may be kept in mind while planning water management for maize crop. Water requirement to produced 1 kg of dry matter is the lowest in case of maize. Thus, a very high water use efficiency can be achieved as compared to other cereal crops. Although maize needs less water than soybean or pea but its water requirement for germination is 10 to 30 times more than that of wheat, berseem and mustard. It is, therefore, imperative that at the time of germination available moisture in case of maize, pea and soybean should be high so that each and every seed may germinate properly.

Availability of moisture at later stages has a direct bearing on the growth and yield of maize. A moisture content in the range of 18 to 21 per cent accelerated the growth rate of maize plant (i. e. 5 mm/hr.) as compared to 1.4 mm/ha, at lower moisture content of 9 to 12 per cent. Adequate soil moisture content throughout the crop growth period resulted in maximum grain yield. The optimum availability of moisture may be maintained by irrigating crop at critical stages. Six critical stages i. e. i) early knee, ii) late knee, iii) flowering, iv) silking, v) milk and vi) dough stages for growth and development have been identified for the prupose of irrigation. However, optimum potential can be maintained by 5 irrigations, missing the first stage. But skipping irrigation at other stages results in significantly lower grain yields. Restricting water supply any further may cause serious water stress and thus lower yields. With four irrigations minimum loss may occur by escaping 1st and 6th stages but the reduction would be much higher if flowering and silking stages are skipped. Maintaining an optimum moisture regime from late knee to silking stage with four irrigation is highly conducive to crop growth as compared to other combinations. For three irrigations, late knee high, flowering and silking stages are the most critical stages. Three irrigations applied at these stages would result in almost as much as that obtained by four irrigations. Any other schedule for a total of 4 irrigations would be highly detrimental to the crop.

Under moisture stressed condition to achieve higher water use efficiency, crop has to be sown in furrows instead of ridges. In low rainfall areas rain water can be retained in depressed zone for a longer period of time. Higher moisture content in furrows near the root zone continues to prevail for a longer period. Yield received from this method found 10 per cent higher than that obtained under ridged method of sowing.

9. WEED CONTROL

If weeds are allowed to grow in maize field they roughly remove equal or even more amount of plant nutrients and water. Weeds generally have a better chance of growing in maize field due to lower competition offered by initial slow growth rate of maize plants and comparatively faster growth of weeds. This leads to almost complete failure of crop if the weeds are not taken care.

By nature weeds are more tolerant to adverse condition and can grow faster then crop plant which are very sensitive to adverse conditions.

Any unwanted plant at a place where it is not required is a weed. Even a crop plant growing in another crop may be considered a weed. Most weeds grow naturally and get propagated from crop field. Weed seed can come mixed with crop seed, through irrigation water, air and through animal excreta. These weeds create serious problem for crop production by reducing crop yield by 5 to 90% under poor or no management conditions. Besides, weeds, harbour diseases and pests which indirectly cause undesirable conditions. Bulk of water and nutrients are also rubbed by weeds. All the dormant inactive weed seeds become viable as soon as first monsoon showers are received. Weeds are genetically better equipped to retard economic crop growth. They have an inherent fast growing and fast multiplying capacity. They pose a serious threat in fields of wider spaced crops like maize. Most common weeds of maize are given below:

9.4.5. Common Kharif Weeds in Maize

Monocot	Dicot
Cyperus rotundus	*Corchorus actangulus*
Digitaria sp.	*Trianthema monogyna*
Digitaria sanguinalis	*Trianthema portulacastrum*
Enchinochloa colonum	*Tribulus terrestris*

All these weeds may be found together at different intensities depending on variable agro-climatic conditions. A weed like *Cyperus rotundus*, which is perennial in nature is found almost all over the country. *Trianthema monogyna* is one of the most common and most abnoxious weeds in this part of country. Dicot weeds spread at a very fast rate and cover the surface area very soon if not controlled timely. Clean seed bed throughout the crop season is a most desirable condition for maize cultivation. Right from the beginning upto 75 days after sowing when the crop reaches the grain filling stage is most desirable. Weeding may be undertaken as soon as the crop germinate and attains 3-4 leaf stage (about 25 days after sowing) and may be continued till 75 days after sowing Any delay in weed removal may cause fall in grain yield. Unweeded crop is likely to give 50 per cent less yields as compared to clean seed bed. Weeds are usually controlled by manual labourers in 2-3 operations. But continuous rain may not allow the manual weeding operation to be undertaken in general. Thus, chemical weed control or the integrated weed management practice become necessary. Atrazine is one of the most efficient herbicides which remains effective for a maximum period of 30 days. It in abundance and so also the dicots. At this stage due to increase in plant height manual weeding becomes difficult. The attempts made in recent years have successfully overcome the problem by the use of paraquat. By its nature this chemical is a non-selective herbicide and harmful to all the plants. It is, therefore, to be directed on weeds plants only. The practice may be advocated where manual labourers are not available.

Atrazine is an effective herbicide for weed control in pure maize crop, but it cannot be used in legume cereal mixture. Among many herbicides tested for such a situation pendimethaline has been found to be the most effective and protects both the legume and cereal crop in pre-emergence application. Oxadizone, another herbicide, is comparatively less effective but has been found better than other chemicals like fluchloraline. Mechanical weed control in such a situation is impracticable.

The practice of chemical weed control may not be considered mandatory for general cultivation. So long maize is a crop of poor farmers in majority, manual weeding has many advantages and may be given first priority over the chemical method. Residues of chemical persist in soil and gradually build up the store and may likely to pollute the soil and water environment. In years to come soil may be unproductive and worthless to produce any plant. Never the less use of chemicals should be restricted. In general, use of chemicals in agriculture may lead to any unknown consequences.

10. PLANT PROTECTION

10.1. Diseases

10.1.1. Maydis Leaf Blight

1. Grow resistant maize cultivars, such as Deccan, Deccan-103, Ganga-5, Deccan-101, VL-42, Ganga-4.
2. Use protective fungicides of thiocarbamate group (dithane-Z-78, dithane M-45, zineb) @ 2.0-3.0 kg/ha in 1000-1200 L of water, as soon as first symptoms of the disease appear at kneehigh stage of the crop growth.
3. Clean slowdown of crop debris may reduce early infection.

10.1.2. Turcicum Leaf Blight

1. Use of resistant maize cultivars is the most effective and economical method of reducing the losses caused by this disease.
2. The disease can also be controlled by two or three foliar sprays with 0.25 to 0.3 per cent mancozeb or zineb at 10-15 days intervals. First spray should be given soon after the disease appears in the field.
3. Application of urea followed by a light irrigation has helped to minimise the disease as this disease sometimes becomes serious under poor management, i. e., under low nitrogen and water stress.

10.1.3. Rust

1. Select resistant maize cultivars for planting in areas where rust becomes problem, e. g., hybrid Ganga 4.
2. The severity of the disease during *rabi* season can be reduced by spraying mancozeb or zineb @ 2.5-3.0 kg/ha. The first spray should be given as soon as rust pustules appear on the foilage and then two or three more sprays at 10-15 days intervals. This method may also be feasible in seed production plots and also on fodder crops.

10.1.4. Seed Rots and Seedling Blights

1. Sow certified seed secured from a reliable source, which is always treated with desired pesticides. If you want to use your own seed, treat it with thiram or any other seed dressing fungicide like bavistin @2-2.5 g/kg.
2. Plant healthy looking and good quality seed taken from disease-free and healthy plants. Also use injury-free seed of high germination.
3. Planting should be done preferably at a time when soil temperature is more than 10° C.

10.1.5. Brown Spot

1. Grow resistant cultivars.
2. Clean slowdown of crop debris.
3. Adequate precautions should be taken to keep the fields free from weeds.
4. Crop rotation of at least two years should be followed to minimise the sources of primary inoculum of the pathogen.

10.1.6. Phyllosticta Leaf Spot

1. Resistant hybrids and varieties offer the best means of control.
2. Fungicide application may be feasible in seed production fields.
3. Clean ploughing or rotation or both, greatly reduces early-season infections and primary inoculum.

10.1.7. Brown Stripe Down Mildew

1. Resistant maize cultivars, such as hybrids Ganga-5, Ganga-9, composites, Kisan, Tarun, Naveen, Sweta, Kanchan and others offer the best means of control.
2. Plant maize at least a fortnight before the onset of rains. Incidence of the disease becomes less if planting is done during dry season.
3. Select well drained fields for maize planting, if this disease is a problem in the area or arrange proper drainage to drain out the excess rain water.
4. The disease can be minimised by spraying the foliage with fungicides such as mancozeb (0.3%). First spraying should be done about a month after planting or as soon as the disease symptoms are seen. Other 3-4 sprays should be done at 10-15 days intervals depending on severity of the disease.
5. Burn the crop residue after the harvest.
6. Practice three-year crop rotation to minimise the initial inoculum load in the field.

10.1.8. Peronosclerospora Downy Mildew

1. Plant disease resistant maize cultivars like Suwan. Amongst the released hybrids, Ganga safed 2 is least susceptible.
2. Provide adequate soil drainage or avoid planting in low, wet spots.
3. Planting should be done at least a fortnight before the onset of rains.
4. Burn the crop residue after the harvest.
5. The disease can be minimised by foliar spraying of fungicides like mancozeb @ 3 kg/ha at frequent intervals, which may be economical to save valuable breeders materials.
6. Treating maize seed with a systemic fungicide, metaxyl @ 3 g/kg as slurry treatment eliminates this group of diseases almost completely.
7. Burn the residues after the crop harvest.

10.1.9. Pythium Stalk Rot

1. Grow resistant maize hybrid Ganga safed 2 or composite Vikram.
2. Plant healthy seed treated with fungicides like thiram @ 2.5 g/kg.
3. The most suitable sowing time to avoid high incidence in the Northern Indian Plains is first fortnight of July.
4. Apply captan or thiram @ 2.5 kg per 100 liters of water per hectare on lower internodes of plants 30-35 days after planting.
5. Arrage proper drainage in the field to drain out excess rain water.

10.1.10. Erwinia Stalk Rot

1. Grow resistant hybrids and composites such as hybrid Ganga safed-2.
2. Apply cultural management practices to avoid flooding.

3. Planting should be done in the second fortnight of June. It helps in reducing stalk rot incidence in northern states of the country.
4. Provide balanced fertilisers, especially potash, which helps in making the plants more resistant.
5. Apply chlorine water irrigation (100-1000 ppm) after a week of planting. if need be, this water should also be applied at pre-flowering stage of the crop. Bleaching powder can be used for generating chlorine.

10.1.11. Late Wilt and Black Bundle Disease

1. Select resistant maize cultivars for planting such as hybrid Ganga 5.
2. Adopt crop rotation with rice or sesame or other non-hosts for these diseases.
3. Apply balanced fertilisers, especially potash, based on soil test. This minimises stalk rot incidence, in general.

10.1.12. Charcoal Rot

1. Sow resistant maize cultivars, such as Ganga-9, Deccan 101, Him 128, Kanchan, Diara Composite, Hunius composite and Sangam, etc.
2. In irrigated areas, charcoal rot cam be minimised by maintaining moist soils, especially during dry periods after tasselling.
3. Apply balanced fertiliser doses and avoid high doses of N and low level of potash.

10.2. Insect Pest

10.2.1. Stem Borer

1. Pull out all dead hearts and destroy them to kill the lingering stages of the pest in the stubbles.
2. Plough up the field soon after harvesting of maize, sorghum and bajra crops, collect and burn the stubbles
3. Grow resistant/tolerant varieties
4. Sowing of crop in first week of July evades borer infestation.
5. Use higher seed rate and remove the plants showing severe border injuries at thinning.
6. Spray endosulfan 35 EC, 0.1 %, lindane 20 EC, 0.05% or carbaryl 50 WP, 0.1% @ 450 l/ha followed by respective granules @ 5-7.5 kg/ha in plant whorls, 10-15 and 25-30 days after germination. Apply carbofuran 3% granules @ 5 kg/ha in 15 days old plant whorls.

10.2.2. White Grubs

1. Hand collection of adults and their killing in kerosinised water help in reducing future generations.
2. Spray carbaryl 50 WP or monocrotophos 36 WSC, 0.05% or quinalphos 25 EC, 0.03% on host trees.
3. Use phorate 10% or quinalphos 5% granules in rows @ 25 kg/ha before sowing the crop in endemic areas.

10.2.3. Termites

- Mix 4% endosulfan dust @ 25 kg/ha in the soil before sowing in termite prone areas.

10.2.4. Cutworms

- Mix 4% endosulfan dust @ 25 kg/ha in the soil before sowing in termite prone areas.

10.2.5. Army Worms

- Dust 2% methyl parathion or 10% @ 25 kg/ha in plant whorls.

10.2.6. Hairy Caterpillars and Climbing Cutworms

- Spray endosulfan 35 EC, 0.1% at the appearance of pests.

10.2.7. Shoot Fly

1. Cover pits of the rubish with soil.
2. Use high seed rate and remove the infested plants at thinning operation.
3. Grow resistant/tolerant varieties.
4. Apply 10% phorate granules @ 15 kg/ha before sowing the seed in shoot fly infested areas.

10.2.8. Thrips

1. Grow resistant/tolerant varieties
2. Sporay phosphamidon 100 EC, 0.03%, malathion 50 EC, 0.05% or endosulfan 35 EC, 0.1% at the appearance of foliar infestation.

10.2.9. Pyrilla

Spray metasystox 25 Ec, 0.03%, monocrotosphos 36 WSC, 0.03%, carbarl 50 WP, 0.1% or endosulfan 35 EC, 0.1% at the time of appearance of pest.

10.2.10. Grasshoppers

1. Trimming the bunds and elevate grounds help in destroying the layed eggs and younger nymphal stages.
2. Dust 4% endosulfarn dust @ 25 kg\ha.

11. QUALITY

Chemical Constituents of Maize

Composition	Ash (%)	Protein (%)	Oil (%)	Sugar (%)	Starch (%)	% by wt. of whole grain
Endosperm	0.31	9.4	0.8	0.64	86.4	81.9
Embryo	10.1	18.8	34.5	10.8	8.2	11.9
Bran	0.8	3.7	10.	0.34	7.3	5.3
Tip cap	1.56	9.3	3.8	1.54	5.3	0.8
Whole grain	1.44	10.3	4.8	1.97	71.5	99.9

Kernel contains about 71 per cent starch, 2 per cent sugar, 9 per cent protein, 5 per cent fat and 1.5 per cent ash. The protein content varies between 6-15 per cent in different strains. In Maize kernel about 80% protein is in the endosperm.

The germ, though constituting only about 1/10 of the grain, contains about 1/15 of the total protein. Maize contains three of protein.

(1) Prolamines soluble in alcohol chiefly Zein; (2) Globulin soluble in neutral salt solution (3) Glutelin

The entire protein in maize is inferior because Zein is deficient in trytophane and lysine which are essential for animal nutrition.

Maize is the richest of the cereals in fat, with the exception of oats, and strains of maize may contains as much as 7 per cent fat in the grains.

More than 70 per cent of the maize kernel is carbohydrate. Starch is mainly in the endosperm, the sugar in the germ and fiber in the bran. Vitonicu is located in germ. About 3/4 of the total mineral are found in the germ and rest in the outer endosperm. Maize is poor in Ca but rich in P and K like other cereals. Mg, Na and Cl are very less. Fe in sufficient.

The mineral composition of maize (85% dry matter) National Research Council (1947)	
Co (%)	0.0239
P (%)	0.2620
K (%)	0.2820
Fe (%)	0.00185
Mg (%)	0.1109
Na (%)	0.00939
Cl (%)	0.0450
S (%)	0.1190
Mn Mg/lb	2.19
Cu Mg/lb	0.734
Cobalt Mg/lb	0.0081
Fluroiral	2.6
10 dine (%)	0.00006
Zn (%)	0.0017

12. POST HARVEST TECHNOLOGY

There are several processes that convert either grain or maize stover into food and chemical products. These processes are: (1) Wet milling and (2) Dry milling.

12.1. Wet Milling and Dry Milling

Wet milling and dry milling are well established industrial processes. In wet milling process maize grain is separated into its four major constituent i.e., oil, starch, fibre and protein which are all sold at higher prices than the raw material (grain).

The major use of maize oil has been for human consumption. It is known to be highly unsaturated and thus much favoured cooking oil.

The mixture of protein, fibre and other extracted ingredients in maize is dried and marketed as the maize-gluten feed, almost exclusively to support meat industry.

Purified maize starch in either native or chemically derived for three major industrial markets i.e., in paper making textile sizing and food processing. Maize starch is being processed by acid and enzyme hydrolyses to produce maize syrups of various kinds. Variety of chemically derived products made from the native granular maize starch. The use of industrial starch in paper, textile and food products. The modified starches are superior to the native polysaccharide in many aspects.

13. REFERENCES

Dayanand (1998) Principles governing maize cultivation during rainy season. Ind. Fmg. 48 (1): 84-87.

De condelle (1886) Origin of cultivated plants. Kegan Paul. Treneh & Co. London.

FAO Production Year Book (1997) Vol. 57.

IARI (1997-98) Annual report.

3. Sorghum

P. S. Rathore

Sorghum is used for food, feed and forage. It is a staple food for millions in Africa and Asia. The grains are used in *chapati* or eaten like rice after boiling or as popped grains. The grains are also used as a feed for cattle, swine, poultry and birds. Green fodder and stover is palatable to milch and draft animals. Industrial uses of sorghum are in fortification of food, preparation of malt corrugating adhesive beverage alcohol. Sweet sorghum varieties (17.8% Brix) are used in extraction of raw sugar.

1. ORIGIN

Vavilov (1926) described Abyssinia as probable centre of origin. Harlan and Dewet (1972) on the basis of occurrence of wild and primitive race bicolor, reported Africa as its centre of origin. Crop was introduced in India around 1500 B.C.

2. DISTRIBUTION

India leads the world with regards to sorghum acreage, but second in production after USA. It is mainly grown in India, USA, Mexico, Sudan and Nigeria. Productivity is highest in USA (4401 kg/ha). The acres and production of in world is given in Table 3.1. India represents about 30% world acreage which is 83% of south east Asia. Among the important states, Maharashtra, Madhya Pradesh, Andhra Pradesh, Rajasthan and Karnataka are major (Table 3.2).

TABLE 3.1: Area and production of sorghum in world (1997)

Countries	Area '000 ha.	Production '000 Mt.	Productivity kg/ha
India	11,200	9000	1414
USA	3800	16727	4101
Nigeria	6589	7297	1107
Sudan	5641	3369	597
Mexico	1909	6276	3288
Total	**45,323**	**64,071**	**1414**

Rajasthan has the lowest productivity among the major sorghum growing states. Crop is grown from extreme south (9° N) to north. West 25° N, in *kharif* and *rabi* season.

TABLE 3.2: Area and production of sorghum in India (1995-96)

States	Area 000 ha.	Production 000 t.	Productivity kg/ ha.
Maharashtra	5557.3	4981.8	896
Karnataka	1983.8	1738.5	876
Madhya Pradesh	1018.0	863.2	848
Andhra Pradesh	851.1	625.6	735
Rajasthan	593.5	139.4	235
Tamil Nadu	505.3	489.1	968
Total	**11444.4**	**9546.9**	**834**

3. CLASSIFICATION

Sorghum bicolor (L.) Moench belongs to family graminae with 2n = 20 chromosome. Harlan and Dewet (1971) divided *Sorghum bicolor* into five basic races, on the basis of spikelet types and grains as bicolor, guinea, caudatum, kafir and durra. The important differences in spikelets and grains of each type is given below:

Different basic races of sorghum

Race	Spikelets	Grains
Bicolor	Persistant, pedicellate	Elliptic - subglobose, attached to panicse & pigmented.
Guinea	Deciduous and persistent pedicellate	Ovate - flattened slight pigmentation.
Caudatum	Sessile, pedicellate deciduous	Flat from one side - round on other side, white or pigmented.
Kafir	Hairy, sessile	Elliptic and broad, flattened.
Durra	Persistent, sessile and pedicellate	Biconvex with broad tip, wedge like base.

4. GROWTH STAGES

Like other crops, sorghum has five distinct growth stages as:

1. **Seedling stage:** Germination takes place in 4-5 days after sowing. Emergence of coleoptile indicate seedling stage.
2. **Flag leaf stage:** 3-4 leaves with fully expended leaf area. Stage reach in 40-55 days after sowing.
3. **Boot stage:** Ear head covered within sheath i.e. under flag leaf stage reaches in 45-60 days after sowing.
4. **Soft dough:** Endosperm filled with watery fluid. It is called milky stage. Stage reaches in 70-85 days after sowing.
5. **Physiological maturity:** Grain with maximum accumulation of dry matter. Stage reach in 85-95 days after sowing.

5. CLIMATE

It is a plant of warm climate. It remains dormant under unfavourable conditions, leaves posses waxy coating and presence of motor cells in leaves rolls the leaves under moisture deficit conditions. It also posses more number of secondary roots/unit of primary roots. Thus crop can withstand drought, better than maize. Crop is grown in areas receiving 800-1000 mm annual rainfall isohytes. High rainfall at heading reduce pollination and give poor yield, though crop is tolerant to water logging. Sorghum is grown both in *kharif* (northern states) and in *rabi* (western and southern states). Minimum temperature required for germination is 7-10° C, germination does not take place if temperature is less than 7° C. Optimum temperature for growth is 25-30° C. Crop is sensitive to low temperature. Temperature below 15° C affects crop growth adversely. It is a short day plant.

FIGURE 3.1: Sorghum ear head.

6. SOILS AND SOIL PREPARATION

Sorghum is raised predominantly in vertisols and to lesser extent in alfisols. Soils having good water holding capacity, rich in humus are best suited. Black cotton soils of central India are categorized as best soils for its cultivation. Crop is grown in pH range of 6.0-8.5.

Crop needs good tilth. Field is prepared by one deep ploughing followed by 2-3 harrowing or ploughing with country plough immediately after onset of monsoon. Planking should be done after each ploughing to break the clods and to level the field.

7. CROPPING SYSTEMS

Ninety five percent of sorghum in India is rainfed. Monoculture is a common practice due to moisture stress. With the development of short duration, improved varieties double cropping is possible. The most common crop sequences associated with sorghum are:

Common cropping system with sorghum	
Sorghum - chickpea	Sorghum - chickpea - Sorghum
Sorghum - mustard	Sorghum - wheat - green gram
Sorghum - wheat	Sorghum - wheat - Ground nut
Sorghum - pea	Sorghum - sunflower
Sorghum - safflower	

Generally it is observed that crop growth and yield is poor after sorghum. This is because of deep root system, more absorption of available nutrients and its high carbohydrate contents in residues (left in field after harvest) in comparison to nitrogen. Plants residue undergoes decomposition, consuming available nitrogen of soil. Thus crop succeeding after sorghum may experience nitrogen deficiency, commonly called soil sickness or yellow effect. Therefore, it is advisable to supplement the succeeding crop with adequate nitrogen to over come its deficiency.

Intercropping of sorghum with pulses or oil seeds, paired planting system (30/60 cm) is good. At national level, maximum returns of soybean, green gram, groundnut in paired rows of 30/90 cm. with two rows of intercrop in 90 cm space. Sorghum in equal distant rows (60 cm) with one row of inter crop in between two rows stand next. In Rajasthan 30/60 cm paired planting of sorghum is recommended. One row of pigeon pea/black gram/green gram is sown interspace of 60 cm between two pairs of sorghum rows. In Maharashtra sorghum-safflower in 6 : 3 or 4 : 2 is better in *rabi* (Rana *et.al.*, 1998). CSH6 with short nature and duration was reported the best variety of sorghum for inter cropping system with pigeon pea. Geno type 'C 11, 'BDN 2' and 'LRG 36' of pigeon pea are ideal. CSH9 sorghum is good for black cotton soils with assured rainfall with 3 : 3 row ratio. (Rana *et.al.*, 1998).

8. MANURE AND FERTILISERS

Application of FYM at 8-10 t/ha prior to sowing not only provide essential nutrients but also improve germination by reducing crust and also increases water retention capacity of soils. It is a heavy feeder of plant nutrients. A grain sorghum crop yielding 50-60 q/ha remove nearly 130-180 kg N, 50-60 kg P_2O_5 and 100-130 kg K_2O/ha. Against such heavy removal of nutrients, the actual addition by farmers is very less. Thus, farmers are harvesting low yield. Response to nitrogen by *kharif* sorghum varies from 10-13 kg grain /kg N. Response is much higher in new cultivar than local ones. Singh *et.al.* (1993) recommended 80 kg N and 40 kg P_2O_5 for sorghum on sandy loam soils of Sumerpur (Rajasthan). Similarly Tripathi and Suraj Bhan (1995) obtained significantly higher yield of sorghum due to 60 kg N/ha over 30 kg N/ha. Therefore, general recommendation is 60 kg N and 30 kg P_2O_5 for rainfed areas and 80 kg N and 40 kg P_2O_5/ha for irrigated areas. For hybrid and fertiliser responsive cultivars like CSH 13 R and CSH 15 R, and Swathi, 100 kg N/ha is beneficial under irrigated conditions. In intercropping system with pulses, *rabi* sorghum is to be supplied with 50% of recommended levels of nitrogen and phosphorus.

Full quantity of phosphate and half quantity of nitrogen should be drilled at 10 cm below at the time of sowing. Remaining 1/2 quantity of N should be given 30-35 DAS at knee high stage of crop. If moisture is scarce, avoid top dressing.

Response to potash is generally low due to high K content of sorghum solis and observed in few pockets. Thus, application of phosphate and potassic fertiliser should be based on soil tests. In zinc deficient area, apply 20-25 kg $ZnSO_4$ or 0.2% $ZnSO_4$ with half quantity of lime to prevent burning of laves. Iron chlorosis may be visible in some sorghum growing regions. Sorghum is sensitive to iron deficiency. Spraying ferrous sulphate with lime is efficient to overcome the deficiency. For fodder crop 80 kg N and 40 kg P_2O_5/ha have been found optimum.

9. BIOFERTILISERS

The term biofertilisers or more appropriately microbial inoculants are the preparations containing live or latent cells of efficient strains of nitrogen fixing, phosphate solubilizing or cellulotytic micro-organisms used for dressing seed, soil or composting areas with an objective of increasing the number of such micro-organisms in soil or rhizosphere and consequently improve the extent of microbiologically fixed nitrogen for plant growth. Application of biofertilisers increase crop yields by one or more of the following reasons:

1. Supplying additional nitrogen to the crop,
2. Improving soil conditions
3. Increasing the solubility of plant nutrients in the soil and
4. Stimulating plant growth through synthesis of growth promoting substance.

Field responses of sorghum var. CSH-5 to inoculation with Azospirillum brasilense was tested at six centres. Simple seed inoculation generally increased the grain yield from 15.2 to 63.6 per cent over uninoculated control. The response of seed inoculation with *Azospirillum brasilense* without nitrogen was equivalent to the application of 40 kg N/ha alone at Delhi, Pantnagar and Dharwar centres of all India Coordinated Projects.

9.1.1. Varieties

Large number of varieties/hybrids with higher yield potential than that of local have been released for cultivation in different states. The varieties recommended for different states are given in Table 3.3 to 3.6.

TABLE 3.3: Recommended hybrids and varieties of sorghum for *kharif* in different states.

State	Areas of adaptation	Hybrids	Varieties
Maharashtra	Light soil areas	CSH 1+, CSH 6+, SPH 468, CSH 5*, CSH 9*, CSH 10*	CSV 10, CSV 11
	Medium to heavy soil areas	CSH 1+, SPH 388*, SPH 468+	SPV 462, SPV 475, AKSV 37
Karnataka	Low rainfall areas	CSH 1, CSH 6, SPH 468	SB 1066, DSV 1+
	Assured rainfall areas	CSH 5*, CSH 10, SPH 468	CSV 10*, CSV 11*, SPV 462, SPV 475
Andhra Pradesh	Low rainfall areas	CSH 6, SPH 468, CSH 5,	CSV 10, CSV 11
	Normal rainfall areas	CSH 9, CSH 10, CSH 11*	SPV 462, SPV 475, CSV 10, CSV 11
Madhya Pradesh	Entire state	CSH 5, CSH 9, CSH 10, CSH 11	CSV 10, SPV 462, SPV 475, SPV 235
Tamil Nadu	Madurai District	CSH 1, CSH 5, CSH 6	-
	Entire state	CSH 1, SPH 468	CSV 11, Co 24, Co 25
Gujarat	South Gujarat	CSH 5, CSH 9, CSH 11	CSV 10, SPV 462, SPV 475, GJ 35, GJ 36
	North Gujarat and Saurashtra	CSH 6, CSH 10, SPH 468	CSV 10, SPV 462, SPV 475, GJ 37
Rajasthan	Semi arid and transitional zones	CSH 1, CSH 6, SPH 468	SPV 245, CSV 10, SPV 96
	Humid and semi humid zones	CSH 5, CSH 9, CSH 10, CSH 11	SPV 462, SPV 475, SPV 245
Uttar Pradesh	Entire state	CSH 9, CSH 10, CSH 11, SPH 468	CSV 10, CSV 11, SPV 462, SPV 475

*Medium and + Early.

TABLE 3.4: Recommended hybrids and varieties of sorghum for *rabi* in different states

State	Areas of adaptation	Hybrids	Varieties
Maharashtra	Dry areas	CSH 13R*, CSH 12R*, CSH 8R*, SPH 677	M-35, CSV 8R*
	Irrigated areas	CSH 13R*, CSH 12R*, CSH 8R*	Swathi+
Karnataka	Dry zone	CSH 12R*, CSH 13R*, SPH 677	M 35-1*
	Transitional belt	CSH 12R*	A-1, CSV 8R
	Irrigated areas	CSH 13 R*	CSV 8R
Andhra Pradesh	Late *kharif* areas of Telangana	CSH 5, CSH 9, CSH 13R, SPH 677	CSV 11*, SPV 462
	Early *rabi* areas of Rayalaseema	CSH 10, CSH 5, CSH 12R	CSV 11, SPV 462*
	Normal *rabi* areas	CSH 12R, CSH 13R	M35-1, CSH 8R, Swathi
Tamil Nadu	Entire *rabi* areas	K. Tall, CSH 10, Co 24, Co 25	-
	Summer irrigated areas	CoH3, CSH 5	SPV 462*, CSV 11, Co 24, Co 25
Gujarat	Entire *rabi* areas	CSH 12R*, CSH 13R+	GJ 9

*Medium and + Early.

TABLE 3.5: Characteristics of *rabi* varieties

Hybrid/variety	Grain yield q/ha	Fodder yield q/ha	Duration (Days)
Hybrid			
CSH 8R	22	37	115
CSH 12R	26	47	115
CSH 13R	33	54	113
CSH 15R	32	56	110
Varieties			
CSV 8R	22	48	120
Swati	22	53	117
CSV 14R	23	55	117
M 35-1	21	61	119

TABLE 3.6: Characteristics of fodder varieties

Variety	Fodder yield (q/ha)		Characteristics
	Green	Dry	
Single cut varieties			
Pusa Chari-9	425	135	If flowers ready for fodder in 80-85 days. It is medium thick slimmed, non-sweet and pithy. Leaves are long and medium broad with white midrib. Panicles are semi-loose, straight, but sometimes goose necked. It natures in 120 days and yields 8 q/ha.
Rajasthan Chari-1 (RC-1)	450	125	Its stem is medium thick, leaves long and dark green with white midrib. Ear-heads oblong fully exerted from the flag leaf. Grains are bold and chalky white. It matures for seed in 110-115 days.
Rajasthan Chari-2 (RC-2)	330	100	Its stem is medium thick, leaves light green having 60-70 cm length and 5-6 cm breadth with white midrib. Plants have 11-13 leaves with drooping habit. Ear-heads are fully exerted from the flag leaf. Grains are bold, flat and chalky white. It matures for seed in 100-105 days.
Rajasthan Chari-3			Plants 2-2.5 meter tall, flower in 65-70 days grees fodder yield 300-350 q/ha, dry fodder yield 100-110 q/ha. Suitable for entire Rajasthan.
HC-171	520	170	Its stem is sweet and juicy and leaves have green midrib. Panicles are semi-compact with small and creamy white seeds. It is highly resistant to most of the foliar diseases. It is highly resistant to mites.
HC-260	450	140	It becomes ready for fodder in 55-60 days and matures for seed in about 85-90 days. It is non sweet and juicy. Leaves are medium broad and long with white midrib. Its panicles are semi compact with white seeds. It is resistant to foliar diseases. It gives good seed. Leaves are narrow and long with green midrib. It is ideally suited for multi-cut due to its high regeneration, faster growth and profuse tillering. The grains are white, small and covered with straw coloured glumes. Panicles are highly laxed. It matures for seed in 90 days.
Jawahar Chari-69 (JC-69)	700	195	It is Sudan type having high regeneration, faster growth and multi-cut ability. Its stem is thin and non sweet and leaves are narrow and long with white midrib. Its panicle is highly laxed with small seeds covered with black or brown glumes.

Variety	Fodder yield (q/ha)		Characteristics
	Green	Dry	
Pusa Chari-23	550	160	It is thin stemmed, non-sweet and non juicy. Leaves are narrow with white midrib. Plants have lateral branches and yield 14 q/ha under north Indian conditions.
HC-136	550	175	It is tall, sweet, juicy and medium thick stemmed. The leaves are very broad and long with green midrib and remain green up to maturity. It has high protein, low toxic constituents and better digestibility. It is also good in regeneration and yields 800-850 quintals green fodder and 260-280 quintals dry mater per hectare in two cuts. It matures for seed in 140 days.
Jawahar Chari-6 (JC)	412	120	Its stem is tall, thick, juicy and non sweet. Leaves are broad and long with white midrib. It is tolerant to leaf spot disease. Panicle is semi laxed. It flowers in 80 days and matures in 130-135 days. Grains are bold, pearly white with reddish brown glumes.
U.P. Chari-2	380	125	It flowers in 75-80 days. It is medium thick stemmed with tan colour and juicy. Leaves are very broad and long having light green colour. Panicle oval shaped, medium in size and contact. Grains are medium bold, round and pearly white in colour. It matures for seed in 110 days and gives 10-12 q/ha seed yield.
Multi-cut varieties			
M. P. Chari	300	95	It takes about 65-70 days to become ready for fodder. Its plants are tall, thin stemmed, leaves medium long and narrow with white midrib. It is non sweet, non juicy and pithy. Panicles are erect and lax with purple or black glumes. Grains are brown coloured. It has better regeneration and hence good for two cuttings. The crop matures for seed in about 110 days.
Meethi Sudan (SSG 59-3)	700	210	It gives a number of cuttings for green fodder during the crop seasons. It is thin stemmed, sweet with profuse tillering and flowers in 55-60 days. It is drought and water logging tolerant and generally gives two cuts. It flowers in 60-65 days. Panicle is laxed with profuse lateral spikes. Seed is purple red coloured with glumes adhered to them. It matures in 95-100 days and yields 10-12 q/ha seed.

10. SOWING

Sorghum is sown thrice in a year. In north it is sown in *kharif* and summer and in west and south besides *kharif*, *rabi* crop is also taken. *Kharif* sorghum is generally sown with the onset of monsoon (Singh *et.al.*, 1976). The experiments conducted in major *kharif* sorghum tracts clearly showed that optimum sowing time under rainfed conditions is with onset of monsoon. Delay after first monsoon, causes substantial reduction in yield (Table 3.7). This is due to decrease in number of seeds per panicle and grain size. Late sown crop is also susceptible to shoot fly and midge. Dry sowing just before onset of monsoon is the best. *Rabi* crop of sorghum should be sown around middle of September. If monsoon persists or moisture is available sowing can be done upto first week of October. Summer sorghum is mainly raised for fodder and crop is generally sown in March-April in north India. In Tamil Nadu, Andhra Pradesh and Karnataka summer crop is sown in Jan.-Feb. under irrigated conditions.

TABLE 3.7: Yield loss due to late sowing in different sorghum tracts

Region	Date of onset of monsoon	Yield kgha	
		Sown with onset of monsoon	Sown late
Saurashtra (Gujarat)	June 18	320	280(9)
Eastern Rajasthan	June 27	390	250(10)
Telengana (A.P.)	June 10	540	460(14)

Figure in parenthesis indicate number of days seeded late (The Hindu survey of Indian Agric., 1999).

Kharif rainfed sorghum is sown at 45 x 12 cm using 9-10 kg seeds/ha and plant population of 1,80,000. *Rabi* rainfed crop is sown either at 45 x 16 or 60 x 12 cm. using 9-10 kg seeds and plant stand should be 1,35,000. While the irrigated *rabi* crop is sown at 45 x 15 cm. using 9-10 kg seeds and total plant stand is 1,50,000. Fodder crop of sorghum is sown at 30 cm. row to space using 40 kg seeds/ha for bold seeded and 20-25 kg for small seeded varieties. Cowpea are often mixed with sorghum to improve nutritive quality. Crop is sown with country plough, however seed cum fertiliser drill is efficient. Sowing sorghum in ridge and furrow system is recommended in low rainfall areas. Generally, the depth of sowing is 3-4 cm.

Crust formation at the time of emergence is a problem in maintaining proper plant stand. Crust is a hard layer on soil surface, resulting by the beating action of falling rain drops. The sandy loam soils with 50-70% sand, 30-50% silt and 20% clay is most crust prone soil, sorghum seedlings fail to emerge because of hard surface soil. Problem occurs when there are rains after sowing sorghum before emergence.

11. IRRIGATION

Sorghum is drought tolerant. High yielding varieties respond well to irrigation. Sorghum has an extensive and deep root system capable of extracting soil moisture from deeper soil layers. Maintaining optimum soil moisture at 3-4 week after sowing helps in development of strong secondary roots. Similar to other cereal crops, water stress affects growth and development of sorghum by affecting:

1. Floral initiation and inflorescence development which is important in deciding grain number.
2. Antithesis and fertilization which decides actual number of grains.
3. Grain-filling decides grain weight.

Based on these, early seedling stage and flower primordial stages are considered most critical for moisture stress (Yadav *et.al.*, 1998). Generally 1 to 8 irrigations are needed, depending on soil and climatic conditions. Irrigation scheduling at 0.8 IW/CPE ratio have been found to be the best criteria. Irrigation may also be decided by available soil moisture. Yadav *et.al.* (1998) recommended irrigations in sorghum when 45-55% available soil moisture is depleted. Depth of irrigation is generally 6-8 cm and water requirement of crop is around 80-480 mm.

In case of severe terminal drought, with chances of total failure, thinning upto 50%, alternate rows before emergence of ear is recommended as a mid season correction (Rana *et.al.* 1998). Summer sorghum fodder needs 4-5 irrigation at 10-15 days interval. Multi-cut varieties require frequent irrigation. Crop should be irrigated after each cut for better regeneration.

12. WEED CONTROL

Sorghum is badly infested with both broad leaf and grassy weeds, because of congenial weather. Mishra (1997) reported 15-40% loss in grain yield due to weeds. He further reported that 15-45 days of crop growth is critical with regard to crop weed competition. *Sorghum halepense* commonly called barru mimics with sorghum, particularly before appearance of inflorescence. Therefore, it is difficult to control this weed. The common weeds infesting sorghum crop are:

TABLE 3.8: Common weeds found in sorghum fields

Broad leaf weeds	Grassy weeds
– *Amaranthus viridis*	– *Euchinochloa colonum*
– *Euphorbia macrocephylla*	– *Setaria glauca*
– *Phyllanthus niruri*	– *Cyperus rotundus*
– *Commelina benghalensis*	– *Dactyctenium aegypticum*
– *Boerhavia diffusa*	– *Sorghum halepense*
– *Trienthema monogynae*	– *Cynodon dactlion*
– *Trienthema partulacustrum*	

Inter row weeds may be controlled mechanically by running blade harrow, but intra row weeds remain. Hand weeding with *khurpi* or hand hoe is most common practice. Two hand weedings at 15 and 30 days after sowing, effectively control the weeds, both in inter and intra row can be controlled by using herbicides efficiently. The herbicides along with their doses and mode of application is given in Table 3.8.

TABLE 3.9: Herbicidal weed control in sorghum

Herbicide	Dose kg/ha.	Mode of application
Trifluralin	0.75-1.0	Pre plant incorporation
Atrazine	0.25-0.75	Pre emergence
Propazine	1.0-1.5	Pre emergence
Simazine	0.25-0.75	Pre emergence
Oxadiazon	0.75-1.0	Pre emergence
Atrazine	0.5-1.0	Post emergence, 1-2 leaf stage of weeds
2,4-D	0.5-0.75	Post emergence, 15-20 days after sowing, direct spray in between rows.

Striga (*Striga lutea*) causes 15-100% loss depending on severity of infestation. Following are the ways to check its infestation and control.

1. Grow *Striga* resistant varieties as CO 20, N-13 work of other resistant lines as SPV 168, SPV 103 and SAR-2 is in progress.

2. Crop rotation with trap crops as cotton, sunflower and groundnut, destroys the seeds and minimize the losses.
3. In standing crop, hand pulling when population is less or spraying 2 kg 2, 4-D sodium salt as directed spray check its infestation and damage.

13. PLANT PROTECTION

Insect pest cause severe economic damage in grain and fodder crop. Adoption of suitable cultural and need based chemical control minimises losses. The major insect pests and diseases of crop are given below:

13.1. Insect Pests

Shoot fly (*Atherigona soccata*): Insect causes damage from 1-4 week after emergence, Maggot feeds on tips resulting into wilting of central leaf. Leaf dries and gives a typical appearance of dead heart. At later stages, infested plant produce side tillers. It is advisable to initiate control, measures only when 10 per cent dead heart or one egg/seedling in 10 per cent plant is observed in field (Singh, 1999). Infestation can be avoided by sowing crop with in 7-10 days of onset of monsoon. In *rabi* sowing in end of Sept. to first week of October, avoid shoot-fly damage. Seed treatment with Furadan 50 SP at 100 g/kg seed. Furadan 3G or Phorate 10G at sowing @.20 kg/ha also check incidence. In case of infestation in standing crop, spray 0.025% Metasystox, one week after germination. Repeat spraying after 15 days.

Stem borer (*Chilo partellus*): Infest crop from 15 days till maturity. Larvae initially feed on leaves. Later on the larvae bore into the main stem causing stem tunnelling leading to breakage of stem. It is advisable to initiate control, measures only when 8 per cent dead heart or 10 per cent plants are affected in the field (Singh, 1999). Other preventive measures are to uproot and burning of the stubles of previous crop. Infestation may be checked by applying Endosulfan 4G or Malathion 10D or Furadan 3G in whorl @8-12 kg/ha in whole at 20-35 days after emergence after any visual symptom of insect.

Sorghum midge (*Contarinia sorghicola*): Cause damage in panicles. Maggot feeds in fertilized ovary thus prevent grain development. In severe infestation, entire panicle appear blasted and red liquid oozes out. It is advisable to initiate control, measures only when 5 midges/ear head or 0.25 adult midge ear is observed in field (Singh, 1999). Early planting, destruction of chaffy panicles, ear heads and weeding of sorghum halepense prevent its infestation in crop. Infested crop may be saved by applying Endosulfan 4D or HSD 10G @ 20 kg/ha or Endosulfan (35 EC) 1.0 litre at pre bloom (50% panicle emergence) stage and repeating it after 4-5 days. Insecticide should only be applied after noting presence of 2-3 midge flies (Thresh hard level).

Head bug (*Calocoris angustatus*) feed and damage the developing grains of sorghum. It causes considerable loss in grain yield. It can be controlled by using insecticides as used for midge control.

13.2. Diseases

Downey mildew: Appearance of vivid green and white stripes on leaves. Stunted growth and failure to produce head occur at advance stage. Seed dressing with metalaxyl (Ridomil 25) at 1 g/kg or Apron 35 SD 1.0 g/kg with thiram control infection from germination. In field, spray of metalaxyl 0.1 g/Lit 40 days after germination check spread of downey mildew. Control measures must be initiated as soon as 4 nymphs or 1 adult per head is seen.

Grain molds: Caused by many fungal species. Grains infected with *Fusarium moniliforme* have fluffly white or pinkish colour, while grains infested with *Curvularia lunata* have black appearance. Discolouration of grain, reduction in grain weight leading to 75-100% damage. The harvested grain are also toxic to animals. Seed treatment with captan 4 g/kg seeds is a preventive way. In field three sprays of aureofungin 200 ppm + captan 0.3% from 50% flowering at 10 days interval is costlier but very effective. Spraying three time dithane M-45, and bavistin 0.2% from flowering at 10 days interval control grain mould.

Ergot: Secretion of creamy sticky liquid from infected florets. Seed setting is adversely affected. Seeds should be cleaned by dipping in 30% sodium chloride and subsequent 3 washing in clean water, spraying with benlate 0.1% three times from 50% grain formation is recommended to control disease. Two sprays of bavistin 0.1% or thiram 0.3% at 50% flowering and second spray after a fortnight is effective against ergot (Rana *et.al.*, 1998).

Rust: Rust pustules invade the lower leaves. In susceptible varieties entire leaf tissues are destroyed by pustules. Use of clean seed, crop rotation and use of resistant varieties are preventive ways against this disease. Effective control in field could be achieved by 4 sprays of Dithane M-45 at 0.2% at 10 days interval from 30 days after germination.

Charcol rot: Lodging and poor grain filling are main symptoms. Fungi infect crown and disorganize the cortical tissues and make them soft and hollow, leading to lodging. Premature drying, stunted growth and lodging cause loss in quality and quantity. Mulching with wheat straw, periodical hoeing reduces chances of charcol rot. Seed dressing with *Trichoderma vivide* (Bioagent) also reduces infection. Soil treatment with thiram also checks its spread.

14. HARVESTING

Crop should be harvested when grains are hard and having 20-25% moisture. Rather crop should be harvested at physiological maturity, as many high yielding and hybrid varieties remain green even after maturity. Crop may be harvested either by cutting ear head or whole plant and then ear heads are collected. Crop is threshed manually by beating with sticks or by running bullocks over heap of ear heads or by mechanical thresher after drying. Produce is cleaned and dried in sun till 10-15% moisture.

15. YIELD

Irrigated crop raised with full improved cultivation practices yield 45-50 q/ha of grain and 100-110 q/ha dry fodder. Rainfed crop provides 20-30 q/ha grain with 75-80 q/ha dry fodder. In general the harvest index is 45-50%. Efforts are going on to develop ideal plant type of sorghum with short stiff stem, erect leaves, deep and extensive root system, loose panicle with more primary branches + having large number of grains capable of producing in moisture stress conditions and tolerant to insect pests and diseases.

16. FODDER SORGHUM

16.1. Cutting Management

Single cut varieties should be cut at 50% flowering stage. In multicut varieties first cut is taken after 50-60 days of sowing. Subsequent cuttings should be taken at an interval of 35-40 days. Crop should be cut 5-8 cm high from ground level for fast sprouting.

16.2. Sorghum Poisoning

Young plants (30-40 days stage) contain cynogenic glucoside 'Dhurrin'. Dhurrin in the stomach of animals is converted into hydrocyanic acid. Thus when young plants (about 5 kg) of sorghum are fed to animals, it causes carcinogenic death of cow. This is known as prussic acid poisoning or sorghum poisoning. HCN content is more in leaves. Concentration of HCN is more in morning, and in summers. The toxic limit of HCN is 200 ppm, concentration decreases after 50 days. To overcome problem, avoid feeding sorghum before 50 days stage. Hay and silage are generally free of HCN.

As first aid treatment affected, animals may be given intravenous injection of sodium thiosulphate to restore capability of oxygen transport of blood. Two full drenches of mollases may be given to provide extra energy to animals.

17. POST HARVEST TECHNOLOGY

Grains are cleaned, dried in sun till 10-12% moisture. Stored grain pests such as rice grain weevil (*Sitrophilus oryzae*) and grain moth (*Cerealella* sp.) damage stored sorghum if moisture is high. Dehulling and pearling is done to improve quality. Grain quality is decided by colour of grain, test weight and shape of grain. Rounded white and bold grains fetch more price. Chemical composition of sorghum and other cereal crops is given in Table 3.9.

TABLE 3.10: Nutrient composition of cereals (per 100 g)

Food grain	Proteins (g)	Carbohydrates (g)	Fat (g)	Crude Fibre (g)	Mineral Matter	Calcium (mg)	Phosphorus (mg)
Wheat	11.8	71.2	1.5	1.2	1.5	41.0	306.0
Rice	6.8	78.2	0.5	0.2	0.6	10.0	160.0
Barley	11.5	69.6	1.3	3.9	1.2	26.0	215.0
Maize	11.1	66.2	3.6	2.7	1.5	20.0	348.0
Sorghum	10.4	72.6	1.9	1.6	1.6	25.0	222.0
Pearl millet	11.6	67.5	5.0	1.2	2.3	42.0	296.0
Finger millet	7.3	72.0	1.3	3.6	2.7	344.0	283.0
Proso millet	12.5	70.4	1.1	2.2	1.9	14.0	206.0
Foxtail millet	12.3	60.9	4.3	8.0	3.3	31.0	290.0
Kodo millet	8.3	65.9	1.4	9.0	2.6	27.0	188.0
Little millet	8.7	75.7	5.3	8.6	1.7	17.0	220.0
Barnyard millet	11.6	74.3	5.8	14.7	4.7	14.0	121.0

18. CONSTRAINTS OF PRODUCTION

Sorghum follows C_4 pathway of carboxylation and PEP carboxylase is very efficient enzyme system. Despite of this fact, farmers obtain low yield. Lack of demand for human consumption, non availability of improved seeds, cultivation of crop on poor soils and low acreage under high yielding varieties are some of the constraints. Moisture stress and poor emphasis on plant protection are the other factors, leading to its poor yield. Efforts should be made to develop short duration, fertiliser responsive drought tolerant cultivars to harvest more.

19. REFERENCES

Anonymous (1992). Technology for increasing sorghum production in India. National Research centre for Sorghum. Rajendra Nagar, Hyderabad.

Anonymous (1999). The Hindu Survey of Indian Agriculture 1999.

Chopra, N.K and Chopra, N.K. (1990). Beware of sorghum poisoning. Intensive Agriculture. 27(11):24-25.

FAI (1997). Fertiliser statistics, Fertiliser Association of India, New Delhi.

FAO. (1997). FAO year book, VVI. 51, 1997.

Harlan, J.R and de Wet, J.M. J (1972). A simplified classification of cultivated sorghum. Crop sci. 12: 172-176.

Katyal, J.C and Das, S.K (1993) Fertiliser News.

Mann, J.A, Kinber, C.T and Miller, F.R. (1983). The origin and early cultivation of sorghum in Africa. Texas A & M University, College Statim, Texas pp 1-21.

Manohar, S.S (1986). Efficient be of fertilisers including biofertilisers for sorghum. Lecture delivered in state level training or sorghum production technology held at SKN college of Agric Jobner. 6-9, May.

Manohar, S.S (1986). Water management in sorghum. Lect.

Mishra, J.S. (1997) Critical period of weed competition and losses due to weeds in major field crops. Former and Parliament. 33(6):19-20.

Rana, B.S., Singh, B.U., Rao, M.H, Indira, S, Rao, S.S. and Kaul, S.L. (1998). Sorghum (*Sorghum bicolor*) research in India. Indian J. Agric. Sci. 68(8):405-22.

Singh, I, Chouhan, G.S. and Choudhary, L.S. (1993) Response of sorghum (*Sorghum bicolor*) to nitrogen and phosphate under western Rajasthan conditions. Indian J.Agric 38(2):305-306.

Singh, S.P. 1999. The Hindu Survey of Indian Agriculture. Pp 178-184.

Tripathi, R.Y and Suraj Bhan (1995). Effect of level and method of nitrogen application and moisture conservation practices on growth and yield of rainfed sorghum (*Sorghum bicolor*) under light textured, eroded soils of central Uttar Pradesh. Indian J. Agric. 40(1):47-50.

Vivilov, N.I. (1926). Studies on the origin of cultivated plants, Bulletin of applied botany and plant breeding. VVI 16(2):

Yadav, R.L., Singh, P., Prasad, R and Ahlawat, IPS (1998). Fifty years of Agronomic Research in India. Indian Society of Agronomy, New Delhi.

4. Pearlmillet

R. C. Gautam

Pearl millet with a popular name *bajra* is an important crop of rainfed areas of India. It provides staple food for the poor in a short period in the relatively dry tracts of the country and that is why pearl millet is cultivated by the economically poor farmers using either no improved production technology or using it only at sub-optimal levels. Under rainfed conditions of Rajasthan state. Pearl millet is generally mono-cropped (100% cropping intensity) but in other pearl millet growing states high intensity sequences (200-300%) are also followed as well distributed rainfall is received in these states and irrigation facilities are also available to provide life saving irrigation to the crop under drought conditions.

Much has been done to increase the per hectare grain yield of pearl millet. Improved disease resistant hybrids/composites are developed and package of agronomic practices are evolved by the scientists of different institutes and agricultural universities. An attempt has been made by the author to compile the research findings of different scientists and to present it in the following pages for the benefit of the farmers/readers who are interested in the new technology of growing pearl millet in India.

In addition to that pearlmillet plant has a better root system with better extraction of soil moisture and its utilization-mechanism in the plant. Their efficient photosynthetic mechanism and rapid transfer of food material from the leaves to the grains have made this plant to withstand acute drought situations.

Approximately 40 per cent of the rural population of the country prefers to take this food as it gives sufficient energy to the field workers at a very low cost. In India pearl millet cultivation in general is preferred in the lands which are poor in fertility, farmers never prefers to give costly inputs like fertilisers to this crop comparing with rice, maize and sorghum crops of the same season. In other words, Pearl millet is grown by the farmers who are economically poor. They grow this crop with lesser inputs and utilize it as a stable food and that's whyPearl millet is called as "Poorman's crop".

1. ORIGIN

Pearl millet is considered to be a very old crop. Most of the scientists believed that the primary centre of its origin is Africa from where it spread to India and other countries. Out of 32 species described by Stapp (1934) only two are known outside the African continent. They are *Pennisetum glaucum* (Pearl millet) and *Pennisetum purpureum* (Elephant grass) and both have a common ancestor. Secondly, a study revealed that a greatest range of genetic variability was found in the strains from Africa. The above facts are enough to establish that the origin of pearl millet is from Africa. Once domestication had occurred, pearl millet was probably quickly introduced throughout the drought prone regions of the Sahara. As the desert continued to expand, the distribution of pearl millet expanded proportionately.

2. DISTRIBUTION

Pearl millet is an annual warm weather coarse cereal covering, annually on 10-12 m hectares of marginal and sub-marginal lands primarily in the states of Rajasthan, Gujarat, Haryana, U.P. and Maharashtra. The rainfall over this area ranges from 200 to 1200 mm received from mid June to end of September. Therefore, major part of pearl millet acreage faces moisture stress during its growing period mainly due to low and ill distributed rainfall. Pearl millet's success is a dependent function of rainfall and or incidence of diseases like downy mildew and ergot. Crop is characterized by the drought evading mechanism. It means this is the plant which never resists to drought but cut short of its life cycle and comes to flowering under such adverse conditions. It grows well in regions receiving 500 mm of rainfall. Pearl millet grows well in areas that have extended dry periods. Pearl millet thrives better than other major cereals in poor soils. It often replaces sorghum on light sandy soils. Pearl millet is therefore, mostly grown in semi-arid and arid climate of tropical and sub-tropical regions. Pearl millet was grown over an area of 22.0 mha in 1990 in the world which falls between 15° W-90° E longitude and 5° S-40° N latitude. Cultivation of crop is confined to drier regions of Southern Asia (mainly, India), East, West, North and Central Africa which cover countries like Nigeria, Niger, Mali, Chad, Tanzania, Sudan and Senegal. In USA, South America, Canada, Japan, Italy and Australia. It is grown in small proportions for fodder purposes.

In India, pearlmillet is grown on 10.4 m ha of cultivated land distributed in most of the states except the states having high rainfall. India has been divided into four ecological zones for Pearl millet cultivation.

1. *Ecological Zone I:* Adequate to abundant, rainfall and fertile soil zone (Punjab, U.P., Delhi, Haryana and adjacent M.P.).
2. *Ecological Zone II:* Limited rains but heavy to light loamy soils (Gujarat, Mahrashtra & M.P.).
3. *Ecological Zone III:* Low rainfall and light soils (Karnataka, North-Central. A.P. and Rajasthan).
4. *Ecological Zone IV:* Limited but well distributed rainfall (T.N., and Coastal A.P.).

Acreage-wise most important Pearl millet growing states are Rajasthan, Maharashtra, Gujarat, Western, U.P. and Haryana. Nearly 87% of the total area lies in these states. Area production and productivity of this crop in these states is given in Table 4.2.

At present in India, the total area under pearlmillet crop is 10.4 m. ha, out of which 94% is rainfed.

TABLE 4.1: Area, production and productivity of Pearl millet in important countries of the world

Country	Area (mha)	Production (000 t)		Productivity (kg/ha)	
	1997	1961	1997	1961	1997
India	10409	3502	7508	329	781
Nigeria	5130	2548	3683	595	716
Niger	3229	490	1463	400	480
Mali	1467	795	1189	661	798
Chad	785	1015	600	842	853

TABLE 4.2: Area, production and productivity of pearl millet

State	Area (m ha)		Production (m.ton)		Productivity (kg/ha)	
	1961	1997	1961	1997	1961	1997
Rajasthan	4.20	5.00	1.04	1.63	247	400
Maharasthra	1.67	1.74	0.43	0.68	258	500
Gujrat	1.21	1.39	0.48	0.79	394	995
U.P.	0.95	0.91	0.38	0.78	426	908
Haryana	0.50	0.63	0.29	0.35	298	780

3. CLASSIFICATION

Stapp divided the genus *Pennisetum* in six sections. Pearl millet cultivated in the world belongs to section *Pennicillana*. Botanically pearlmillet is *Pennisetum glanccum* (L.) R.Br. emend Stunntz) belongs to family gramineae and diploid chromosome no is 2n=14.

Pearl millet is an important rainfed crop. Major part of its acreage faces moisture stress at its growing time due to low and poor distribution of rainfall. But being a drought evading crop, it can cut short of its life cycle and comes to flowering and is having the characteristic of maturing within 40 to 60% of its actual life cycle period. Because of drought evading mechanism, pearl millet crop is preferred in states like Rajasthan and Gujarat where the *kharif* season is more prone to drought like conditions. Farmers of these states may hope to harvest the crop even in acute drought seasons.

Before flowering, plant of Pearl millet is in growth stage. The stem is divided into nodes and internodes. One leaf and one auxillary bud emerge from each node. The auxillary bud may grow into a tiller or shoot. The inflorescence comes out from the upper most node. The upper most nodes are rather much enlongated than the lower most ones. The internodes may be hairy or glabrous. The seed on germination gives rise to radicle which grows into primary root. After the emergence of 2-3 leaves, the adventitious roots appear from the first node at an angle of 90° from the first auxiliary bud. About 14-25 roots per plant each measuring on an average, 34 on have been recorded. At the second node, the second pair of secondary roots also appear in the same manner. The nodes above the soil level produce thick strong roots which serve as proof roots, and prevent the plant from lodging. Deeper root system, efficient photosynthetic mechanism and rapid transfer of food material from leaves to the grains has made this plant to withstand, serious drought conditions in rainfed areas.

FIGURE 4.1: Pearlmillet ear.

4. CLIMATE

Pearl millet is a rainfed crop and therefore, it is mostly grown in the area where the annual rainfall is between 400 mm and 750 mm. In India rainfall in pearlmillet growing arid and semi-arid areas ranges between 190 mm to 750 mm at Jodhpur and Nandyal, respectively. This plant grows well under warm climatic conditions with 500-600 mm rainfall, during the entire growing period. During the vegetative growth of the crop, moist weather and light to medium rainfall is very useful. But at the time of flowering and grain formation there should be no rain and the weather should be clear and dry. Grain formation is adversely affected and results in poor grain setting if rains continue till that stage. If there is rain at the grain maturity period, the grains are infected with diseases like ergot which appear due to high humidity and low temperature. Waterlogged conditions are not congenial for this plant to grow, therefore, this crop should not be put under low lying areas. Experimental findings revealed that the temperature of 28-32°

C is considered to be most optimum for the vegetative growth of the crop. High temperature at this stage induce early flowering. The low temperature increases the chances of ergot infection. It is, therefore, recommended to sow the crop at optimum sowing time. This crop also protects the crop from such incidences.

5. CROPPING SYSTEM

5.1. Cropping Sequences

In northern India many rotation with Pearl millet crop are adopted. In states like Rajasthan and Gujarat where rainfall is scanty most part of Pearl millet area is covered with 100% cropping intensity which means only one crop is grown in one calendar year and in these areas therefore, Pearl millet is followed by Pearl millet. This cropping sequence is proved to be inferior to the sequence preceded by a legume crop as reported by Giri and De (1979). In experiments conducted over 3 years, the grain yield of Pearl millet was increased by 23% when grown after groundnut, 24% after cowpea and 12% after pigeonpea. This increase was calculated to be equal to 60 kg N/ha. It is therefore, recommended for Rajasthan and Gujarat farmers to take Pearl millet crop in the alternate year and grow a legume in between.

In assured rainfall areas Pearl millet is followed by *rabi* crop and general rotations are: Pearl millet-barley, Pearl millet-wheat, and pearlmillet-toria etc. In these rotations, pearlmillet is grown as rainfed crop and *rabi* crop is grown on conserved moisture. One or two irrigations recommended to the *rabi* cros if conserved moisture is not sufficient for the entire crop season. Such rotations (200% intensity) are most popular in Western U.P., north M.P., Maharashtra and in Haryana states. Total productivity is increased considerably in such rainfed areas by following the above mentioned rotations.

5.2. Intercropping

Intercropping is defined as growing of 2 or more crops simultaneously in the same field in rows of definite geometrical pattern. In case of Pearl millet a paired-row of planting is adopted in place of uniform rows. Spacing of 30/70 cm is the most appropriate spacing where Pearl millet is sown in paired rows at a distance of 30 cm and the spacing of 70 cm is kept between two paired rows. The inter crop is sown in the centre of 70 cm. In rainfed areas intercropping is an insurance against bad weather. Selection of inter crop is generally from oil seeds and legume crops. Most suited intercrops from the legume group are: green gram, black gram, pigeon pea, cowpea, dew gram, and cluster bean. Inter crops from oilseed group are groundnut, seasame castor and soybean. Intercropping in Pearl millet generally increases the land-equivalent ratio (L.E.R.) and monetary returns considerably.

Experimental findings revealed that intercropping not only stabilize yields but also maximise total productivity/unit area under normal climatic conditions. The basic concept of intercropping is that 2 or more species can exploit the environment better than either of the species grown separately.

TABLE 4.3: Ideal companion crops identified at different centres

Region/Centre	Suitable inter crops
Mandor, Jaipur Fatehpur	Pearlmillet + mothbean or cluster bean
Hisar, Jamnagar and Madhuri Kund (UP)	Pearlmillet + greengram or sesame.
Delhi	Pearlmillet+ Groundnut or pigeonpea or castor
Gwalior	Pearlmillet + Blackgram or soybean
Bijapur, Pune, Aurangabad & Coimbatore	Pearlmillet + pigeonpea, Pearl millet + groundnut or cowpea or sunflower

TABLE 4.4: Net returns in the intercropping system

Intercropping system	Data of 4 years at following no. of centres	grain yield			Mean	
		Pearl millet	Inter-crop	Pearl millet equivalent	Input in intercropping (Rs./ha)	Net return from intercropping (Rs./ha
Pearl millet + cowpea	6	25.0	3.5	34.6	6055	1680
Pearl millet + mungbean	7	26.0	2.7	34.0	5950	1404
Pearl millet + urdbean	4	24.9	1.3	28.4	4970	624
Pearl millet + Castor	3	26.4	2.3	31.6	5530	920
Pearl millet + sesame	4	22.0	1.7	29.0	5075	1241
Pearl millet + ground nut	5	4.9	3.3	32.9	5757	1419
Pearl millet + pigeonpea	3	23.6	2.4	32.3	5635	1512
Pearl millet alone	8	26.4	-	26.4	4620	-

All India Coordianted data on intercropping presented in Table 4.5 revealed that Pearl millet yield did not very much due to intercropping. Promising intercrops in the intercropping were groundnut, mungbean, pigeonpea and castor which gave 2.3-3.5 q/ha additional yields to the growers. Most promising and remunerative mixtures in different agro-climatic regions were Pearl millet + cowpea, pearl millet + pigeonpea, Pearl millet + groundnut, Pearl millet + mungbean and Pearl millet + sesame in terms of net returns.

6. SOIL AND FIELD PREPARATION

Pearl millet is successfully cultivated in different types of soils. Sandy loam to loamy soils which are well drained and non-saline and non-alkaline in nature. In Central and peninsular India, Pearl millet is grown in black cotton soils and laterite soils. Heavy soils are not suitable for Pearl millet cultivation and gives fairly good yield.

In moisture deficit areas, the tillage operations before and after seeding the crop are extremely important as they are associated with the conservation of rain water. In such cases summer ploughing every year and deep ploughing once in 3-4 years are recommended. The deep ploughing once in 3-4 years are recommended. The deep ploughing, is capable of releasing major nutrients and greater retention of moisture in the soil. Generally, pearl millet requires a fine and smooth seed-bed which is free from clods. Nearly 2-3 ploughings are given to prepare a fine seed bed. To minimise damage due to termites, application of 5-8 kg. Furadon per hectare is recommended. The water holding capacity of both the coarse-textured desert soils in Western India and the red sandy-loam soils of southern India is extremely low. Subramaniam *et. al.* (1973) reported that organic material like rice-husk or FYM. (25 tons/ha) if incorporated to a depth of 20-45 cm, 8 weeks before sowing improved the water-holding capacity of the soil and there by the yield of Pearl millet. Ali and Prasad (1974) conducted field trials for 2 years to study the influence of type of seed bed and mulch on the yield and nutrient up take by Pearl millet under semi-arid conditions. A flat seed bed stored more moisture, gave significantly higher grain yield and also increased the uptake of NPK compared with ridge and furrow beds.

7. MANURES AND FERTILISERS

Fertiliser application to field crops is necessary to realize maximum yield potential of a crop and its variety and Pearl millet is no exception. The major nutrients viz., N, P & K are required by the crop plant to provide balanced nutrition in conjunction with micro-nutrients, if needed. At the present productivity level, Pearl millet, removes 72 kg. N+P+K/ha/annum, whereas only 10-11 kg. of these nutrients are being actually applied. It is estimated that for the production of 100 kg grain, Pearl millet requires 3.6 kg N, 0.8 kg P and 3.4 kg. K. Fertiliser not only increase above-ground growth but also enhance below-ground root proliferation. In this way, replenishment of the soil nutrients encourage more efficient water use. The utilization of the applied N, P and K from fertiliser by this crop amounts to 36.4, 20.5 and 126.3 per cent, respectively. These figures may change with the texture of the soil especially for nitrogen but these can be held nearly constant by increasing the number of split doses for nitrogen in light textured soils.

Of the 3 macroelements, response to N/has almost been universal in all regions of the country where Pearl millet is grown. The optimum rate of N fertilization in experiments averaged over 5 years and 30 locations ranged from 92 to 137 kg/ha at a response level of 6.0-10.5 kg grain/kg N applied. The output: input ratio of nitrogenous fertilisers ranged between 0.8 and 2.7 in different zones (De and Gautam, 1987). The studies aimed at determining response to N fertilization of newly developed hybrids and populations under varying time of sowing showed that most of the cultivars were found responsive to N fertilization. Application of 40 kg N/ha in zone 'A' (Rajasthan & Gujarat) and 60-80 kg N/ha in zone 'B' (Haryana, Western UP., North M.P. and Maharashtra) were found optimum doses for both hybrids and composite varieties under rainfed conditions. Highest response has been reported upto 80 kg N/ha but did not differ significantly with 60 kg N/ha at IARI, New Delhi (Gautam, 1997).

Studies also revealed that Pearl millet responds to phosphorus also. But the responses are recorded at the few centres where there was deficiency of phosphorus in the soil. At IARI, it was observed that N could not be fully utilized without phosphorus. As much as 800 kg of additional grain yield was obtained merely by adding 20 kg P_2O_5/ha. In another trial, yield response to applied N increased considerably with the increasing rate of phosphorus application. Increase in nitrogen beyond 80 kg N/ha without phosphorus reduced the yield. The recommended levels of phosphorus for Pearl millet crop are 20-24 kg P_2O_5/

ha. In case of potassium, though the crop is heavy user of potash than maize or sorghum at comparable yield levels. Yet no worthy response to applied potash has been observed. At places like Raichur, Junagadh, Coimbatore and Jammu, a very good response of 12 q/ha was obtained due to the application of 90 kg K_2O/ha while it ranged between 1.9-2.2 q/ha at Rohtak and Tonk. Application of P and K fertilisers, therefore, be based on soil test values of these nutrients. From farmer's point of view the general recommendations of fertiliser application to Pearl millet crop are as follows:

- **A Zone** = 40 kg N + 30 kg N + 30 Kg P_2O_5 + 30 kg K_2O/ha
- **B Zone** = 60 kg N + 40 kg P_2O_5 + 40 kg K_2O /ha

Among the fertilisers urea, calcium ammonium nitrate (CAN) and D.A.P. are most suited to pearl millet. Full dose of N should be applied in 2 equal splits i.e. half at planting and other half as top-dressed 3-4 weeks after sowing, subject to the availability of soil moisture. Split application saves N losses due to leaching particularly in light soils. In case of P and K fertilisers, full dose should be applied (band placed) before sowing. The placement of these fertilisers increases the efficacy of applied fertilisers. Pearl millet takes up NO_3 nitrogen more rapidly than NH_4 nitrogen. The uptake of P_2O_5 and K_2O is positively correlated with N and grain yield. There is a decline in N uptake after flowering in poor soils. But it continued at all stages upto maturity in N rich soils. Nitrogen content in plant at all growth stages was related to grain yield. Some observations revealed that N, P_2O_5, K_2O, Ca and Mg were transferred to grain during dough to maturity stages.

The field experiments started in 1992 at several coordinated centres, On the permanent site with the objective to determine the long term effect of organic manures like FYM. on N economy in pearl millet cultivation and also on soil physical condition revealed that application of 2.5 tons of FYM. for A Zone and 5.0 tons of FYM. for B Zone every year in combination with 40 kg N/ha was more economical and helpful for yield stability in pearlmillet.

8. BIOFERTILISERS

Azospirillum brasilence is a free-living bacterium which shows associated symbiosis when present in the rhizosphere. These bacteria are capable of fixing 10-28 mg/g N of calcium malate in pure culture. In experiments conducted at 7 locations over 3 years, inoculation of Pearl millet seed with the local strains of the bacterium increased the yield markedly. Field studies aimed at quantifying the response of Pearl millet to inoculation with *Azotobactor* and phosphate for economizing nutrient solubilizing bacteria revealed that part of inorganic N could be saved with *Azotobactor*. Seed inoculation with *Azospirillum* in conjunction with 10-20 kg/ha and fertiliser was found to save N to the extent of 10-20 kg/ha. In other study, sole application of phosphobacteria or with P_2O_5 did not produce as much visible effect as sole *Azotobacter* or *Azotobacter* with nitrogen.

9. VARITIES

- ***Hybrids:*** Pusa-23 (MH-169), Pusa-322, ICMH-451 (MH-179), ICMH-356, HHB-60, HHB-68, HHB-50, RHB-30, GHB-235, X-5, Nandi-30, MLBH-285, RHB-58 (MH-320), RHB-90, MH-605 (PUSA-606) MH-790, MH 782

- ***Composites:*** ICTP-8203, HC-4, WC-C-75, ICMV-221, ICMV-155, RAJ-171, HP 8601, Pusa Bajari-266, MP 334.

Characteristics of some of the varieties are given in Table 4.3.

Variety	Maturity (Days)	Yield (q/ha)	Recommended for
BJ 104	85	35-40 q	Gujarat, Maharashtra, Delhi, Karnataka, Rajasthan, UP & Haryana
CJ 104	82	20-35	Gujarat
BX 560-230	85	35-40	Whole India
MBH 110	85	35-40	Maharashtra
MBH 118--WCC 75	85-90	15	Maharashtra and Rajasthan
MH 179	80-85	15-20	Maharashtra and Rajasthan
RHB 30	70-75	20-25	Rajasthan
HHB 67	65-70	15-25	Rajasthan
RCB 2	80-85	15-20	Rajasthan
Raj 171	-	18-20	Whole Country
RHB 30	-	17-22	Arid & Semi and regions of Rajasthan
RCB 911	70-85	20	Rainfed, low fertility soils of Rajasthan
Raj Bajra Chari 2	60	300 q/ha green fodder	Rajasthan

10. SOWING

10.1. Time of Sowing

Optimum time of sowing plays an important role in determining the performance of a crop. In rainfed areas the sowing of Pearl millet should be commenced with the onset of monsoon showers. Sowing at the optimum time gives maximum grain yield and net profit. Chances of disease incidence are also reduced. But delay in sowing lower the grain yield consistently. The rate of decline varied from 4-80 kg/day/ha under rainfed conditions. A maximum temperature range of 28°-32° C is found to be optimum for proper growth in the initial stages and to achieve this, the crop should be sown by the first fortnight of July. Late sowings in the month of August cause poor stand due to high rate of mortality of the seedlings, restricted vegetative growth, poor grain setting and more incidence of diseases due to comparatively low temperature during the period.

10.2. Transplanting

For delayed sowing, transplanting is recommended in pearl millet as it is found to be better substitute. For this a nursery measuring 500 m^2 area is sown in advance and 3 week old seedlings are transplanted in the main field in the month of August. A row to row spacing of 50 cm and plant to plant spacing of 10-12 cm is kept to have 1,75000-200,000 plants/ha. Rainy day is generally preferred for transplanting.

10.3. Seed Rate and Plant Population

Recommended seed rate is 4-5 kg/ha if the seed is sown by *Kera method* (behind the plough) and row to row spacing is kept 45 cm. In case of pearlmillet, plant to plant spacing is maintained after 3 weeks of sowing at the time of thinning and weeding. The optimum plant population was found to be 1,75,000 to 200,000/ha. However, plant population of 160,000/ha should be maintained in erratic rainfall area. Widening the row distances decreases the pre-flowering moisture use but in narrow spacing conserved water can be utilised for later growth stages as well. Seeds of Pearl millet should be sown about 2-4 cm deep preferably by drill or with plough as *Kera method*.

11. IRRIGATION

As pearlmillet is a rainfed crop, there is hardly any need of irrigation provided rainfall distribution satisfactory. Water requirement of this crop is much lower (250 mm - 350 mm) than maize, sorghum and finger millet (500 - 600 mm). It has been estimated that pearl millet requires, on an average, 100 mm of water per 7 quintals of Pearl millet grain yield/ha. This calculation is based on the facts that soil moisture is properly utilized and there is no disease or pest incidence on the crop, nutrient losses and washout of the pollens by the late rains at the development stage. But in view of the various kinds of water losses under field conditions, it is generally assured that utilization of rain water varied with the type of soil, organic matter content in the soil, leveling of the field and moisture conservation practices followed in that field. But in spite of several precautions, there is nutrient loss, specially of N, associated with the water lost in drainage or run-off. Thus 500 mm of rain received during the crop season would be able to produce the grain yield varying from 7-35 q/ha. On an average 25 q/ha of grain yield is obtained with 500 mm of rainfall. Though Pearl millet is chiefly a rainfed crop irrigation at anthesis of flowering stage would be beneficial in irrigated areas. As many as 3 to 4 irrigations are required in the event of complete failure of rainfall.

11.1. Moisture Conservation Practices in Rainfed Areas

1. Deep ploughing once and 3-4 ploughings before sowing of the crop.
2. Ridge a furrow system of planing for better water harvesting.
3. Application of FYM. at 5 tons/ha to increase the water-holding capacity of soil by 3-4% and reduces of soil temperature by 4° C.
4. Use of different kinds of mulches to check the evaporation losses e.g. wheat straw, FYM., stover etc.
5. Use of different kinds of transpiration - suppressants e.g. kaolin and atrazine, etc.
6. Removal of 1/3 upper part of the seedlings to minimise the transpiration losses.
7. Seed treatment with NaCl or KNO_3 (0.2%) to give hardiness to the crop plant.
8. Reduction of plant stand (mid season correction) according to gravity of drought.
9. Appropriate weed control measures.
10. Breaking of upper crust of the soil layer to check the evaporation losses from Pearl millet field.
11. Use of short duration and drought tolerant varieties/hybrids.

12. Adoption of intercropping under dryland conditions in the paired row planting system helps in reducing the evaporation losses.

12. WEED CONTROL

As Pearl millet crop is grown pre-dominantly in the warm and submitted rainy season, weeds of different kinds deprive Pearl millet of vital nutrients, moisture, light and space. Like other rainy season crops, Pearl millet also faces severe weed competition leading to heavy reduction in grain yield. Competition with weeds could reduce yield by 25-50%.

Important weeds grow with pearlmillet plant are as follows:

- Sathi - *Trianthema partulacustrum*
- Kewai - *Digitaria sanguinalis*
- Makra - *Dacteloctenium aeqyptium*
- Anzan ghas - *Cenchrus ciliaris*
- Sama ghas - *Echinochloa colonum*
- Doob ghas - *Cynodon dactylon*
- Motha - *Cyperus rotundus*

As manual weeding in Pearl millet fields is a costly affair and also at times becomes difficult because of soggy nature of soil during the crop season. Experiments conducted at IARI, New Delhi and other centres revealed that chemical weed control is an effective and economical method in case of pearlmillet. Gautam and Kaushik (1980) reported that triazine compound atrazine when applied at the pre-emergence stage controlled 71-96% of the broad-leved weeds and increased the yield by almost 30-40% over the un-weeded control. In Pearl millet atrazine at 0.5 kg /ha if applied as pre-emergence in 600 L of water/ ha was recommended as the safer dose. In addition to that one manual weeding was also recommended to up root the perennial weeds like *Cynodon dactylon* and *Cyperous rotundus*.

13. PLANT PROTECTION

13.1. Disease

Downy mildew: Considerable information is generated on the biology, epidemiology and control of downy mildew. This fungus can be transmitted as oospores on the seed surface or as the mycelium in embryonic tissues of seed. After germination the downy growth of the fungus is seen covering mostly the undersurface of the leaf. Affected plants sometimes do not produce ear-heads. But appearance of this disease at ear-emergence is known as Green-Ear' disease, as small green leaves replace grains in the earheads.

Infected seedlings should be rogued out and destroyed by burning. Disease resistant hybrids should be grown. Spray of dithane Z.78 (0.2%) or copper oxychloride (0.35%) proved effective in controlling the disease. At present the hybrids and composites developed by the breeders are found to be highly resistant to downy mildew disease, therefore, the incidence is not a serious problem.

Ergot: Sclerotia in the soil germinate to produce ascospores which become air borne and infect individual florets through fresh stigma. Secondary infection occurs by rain, or insects or by physical contact among earheads. The optimum conditions for the spread of the diseases are high atmospheric humidity, low temperature range (20°- 25° C) and existence of a high susceptible host. The honey like secretion dry-up after few days and become hard. This is called as ergot and contain a poisonous substance which is harmful to man and cattle.

Cultural methods including intercropping, early sowing, and deep ploughing to bury sclerotia are proved to be effective to prevent the disease. Fungicidal control (Ziram 0.2%) was also suggested but not found economical. Host-resistance is the best way to control the disease. The cultivars presently in practice possess a good level of resistance against ergot.

Smut: This disease also attacks earheads which produce powder in place of grain. In the beginning it is green in colour and turns to black in later stages. Secondary spread of inoculum may also occur from early flowering panicles to late flowering ones.

Some fungicides (lantavax or vitavax, 0.25%) were found to reduce smut infection when sprayed on to panicles at bootleaf stage. However, this is not considered a practical method of control. The most economical measure is the host-plant resistance.

Rust: The uredinial and telial stages occur on Pearl millet and the spermagonial and aecial stages on alternate host, *Solanum melongena*.

Control of rust was attained with varying degree of success through fungicides (zineb 0.15%) and cultivation methods like adjusting sowing time and fertiliser application. Host resistance is the best remedy in case of rust.

13.2. Insect and Pest

Unlike the diseases, the gravity of problem of insect pests in pearl millet is less severe. Important insects pests of Pearl millet are shootfly, White grub, grey weevil, chaffer beetle and midge. The insects like White grub, shootfly and grey weevil are more damaging during early growth stages. The seed breeding material, hybrids and composites are screened against whole spectrum of insect pests. This has helped in identifying the sources of resistance and also in developing suitable control.

Shootfly: Early planting with the onset of monsoon is effective in controlling shoot fly incidence. Two dustings of malathion (5% dust) @ 25 kg/ha at 10 & 20 days after germination or 2 sprays of neem oil (in 4 g soap/litre water solution) 0.05% or 2 sprays of endosulfan (0.07%) at 10 and 20 days after germination is recommended for controlling shootfly. White grub: Intercropping of Pearl millet with pulses like green gram, clusterbean and cowpea significantly reduced the white grub damage on Pearl millet. Mixing of phorate or carbofuran @ 12 kg/ha with Pearl millet seed and applying in furrows at the time of sowing or application of FYM + endosulfan dust 4% (in 3 : 2) @ 100 kg/ha is also quite effective.

Grey weevil: Quinalphos 1.5% methylparathion 2% and fenvalerate 0.4% dust @ 25 kg/ha effectively controlled grey weevil.

Midge: For controlling midge 250 cc of metacid (50 E.C.) or 625 cc of thiodone (50 EC) in 250 L of water/ha is sprayed on pearl millet crop.

14. HARVESTING

Harvesting at physiological maturity or at little later stage when grains are hard enough & having nearly 15-20% moisture are the ideal stages. Harvesting is done by cutting the entire plant or removing the earheads first and cutting down the plants later on. The earheads after harvesting should be sun dried at the threshing flower for 10-20 days. The threshing is done thereafter either by beating the earheads with sticks or under the feet of bullocks. The threshed grain should be cleaned and dried in Sun to bring down the moisture content of grain by 12-14% for safe storage.

15. YIELD

By adopting improved technology, Pearl millet crop can produce grain yield of the order of 25-35 q/ha. Hybrids produce more yield and respond well to N fertilisers as compared to composites/ synthetics. Rainfall distribution also determine the level of grain yield in that particular zone. About 75-85 q/ha of dry fodder is also harvested from pearlmillet fields which is very useful for milch cattle and drought animals. Generally grain to stover ratio is 1 : 3 in case of Pearl millet.

16. QUALITY

Despite genetic improvement, agronomic factors also affect the quality of Pearl millet. Use of nitrogenous fertilisers helps in realizing maximum protein potential of Pearl millet. Application of 80-120 kg N/ha enhanced protein content by 50%, methionine and tryptophane by 30-50% but did not affect the leucine content. Biochemists at IARI reported that 30-90 kg N/ha increased prolamine but decreased glutenin content of Pearl millet (Anonymous, 1974). Similar study also revealed that with an increase in the level of N fertiliser, the quantity of each of the 6 amino acids in the grain was enhanced significantly but their quantity in protein remained constant. In Pearl millet grain, lysine, threonine and methionine were inversely correlated and leucine, insolucine and tryptophane were positively correlated with the protein content.

From quality point of view Pearl millet grain is rich in minerals (2.0-3.5) and fat content (4.0-8.0%) Pearl millet grains posses a higher (10.5 to 14.5%) protein content with more desirable levels of essential amino acids. The popular hybrid and composites of pearl millet, on an average, contain 11.6% protein which is higher than the protein content of sorghum and maize. The grains of pearlmillet posses a biological value similar to what rice and imparts substantial energy to the body. Due to its above qualities, pearlmillet is considered to be an excellent nutritious food particularly for those who are habituated to this food.

At IARI, it was observed that N could not be utilized fully without phosphorus application. A dose of 20 kg P_2O_5/ha helps in getting maximum recovery from N fertiliser. At this rate of phosphorus 800 kg/ha of additional grain yield was obtained and the protein content improved by 5-8%/

Potassium plays a vital role in the synthesis of amino acids and protein from ammonium ions which are absorbed from the soil. As potassium is essential nutrient for the photosynthetic activity of the plant leaves, resistance to lodging and diseases, it is directly involved in the protein improvement.

17. REFERENCES

Ali, M. and Prasad, R. (1974). Effect of mulches and type of seed-bed on Pearl millet under semi-arid conditions. Experimental Agriculture 10(4):263-272.

Anonymous (1974). Annual report of Division of Biochemistry, IARI, New Delhi-110012.

Bhatnagar, S. K., Yadav, O. P. and Gautam, R. C. (1998) Research achievements in pearlmillet (*Pennisetum glaucum*). Indian Journal of Agricultural Sciences 68(8) (Special issue): 423-430.

De, Rajat and Gautam, R. C. (1987). Management practices to increase and stabilize pearlmillet production in India. Proceedings of International Pearl millet Workshop held during 7-11 April, 1986 at ICRISAT, Hyderabad, pp: 247-253.

Gautam, R. C. (1985). Intercropping in Pearl millet: A remunerative practice of rainfed farming. Indian Farming 45 (4): 21-22.

Gautam, R.C. and Kaushik, S.K. (1980). Spray atrazine or propazine to control weeds in *bajra*. Indian Farming 29(2): 21-25.

Subramanian, S., Lognathan, S.; Ravi kumar, V. and Krishnamurthy, K.K. (1973). Effect of tillage and organic amendment on the physical properties of soil and yield of *bajra*. Madras Agriculture Journal 62:106-109.

5. Finger Millet

P. S. Rathore

Finger millet (*Eleusine coracana*) commonly known as Ragi or Mandua, it is an important minor millet grown in India. It is staple food crop in southern and hilly area of our country. It is a *Kharif* season crop grown both for grain and forage. It contain 9.2% protein, 1.29% fat, 76.32% carbohydrate, 2.24% minerals and 3.90% ash, besides vitamin A and B. The green straw is suitable for making silage.

It can be grown easily in areas having scarcity of water and on comparatively poor soils. Grains are rich in protein as compared to wheat and rice thus provide proteins, minerals and vitamins to weaker section of society.

1. ORIGIN

Ethopian region of Africa is considered centre of its origin where it was developed from *Eleusine corocana* sub-species *africana* probably in Ethopian region. It is a tropical crop and grown in large area as small millet in India and Africa.

2. DISTRIBUTION

Finger millet is widely cultivated in India, Africa, Sri lanka, Malaysia, China and Japan. In India total area under finger millet in different states is around 2.5 million hectares. Karnataka, Tamil Nadu, Andra Pradesh, Orissa. Maharashtra, Uttar Pradesh, Bihar Gujrat, Madhya Pradesh accounts for nearly 98% of the total area under this crop.

3. CLIMATE

Finger millet is crop of tropical and subtropical climate and can be grown in hills as well as plains as a rainfed crop. The crop can be grown in areas receiving annual average rainfall between 500 to 1000 mm.

4. SOIL AND SOIL PREPARATION

Crop can be grown on a wide variety of soils ranging from very poor to very fertile soil. It thrives best on well drained loam or clay loam soils. However, crop can also be raised on saline soils.

The seeds of *ragi* are very small. It needs well prepared field. One ploughing with mould board plough should be done soon after harvesting of the previous crop. The field should be prepared with onset of monsoon. Two-three harrowing or ploughing with local plough and finally the field should be levelled.

5. CROPPING SYSTEM

Finger millet (*ragi*) under unirrigated condition is grown mixed with sorghum, pearlmillet, oil seeds and pulses. Under irrigated condition it can be grown in rotation with tobacco, chillies, vegetables chickpea, linseed, mustard and barley. Common crop rotation are given below:

North India	South India
Ragi-Chickpea	Ragi-Tobacco
Ragi-Barley	Ragi-Sugarcane
Ragi-Mustard/Taramira	Ragi-Potato-Maize
Ragi-Linseed	
Ragi-Tobacco	
Ragi-pigion pea	

Seetharam (1998) identified profitable intercropping system with finger millet as given in Table 5.1.

TABLE 5.1: Intercropping system with finger millet

System	Recommoned for state(s)
Finger millet + Pigeon Pea 8:2	Karnataka
Finger millet + Pigeon Pea 6:2	Bihar
Finger millet + Soybean	Uttar Pradesh and Orissa

6. MANURES AND FERTILISERS

Finger millet responds well to fertiliser application. In absence of soil test values generally 50-60 kg N, 30-40 kg P_2O_5 and 20-30 kg K_2O per hactare is recommended for the crop. Full quantity of phosphate and potash and half dose of N is to be applied at sowing. Remaining half dose of N is applied after 40-45 days of sowing depending upon moisture availability. In addition to inorganic fertilisers FYM at 8-10 tonnes per ha be applied. This should be incorporated 2-3 weeks before sowing.

Seetharam (1998) recommonded N.P.K. for different states for finger millet given in below:

States	Fertilisers Doses NPK Kg/H	
	Irrigated	Rainfed
Andhra Pradesh	90-30-20	40-20-20
Bihar	60-30-20	40-20-20
Gujarat	80-40-0	-
Himachal Pradesh	-	40-20-00
Karnataka	100-50-50	50-40-25

States	Fertilisers Doses NPK Kg/H	
	Irrigated	Rainfed
Maharastra	50-25-0	25-20-0
Madhay Pradesh	-	40-40-0
Tamil Nadu	90-45-45	60-30-30
Uttar Pradesh	-	60-30-0

Seed inoculation with mixture of *Azospirillum brasilence* and *Aspergillus awamuri* increases the yield. The use of *Aspergillus awamuri* increases phosphorus availability.

7. VARIETIES

Many varieties with high yield potential have been released for cultivation in different states. Varieties recommonded for each state are given below-

1. **Andhra Pradesh:** Godavari, Goumati, Padmavati, EC-4840, HR-374, IE-28, AKP-7, AKP-2
2. **Bihar:** BR-2, RAU-1, RAU-2, RAU-8, A404, BM-2
3. **Gujrat:** GN-3
4. **Karnataka:** Godavari, Indof-1, Indof-5, HR-374, Hamsa, EC-4840, IE-28, MR-1, MR-2, GPU-28.
5. **Maharastra:** BR-407, HR-374, PES-176.
6. **Madhya Pradesh:** JNR-1008, JNR-852, PR-202, BR-407
7. **Orissa:** HR-374, PES-176, PR-202, BR-407
8. **Tamil Nadu:** Indof-5, EC-4840, Godavari, IE-28, Sarda, BR-407
9. **For all states:** VL-149.

8. SOWING

- **Time of Sowing:** Unirrigated *ragi* crop in Karnataka, T.N., AP is sown during June-July and MP, Gujarat, Orissa, Bihar and UP in June. In hills, the crop is sown in April-May. The irrigated crop in Karnataka, T. N. and A. P. is sown in September and October.
- **Seed Rate:** The quantity of seed required would be between 8 and 10 kg/ha in case of line sowing. By transplanting 4 kg seed is sufficient for raising nursery
- **Method of Sowing:** Crop is sown in lines 3-4 cm deep. Line sowing ensures better germination, reduces seed requirement and facilitates inter-culture operations. Line to line distance of 20-25 cm and plant to plant distance of 8-10 cm are ideal for *ragi* crop. In regions of adequate soil moisture, it can be raised by the transplanting.

9. IRRIGATION

Depending on the soil and weather conditions and duration of the varieties, 8 -12 irrigation are necessary. The crop should be irrigated once in 8 days in light soils and 15 days in heavy soils. There should not be moisture scarcity at ear formation.

10. WEED CONTROL

Timely weeding and interculture is important. It is difficult to distinguish grass weed from *ragi* seedlings in early stage. When crop is 35 days old, two inter-cultivations are necessary. Extra plants may be removed at the time of inter-cultivation, this will provide desired plant to plant space.

11. PLANT PROTECTION

Blast helminthosporium, blight, root rot, green ear grain smut and virus diseases occur in finger millet. Insect pest are not causing much damage to *ragi,* but stem borer, aphids and ear caterpillers may cause some damage. The following measures may be adopted to control these diseases.

1. To control blast and brown spot diseases, spray 0.2% dithane M-45 or 0.1% kitazine twice. First spray at flowering and second after 15 days of first spray.
2. Virus diseases (motile steak and streak virus) is transmitted by leaf hopper. Leaf hopper is an important measure in the control of this disease. On appearance of first symptoms, rogue out the infected plants and spray 0.5%, rogor or any other systemic insecticide. Repeat the spray if required.
3. The ear catterpillers and stem borer can be controlled by dusting 4% malathion at 15 kg/ha or by spraying 0.05% endosulfan.
4. The aphid can be controlled by 4% endosulfan or 4% malathion at 15 kg/ha.

12. HARVESTING AND THRESHING

The crop matures in 120-135 days. Harvesting is generally done in two steps. First earheads are harvested with sickle and straw is cut close to the ground latter. The earhead are staked for three or four days to dry. Threshing is done by beating with sticks.

13. YIELD

With improved package it is feasible to harvest 20-25 quintals of grain and 50-100 quintals of fodder/ha.

14. CONSTRAINTS IN PRODUCTION

1. Crop is mainly grown under rainfed condition in poor and marginal soils of poor fertility.
2. Subsistence level of farming still adopted in large area.
3. Unavailability of improved varieties to farmers improved technology is not popular among the farmers.
4. No attempt is made in plant protection measures to control pest and diseases.
5. Low input uses, particularly fertiliser.

15. REFERENCES

Seetharams A. (1998) Small millets research: achievements during 1947-97. Indian J. of Agric. Sci. 68(8):431-437.

6. Kodo Millet

P. S. Rathore

Kodo millet is drought resistant crop. The grain remains under horney seed coat which should be removed before cooking. Kodo grain contains 8.3% protein 1.4% fat, 65.6% carbohydrate and 2.9% ash. Straw is very poor in quality.

1. ORIGIN

Kodo millet is cultivated as a cereal in India. Based on investigation Kodo probably originated in the South East Asia.

2. DISTRIBUTION

Kodo is grown mostly in A.P; Maharastra, Karnataka, Tamil Nadu and U.P. For forage kodo is grown in America, Australia, and South Africa.

3. CLIMATE

Kodo requires warm and dry climate. It is highly drought tolerant hence can be grown in areas of scarce and erratic rainfall. It is generally grown in areas receiving 40-50 cm rainfall annually.

4. SOIL AND SOIL PREPARATION

Sandy loam to loam soils are best for Kodo. However it can be grown from poor upland to gravelly stoney soils. One ploughing with mould board plough followed by 2-3 harrowings and planking to level the field are sufficient for the crop.

5. CROPPING SYSTEM

It is grown as a pure crop. It is also grown mixed with red gram, sesamum and sorghum. The common rotations are given below:

Kodo	Linseed
Kodo	Chick pea
Kodo	Mustard/Taramira
Kodo	Barley

6. MANURES AND FERTILISERS

Five to seven tonnes FYM/ha is recommended for the crop. Seetharam (1998) recommended the following doses of fertiliser for different states:

States	Doses N. P. and K kg/ha
Karnataka	20-20-20
Madhya Pradesh	40-20-0
Tamil Nadu	40-20-0
Uttar Pradesh	20-20-20

7. VARIETIES

TABLE 6.1: Varieties of kodo and their characteristics

Varieties	Characteristics
Niwas-1	Evolved from M.P. takes about 115 days to mature and yield about 18 q/ha
Dindori: 73	Mature in 112 days. Tolerant to insect pest and diseases.
Pali	Mature in 110-112 days. Tolerant to insects pest and diseases, yield potential 18-20 q/ha.
J.K. 76 and J.K.62	Recommended for M.P.
GPUK-3	Recommended for all koda growing states.
APK 1 and KMV-20	Recommended for T.N.
JK-2	Recommended for Gujarat region

8. TIME OF SOWING

Crop is sown with the onset of monsoon. Sowing time extends from middle of June to end of July.

9. SEED RATE

The seed rate for line sown crop is 10 kg/ha and 15 kg/ha for broadcast sown crops. However line sowing is better than broadcast.

10. SPACING

Crop is sown at row to row space of 45 cm and plant to plant space of 10 cm in Tamil Nadu and Karnataka but at other places it is grown at 25 cm row to row and 10 cm plant to plant.

11. IRRIGATION

Kodo is sown is *kharif.* It does not require any irrigation. It is mostly grown as a rainfed crop. If irrigation facilities are available one or two irrigation may be applied if long dry spell occur.

12. WEED CONTROL

It is essential to control weeds. In the initial stage, field should be kept weed free upto 35-40 days of sowing. Two weeding at 15 and 30 days after sowing control the weeds.

13. PLANT PROTECTION

13.1. Diseases

Following are the main diseases of Kodo:

- **Ergot:** It is a fungus disease. Consumption of affected grain is injurious to human and cattle.
- **Smut:** It is a seed borne disease. It is controlled by seed treatment with agrosan G.N or ceresan @2.5 to 3 g/kg of seed.

13.2. Insects Pest

Main insects of kodo millete are white ants and stem borers. White ants may be controlled by applying 20-25 kg endosulfan 4% dust in soil, thimet 4% granules @ 15 kg/ha may control the stem borers.

14. YIELD

The average yield of kodo millet ranged form 10-15 q of grains and 20-35 q of fodder. Moisture in grain for storage should not be more than 10-12%.

15. REFERENCES

Seetharam, A (1998). Small millets research: achievements during 1947-97. Indian J of Agric. Sci. 68(3):431-437.

7. Wheat

S. N. Sharma

In most area of the world, it is the principal food of man. The properties of gluten in wheat are such that it produces bread stuffs, generally superior to those from any of the other cereal grains (Clark, 1936). Because of its excellent baking quality, wheat has become the most important source of carbohydrates in the majority of countries in the temperate zones. A small amount of wheat is also used in the manufacture of dextrose, alcohol, and certain breakfast foods.

The by products derived from the milling of wheat into white flour include wheat bran, standard middlings, flour middlings, red dog, brown shorts and grey shorts. All of these by-products are higher in protein content than wheat itself and serve as a valuable protein supplement in many livestock rations. The durum wheats are processed into semolina from which macaroni and related products are made.

Wheat straw is used as excellent feed for livestock in undeveloped countries whereas in developed countries it is used for bedding on farms, and some manufactured into paper. Much of the straw in the leading wheat producing areas is wasted, usually being burnt following threshing

1. ORIGIN

The growing of wheat began very early in pre-historic time, so long ago that the origin of wheat is still a matter of speculation. Records from ancient China show that it was cultivated there by 2700 BC and it was known also to the Egyptians and to inhabitants of Switzerland as early as the stone age.

De Candolle believed that wheat originated in the valley of Euphrates and Tigris and spread from there to China, Egypt and other parts of the world. Some American investigators believed that wheat probably originated in Syria and Palestine mainly on the evidence of Aaronshon, who brought a type of wild wheat to the USA from that area. Vavilov (1950) after extension studies of the origin of cultivated plants, came to the conclusion that the origin of the durum wheat was probably in the region of Abyssinia, whereas the whole group of soft wheats probably originated in the region of Western Pakistan, South-western Afghanistan and the southern parts of mountainous Babohara. In India, evidences from Mohan-jo-daro excavations indicate that wheat was cultivated here more than 5000 years ago. The locus of domestication of wild wheat undoubtedly is its area of natural distribution.

1.1. Diploid (7-Chromosome Group)

Cultivated einkarn - probably the ancestor of all the other cultivated wheat (Mangelsdorf, 1953) - undoubtedly was domesticated from wild einkarn (*Triticum aegilopoidis*) by selection of plants with slightly larger kernels, a slightly tougher stem, and heads that shatter less readily.

Einkarn may have originated somewhere in the near east, since wild enkarn as well as a large number of Aegilops species were found there. Cultivated einkarn possibly originated in the mountains of northeastern Turkey or the Southwestern Caucasus (Vavilov, 1950), or slightly farther south in the uplands of eastern Iraq (Mangelsdorf, 1953), and then spread to Asia minor (Helbaek, 1959) and Europe.

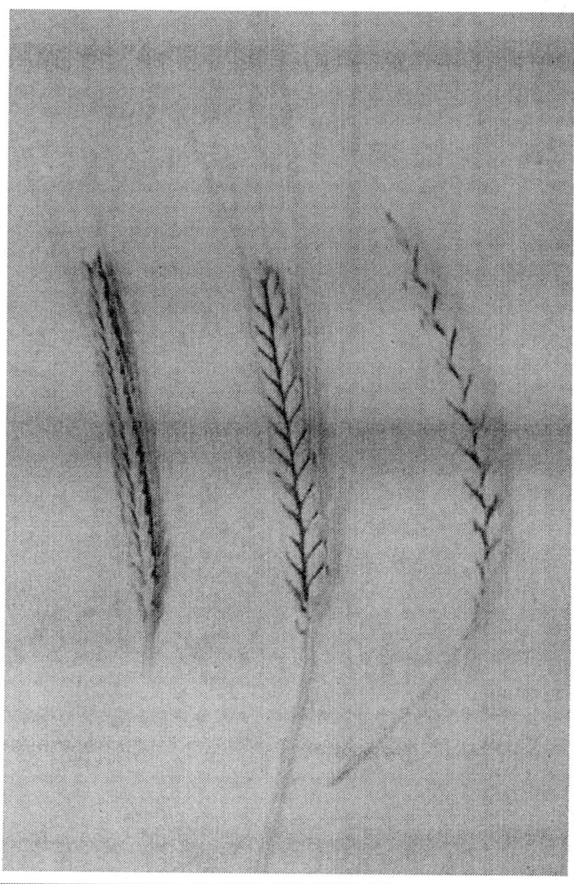

FIGURE 7.1: Cultivated species of *Triticum monococcum*, *T. dicoccum* and *T. spelta*.

1.2. Tetraploid (14 Chromosome Group)

The tetraploid wheat probably arose through hybridization of a 7-chromosome wheat with a 7-chromosome related grass. The parent involved undoubtedly was einkarn or its wild relative, while the wild grass species may have been *Aegilops speltoides* (Riley *et. al.*, 1958). Emmer possibly was derived directly from wild emmer, although it too may have been a product of hybridization (Mangelsdorf, 1953).

Wild emmer (*T. dicoccoidis*) is indigenous to southern Armenia, northeastern Turkey, Western Iran, Syria and northern Israel and was found in the Jarmo site in eastern Iraq. Cultivated emmer appears to have had its origin somewhere in Asia minor, but there is a possibility that it was domesticated in the uplands of Iraq. By 4000 BC emmer, the earliest of the 14-chromosomes wheats, was widely cultivated from the lower plain of Iraq to Asia minor to the Rhine river in Europe.

1.3. Hexaploid (21-chromosome Group)

Haxaploid wheats appear to have evolved most recently. No wild forms are known. These wheats seem to have originated by the hybridization of a 14-chromosome wheat with a wild 7-chromosome some relative of wheat, probably from the genus algilops (Mangelsdorf, 1953).

FIGURE 7.2: Cultivated species of hexaploid wheat *Triticum aestivum vulgare*, *T. aestivum sphaerococcum* and *T. aestivum compactum*.

2. HISTORY

Simple selection exercise by wheat farmers and plant breeders since many centuries resulted in the development of high yielding cultivars. Traits such as increased number of fertile florets, length and density of spike, reduction in shattering, disease resistance and greater responsiveness to fertilisers without lodging have been genetically introduced by recent tools of mutation and hybridization. Release of semi-dwarf variety, "Grain" in 1961 by crossing the Japanese cultivar Norin-10 with American winter wheat "Brevor" opened the era of modern wheat production. The international maize and wheat centre (CIMMYT) in Mexico is the major germplasm repository of the world, while the International Center for Agriculture Research in Dry Areas (ICARDA), Syria collects and disseminates the "durum" germplasm.

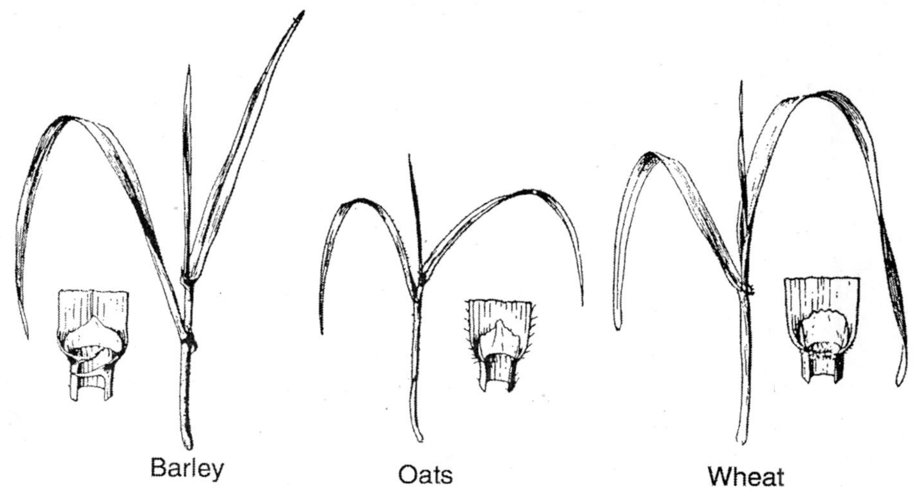

FIGURE 7.3: Morphological characteristics of barley, oats and wheat.

2.1. History of Indian Wheat Improvement

Earliest systematic efforts for improvement of wheat germplasm in India began at the Indian Agricultural Research Institute (IARI), Pusa, Bihar (Howard and Howard, 1906). They released local *T aestivum* based cultivars, Sarbati, Dara, Hard red Calcutta, Bansi, Malvi and Kandula of the durum type. Kapli, Samba and Poptia of *T. dicoccum* were distributed in several parts of the Indian subcontinent. Several other significant achievements were made at IARI, New Delhi by the wheat improvement programme led by Dr. B.P. Pal. The introduction of semi-dwarfs in 1963 provided a spurt in wheat production in India. Mexican dwarf varieties Sonora-64 and Lerma Rojo, introduced in 1963, possessed higher yield levels, non-lodging trait, coupled with greater response to fertiliser nitrogen. This enthused the Indian Government to import 1,800 tonnes of seeds of Lerma Rojo and Sonoera 64 in 1965. Further genetic advancement in these varieties resulted in Indian versions of dwarf and semi-dwarf genotypes named Kalyan Sona, Sonalika, UP 301 etc. (Alexander, 1985). The rapid increase in wheat yield in late sixties is generally termed as the *green revolution*. It is a highly conspicuous example of alleviation of food shortage, related economy, and living standards of a country through inputs on agricultural research.

3. DISTRIBUTION

Wheat (*Triticum spp.*) is the most widely cultivated of all cereals. It is grown under various climatic conditions between 47°S and 57°N latitudes. Hence wheat crop is almost sown or harvested, all round the year in one country or the other. A systematic zonation of wheat culture on the globe includes the north American continent, Latin American including southern cone, Europe, west Asia and North Africa, South and East Africa, South Asia and Far East and Australia.

Wheat is cultivated on about 226945 thousand hectare area with production of about 586036 thousand million tonnes. Area and production of important country is as follows:

TABLE 7.1: Area and production of wheat in world (1997)

Countries	Area 000 ha	Production 000 Mt.	Productivity kg/ha
USSR	25000 F	44180	1767
China	30000	122600	4087
USA	25729	68761	2673
India	26000	69000	2654
Kazakistan	11512	8955	778
Australia	10839	18554	1712
World	226945	586036	2686

Wheat is the main cereal crop of India after rice. Uttar Pradesh, Punjab, Madhya Pradesh, Rajasthan, Bihar and Haryana are the main wheat producing states. Maximum area is in Uttar Predesh but productivity is highest in Punjab. Areas and production of wheat in different states during 1995-96 is given in Table 7.2.

TABLE 7.2: Area and production of wheat in India (1995-96).

States	Area 000 ha	Production 000 Mt.	Productivity kg/ha
Punjab	3282.0	12724.0	3877
Uttar Pradesh	9052.3	22202.6	2453
Madhya Pradesh	4007.6	6467.9	1614
Rajasthan	2196.2	5493.1	2501
Bihar	2071.9	4180.5	2018
Total	26121.8	62620.1	2493

Wheat cultivation in India spans between 10°N in the peninsular India and 35°N in the northwestern Himalayan foothills. For co-ordination of wheat research in India, its cultivation zone have been demarcated (Tandon and Sethi, 1987) as follows:

- **North hill zone:** Foot hills of Himalayas in Jammu and Kashmir, Himachal Pradesh and Uttar Pradesh forms the region. Nearly 4% of area and 3% of production of the Indian wheat is contributed by this zone. Crop durations are long starting from October/November to May/June.

- **Northern Plains:** The Indo-gangetic alluvial area contributes nearly 45% production from 30% of total area under wheat cultivation in India. Predominantly hexaploid wheat is sown in November/December and harvested in April.

3.1. North-Western Plain Zone

This zone contributes nearly 10% area and 11% production of the country. Growing season is relatively short starting from November to March. *Triticum aestivum* dominates in this zone

3.2. North-Eastern Plain Zone

This zone contributes about 24% of total production of wheat from 27% area. Crop season spans from November till March/April, however, cool seasons are relatively short in this zone. Sowing is often delayed due to higher moisture conditions in the fields. Most areas sow the hexaploid *T. aestivum*.

1. **Far-eastern zone:** It comprises a small wheat zone of 2% area and an equal share of wheat production of the country. Shorter growing seasons last from November to February/March.
2. **South-Eastern zone:** Sub-humid upland of eastern India and parts of peninsular India contribute nearly 1.5% of wheat production of the country. It is predominantly a rice zone and wheat cultivation has least importance.
3. **Southern hill zone:** This zone comprises hilly areas of Tamilnadu and Kerala. Wheat has a minor importance with respect to area and production. Though *T. aestivum* is the predominant species, *T. dicoccum* is also grown in some areas. Wheat is grown twice a year in this zone, May to September and October to April. Stem rust perpetuates in this zone.
4. **Central zone:** Parts of arid western India, central India, and Vindhyas contributes nearly 17% area and 13% production. Crop is sown during October and harvested by March. *Triticum durum* is preferred in this zone.
5. **Peninsular zone:** The deccan plateaus, western ghats and Karnataka plateau contributes 6% area and 2.5% production of wheat. The crop is sown during later half of October and harvested by March. *Triticum aestivum, T. durum* and *T. dicoccums* are cultivated in this zone. Thermo sensitive varieties are preferred.

4. CLASSIFICATION

There are several bases that may be used to classify wheat, but the most important are the groupings based on botanical characteristics. Wheat belongs to the grass family Gramineae (Poaceae) and the genus *Triticum*. The species of *Triticum* can be classified into three distinct groups; dilpoids, tetraploids and hexaploids, with 7, 14, and 21 pairs, respectively, of chromosomes in their cells. Although as many as 18 species of wheat have been described and recognized by Percival (1921). The 12 cultivated species of *Triticum* are as follows:

TABLE 7.3: Cultivated species of genus *Triticum*.

Group	Chromosome No. (2n)	Species	Genome formula	Common name
Diploid	7	*T. monococcum*	AA	Einkarn
	14	*T. dicoccum*	AABB	Emmer
		T. durum	AABB	Durum
		T. persicum	AABB	Persian
		T. turgidum	AABB	Poulard or Rivet
		T. polonicum	AABB	Polish
Hexploid	21	*T. vulgare (aestivum)*	AABBDD	Common
		T. compactum	AABBDD	Club
		T. speltra	AABBDD	Spelt
		T. sphaerococcum	AABBDD	Shot
		T. macha	AABBDD	Macha
		T. vavilovi	AABBDD	Vivilovii

These species have been described by Bayles and Clark (1954).

1. Einkarn wheat (*T. monococcum*) is a one-grained wheat. The spikes are awned, narrow, slender and laterally compressed. The spikelets usually contain only one fertile floret. The terminal spikelets are aborted. The palia splits into two parts at maturity. The kernels remain enclosed in the spikelets after threshing; they are pale red, slender and very much compressed. The kernel crease is very slight.
2. Emmer wheat (*T. dicoccum*) may be of either winter or spring habit. It is usually awned. The culms often are pithy within, while the leaves usually are pubescent. The rachis is brittle, while the spikes are very dense, laterally compressed, and narrow. The pedicel is short, narrow and pointed. The spikelets usually contain two flowers. The kernels are red or white, long and slender with both ends acute. Emmer is distinguished from spelt by the shorter, denser laterally compressed spikes.
3. Durum wheat (*T. durum*) plants are of spring habit. The peduncle is pithy, at least in the upper part. The spikes are laterally compressed; the glumes are sharply keeled and the lemmas are awned (except in a few awnless forms which are not grown commercially). The awns are long and coarse, and white, red and usually rather long and pointed; they are very hard and translucent, which makes the white kerneled forms appear amber-coloured. The kernels always have a short brush and angular cheeks.
4. Persian wheat (*T. persicum* or *T. carthlicum*) is a fungus-resistant early maturing spring wheat. It has a strong yellow or light red straw with mostly solid nodes. The spikes are very flexible and lean over. The rachis is very narrow; the spikelets usually are three kerneled. This species differs from others in that the glumes have long awn like projections instead of beaks. Since awns also occur on the leminas, the spike has a bushy appearance. The kernels are oval, almost always red and hard in texture.
5. Pouard wheat (*T. turgidum*) may be of either winter or spring habit. The plants are usually tall with broad leaves. The culms are thick, usually solid and sometimes pithy. The spikes are long and occasionally compound or

branched. The spikelets are compactly arranged on the spike. The glumes are short and compactly keeled. The kernels are thick, humped, and mostly hard, but usually very starchy. The poulards are most closely related to the durums.

6. Polish wheat (*T. polonicum*) has a spring habit, tall stems and a pithy peduncle. The spike is awned, large and lax. The glumes are paperly, very long and narrow. The kernel is long and narrow, sometimes it is nearly a half-inch long. It is hard and somewhat similar in shape to a rye kernel.
7. Common wheat (*Triticum vulgare/aestivum*) is distinguished from club wheat by a spike that is long in proportion to its thickness. The spike is usually dorsally compressed. The spikelets are 2-5 flowered, far apart, slightly overlapped, pressed close to the rachis, and nearly erect. The glumes are keeled only in the upper half, shorter than the lemmas, firm and either glabrous or pubescent. The lemmas are awnless or bear awns less than 10 cm long. The palea, as long as the lemma, leminas entire at maturity. The culm of the plant is usually hollow, but occassionally pithy within and it varies in strength and height. The blades of the leaves are usually narrower than those of the durum and poulard wheats. The kernels may be either soft or hard, white or red. There are a number of both winter and spring varieties grown world over. Common wheat may be subdivided into:
   a) *Hard red winter wheat:* Hard red winter wheats are the most important commercial class of wheat. The grain is hard and generally high in protein content. In quality, as a bread wheat, it ranks second only to hard spring wheat.
   b) *Hard red spring wheat:* is grown where winters are too severe for winter wheat. It is noted for its high protein content and excellent bread making characteristics.
   c) *Soft red winter wheat:* is grown mostly under humid conditions. The grains are soft of low protein content which produces a flour most satisfactory for pastries, such as cakes, cookies and pies.
   d) *White wheat:* white wheats are mainly used for pastry purposes, but some of them go into shredded wheat and bread.
8. Club wheat (*Triticum compactum*) Plants may be of either winter or spring habit and either tall or short. The stems usually are stiff and strong. The spikes usually awnless but sometimes awned-all elliptical, oblong, or sometimes clavate or club-shaped, short (usually less than 6.5 cm in length) very compact and laterally compressed. The spikelets usually contain five fertile florets and spread at nearly a right angle to the rachis. The kernels of club wheat are small and laterally compressed because of crowding in the compact spikes. The grain may either white or red. The club wheats are distinguished from common, wheats by the shorter and denser laterally compressed spikes.
9. Spelt wheat (*T. spelta*) may be of either winter or spring habit and awnless or awned. It has a long, narrow, lax spike with a brittle rachis. The pedicle, or internode of the rachis, is long and wide. Spikelets are of two kerneled, arched on the inner side and closely appressed to the rachis. The kernels, which remain enclosed in the glumes after threshing, are pale red, long and laterally compressed. They have an acute tip as well as a narrow, shallow

crease. Spelt is grown commercially only to a slight extent in the united states.

10. Shot wheat (*T. sphaerococcum perc.*) differs from other hexaploid wheats in its short or dwarf stature (60-70 cm). It has small heads, small rounded grains and hemispherical inflated glumes. It has a characteristic tufted appearance when growing, usually being profusely tillered. It has erect stiff straw which is thick in proportion to its height. It may be awned or awnless. It has red or white and glabrous or pubescent chaff, and may have red or white grain. This species was grown to a minor extent in northern India and west Pakistan for local consumption (Percival, 1921; Ellerton, 1939).
11. Macha wheat (*T. macha*) is a late maturing winter type. The plants have prostrate growth habit. The spelt like spikes are lax or dense, fragile and with rachises that break readily. The kernels are red and semihard. This species is cultivated in Georgia in Soviet Russia.
12. Vavilovii wheat (*T. vivilovi*) has a considerable resemblance to splet, except that it has a compound spike with glumes covered with bristlelike spines, a non fragile rachis, black awns and white grains. It too is grown in Soviet Russia.

5. GROWTH STAGE

The growing wheat plant passes through a vegetative phase followed by a reproductive or fruiting period.

5.1. Vegetative Phase

Vegetative phase includes germination crown growth, tillering, jointing, culm elongation and heading.

Germination: When the wheat grain is exposed to moisture, water is absorbed and the grain swells. At appropriate temperatures the embryo begins to push-out both the coleorhiza (a terminal primary root enclosed within a cone-shaped root sheeth) and Colioptile (an axis with a terminal plumule completely enclosed in a tubular cone-shaped sheath) The seed coat is ruptured, the coleoptile turns upward, and the primary root pierces the coleorhiza and turns downward. Shortly after the primary root pierces the coleorhiza, the lower pass of lateral roots emerges through the axis cortex, followed soon after by a second pair just above the first and sometimes by a third pair. This seminal root system remains functional throughout the life of the plant unless destroyed by disease or in some other way. As the coleoptile continues to elongate, the first foliage leaf of the plumule pushes through the vent. As growth continues, the second foliage leaf pushes out.

Crown root development: Coronal roots arise from the node of the first foliage leaf and this becomes the first node of the crown. Root initiation occurs in the dwarf wheat about 20-25 days after sowing. The crown root consists of several nodes and internodes. Two or three more coronal roots are formed at each crown node. These crown roots, together with their branches are principal supply roots of the plant.

Tillering: A bud primordium lies in the leaf axil at each crown node. The bud primordia develop into tillers (Mc Call, 1934). Tiller production in wheat often starts about 15 days after sowing with a new tiller added every 4-5 days until 45 days after sowing. Though tillering is a varietal character, it is also affected by environmental factors viz., temperature, light intensity, day length, soil moisture, fertility status, depth of sowing, spacing between plants and purpose of preceding crop etc.

Jointing stage: This is the peak plant growth stage starting from 45 to 60 days after sowing. The upper and intermediate nodes expand during this period. The internodes become progressively longer from the base to the top. The uppermost internode is the longest among all the others. The existing leaves are pushed upwards and they increase in size.

5.2. Reproductive Phase

The reproductive phase commences with the tip of the growing point, differentiated into a spike. Reproductive phase consists of three distinct stages.

Booting/flag leaf stage: It is characterised by the swelling of the upper most internodes. The spike is hold inside the last leaf. This last leaf is called flag leaf. The flag develops completely at this stage. The maximum leaf area of plant is observed at this stage. This stage reaches 70-75 days after sowing.

6. CLIMATE

Wheat is grown over a wide range of latitudes ranging between 60°N and 60°S and altitudes ranging from the sea level to an elevation upto 3500 metres in the tropics and subtropics. Normally, the most ideal conditions for wheat growth are cool and moist weather during the vegetative growth period and warm and dry weather during grain formation. Because of this reason the great wheat regions of th world are found in the temperature zones between 30-60°N and 25-40°S. Growth and reproduction of the wheat plant is influenced by temperature, rainfall and light.

The optimum temperature for the germination of wheat is between 20-22°C, though wheat can also germinate at 4°C. At temperatures considerably above the optimum, the grains germinate irregularly and the embryo frequently dies. At temperature as high as 35°C, the endosperm may undergo decomposition due to the activities of bacteria or fungi (Percival, 1921).

The optimum temperature for vegetative growth ranges from 16 to 22°C. Green house studies with controlled soil temperatures indicate that plant height, root length and tiller number decreased as temperatures raise from 22°C to 42°C. Leaves were largest at 22°C and became lighter in colour as the temperature rose above 32°C. Heading was accelerated as much as 11 days as the temperature was increased from 22 to 34°C, but it was retarded or prevented at temperature above 34°C.

During the grain development period wheat requires a mean maximum temperature of about 25°C for atleast four to five weeks. The temperatures above 25°C reduce the test weight of grain due to hastened maturity.

Wheat crop thrive well in those areas where annual rainfall ranges between 25 and 180 cm. However, heavy rains are detrimental to wheat growth as it causes delay in sowing, waterlogging, leaching of nutrients, lodging, infestation of insect-pests and diseases. Crop can not withstand extended periods of soil moisture stress. Rainfed wheat requires a minimum evenly distributed winter rainfall of 15 to 20 cm. Low as well as high rainfall is detrimental to wheat yield.

Wheat normally flowers in the long days of spring or early summer. Short days increase the vegetative growth of the wheat, whereas longer days hasten the formation of inflorescence. Spring wheats usually flower at any day length from less than eight hours upto continuous light. But the process is accelerated with an increase in day length. Winter wheats generally complete their life cycle most rapidly when given low temperatures during the early stages of growth but a long day and high temperatures during the later stage of growth.

7. SOIL AND SOIL PREPARATION

Deep, well drained sandy loam, loam or sandy clay loam with nearly neutral pH are most suitable for wheat cultivation. However it can be grown on a variety of soil ranging from sandy to clayey soil with a pH range from 6.0 to 8.5. On light soils, crop suffers with excessive drainage and low water holding capacity and requires frequent irrigations. On the other hand on heavy soils crop may suffer due to poor drainage. Salinity and/or alkalinity of the northern Indian wheat cultivation zone reduces crop stand and yield considerably (Singh, 1990). The threshold salinity level for wheat is 6 ECe ds/m, yield reduces by 50% at 14 ECe ds/m.

The wheat growing areas of the country can be divided into five soil divisions: (1) the Gangetic alluvium of Uttar Pradesh and Bihar, (2) the Indus alluvium of the Punjab and Haryana (3) the black soils of central and southern India (4) the hilly soils of the Himalaya and else where and (5) the desert soils of Rajasthan. Out of these five division first two categories are most suitable for wheat cultivation.

Traditional system of land preparation for wheat includes one deep ploughing followed by two-three harrowing. After the harvest of *kharif* crops the field is irrigated and at an optimum workable moisture content. The field should be ploughed with disc or mould board plough followed by two to three harrowing and planking. Under late sown conditions when wheat follows sugarcane pigeonpea delay in sowing causes drastic reduction in yield. The pre-sowing irrigation for wheat may be applied in previous standing crop so that field may be prepared immediately after the harvest of previous crop. Alternately, a no tillage system may be followed and wheat is sown directly using zero-tillage seed drills. This system has been practiced consistently for over 15 years, by the farmers in the Panama State of Brazil (Vieira and Kachan, 1990). In India zero-tillage seeddrills have recently been developed by Punjab Agriculture University Ludhiana and G.B. Pant University of Agriculture and Technology, Pantnagar. Zero-tillage system improves soil fertility besides maximum possible use of the short turn around period available in intensive cropping systems or under high soil moisture situations. But weed control by pesticides is a prime requirement of zero-tillage system.

In rainfed areas, collection and conservation of soil moisture and timely cultivation are key practices for the success of wheat crop. One deep chiselling once in three years should be done to break the hard pan created by the continuous use of the country plough. Two to three summer ploughing have been found useful for soaking the rainwater in deeper layer. After cessation of monsoon all the possible efforts should be made to conserve moisture. The ploughings (3-4) should be done in the evening time, furrows should be kept open whole night and planking should be done after each ploughing early in the morning.

8. CROPPING SYSTEM

The development of short duration and photo-insensitive varieties of various crops and assured irrigation had made a distinct change in the cropping systems in our country. The flexibility in sowing time of high yielding dwarf wheat varieties has made wheat a very common component of various cropping systems. The cultivation of wheat has also been spread to traditionally non-wheat growing provinces like West Bengal, Assam, Orrisa, Andhra Pradesh, Karnataka and Maharashtra. Some of the most common cropping systems are given below:

TABLE 7.4: Common cropping system.

Cropping System	Duration
For traditional wheat belt of northern India:	
Rice-wheat	One year
Rice-wheat-greengram	One year
Maize-wheat	One year
Maize-wheat-greengram	One year
Maize-potato-wheat-greengram	One year
Maize-toria-wheat-greengram	One year
Pearlmillet-wheat	One year
Cotton-wheat	One year
Pigeonpea-wheat	One year
Sorghum (grain or fodder)-wheat	One year
Maize-wheat-cotton-perseem	Two years
Maize-wheat-sugarcane-ratoun	Two years
For Non-traditional wheat belt	
Rice-rice-wheat	One year
Jute-Rice-wheat	One year
Rice-wheat-maize	One year

Wheat can also be intercropped with a number of crops like mustard, chickpea, lentil, safflower and lineseed. Mustard is most commonly intercropped with wheat, row of mustard is alternated with 8-10 rows of wheat. This system may not be economically viable in irrigated areas but it meets the requirement of oil and fuel of the farmers.

9. MANURES AND FERTILISERS

Nutrients absorbed by wheat plant include nitrogen, phosphorus, potassium, calcium, chlorine, silicon, sulphur, and small amounts of other essential elements.

9.1. Nitrogen

Wheat plant absorbs much of its nitrogen by the time of flowering, but absorption of nitrogen continues till the maturity. The critical concentration of nitrogen in wheat leaves is reported to be 2.5% below which deficiency symptoms appear. In the absence of adequate supply of nitrogen, wheat plants turn yellow, and growth becomes stunted. There is poor tillering and few tillers bear small earheads. Grains are also not fully developed. All these results in very low yield. Our soils are universally deficient in nitrogen which calls for adequate nitrogen application. Generally wheat crop responds upto 120-150 kg N/ha. in irrigated areas and 40-60 kg N/ha. in rainfed areas.

Nitrogenous fertilisers should be applied in two-three splits. In medium and heavy soils two third or half of the dose should be applied at the time of sowing and remaining one-third or half dose should be top dressed at first irrigation. On sandy soils top dressing should be made in two splits, first at first irrigation and second at second irrigation. Under rainfed conditions full dose of nitrogen should be placed before sowing. If there is good rainfall during crop growth an addition dose of nitrogen may be top dressed for harvesting of potential yield.

Under normal soil, all-nitrogenous fertiliser presently available in market are equally effective. But in calcareous soils and in strongly alkaline soil, 10-30% of top dressed nitrogen may be lost through ammonium volatilization. In such cases calcium ammonium nitrate should be preferred over urea.

Use of farm yard manure, green manure or nitrogen fixing micro-organisms such as *Azotobacter*, *Azospirillum* etc. reduces nitrogen fertiliser requirement by one-third to half. Organic manures also improves physical and biological properties of soil.

9.2. Phosphorus

Phosphorus absorption takes place at a rapid rate throughout the crop growth period and its accumulation in the plant reaches the maximum about two weeks before the maturity. Phosphorus concentration below 0.1% in dry leaf tissue is considered deficient. Phosphorus deficient wheat plants bear less as well as smaller ears. Phosphorus deficiency also reduces the number and size of grains. Soil P concentration upto 8 ppm (Olsen's P) is often considered low. It necessitates external P application. The crop yield responses have been observed upto 60 Kg P_2O_5/ha. However response to P depends on the source and concentration of soluble P in the fertiliser in addition to many other relevant soil factors. On neutral and alkaline soils, phosphatic fertiliser containing atleast 50% of total P in water soluble form is more efficient as compared to those fertiliser which contain less water soluble P. Sharma and Prasad (1996) have recommended mixing of rock phosphate with pyrite for higher efficiency on neutral soils of Delhi. On acid soil (pH less than 6) rock phosphate and basic slag are more profitable than super phosphate or diammonium phosphate (Prasad,1975).

Water soluble phosphatic fertilisers such as diammonium phosphate and super phosphate are more efficient when applied in granular form and are placed near seed band whereas citrate soluble and insoluble fertilisers are more efficient in their finely powdered form and when they are mixed with large volume of soil. Phosphatic fertiliser should be applied at the time of sowing.

9.3. Potassium

Most of the potassium in wheat is absorbed by the seven weeks before the maturity of the crop. Potassium is essential for carbohydrate synthesis. Potassium deficiency adversely affects the grain productivity as well as quality. The response to potassium largely depends on its soil status. Therefore the correct level of its application should be decided on the basis of soil test values. In recent years wheat crop has been found to respond to potassium upto 40-50 kg K_2O/ha in the soils having low to medium level of potassium. Muriate of potash (potassium chloride) and potassium sulphate are only potassic fertilisers presently available in the market. Both are equally effective for wheat. Potassic fertiliser should be applied at the time of sowing however it can also be top dressed at the time of first irrigation.

10. MICRONUTRIENT

Deficiency micronutrients lke Zn, Fe, Cu, Mn and B are notices in one part and other. Zinc deficiency leads to retardation of stem elongation, necrosis and chlorosis of leaves. The interaction behaviour of Zn with P availability also complicated Zn fertilization. Higher levels of P in soil often interferes with Zn availability in soil. Zinc is commonly applied in the from of commercial grade of Zinc sulphate @25 Kg/ha. Zinc sulphate should preferably placed in soil at the time of sowing. However, its deficiency if noticed during the early growth stages could be corrected by a foliar spray of $ZnSO_4$ (0.5%). For this 5 Kg of $ZnSO_4$ and 2.5 Kg of slaked lime is dissolved in 1000 litres of water to spray an area of one hectare.

Manganese deficiency symptoms are seen as necrotic streaks on the leaf lamina. Such streaks are highly conspicuous on the flag leaf. Foliar spray of 0.5% $MnSO_4$ is recommended to correct Mn deficiency. Similarly, earliest symptoms of Fe deficiency of chlorosis of wheat leaves, that appears within two weeks after sowing, can be corrected by foliar spray of Fe. Boron deficiency results in improper development of inflorescence, very short awns, and reduction in yield and grain quality. Copper deficiency begins as limpness or wilting with resultant loss in turgar. The tips of leaves become pale yellow. Reduced stem elongation, improper heading and spikelets of copper deficiency. The Cu deficiency can be corrective by a foliar spray of $CuSO_4$.

11. VARIETIES

Wheat is grown in diverse agroclimatic conditions. In India wheat growing areas have been divided into nine wheat zones on the basis of agroclimatic conditions. The prevailing and recently released varieties for each wheat zone are given below.

11.1.1. Northern Plain Zone

Plains of Punjab, eastern Haryana, Alwar, Bharatpur, and Sriganganagar districts of Rajasthan, western Uttar Pradesh, Jammu region of Jammu and Kashmir and Gwalior, Bhind and Morena districts of Madhya Pradesh comprise this zone Triticum aestovum exclusively dominates this zone. The varieties are recommended for zone are given below:

- **Timely sowing irrigated:** HD 2329, HD 2204, HD 2177, VP 368, VP 2003, PBW 343, WL 711.
- **Late sowing irrigated:** HD 2285, Sonalika, VP 115, Raj 3765.
- **Rainfed:** C 306, IWP 74, WL 410.

11.2.2. Northern Hill Zone

It comprises of the humid western Himalayan region, hilly areas of Kashmir, Uttar Pradesh (except tarai) and Himachal Pradesh (except Una district and Paonta village). Only *Triticum aestivum* is grown in this zone. The recommended varieties for this zone are listed below:

- **Irrigated wheat:** Girija, Sonalika, VL 738
- **Rainfed wheat:** VL-421, VL-401, VL-404, Ridley.

11.3.3. North-Western Plains Zone

This zone covers the arid western plains consisting of Kota and Udaipur divisions of Rajasthan, western districts of Haryana and northern Gujarat. *Triticum aestivum* is the predominant species. However, *Triticum durum* is gaining ground in irrigated areas. The varieties recommended for this zone are given below

- **Timely sowing irrigated wheat:** PBW 343, HD 4530, HD 2687, WH 147, Raj 911, Jairaj.
- **Late sowing irrigated wheat:** Sonalika, Raj 3765, PBW 373, PBW 435, UP 2425, Shera, Raj 821
- **Rainfed wheat:** PBW 396, A-9-30-1, Mukta, Meghdoot, C 306, D 134.

11.4.4. North-Eastern Plains Zone

It includes the sub-humid sutlej-Ganga alluvial plains and the whole of eastern Uttar Pradesh except the Bundelkhand region, and entire state of Bihar excepting the Chotanagpur plateau. *Triticum aestivum* is exclusively grown in this zone. Recommended varieties are given below.

Triticum aestivum is exclusively grown in this zone. Recommended varieties are given below.

- **Timely sown irrigated wheat:** HP 1731, HP 1761, K-9107, Janak, UP 262, HD 1102, Malavia-12, Poorvi
- **Late sown irrigated wheat:** HD 2643, HP-1744, HP 1209, Sonalika, UP-115
- **Rainfed wheat:** C 306, K 68

11.5.5. Far Eastern Zone

It consists of west Bengal, Chotanagpur region of Bihar, Assam, Sikkim, and other north-eastern states. *Triticum aestivum* is the only species grown in this zone. Suitable varieties of wheat are listed below.

- **Timely sown irrigated wheat:** Janak, VP 262, HP 1102, HUW 12, K 7410
- **Late sown irrigated wheat:** Sonalika, VP 115, HP 1209
- **Rainfed wheat:** NP 852, C 306

11.6.6. South Eastern Zone

This zone covers Orrisa, coastal Andhra Pradesh, Chattisgarh region of Madhya Pradesh and Chandrapur and Bhandara districts of Maharashtra. *Triticum aestivum* is the predominant species in this zone.

- **Timely sown irrigated wheat:** Janak, VP 262, VP 301, MD 2189
- **Late sown irrigated wheat:** Sonalika, VP 115, HP 1209.
- **Rainfed wheat:** Bijaga yellow, Bijaga red, C-306.

11.7.7. Central Zone

The zone comprises Madhya Pradesh except its south and south-eastern parts, a few northern districts and Kota and Udaipur division of Rajasthan, Baroda, Kroach, Kito, Surat, Bulsar districts and Kathiawad region of Gujarat and Jhansi division of Uttar Pradesh. Besides *Triticum aestivum, T. durum* is also grown in this zone. *T. dicoccum* is grown only in some localized pockets of Gujarat. The recommended varieties are listed below.

- **Timely sown irrigated wheat:** GW 139, GW 199, Lok-1, J-24, WH 147, HD 4502, HD 4530, Raj 911, P 215
- **Late sown irrigated wheat:** Swati, GW-173, DL 788-2, Sonalika and HD 2327
- **Rainfed wheat:** JWS 17, HW 2004, Sujata, Mukta, Harbada-4

11.8.8. Peninsular Zone

This zone comprises of Maharashtra except Bhandara and Chandrapur districts, Andhra Pradesh except its eastern regions, Karnataka, Goa and Tamilnadu except its hilly areas. In this zone *Triticum aestivum* and *T. durum* and *T. dicocuum* species are commonly grown. Following varieties are recommended for this zone.

- **Timely sown irrigated wheat:** DDK 1001 (T. dicoccum), MACS 2846 (T. durum), DWR 1006, CC 464, HD 2189, Malvika.
- **Late sown irrigated wheat:** DDK-1001 (*T. dicoccum*), NIAW 34, Sonalika.
- **Rainfed wheat:** NIAW-15, N-59, NI-747-19, NI-5439, HW-657, Bijaga yellow, Bijaga red.

11.9.9. Southern Hill Zone

The zone comprises hilly areas of Tamilnadu and Kerala. Though *Triticum aestrivum* is the predominant species, *T. dicoccum* is also grown in some areas. The recommended varieties for this zone as given below-

- **Timely sown irrigated wheat:** HW-517, NP 200, HD 2135
- **Late sown irrigated wheat:** HD 2135, HW 517, NP 200
- **Rainfed wheat:** HW 517, NP 200, HD 2135

12. SOWING

12.1. Selection of Seed

Seed should be healthy and of recommended variety for a particular area. There should not be any mixture of seeds of other crop/variety or weeds. The seed should be purchased from a reliable source. Seed should be treated with vitavex, thiram or agrosan G.N. @ 2.5 g/Kg seed.

12.2. Time of Sowing

Time of sowing is one of the agronomic practice which affect the yield most. Time of sowing depends on the climatic conditions, soil conditions and other available facilities like irrigation. In irrigated areas long duration (135-145 days) high yielding semi-dwarf varieties like HD-2329, CPN 3004 HP 368, UP 2003, WL 711 should be sown between 10th and 30th November. Under late sowing conditions, short duration (120-125 days) varieties like sonalika HD 2285, HD 1553, UP 115, UP 262 should be sown from December 1 to December 15. However, if the sowing of wheat is delayed beyond December 5 there is drastic reduction in yield inspite of the development of the best production technology.

Rainfed wheat is generally sown from second fortnight of October to early November to realise the maximum advantage from the residual conserved soil moisture.

12.3. Method of Sowing

Sowing of wheat is generally done by following methods:

- **Broadcasting:** It is the most prevailing and simple method of wheat sowing in India. Seeds are broadcasted on soil surface, then worked in by harrowing and covered by planking. However all the broadcast seeds do not get proper contact with moist soil. Secondly some seeds left uncovered and are eaten away by birds. This results in poor and uneven germination. To overcome these impediments, 25% higher seed rate is recommended.
- **Behind the plough:** In this method, the seeds are dropped by hand into the furrows that have been opened with local plough. When seed is dropped in furrows by hand, it is called *kera* method and when seeds are dropped through a *pora/nai/hazara* a special attachment with local plough it is called *pora* method. In this method seeds are dropped at a desired depth and germination is satisfactory.
- **Drilling:** Drilling of seed with the help of seed drill or seed-cum fertiliser drills is the best method of sowing. Seed drills are calibrated at a optimum seed rate. These seed drills dropes the seed at a optimum depth an d spacing while running over the field with the help of a tractor or bullocks. The field should be levelled and free from clods, weeds and residues of the previous crop to ensure uniform germination.
- **Dibbling:** It is used when quantity of seed is limited. Sowing is done with the help of a small implement known as *dibbler*. In this method dibblers are pressed into the soil to make the holes and one or two seeds are dropped by hand in each hole and then seeds are covered. Though the method greatly economies the seed, it can not be commercialized because it is time

consuming and labour intensive.

12.4. Seed Rate

Seed rate of wheat varies with varietal characters like seed size tillering ability and germination percentage, time of sowing, moisture content into the soil at sowing time and method of sowing. Under normal conditions a seed rate of 100-125 kg/ha is sufficient. A variety of higher test weight and shy in tillering requires higher seed rate. Under late sown conditions seed rate should be increased by 25 per cent. Similarly when soil moisture is less at the sowing time seed rate should be increased by 25 per cent. Method of sowing also effects the seed rate of wheat. Broadcasting requires 150 kg/ha while sowing with local plough and seed drill requires 125 and 100 kg seed/ha, respectively under normal condition. Where seed is sown with dibbler a seed rate of 25-30 kg/ha is sufficient.

12.5. Spacing

Spacing varies with variety time of sowing and irrigation. A wheat variety that tillers excessively is spaced widely. Irrigated wheat is spaced 22.5 cm between rows and 8-10 cm between plants. Rainfed wheat is spaced 25-30 cm between rows and 5-6 cm between plants. When sowing is delayed a closer spacing of 15-18 cm between rows is recommended. Narrow spacing coupled with higher plant density generally helps in the suppression of weeds.

13. IRRIGATION

Irrigation is a costly input, requiring major expenditure. It is often a scare commodity in parts of India. However dwarf wheat varieties are highly responsive to irrigation. The response of wheat to irrigation depends upon the soil and climatic conditions. In north-western India about 4-6 irrigations are required for wheat. The optimum depletion limit of available soil water for wheat has been observed to be about 40-50 per cent in the surface 60 cm depth of soil layer on the sandy loam soil of Delhi. (Bathkal and Dastane, 1968; Gautam *et. al.*, 1968). Similar observations have also been recorded on the sandy loam and loam soils of Uttar Pradesh, Punjab and Haryana (Modgal *et. al.*, 1968, Mishra *et. al.*, 1969; Sukla *et. al.*, 1972, Yadav, 1972, Singh *et. al.*, 1973). On the clay loam soils of Madhya Pradesh, however, 80 per cent of available water in 15-20 cm soil layer could be allowed to deplete (Jethmalani and Khurchania, 1967, Motiramani, 1968). This required about 4 irrigations. In Andhra Pradesh, 25 per cent depletion of available soil moisture was found optimum on light soils (APAU, 1973, Katama Reddy *et. al.*, 1975). In Rajasthan irrigation at 20-25 per cent depletion of available moisture in the surface 30 cm depth of soil layer was found optimum on the sandy soils (Bajpai and Singh, 1972; Rathore and Singh, 1973; Bajpai, 1975) and this required about 6-8 irrigations.

The most practical criterion for scheduling of irrigations for wheat is the one based on the physiological growth stages. The crown root initiation stage (20-25 days after sowing) is considered as the most critical stage for irrigation (Bhardwaj and Wright, 1968). The second important stage is flowering which takes 80-85 days after sowing. Among the other stages, jointing and milking ranked third followed by tillering stage and then dough stage. The most critical stages of wheat crop as per the availability of irrigations are given below:

TABLE 7.5: Critical stages of wheat crop for varying numbers of irrigations

Number of available irrigation	Critical stages
1	CRI
2	CRI+LJ
3	CRI+B+M
4	CRI+LT+F+M
5	CRI+LT+LJ+F+M
6	CRI+LT+LJ+F+M+D

CRI: Crown root initiation (21), **LT**: Late tillering (42), **LJ**: Late jointing (60), **F**: Flowering (80), **M**: Milky (95), **D**: Dough (115). Figures in parenthesis are the number of days after sowing.

Climatic parameters especially pan evaporation data can also be used for scheduling irrigation in wheat in areas where water resources are in abundance. Dastane and Singh (1964) at Delhi recommended irrigations at 99 mm cumulative pan evaporation (CPE) irrespective of the variety. Kuspe (1975) recommended irrigation at 60 mm CPE at Parbhani in Maharashtra. Bhan (1975) at Kanpur recommended irrigations at 100 mm CPE.

In recent years, ratio between irrigation water (IW) and CPE has also been used for scheduling irrigation in wheat. The appropriate IW: CPF ratio for a irrigation in wheat ranged between 0.7 and 0.9. However, from the operational point of view, this method has limited applicability because CPE measurements are beyond the means of a common farmer. Moreover, the yield of wheat due to irrigations scheduled on the basis of different criteria did not differ significantly but the quantity of irrigation water needed was least where irrigations were scheduled on the basis of critical stages. Parihar *et. al.* (1975), however, observed that irrigation based on IW/CPE ratio economised the irrigation water as compared to irrigations based on growth stages.

Among the methods of applying irrigation, the border method has been found to be the most suitable for wheat. It ensures uniform distribution and high water application efficiency. The application efficiency is influenced by the width, length and slop of the borders and size of the irrigation streams.

14. WEED CONTROL

It has been generally observed that semi-dwarf wheat varieties are not as competitive as taller wheat varieties with weeds (Fischer, 1982). Weeds emerge with the crop and if not controlled in the early stages of crop growth there may cause reduction in yield varying from 10 to 40 per cent depending upon the intensity and stage of weed infestation. In dwarf wheat, the first 30 to 40 days are regarded highly critical from the point of view of crop weed competition.

Wheat crop is generally infested with both dicot and monocat weeds. The major dicot weeds are: *Chenopodium album* (bathua), *Fumaria parviflora* (gajri), *Cirsium arvense* (Kotili), *Anagallis arvensis* (krishan neel), *Vicia sativa* (Ankeri), *Melilotus alba* and *Melilotus indica* (Senji), *Lothyrea* spp. (Chatri matu), etc. Monocot weeds include: *Phalaris minor* (little canary grass), *Avena fatua* (wild oats), *Cyperus rotundus* (matha), *Cynadon dactylon* (doob).

Generally, weeds are controlled by hand weeding with hoe or *khurpi,* etc. First hand weeding is done with the appearance of a thick flush of weeds. Second weeding may be repeated after interval of two weeks. This type of schedule of hand weeding has been found very effective to control annual weeds in dwarf wheat. However, now-a-days hand weeding schedules have become not only difficult but also impossible due to high cost and scarcity of labcur. Hence, use of herbicides to control the weeds has become inevitable. Dicot weeds can be controlled by the application of 2,4-D (ethyl ester) at 0.3-0.4 kg a.e./ha in 700-800 litres of water about 35 days after sowing. If 2,4-D is applied between 25 and 30 days after sowing, it results in the earhead malformation. On the other hand its application after 40 days of sowing leads to the unsatisfactory control of weeds.

Introduction of semi dwarf varieties in India has complicated the problem of little canary grass and wild oats in wheat. Both the weeds normally germinates along with wheat crop. During the initial growth stages their morphological features closely resemble the traits of wheat seedlings. Therefore it is difficult to recognize them. Few distinguishing features by which these weeds can be differentiated from wheat plants are given in Table 7.4.

Monocot weeds like little canay grass and wild oats can be controlled by post emergence application of isoproturon at 1-1.5 Kg or methabenzthiazuran at 1.5 Kg or metoxuron at 1.5 Kg per hectare in 700-800 litres of water within 30-35 days after sowing. Pre-emergence application of pendimethalin at 1.0 Kg per hectare in 900-1000 litres of water gives a broad spectrum control of weeds in wheat. Spraying a mixture of 2,4-D (ethyl ester) either with isoproturon or with methabenzthiazuron or metoxuron also provide satisfactory control of both monocot and dicot weeds commonly found in wheat crop.

TABLE 7.6: Distinguishing features of little canay grass, wild oats and wheat

Plant part	Little canay grass	Wild oats	Wheat
Stem	Solid, branched, pink or purple at the base upto 50 days of growth	Hollow, unbranched normally green	Hollow, unbranched, dark green
Leaves	Smooth, pale green with purplish green spots at the junction of lamina and leaf	Green but no serration and hair are present at the margins, slightly rough	Dark green with serrated margins, pubscent and rough
Ligule	Very large	Large and serrated	Very small
Auricles	Absent, hairs present at the junction of leaf sheeth and lamina	Absent, no hair at the junction of leaf sheath and lamina	Two in number and mostly hairy.

15. PLANT PROTECTION

15.1. Diseases

Wheat crop is infected with rusts, alternaria, leaf blight, loose smut, Karnal bunt and powdery mildews.

Rusts: Wheat is infected by brown yellow and black rusts. Brown and yellow rusts are major diseases in north-western part of the country, whereas black rust is severe in the southern, central and eastern part of India.

Brown rust is caused by a fungus known as *Puccinia recondita tritici*. The small, round-oval uredial pustules of brown rust develop mainly on the leaves, but in severe attacks, these appear on sheath and stem as well. The uredial pustules are bright orange in colour and appear either in small clusteres or irregularly scattered. In case of severe attack, plants mature early, produce light and shrivelled grains and the most damaging in several areas.

Yellow rust is caused by the fungus, *Puccinia striformis*. In the mild attack, uredia are appeared mainly on the leaves but in case of severe attack, these appear on leaf sheaths, stalks and glumes as well. The uredial pustules are oval in shape and lemon-yellow in colour and are formed in long streaks or rows.

Black rust of wheat is caused by the fungus, *Puccinia graminis tritici*. The uredial pustules reddish-brown in colour appear on stem, leaf sheeth, leaves and earheads in elongated shape. Black rust is a killer disease and in severe infection there is heavy crop failure.

For controlling rusts, following measures are suggested:

1. Grow rust-resistant varieties recommended for a particular area.
2. Avoid late sowing because late sown crop is more susceptible to rust.
3. High nitrogen dose favours rust infection, whereas high potash dose reduces rust infection. Use balanced fertilisers.
4. Spray 0.2% zineb or dithane M-45 and 0.1% Sandovit (spreader-sticker) after appearance of pustules. The pustules normally appear during the last week of January. Second and third spray may be repeated with an interval of 10-15 days.

Loose smut: Loose smut is caused by a fungus *Ustilago nuda tritici*. Ternimal symptom of loose smut is the production of black powder in place of wheat grains in the ears. Black powder is a mass of olive-green microscopic spores of smut fungus. When spores are blown to flowers of healthy plants, they germinated on the female organ stigma and lead to infection of developing seed. The smut fungus thus reaches in the matured grain and remains dormant until the germination of the grain. When infected grains germinate, the fungus inside the seed becomes active and grows inside the plant, keeping the pace with the apex of the growing shoot. As the ear formation starts, fungus accumulates in the floral parts which are completely destroyed due to the formation of the black powder.

Loose smut can be controlled by:

1. Growing loose smut resistant varieties.
2. Treating the seed with vitavex @ 2.5 g/kg seed before sowing.
3. Uprooting the infected plants, burying it underground or burning it.

Karnal bunt: Karnal or partial bunt was first observed in Karnal (Haryana) in 1930, and is caused by the fungus *Neovossia india*. A portion of infected grain along its groove is converted into a black powdery mass. The black powder gives a foul smell due to presence of trimethylamine.

Since spores of the Karnal bunt fungus are soil-borne as well as seed borne and infection occurs at the flowering time by air-borne sporidia, the disease is difficult to control in the absence of disease resistant varieties. However the following measures can minimise occurrence of Karnal bunt:

1. Do not grow highly susceptible wheat varieties like HD-2009, UP 319, WH-711 and HD 1982.
2. Use disease free seed and treat the seed with mercury fungicide (cerasan or agrosan G. N.) @2.5 g/Kg seed.
3. Avoid continuous cropping of wheat in the same field
4. Avoid excessive irrigation particularly during flowering of wheat.

Alternaria leaf blight: Alternaria leaf blight is caused by the fungus, *Alternaria triticina*. The disease first makes it appearance as small, oval, discoloured lesions on lower most leaves. The spots become irregular in shape take up dark brown to gray colour and gradually spread to the upper leaves. A bright yellow marginal zone is sometimes seen around the spots. In case of severe attack, leaf sheaths, awns and glumes are also infected. Black powdery spores are disseminated by wind and cause disease on healthy leaves and plants.

Alternaria leaf blight can be controlled by:

1. Treating seed with vitavax @ 2.5 g/Kg seed.
2. Applying adequate fertilisers and irrigation.
3. Spraying zineb or dithane M-45 as recommended in case of rusts.

Pouderg mildew: This disease is caused by a fungus *Erysiphe graminis tritici* and is characterised by formation of white powdery growth of the fungus on the upper surface of leaves. Later on, the colour of powdery growth becomes gray or brown under severe infection, leaf sheaths, stems and ear-heads may also be covered by the fungus, plants become stunted and produce poor yield.

The disease can be controlled by

1. Growing resistant varieties
2. Spraying a mixture of dithane m-45 and karathane in the ratio of 4 : 1 @ 2 kg/hec. The spray should be repeated twice at an interval of 10-15 days.

16. INSECT PESTS

Wheat crop is attacked by a number of insect pests and rodents, which are given below.

Termites: The damaged plants dry up completely and are easily pulled out. Infestation is heavy under un-irrigated conditions and in the field where un-decomposed farm yard manure is applied before sowing. Termites can be controlled by the use of 4% endosulfan dust @ 20-25 Kg/ha.

Gujhia weevil: Gujhia weevil attacks germinating seedling and plants die out. Its control measure is same as of termites.

Army worms: Caterpillars of army worms feed on the plants during the night and pick during the day time. They may be controlled by dusting 4% endosulfan dust @ 20-25 Kg/ha or by spraying of carbaryl (Sevin 50 WP) @ 2.5 Kg/ha in 800 litres of water.

Field rats: They also cause heavy loss to wheat crop in field. For the control of field rats fumigate the burrows of rats with aluminium phosphate @ 1 tablet of 0.5 g/small burrow and 3 g/large burrow.

17. HARVESTING AND THRESHING

Yellow and dry straw is an important visual indicator for readiness to harvest wheat crop. Shredding, breaking of spikes and shattering grains are common if the standing crop is allowed to over ripe. The most suitable stage for harvesting wheat is when the grain becomes hard and contains 20-25 per cent moisture. Harvesting is generally done manually with serrate edged sickles. Bullock-driven reapers are also used occasionally. Combines are used by big farmers to do harvesting, threshing and winnowing in single operation. Wheat crop harvested manually or by reapers is dried for three four days on the threshing floor and then threshing is done by power driven stationary threshers.

18. YIELD

The potential yield of semi-dwarf wheat varieties varies between 60 to 75 quintals per hectares. A grain yield of 40 to 50 quintals per hectare of these high yielding varieties has become a common features at farmers field. Though the maximum yield of 60 q/ha or more has been obtained, the national average is about 25 q/ha. Low fertiliser use and delay in sowing are the main reasons for low national average. Short growing season in central India is another reason for low yield.

19. QUALITY

The physical and nutritional aspects of wheat grain determine its quality. Purity, moisture content, appearance (colour, size and plumpness) and hardness of kernel decide the physical quality of wheat grain. Purity of grains is determined by calculating the percentage of pure wheat grains in a bulk sample. The moisture content of wheat grain is important from the milling and storage view point. The optimum moisture content in the grain is about 12 per cent. Grains which are large, plumed and amber in colour are considered to be better. The hardiness is determined by the pearling index. A higher value of the pearling index indicates less kernal hardiness. The pearling index is determined as follows:

$$\text{Pearling index} = \frac{x - y}{x} \times 100$$

where, x = initial sample weight which is normally 20 grams, y = weight of the material left on a 30 mesh sieve.

The nutritional aspect of wheat grain is determined by chemical analysis of wheat grains. Normally wheat grain contains 60-68% starch, 8-15% protein, 1.5-2.0% fat, 2.0-2.5% cellulose and 1.5-2.0% minerals. Higher values of above mentioned parameters indicate the better quality of wheat grain.

Milling quality, *chapati* (leavened bread) making quality, bread (unleavened) making quality and biscuit making quality are also significant from the industrial and consumers point of view. The milling quality is judged by the following standards:

1. The maximum yield of flour free from germ and bran particles.
2. The relative ease with which a high turnover of flour is obtained.
3. The low ash content indicating a thorough separation of bran from flour.

Chapati making qualities comprise water absorption, dough colour, dough properties (mixing and handling), puffing, texture, preservable quality, taste and aroma. High water absorption, gas production capacity and gas retention capacity determine bread making quality. Soft and weak flour of wheat having low water absorption capacity and high protein content is considered ideal for biscuit making. Gluten should have plasticity so that the flour can be stretched without much shrinkage.

An ideal durum wheat for macaroni products should have excellent grain qualities like bold and plump grains, low yellow berry, high carotene content, high semolina yield and excellent quality for making vermicelli.

20. POST HARVEST TECHNOLOGY

20.1. Seed Storage

Seed storage is required for consumption in subsequent months and seed for next sowing. Seed can be stored in metallic drums, polythene bags and earthen pots. The critical moisture content of wheat seeds at which the rate of respiration increases is 14.6%. Seeds containing higher moisture respire faster, depleting seeds food reserve (Clements, 1988). The monsoon climate is ideally suited for insect proliferation. The major storage pests are rice weevil (*Sitophilus aryzae* L.), grain moth (*Rhizopertha dominea*) and the Khopra beetle (*Trogoderma graversium*). *Aspergillus* and *Penicillium* can also damage wheat grains stored in moist conditions.

20.2. Milling

Dried grains are first cracked or crusted gradually, through a series of chilled iron break-rolls. Purified middlings are then passed through reduction rolls, wherein finer and whiter fractions are combined into potent flour (commercial flour). The remaining darker grades are marketed for special baking uses or mixed with bran. Bleaching is done to remove xanthophyll pigments. Benzoyl deroxide or nitrogen trichloride in small quantities has bleaching effects, but higher doses are detrimental. Flour yields from milling are approximately 70 to 74% (Martin and Leonard, 1950).

20.3. Cooking

A wide range of food products like bread, tortillas, moulded macaroni, cookies and crackers are cooked from wheat flour. The stuffed or unstuffed *chapaties* used by most population in the Indian subcontinent are also cooked from wheat flour.

21. REFERENCES

A.P.A.U. 1973. Annual reports, Deptt of Agronomy, S.V. Agril College, Andhra Pradesh Agril. Univeristy, Tirupati, Andhra Pradesh.

Alexander, T.M. 1985. Wheat. Fertiliser Association of India, New Delhi, pp. 21-59.

Bajpai, M.R. 1975. Effect of irrigation, mulches and modes of nitrogen application on yield attributes and yield of wheat (*Triticum aestivum L.*) Var. Rs 31-1, S.K.N. College of Agri.University of Udaipur, Jobner, Rajasthan (Personal communication)

Bajpai, M.R. and Singh, R.N. 1972. Water requirements of wheat and barley on sandy loam soils of semi-arid region. Proc. Symp. 1969, Soil and Water Management, I.C.A.R. spl. Publication; 77-85

Bathkal, B.G. and Dastane, M.G. 1968. Application of penman and thornwaite formulae for estimation of water use rate by crops under Indian conditions. Proc. Symp. 1966. Water management monograph, Indian Soc. Agron.; 77-85.

Bayles, B.B. and Clark, J.A. 1954. Classification of wheat varieties grown in the united states in 1949. USDA Tech. Bul. 1083.

Bhan, S. 1975. Water management research at U.P. Institute of Agricultural Science, Kanpur, Uttar Pradesh (Personal communication).

Bhardwaj, B.L. and Wright, B.C. 1968. New Agronomy for dwarf wheats. Indian Fmg. 17(5): 34-39.

Clark, J.S. 1938. Correlation between annual precipitation and yield of spring wheat in the Great Plains. USDA Tech. Bul. 636.

Clements, D.J. 1988. Wheat seed storage under tropical conditions. *In* wheat Production constraints in tropical environment (ed.) A.R. Klatt UNDP/CIMMYT, Mexico, pp. 360-365.

Dastane, M.G. and Singh, M. 1964. Assessment of irrigation requirements of crops on climatological basis. Proc. 5th NESA irrigation practice seminar, New Delhi; 403-405.

Ellerton, S. 1939. The origin and geographical distribution of *Triticum sphaerococcum* Perc. and its cytogenetical behaviour in crosses with *T. vulgare* Vill. J. Gen. 38; 307-324.

Feldman, M. 1970. Wheat, *Triticum* spp. *In:* Evolution of crop plants (ed.) N.W. Simmonds. Longman Scientific Co. Edinburgh, Scotland. pp. 120-128.

Fisher, R.A. 1982. Developments in wheat agronomy *In* wheat science today and tomorrow (Eds) L.T. Evans and W.J. Peacock, Cambridge University Press, Cambridge. pp. 249-271.

Fisher, R.A. and Byerlee, D.R. 1990. Trends of wheat production in the warmer areas: Major issues and economic considerations. In wheat for the non-traditional warm area (ed.) D.A. Saunders UNDP/CIMMYT, Mexico, pp. 3-27.

Gautam, O.P., Bhardwaj, R.B.L. Singh, Y. and Singh, R.P. 1968. Irrigation studies with tall and dwarf varieties of wheat. Proc. Symp. 1966. Water Management Monograph, Indian Soc. Agron. 110-117.

Helback, H. 1 959. Domestication of food plant in the old world. Science, 130(3372): 365-372.

Howard, A. and Howard, G.L.C. 1906. Wheat in India-Production, varieties and improvement Thucker Sprinkler & Co., Calcutta, India, pp. 72.

Jethmalani, S.C. and Khurchania, S.P. 1967. Research on irrigation water use in Madhya Pradesh. Proc. Symp. Irrigation Water Management. 40th Annual meeting central Bd. Irrigation and Power, New Delhi.

Katana Reddy, B.C., Sankara Reddi, G.H. and Bapi Reddy, T. 1975. Effect of moisture regions and fertility levels on growth and yield of wheat. A.P.A.U. J. Res. 1(4):13-56

Khuspe, V.S. 1975. Research work on water management under irrigated conditions. Marathwada Krishi Vidhyapeeth Parbhani, Maharashtra (Personal communication).

Mangelsdorf, P.C. 1953. Wheat Sci. Am. 189(1): 50-59.

Martin, J.H. and Leonard, W.H. 1950. Principles of field crop production. The McMillan Company, New York, pp. 453-522.

McCall, M.A. 1934. Developmental anatomy and hormologues in wheat J. Agric. Res. 48(4): 283-321.

Misra, R.D. Sharma, K.C., Wright, B.C. and Singh, V.P. 1969. Critical stages in irrigation requirement of wheat variety Lerma Rojo. Indian J. Agric. Sci. 39: 898-906.

Modgal, S.C. Joseph, A. and Goel, S.S. 1968. Effect of soil moisture regions and nitrogen levels on the performance of Indian and Mexican wheats under Tarai conditions of Uttar Pradesh Proc. Symp. 1966. Water Management Monograph, Indian Soc. Agron: 118-136.

Motiramani, D.P. 1968. Water management in black soils. Indian Fmg 17(10): 23 and 47.

Parihar, S.S., Sandhu, B.S. and Singh N.T. 1975. A critical appraisal of research on irrigation scheduling of crops. Water for human needs. Proc. second world congress on water resources, I.W.R.A., New Delhi, Vol. 1: 343-349.

Percival, J. 1921. The wheat Plant. Genold Duckworth and Co. Ltd. London.

Rathore, S.S. and Singh, R.M. 1973. Effect of different soil moisture regimes and fertility levels on the growth and yield of wheat, Sonara-64. Indian J. Agric. Sci. 43: 712.

Riley, R.J., Unrau and Chapman, V. 1958. Evidence on the origin of the B genome of wheat J. Hered. 49(3): 91-98.

Saurders, D.A. 1990. Wheat for the non-traditional warm area. UNDP/CIMMYT, pp. 549.

Shukla, M.P. Lal, R.B. and Lal, R.N. 1972. Effect of levels of soil moisture nitrogen and phosphorus on the yield and quality of wheat. Indian J. Agric. Res. 6:309-314

Singh, A. Bajura, M.S. and Singh, M.T. 1973. Efficiency of moisture and nutrient use by wheat under different moisture regimes and fertiliser schedules. Indian J. Agric. Sci. 43: 133-137.

Singh, K.N. 1990. Overcoming salt stress through breeding for tolerance. *In* Wheat for the non-traditional warm areas. (Ed.) D.T. Saunders UNDP/CIMMYT, Mexico, pp. 422-428.

Tandon, J.P. and Sethi, A.P. 1987. Twenty five years of co-ordinated wheat research. Wheat Project Directorate, Indian Council of Agricultural Research, New Delhi pp. 270.

Vavilov, N.I. 1950 The Origin, Variation Immunity and Breeding of cultivated plants. Chron. Bot. 13(1-6) (Trans by K. Starr chester) 1949-50.

Vieira, M.J. and Kochan, R.A. 1990. Tillage practices and soil physical degradation in the wheat cropping systems of the warmer areas of Latin America. In Wheat for the Non-traditional warm areas (Ed.) D.T. Saunders UNDP/CIMMYT, Mexico. pp. 242-256.

Yadav, J.S.P. 1972. Annual progress report of I.C.A.R. All India Coord. Scheme for Research on water management and salinity, Central Soil Salinity Research Institute, Karnal.

8. Barley

B. N. Mishra and B. G. Shivakumar

Barley is one of the leading cereal crops occupying about 9.4 % of the total cereal acreage with about 7.8 % of the total cereals production in the world. It ranks next only to wheat, rice and maize among cereals in its spread and cultivated in all the continents except the Greenland. In India barley is the most important cereal crop in winter after wheat in both area and production. It is very old crop in India having found its mention in the ancient Sanskrit texts. The term '*yava*' in Sanskrit is the most frequently mentioned cereal in the earliest records of Indo-Aryans and was probably the principle staple food during the *Vedic period*.

Barley is a very hardy crop which can be grown in diverse agroclimatic conditions like drought, salinity, alkalinity, rainfed or irrigated condition, plains or hilly areas where other crops can not with stand. It can be a good substitute for wheat in areas where owing to some agro-climatic and managerial constraints wheat can not be successfully grown. It is further a highly suitable crop under late sown condition. Being a less resource demanding crop it can be raised where there is limitation for inputs.

In some areas of U.P., Rajasthan and Haryana parched barley grains are also eaten. Barley is believed to have cooling effect on human system and in several regions it is consumed in preference to wheat and other cereals particularly during summer.

Although about 90 % of the produce is used as human food in India, most of the produce is used as cattle feed in USA and Europe. With the development of scientific dairy farming the demand for barley as feed is also increasing. Barley has proven to be a palatable and nutritious feed for live stock. Further the straw is also used as good forage for making hay and silage. Besides the grains are also used in making the feed for piggery and poultry. The quality of pork was found to improve upon feeding the barley feed to pigs in Europe.

Barley has been quite famous commodity in malting and brewing industry since time immemorial. The best quality seeds are used for malting which is a midway product used in many bakery preparations and energy rich foods for human consumption. The end products are usually used in breweries for the manufacture of alcoholic drinks viz., 'beer', 'whisky' etc. There are many breweries wholly dependent on the supply of barley as major raw material.

Malting consists of germinating barley seeds for a limited period and then drying and lightly cooking the pregerminated seeds. The product from which the dried rootlet (culms, coombes, cummins, sprouts) has been separated is called malt. Malt is an intermediary in several manufacturing processes. Malt types differ depending on their intended use e.g. the manufacture of malt flours, malt extracts, cereal syrups, various types of distilled spirits, beers and malt vinegar. The malt culms and the residues from these manufacturing processes are used in animal food stuffs (Briggs, 1978).

There is a great demand for quality raw seeds and malt in the international market. A well thought and executed barley production plan may lead to earning a lot of hard currencies in the form of exports. Thus there is good untapped potential for barley and its processed products.

In addition there have been many location-specific uses of this crop developed and appreciated by farmers over a period of time.

1. ORIGIN

Barley is one of the oldest crop being cultivated since the beginning of agriculture. However there is no definite knowledge about its origin. Now the most accepted view based on the available evidence indicate that the Near East region was the probable place of origin of barley around 8000 BC (Harlan, 1968). Archeological remains dating back to 8000 BC at Mureybat on the Euphrates in Syria, 6800 BC at Beidha in southern Jordan, 6000 - 7000 BC at Alikosh (Iran), Ramad (Syria), Jericho (Palestine) and Catal Huyuk and Halcier in Turkey have been found (Harlan and Zohary, 1966). Recent archeological evidence indicate that barley was used as a food source near the site of Aswan dam in Egypt about 16000 BC much earlier than supposed previously (Wendorf *et. al.* 1979). There is no evidence of whether the inhabitants of the settlement planted and cultured the barley or harvested it from wild stands.

Although barley may have had one center of origin. Vavilov proposed two main centers of diversity.

1.1. Northeast Africa and Mountainous Regions of Abyssinia

Variations found here are seen among barley types in Ethiopia, Western Asia, Europe, North America and most of India. Barley found in this whole region are classified as *occidental type*. The typical characters of this are broad empty glumes, two rowed with lateral spikelets, pubescent palea and brightly colored stem.

1.2. Southeast Asia, China, Japan and Regions Adjoining Tibet

These barleys are grouped as *oriental types*. In this region various kinds of naked, short awned, awnless or hooded barleys are found. These barleys also carry resistance to certain races of mildew. These are identified also by extremes of earliness or lateness.

All barleys, wild and cultivated belong to the same potentially interfertile population, and are grouped under one species *Hordeum vulgare* L. emend. Bowden. The two rowed barleys are the oldest known types so far. Various genetical, ecological and taxonomic investigations support the conclusion that cultivated barley had a monophyletic origin from a two rowed species resembling *Hordeum vulgare* sp. *spontaneum.*

From the archeological evidence it is concluded that barley was domesticated around 7000 BC; and by 6000 BC six rowed naked type had become an important cultivated crop.

Barley culture in India appears to have come from the Near East region around 3000 BC It spread slowly across northern India in to the Gangetic plain and very slowly southward reaching central India about 1000 years after Mohenjo-Daro (2300 BC). It is believed that Indo-Aryans brought their staple food with them and also used the barley material of Indus people which led to generation of new variability required for intensive cultivation. It is also believed that barley probably came to India not once but on several prehistoric occasions, either through the trade routes or with the immigrating people (Anon, 1973).

2. DISTRIBUTION

Barley is one of the premier crops spread across the globe in many countries with wide variation in agroclimatic conditions and traditions. Its cultivation extends from the agricultural frontiers like plateaus of Tibet and Ethiopia, to Oasis of Sahar. Barley has been the most tolerant cereal crop particularly where alkali soils, frost or drought are encountered. Thus it is cultivated in many countries with varying acreage. The total area, production and productivity of barley in the world along with a few major countries have been furnished in the Table 8.1.

TABLE 8.1: Area, production and productivity of barley (1998)

Country	Area (mha)	Production (m tonnes)	Productivity (kg/ha)
Russian Federation	13.000	20.000	1538
Canada	4.720	13.647	2281
Turkey	3.700	8.200	2216
Ukraine	3.615	8.071	2233
Spain	3.577	8.590	2401
World	66.290	156.617	2363

TABLE 8.2: Area, production and productivity of barley in India (1998)

State	Area (lakh ha)	Production (lakh tonnes)	Productivity (kg/ha)
UP	3.323	6.688	2013
Rajasthan	1.885	3.780	2005
MP	0.880	0.959	1090
Bihar	0.421	0.496	1178
Haryana	0.340	0.880	2588
All India	7.604	14.358	1888

In India barley occupied an area of 7.604 lakh hectares with a total production of 14.358 lakh tonnes during 1997. The average national productivity was 1888 kg/ha. Barley was the seventh important crop in India after rice, wheat, jawar, bajra, maize and ragi in terms of area and production. But the productivity -wise it was second among all the crops after the wheat, and it was higher than the national average of 1825 kg/ha for all the cereal crops together. The data on area, production and productivity in important states growing barley has been furnished in Table 8.2.

Uttar Pradesh topped the barley area with little less than 50 % of the total area in India. It was followed by Rajasthan, M.P., Bihar and Haryana.

3. CLASSIFICATION

Barley is one of the highly acclaimed member of the family Graminae and tribe Hordea. It has been classified into different types by different authors depending upon one or many morphological features. Often these classifications overlap on one another. However these help in easy identification of different types of barley. Some of the popular and commonly used classifications are outlined in the following paragraphs.

3.1. Classification Based on Fertility of Lateral Spiklets

On the basis of fertility and arrangement of spikelets on rachis barley has been classified into three different types

1. **Two rowed type: *Hordeum distichon* L:** In the two row barley the two lateral spikelets are sterile, thus each spike has only two vertical rows of kernels. Generally each spike produces 15-30 kernels.
2. **Six rowed type: *Hordeum vulgare* L. emend:** In six row barley both the lateral florets are fertile but their size may vary from slightly to moderately smaller than the central floret. In this type each spikelet consists of one floret with three fertile florets attached to one side of the rachis and three on the other, six vertical rows of kernel are usually produced on each spike. About 25-60 kernels are usually produced on each spike (Delorit *et. al.* 1984).
3. **Irregular type: *Hordeum irregulare* L.:** Irregular barley has fertile central florets while some of the lateral florets are fertile and others infertile. This results in the irregular distribution of florets and irregular rows of kernels.

However in terms of botanical classification it is generally accepted that cultivated barley regardless of kerenel row number in the spike, is all one species namely *Hordeum vulgare* L. (Fedek, 1994). All the truly wild forms of *Hordeum* are two rowed that is of the four spikelets at each node of the ear, the lateral ones are female-sterile and only the central one develops a grain. Under domestication six rowed races appeared in which all these spikelets produce grains. There are two genes involved, both with multiple allelic series, but a single recessive mutation (*vv*) is adequate to cause two rowed barley to become six rowed. six rowed genotypes with fragile ears are known but do not appear to be truly wild plants and are probably derived from six rowed cultivars (Harlan, 1976). Within each of these species there is a number of morphologically distinct forms and these have been given the status of varieties. Whether these should be taxonomically designated variety or form is a moot point, but the natural relationship is indicated by the fact that there is easy exchange of genes between recognizably distinct forms.

3.2. Based on Awns

Depending upon the presence or absence of awns in grains it has been grouped into

1. Awned type
2. Awnless type

Again on the basis of nature of awn present they are sub grouped into

1. Smooth awned
2. Rough awned

Awnless types are also called hooded because of the hood shaped structure that develops in place of awn.

3.3. Based on Seed Coating

on the basis of presence or absence of seed cover both two row and six row barleys have been sub divided into:

1. *Hordeum vulgare* var.
   a) *nudum* -naked six rowed
   b) *trifurcatum* -hooded six rowed
   c) *inerme* - awnless six rowed
   d) *hexastichon*
2. *Hordeum distichon* var.
   a) *nudum* -naked two rowed
   b) *trifurcatum* -hooded two rowed
   c) *inerme* -awnless two rowed
   d) *zeocrithon* -fan-tail barley
   e) *deficiens*

In the varieties *nudum* the caryopsis falls free, in *trifurcatum* the long awn is replaced by an inverted sterile floret which forms a hood, shaped like a *fleur delys* (sometimes these forms are called Himalayan barley) and in *deficiens* the lateral spikelets are gone except for the glumes. In the *distichon* types other than *deficiens* the lateral spikelets are represented by sterile, club-shaped structures and the glumes (Berrie, 1977).

3.4. Based on Adherence of Chaff to Grains

Here barley are grouped into two types as

1. Hulled (*syn.* husked) type
2. Hull-less (*syn.* naked) type

In case of hulled barley the husk i.e. floral glumes also called chaff remains attached to grains resulting in poor flour making. It also down grade the quality and nutritive aspects of grain. while in the hull-less types the husk readily falls after threshing and naked grains free of chaff can be collected.

4. GROWTH STAGES

Barley has a well defined stages of growth and development similar to wheat.

Expression of these different stages of growth and development is regulated by climate, soil moisture, cropping system, prevalence of weeds, pests and diseases and agronomic practices. In the beginning, from germination to seedling establishment the rate of growth will be slow but still it will be vigorous compared to other cereals. Growth picks up during grand growth stage involving elongation of stem and tillering. All the tillers produced during this period may not last and succeed in producing earheads. The weak one may whither away in the process. The emergence of earhead and grains formation results in termination of growth and drying of plant starts. A few stages of agronomic importance have been clearly identified.

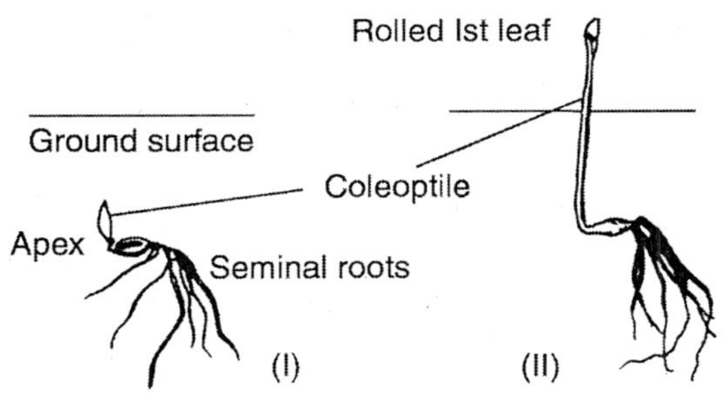

FIGURE 8.1: (I) Germinating grain, (II) first leaf emergence.

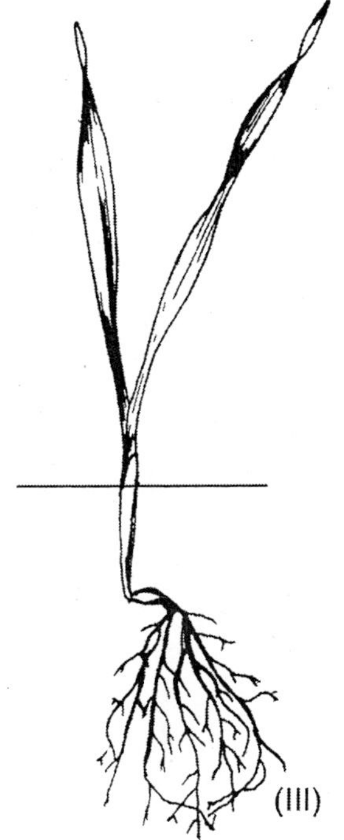

FIGURE 8.2: Barley. Second leaf emergence.

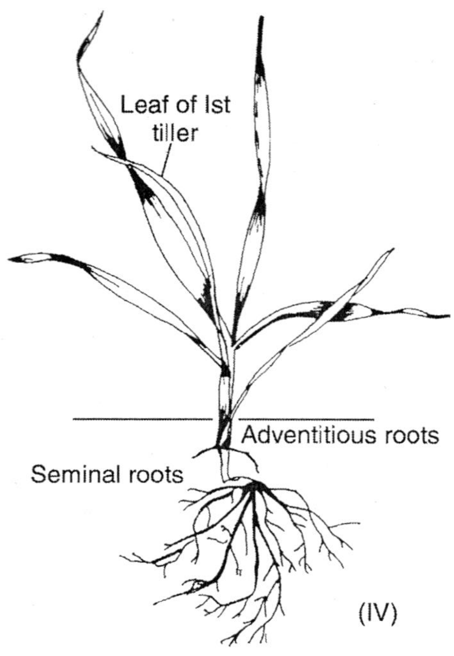

FIGURE 8.3: Barley. First tiller leaf.

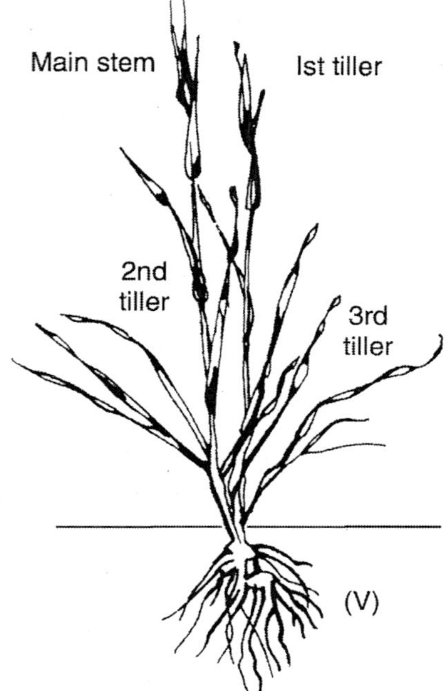

FIGURE 8.4: Barley. Jointing and shooting.

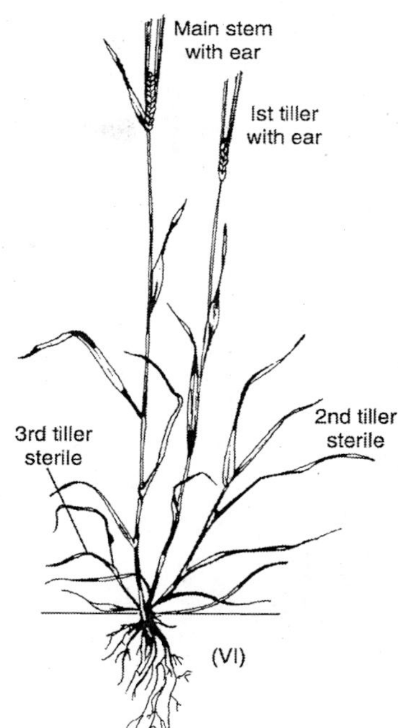

FIGURE 8.5: Barley. Emergence of ear.

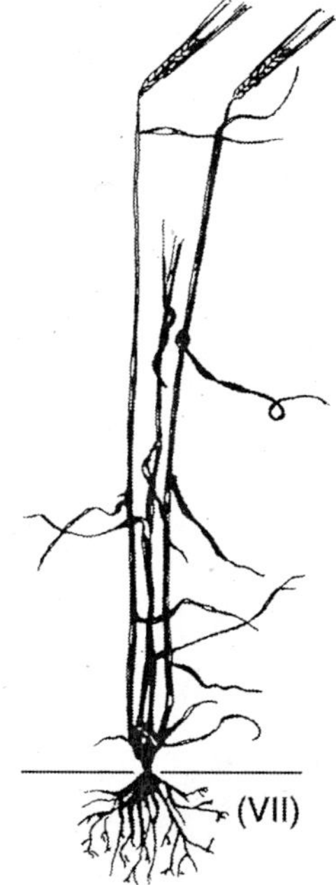

FIGURE 8.6: Barley. Maturity stage.

FIGURE 8.7: Barley. (A) spike of six row (B) spike of two row.

1. **Germination and initial seedling establishment:** Here after germination below soil surface the coleoptile emerges on the soil surface. With the exposure to light plants become green and starts photosynthesis which accelerates the rate of growth and development of leaves. This stage is marked by the exhaustion of endosperm and initiation of crown root. This stage lasts up to 20-25 days from sowing (Pederson, 1950).
2. **Tillering:** The seedling after initial establishment of crown root system starts tillering. The tillers emerge from the crown and grow along with the main stem. A plant may produce up 4-5 productive tillers depending on the variety and environment. There may be some tillers which will not develop and produce earheads. In general two row barley produces more tillers than the six row barley. This phase of growth may continue upto 30-35 days after sowing.
3. **Jointing:** It is also called shooting. Here the plant develops its vegetative part. The stem becomes more visible, nodes start multiplying and internode distance become longer. It shows the emergence of last leaf in the form of flag leaf covering the still unemerged ear called booting. In the later part of jointing stage the lower leaves start withering away while the younger leaves proliferate and grow to their maximum size. Jointing may last up to 55-65 days after sowing.
4. **Heading:** This stage is also known as 'earing' and 'ear head emergence'. The flag leaf gives rise to emergence of ear and the tips of awn become visible in the beginning. Subsequently the full earhead comes out and anthesis begins in the central florets, often with the ear in the boot. Stamens may or may not

be excerted. This stage ends with the production of watery ripe grains. The crop will be in this phase upto 75-85 days after sowing.

5. **Ripening:** This stage includes the sub-phases of post-milking, grain filling and grain development. The soft dough gradually turns to hard dough. This is most important stage of the crop wherein any stress result in substantial reduction in the yield of the crop. Here the neck of the ear continues to lengthen and angle of head to peduncle may change. This stage lasts upto 90-100 days after sowing.
6. **Maturity and drying:** Once the grains pass the hard dough stage they start loosing moisture and plants begin to dry. The plants turn yellowish, loose stiffness and become droopy. The physiological maturity of the seed is completed and ready for the harvest. Any further delay in harvesting results in loss of grains shedding of ears, shattering and collapse of stem. This stage is the final stage in the crop life and it depends on the duration of the variety and agronomic practices.

5. CLIMATE

Barley is cultivated under wide variations of agro-climatic conditions. In India it is grown mainly as a *rabi* (winter) season crop, except in the high altitude areas of Ladakh and Lahaul, where it is raised during spring season. Although barley is a crop of cool climate it may be successfully grown in subtropical and tropical conditions. Barley requires a mean daily temperature of about 12-15^0 C during the growing season and about 30 ^{0} C during ripening and maturity stages. Higher temperature coupled with dry winds during later stages of crop result in poor grain filling and chaffy seeds resulting lower yield and poor quality grains. The warm and humid climate is not very suitable for the barley cultivation. Barley is highly drought tolerant and can be grown with lesser soil moisture. An annual average rainfall of 40-50 cm will result in increased yield. Even a well distributed and assured rainfall of as low as 20-25 cm is sufficient for realizing a good crop of barley in rainfed conditions. Barley is basically a long day plant. Thus a light period of 10-12 hours during vegetative growth and 12-14 during reproductive growth would be ideal. Of late with the release of photo-insensitive varieties the criteria of photo-period requirement is minimized.

6. SOIL AND SOIL PREPARATION

Barley can be grown on variety of soils ranging from light sandy to heavy clay soils. It performs well on properly drained loamy soils having a pH range of 7-8. Excessive available N in the soil results in heavy growth and lodging. Hence a medium fertile soil will be ideal for the cultivation of barley. As it is susceptible to water logging effort should be made to provide proper drainage facility in the field. Although barley can withstand milder salinity and alkalinity its performance will be poor in acidic soils. Here under acidic pH, the solubility and availability of aluminium will be high resulting in poor root development and proliferation. It is therefore better to avoid acidic soils. Liming is found to mitigate the problem of aluminium toxicity in such soils. There are incidence of boron toxicity in some cases. So soils having higher available boron content should not be preferred for barley. Barley may be grown on a rough seed bed and hence one cross cultivation with soil turning plough followed by planting, is sufficient to get a good seed bed for sowing. Conservation of late monsoon rains received during August and September is very important. In moderately saline-cum-high water table conditions, the last preparatory cultivation should be given a week to 10 days before sowing to allow the salts to accumulate near the surface to facilitate sowing of seeds in a relatively less or non-saline soil layer.

7. CROPPING SYSTEM

Barley has been found suitable for multiple and mixed cropping systems in variety of conditions. As it is a short duration crop it can be fit in cropping sequences involving a *kharif* crop in the previous season and a short duration crop in succeeding spring and/or summer season. Depending on the assured rainfall or irrigation facility it can be grown in the following sequential cropping systems.

TABLE 8.3:

Cereal - Cereal System	Legume - Cereal System	Miscellaneous System
Rice - Barley	Arhar - Barley	Groundnut - Barley
Maize - Barley	Greengram - Barley	Cotton - Barley
Pearlmillet - Barley	Blackgram - Barley	Sunhemp - Barley
Sorghum - Barley	Cowpea - Barley	Potato - Barley
	Sugarcane - Barley	

In some cases even the delayed harvest of *kharif* crop can be overcome by successful late barley in many areas. Besides the above sequences vegetables-barley is also another important sequence in *diara* lands in eastern UP and Bihar. Further wherever there is irrigation facility the growing of short duration summer crops like mung, cowpea, fodder or forage crops is also possible before going for *kharif* crop.

In addition barley is also suitable for mixed or intercropping with *rabi* crops like field pea, chickpea, rajmash, mustard and rapeseeds, linseed etc.

8. MANURES AND FERTILISERS

Like any other crop barley too needs all the essential nutrient elements for satisfactory growth and development. Nutrients in too low or high concentrations result in either deficiency or toxicity symptoms respectively. Most of the major nutrients usually become scarce over seasons and if not replenished through manures and fertilisers, are likely to have adverse impact on plants and make them to show deficiency symptoms. Mineral deficiencies probably can not be reliably diagnosed from plant appearance alone, abnormal growth may be caused by many other factors. Symptoms of N, P and Mg deficiency are first noticed in older leaves while symptoms of Ca, B and Cu are noticed in younger tissues first and spread to older parts later. Some of the deficiency symptoms of major elements in barley are as follows.

1. **Nitrogen:** Deficiency results in restricted root growth, reduced tillers, thinner and small stem, pale green leaves and smaller ears and premature ripening of grains.
2. **Phosphorus:** Restricted growth, bluish-green leaves and brown tip, anthocyanin pigments on leaves in severe deficiency and delayed ripening of grains.
3. **Potassium:** Dwarfed plants, shortened internodes, increased tillers, smaller ears with fewer grains, bluish-green leaf, general mottling with brown-purple spots on leaves.
4. **Calcium:** Stunted growth, high seedling mortality and defective growing tips.
5. **Magnesium:** Chlorotic leaves, older leaves turn yellow and die, and dwarfed plants in extreme deficiency.
6. **Sulphur:** Pale yellow leaves, stunted growth and delayed maturity.
7. The micronutrients availability to plants is pH dependent in soil. As these are required in very small quantities the range of sufficiency and toxicity is very narrow.
8. **Iron:** Interveinal chlorosis or totally bleached leaves; may be lime induced and excessive supplies of P, Cu, Mn or Zn cause this deficiency.
9. **Manganese and Zinc:** Deficiency mostly in calcareous soils.

Acid soils may result in *manganese* and *aluminium* toxicity. In the former leaves show brown spots between veins and symptoms similar to phosphorus deficiency are shown in the later.

8.1. Manures

Since barley is grown in soils low in fertility, afflicted with salt problems and under dryland conditions, application of manures *viz*, FYM, organic manures, crop residues etc. improves the water holding capacity and fertility status of the soil. Besides it also increases the buffering capacity of the soil. Application of well decomposed and friable manure free from pathogens, insects' eggs and toxic substances 10-15 tonnes per hectare about a month prior to sowing and mixing in the soil will improve the soil condition resulting in higher productivity of barley.

8.2. Fertilisers

The improved high yielding, dwarf and semi-dwarf varieties of barley are quite responsive to fertiliser management. Hence a fertiliser package tailored on soil fertility status, irrigation facility and targeted yield results in highly satisfactory producutivity in barley (Misra *et. al.* 1979). The following fertiliser dose has been found economic in many barley growing areas.

Rainfed condition	: 30-40 kg N and 20 kg P_2O_5/ha
Irrigated condition	: 60-80 kg N and 30 kg P_2O_5/ha
Late sown condition (Irrigated)	: 60-80 kg N and 30 kg P_2O_5/ha

In areas where potassium deficiency has been noticed application 30 kg K_2O will improve the yields. Further if yields are to be enhanced, application of little higher doses of these fertilisers may be tried.

One third of total nitrogen and full quantity of phosphorus and potassium should be given at the time of planting either through draw tube (*pora*) attached to seed drill or behind plough and mixed in the soil. The remaining one third of nitrogen should be given after first irrigation which helps in the tiller growth and development. The last dose of nitrogen should be given when the crop has entered in the flowering stage. Top dressing of nitrogen should always be given in the form of urea. The high yielding hull-less varieties need higher doses of nitrogen as compared to hulled varieties.

Micronutrients in the form chelated products or simple compounds containing the needed element may be applied to the soil before sowing, if soil analysis showed deficiency prior to sowing or through liquid sprays, if found out by plant diagnosis. But usually deficiency for micronutrients on large scale basis has not been noticed in barley growing area.

9. VARIETIES

The barley growing area in India has been divided into four major zones based on the agro-climatic conditions prevailing in the area (Ram, 1982). They are

1. **Hill zone:** This includes Himachal Pradesh and hilly areas of Uttar Pradesh
2. **North-western plain zone:** This includes Punjab, Haryana, Delhi, western and central parts of Uttar Pradesh
3. **North-eastern plain zone:** This includes parts of eastern Uttar Pradesh, Bihar, West Bengal and Orissa
4. **Central zone:** This includes Rajasthan and Madhya Pradesh

As the situation in each zone varies from other, varieties have been evolved suiting to the needs of each zone. However there are some varieties which can be used across the zones without sacrificing the yield or quality. Varieties which were in use prior to constitution of All India Co-ordinated Barley Improvement Project by the ICAR in 1967, were mostly tall and low yielding. But later, high yielding varieties having short stature, resistance to pests and diseases and suitability to stressed conditions like late sowing and saline and alkali soils have been evolved. A few varieties have been either introduced or evolved to meet the specific needs of users like for malting. However some old varieties are also popular even now. There is a big list of varieties available for cultivation in different zones. A few varieties have been given below with important characters and the areas most suited for their cultivation.

1. **Amber:** This is a tall. good tillering, early maturing variety recommended for eastern UP under rainfed conditions suitable for brewing and malting.
2. **Azad:** This is a tall, culm green with light bloom, moderately tillering, early maturing variety recommended for north-eastern plain zone under non-irrigated, saline -alkali soils. It is suitable for fodder purpose.
3. **BG 105:** It is a semi-dwarf lodging resistant, medium duration variety with high tillering capacity recommended for late sown conditions in north-western plain zone. It is highly resistant to yellow rust, leaf spot and smut diseases.

4. **BG 25:** This is a tall, high tillering, medium to long duration variety recommended for irrigated cultivation in north-western plain zone. It is highly resistant to yellow rust.
5. **BH 39:** This is a tall variety with high tillering ability, medium to long duration variety recommended for irrigated normal sown conditions of north-western plain zone. It is highly resistant to yellow rust and smut.
6. **BHS 46:** This is a medium tall variety with high tillering ability, recommended for hill zone. It is resistant to rust, smut and leaf spot diseases.
7. **Bilara 2:** This is a medium tall variety released for cultivation in saline soil areas in Rajasthan.
8. **Clipper:** This is a two-row barley introduced from Australia, suitable for irrigated conditions of UP, Delhi, Haryana and Rajasthan. It is highly suitable for malting and brewing purpose.
9. **DL 36:** This is a medium dwarf, mid duration variety highly suitable for both timely sown and late sown irrigated conditions.
10. **DL 70:** This is too a medium dwarf, mid duration variety highly suitable for both timely sown and late sown conditions.
11. **DL 85:** This is a medium duration semi dwarf variety recommended for *diara* lands of north eastern plain zone.
12. **DL 88:** This is a medium tall, medium duration variety for late sown conditions of northern plains.
13. **Dolma:** This is medium tall hull-less variety recommended for hills of HP It is resistant to yellow rust.
14. **HBL 3:** This is tall, mid duration, hull-less variety recommended for noth-western plain zone.
15. **Himani:** This is a medium early maturing variety for timely sown rainfed conditions of hills of HP, UP, WB and J&K. It is highly resistant to yellow rust.
16. **IB 65:** This is a tall, hull less variety having very good and attractive grain quality similar to wheat. Recommended for northern plain zones.
17. **K 18:** This is a dwarf hull-less variety suitable for saline, diara soils of north-eastern plain zone and north western plain zone.
18. **K 19:** This is similar to K 18 except that this is high yielding than that.
19. **Jyoti:** This is a erect, medium tall, good tillering, medium to late maturing variety recommended for north-western plains.
20. **K 141:** This is a erect, tall, leaf broad light green, moderately tillering, early maturing variety recommended for north-eastern plain zone under rainfed conditions. It is moderately resistant to rust and smut diseases.
21. **K 169:** This is tall, moderately tillering, early maturing variety suitable for rainfed conditions in north-eastern plain zone.
22. **K 226:** This is medium tall, early maturing variety suited for north-western plain zone under unirrigated conditions.
23. **K 252:** This is erect, medium tall, moderately tillering, early maturing variety suitable for *diara* lands of UP Bihar and sunderban delta of West Bengal.
24. **K 257:** This is a tall, erect, medium duration variety recommended for irrigated areas in central zone.
25. **K 287:** This is a medium tall, mid duration variety recommended for north-western plain zone in irrigated condition.

26. **P 103:** This is a semi tall, medium to long duration variety released for irrigated timely sown conditions of Bihar.
27. **PL 56:** This is a medium tall, mid duration variety recommended for rainfed conditions in Punjab. It is moderately resistant to yellow rust.
28. **Rajkiran:** This is medium tall variety released for general cultivation under high fertility irrigated conditions. It is highly resistant to molya disease caused by nematode. It is known as RD381.
29. **Ratna:** This is a medium tall variety recommended for north-eastern plain zone. It is tolerant to saline-alkali soils.
30. **RD 117:** This is a medium tall variety suited for late sown areas.
31. **RD 137:** This is a medium tall variety released for rainfed areas of central zone.
32. **RD 31:** This is a tall variety suited for cultivation in rainfed areas of Rajasthan.
33. **RD 57:** This is a medium tall variety released for cultivation in low fertility irrigated areas of central zone.
34. **RDB 1:** This is a medium tall variety released for cultivation in Rajasthan under high fertility irrigated conditions.
35. **RS 6:** This is a medium tall variety released for general cultivation in Rajasthan. It is suitable for malting and brewing.
36. **Vijaya:** This is a medium tall, good tillering, medium to late maturing variety recommended for western UP, Delhi and MP. It is suitable for saline and alkali soils.

10. SOWING

After land preparation and selection of appropriate variety, sowing is an important agronomic practice which determine the good germination, population and crop stand. For higher productivity it is obvious to have optimum seed rate, spacing, depth of sowing and time of sowing. The details of sowing operations for different conditions has been furnished in Table 8.4.

TABLE 8.4: Sowing methodology for different conditions

	Rainfed	Irrigated	Late sown (irrigated)
Time of sowing	October last week to November first week	November middle to December first week	December middle to January first week
Seed rate	100 kg/ha	75-80 kg/ha	100-120 kg/ha
Row spacing	22.5-25 cm	22.5 cm	20-22.5 cm
Depth	6-8 cm	4-5 cm	4-5 cm
Method of Sowing	Line sowing by bullock drawn *Desi* plough / seed drill	Line sowing by bullock drawn *Desi* plough / seed drill / tractor mounted seed drill	Same as in irrigated condition

The depth and method of sowing may change depending upon the soil condition and moisture content in the soil.

11. IRRIGATION

Basically barley is a winter season drought tolerant crop having limited water requirement. As barley has initial vigorous growth it proliferates its root system deeper into the soil and extract moisture. However providing irrigation is quite useful in increasing yield from fertiliser responsive short statured dwarf and semi-dwarf varieties. It performs much better than wheat with lesser water under stressed conditions. The quantity and frequency of irrigation to barley depends upon the late rains, soils and climatic conditions. The light textured soils need irrigation more frequently than heavy soils which retain more moisture and for long time. The moisture availability determine the pattern of root growth in barley as mesophytic to xerophytic with intermediary types (Figure 8.3) (Bose and Dixit, 1931).

If barley is grown under irrigated condition a pre-sowing irrigation will result in better seedling establishment. It is reported that about 50 % available soil moisture gives satisfactory germination and irrigation at 25 % available moisture in 15-30 cm soil layer resulted in higher yields. Irrigation based on critical stage approach has been found to be economical and effective (Singh and Sandhu, 1981). The stages Active tillering (About 30-35 DAS), Flag leaf / booting (About 60-65 DAS) and Milking / grain filling (About 80-85 DAS) have been identified critical for irrigation.

In case of limited irrigation; if only one is available should be given at active tillering and if two are available give at active tillering and milking stages. Further irrigation at 0.45-0.50 atmosphere soil moisture tension has been reported to give higher yield of barley in late sown conditions. In dry land conditions too, application of one irrigation at most critical stage will increase yield substantially. Frequent irrigation in saline soils counteract the effect of salinity on crop and help increase productivity. The consumptive use, moisture and nutrient extraction pattern and water use efficiency are mostly influenced by the soil moisture status in the crop root zone, stage of plant growth and crop productivity.

12. WEED CONTROL

Barley exhibits early vigorous growth compared to other winter crops. It almost covers its growing area by active tillering stage. This initial fast growth gives it upper hand in competing with weeds and under ideal conditions of growth it almost smother all the weeds and survives from competition. Also the number of weeds in winter is less compared to other seasons. Thus the weed problem is less in barley. Nevertheless with the improvement in barley growing with higher doses of fertilisers under irrigated conditions have prompted weed competition too. The most common weed species identified in barley are *Avena fatua* (wild oat), *phalaris minor* (canary grass), *Fumaria parviflora, Chinopodium album* (bathua), *Cirsium arvensis* (kateli), *Anagalis arvensis* (krishna neel), *Melilotus alba* and *Melilotus indica* (senji), *Convolvulus arvensis, Ageratum conyzoides* etc.

A good cultural practice like one hoeing and one hand weeding after first irrigation at 30-35 days after sowing gives effective control. If weeds still persist chemical weed control may be contemplated. All broad leaved weeds can be controlled by spraying 2,4-D 0.75 kg a.i./ha dissolved in 750 liters of water at 40-45 DAS. Under dryland conditions the dose may be reduced to 0.50 kg. Addition of 3 % urea solution to 2, 4-D at spray mixture improves its efficacy.

Application of isoproturon at 1-1.5 kg a.i./ha or methabenzthiozuron at 1.5 kg a.i./ ha or metaxuron at 1.5 kg a.i./ha dissolved in about 750 liters of water as post-emergence spray at 30-35 DAS effectively controls other grassy weeds.

13. PLANT PROTECTION

Like any other cereal, barley too is susceptible to many insect pests and diseases. These cause damage ranging from moderate to heavy depending upon the variety, method of cultivation and agro-climatic condition. There is thus an obvious need to take up effective plant protection measures to ward off any loss of yield and quality due to these pests and diseases.

13.1. Insect Pest

The following insects pests have been reported to cause damage in barley in many areas warranting preventive and control measures.

1. **Termites:** There are two species of termites noticed in barley. *Microtermes obesi* Holm., *Odontotermes obesus* Rhamb. These attack the root system of plant and plants start showing wilting symptom and later on dry up. Deep ploughing during fallow period and exposing the termites to sunlight limits the attack of termites. Further application of insecticides like endosulfan 4% dust at 20-25 kg/ha prior to sowing and mixing in the soil brings bout effective control of termites.
2. **Aphids: *Rhopalosiphum maidis* Fitch.:** These are small green to dark coloured, spiny insects swarming around all parts of plant sucking the sap. Subsequently infested plants turn pale green in colour and remain stunted. They also serve as vectors for virus causing barley yellow dwarf disease. The aphids in barley may be controlled by spraying of Methyl parathion at 0.025 % or Dimethoate at 0.03 %. The spray may be repeated in case there is reoccurrence or incomplete control at 10-15 days interval.

14. DISEASES

The following diseases have been commonly observed in barley (Suryanarayana, 1979).

Among these yellow rust, smut and stripe diseases are important in plains and powdery mildew is important in hills and foot rot is important in non traditional saline areas of Bihar and West Bengal. All are fungal diseases except molya disease which is caused by a nematode and has been more serious in Rajasthan.

14.1. Rusts

The yellow rust (*Puccinia strifomis*) is identified by parallel rows of pale to orange-yellow pustules occur in rows on the foliage, stems or ears and may be close enough to appear as continuous stripes. Leaves may wither early; grains are shriveled and small.

The black stem rust (*Puccinia graminis tritici*) is identified by reddish brown rows of spots on stems and leaves.

The leaf rust (*Puccinia hordei*) is identified by numerous small round yellow to orange-brown pustules of the summer, uredial stage, occur scattered over leaf blades and sheaths. The uredospores spread the disease.

The rust diseases are controlled by growing resistant varieties, removal of alternate hosts of causal organisms, dusting sulphur and spraying of fungicides like zineb at 2 kg/ha or any systemic fungicide at appropriate dose.

14.2. Smuts

In loose smut (*Ustilago hordei*) the entire ear except the rachis, is replaced by brown, enclosed spore-masses. Ultimately the spores are liberated and blow away and the naked rachis remains. When the spores reach new ears they infect the developing grain.

In covered smut (*Ustilago nuda*) the disease in noticeable only when the heads emerge from the boot when it is seen that the grains are replaced by hard black spore masses, each covered with a thin membrane. Membrane begin to split after emergence and spores are blown to healthy ears. Others breakup and spread during threshing contaminating the grain.

These smut diseases are controlled by growing resistant varieties, seed treatment with systemic fungicides like vitavex at 2.5 g /kg seed. Removing and destroying the affected ears will reduce the build up of inoculum of fungus for the next season and further spread.

14.3. Leaf Spots

Net blotch (*Helminthosporium teres*) results in lesions on leaves as brown, net like blotches of dark brown lines in lighter brown zones or long, narrow rectangular or irregular stripes. Infected stems show brown discolourations, light brown blotches occur on the kernels. Grains are reduced in size.

Spot blotch (*Helminthosporium sativam*) is identified by dark brown, rotting lesions on the roots and seedling mortality. Often the plants grow, and the fungus spreads slowly through the tissues forming dark brown, elliptical spots and blotches on the leaf blades, stalks, and glumes. Diseased plants may be stunted and excessively tillered. Leaves tend to dry and wither away early.

Barley stripe disease (*Helminthosporium gramineum*) is noticed by long narrow, brown or yellow stripes, as many as seven stripes may run nearly the length of the plant. Leaves may split fray and collapse. Even if the plant is not killed there will be reduction in vigour. Ear head emergence may be affected or emerge with no grain. If grains are formed they will be discoloured and infected.

Control Measures: Use of resistant varieties, clean and treated seeds with organo-mercurial compounds like ceresan or agrosan at 2 g/kg seed, crop rotation reduce the incidence of these blotch and stripe diseases.

14.4. Mildew

The first signs of infections in powdery mildew (*Erysiphae graminis*) are the appearance of small white/grey fluffy spots of mycelium on leaf sheaths, leaves or on the ears. These grow darken and become powdery with spores. Later small, round, black reproductive bodies, the perithecia, may be formed on the mycelial mats. Disease development is favoured by rapid lush growth of the host.

Control measures include growing resistant varieties, dusting sulphur at 20 kg/ha and sprays of systemic fungicides like Karathane (0.2 %) at the interval of 15-20 days.

14.5. Foot Rot

Foot rot disease (*Fusarium culmorum*) is identified by discoloured and rotten roots and stunted growth with yellowish leaves. Plants may fail to produce ears, may lodge, may produce dwarfed or distorted heads, which may contain small grains or be empty. white or pinkish spore masses may occur.

Control is difficult. Use of clean and organo-mercurial compound treated seeds, early sowing and good drainage reduce the incidence of foot rot disease.

14.6. Molya Disease

Molya disease (*Heterodera avenae*) affected plants occur in patches, and are stunted and yellowish. Tillering is reduced and ears may be shriveled. Attacked plants have many dense, short, thick, highly-branched roots, matted together. The nematode attacks each root near the growing point, growth is arrested and laterals are formed.

Control of this nematode infection is achieved by crop rotation and growing resistant varieties. Fumigation with methyl bromide or chloropicrin, ethyl dibromide etc. also kill the nematode present in the soil.

15. HARVESTING

Barley will be ready for harvest once the plants turn yellow and dry up. The grains become hard, moisture content will be around 18-20 per cent and plants become brittle and rigid. Stem breaks down with slight pressure. Fast winds make plants lean and lodge. Any delay in harvest result in shattering of ears and loss of grains. Lodging of plants deteriorates the fodder value of crop and at times the quality of the grains gets impaired. The crop is harvested by cutting at ground level by using sickle. The harvested product is bundled and moved to threshing yard. It is further dried there and threshed by allowing bullocks to trample or by using stone rollers or mechanical threshers. The grains so threshed are winnowed to separate them from chaff and other admixture of plant parts. Alternatively the crop is harvested by using the combine harvester wherein the cutting, threshing and winnowing is done at one go. However, the small plots having bunds are difficult to harvest by combine harvester as it requires large area for its mobility. The cleaned grains are dried and moisture content of about 12 % is moisture before there are bagged in gunny bags and moved to storage godown. The straw collected and kept as 'hay' and is used as animal feed in some areas. It is also used for making farm yard manure.

16. YIELD

Grain is the economic yield in barley unless and otherwise it is grown for fodder purpose. The straw left over after threshing is a by-product which may be used for feeding animals or other purpose. The yield of barley varies depending upon the variety used, soil and agroclimatic conditions, irrigation, fertiliser application, plant protection measures and other agronomic practices. Other things being an optimum the yield of the barley mainly depends on variety. Recent high yielding varieties have the yield potential of 40-45 q/ha. However the straw yield varies depending upon the plant type and duration of the crop. A tall variety usually produces higher straw as compared to dwarf and semi-dwarf varieties. Further the same variety under late sown condition may yield lower straw as compared to timely sown crop. Pests, diseases and weed competition also reduce the straw yield. A good crop under normal circumstances may yield about 4-5 tonnes of straw per hectare. The grain to straw ratio i.e. harvest index, depends mainly on three factors *viz*., population density, water availability and nitrogen supply (Verma, 1979). On an average, grain to straw ratio of 0.3 to 0.5 has been observed in barley.

17. QUALITY

It has been a source of staple food for quite long since the pre-historic period. In India about 90 % of the total production is still consumed as human food in the form of *chapati*, *sattu* etc. Further the malt extracts are used as supplements in energy rich drinks like Bournvita ™ , Boost ™ , Horlicks ™ , etc.

Barley is a good source of vitamin B complex. Consumption of barley is found to curtail cholesterol levels in liver and plasma by inhibiting the rate limiting hepatic enzyme β -hydroxy, β -methyl glutaryl CoA reductase. It also stimulates fatty acid synthesis in liver (Foster and Prentice, 1987).

Quality aspects of barley consists of two major parts:

- Physical appearance and colour of the grains
- Chemical composition of grains

Traditional barley grains consist of 15-20 % husk. Because of husk, consumers do not like to eat barley as a food grain. Husk deteriorates the quality of the chapati. As animal feed too husked grain has poor market because of high crude fiber percentage (8-10%). The hull-less amber coloured barley types will be attractive for human consumption and as animal feed (Ram, 1982).

TABLE 8.5: Composition of barley grains on percentage basis (Harris, 1962)

Starch	63-65	Amino acids : Peptides	0.5
Sucrose	1-2	Nucleic acids	0.2-0.3
Reducing sugars	0.1-0.2	Phosphate	1
Other sugars	1	Silicate	0.9
Gums	1-1.5	Potash	0.7
Hemicellulose	8-10	Magnesia	0.3
Cellulose	4-5	Other minerals	5-9-6
Protein : Albumino	0.5	Lipids	2-3
Globulin	3	Crude protein	8-11
Hordein	3-4		
Glutelin	3-4		

Barley is highly digestible because of negligible gluten content. In protein and lysine values, it is more or less at par with wheat and superior to maize and millets. But higher content of protein deteriorates the malting value of barley. Thus for malting, barley needs entirely different set of quality parameters.

Uniform kernel size, free from insect or fungal infection and cuts or blemishes, is an important factor to the malting industry. The thin kernels are not suitable for malting purpose. A high protein content is undesirable for malting and reduces the value of malt. The preferred protein limits in USA is 10-12 % for two row barley and 11.5-13 % for six row barley but in India barley having protein content >9 % is not preferred for malting. High protein prolongs the time required for stepping and causes uneven germination, produces malt with low extract which result in a decreased yield of beer per unit of barley, produces malt with excessive enzyme activity, reduces the mellowness of malt, causes greater losses during malting due to abnormal growth of the kernels and produces worts with a higher content of soluble protein which may be because of changes in processing and in turn character of the beer.

Higher protein is associated with high soil N, high temperature, low soil moisture, diseases or insects. Situations which cause stress in the plant tend to increase grain protein percentage. The proper C/N ratio forms in the grain in the absence of stress. However if stress occurs during grain filling, translocation of carbohydrates to the developing grains is affected, more than translocation of amino acids, resulting in higher grain protein. Hence judicious application of N fertiliser coupled with proper water management is imperative in production of barley especially for malting purpose.

18. HULL-LESS BARLEY

Since hull-less varieties require about 10 days more than the hulled type, they require about 20 kg higher nitrogen. During the maturity due to longer duration they will be caught in the hot winds in the hilly areas. This will result in the poor grain filling in the upper portion of earhead and reduction in the grain yield. Therefore it is essential to give one more irrigation at grain filling stage to avoid the effect of hot winds at the time of grain filling.

19. FODDER AND FORAGE BARLEY

When barley is grown for fodder and forage purpose, the first cut or harvest should be done at about 60-65 DAS. During cutting care should be taken to leave about 2.5-3.0 cm above ground to give better regeneration power to crop. Further application of 20 kg N /ha and giving one irrigation is essential to have better and vigorous growth after harvest.

20. MALT BARLEY

Barley grown with the purpose of using grains for malting needs a different management strategy. To maintain the protein content at required level both fertiliser and irrigation should be given cautiously. Nutrients should be balanced like 50 : 25 : 25 kg NPK/ha and all the nutrients should be given at once at the time of sowing. And no top dressing should be attempted, as it results in higher protein content in grains. Frequent irrigations should given to counter the effect of nitrogen fertilisers. At no stage of crop stress condition should not be allowed set in.

21. POST HARVEST TECHNOLOGY

The important post harvest operations consists of drying, cleaning and storing. Further based on the intended use, processing can be done like, grading, polishing, flouring etc.

- **Drying:** Soon after the harvest of crop, the moisture content in the grains will be high. It is pertinent to bring the moisture percentage in the seeds to lower level. As the higher moisture condition is suitable for the development of moulds and other insect pests. High moisture content also result in higher humidity and biochemical activity inside the seed resulting in the poor storability. The desired moisture content in seed and temperature depends on the ambient temperature and length of storage. On an average moisture content should not exceed 12 -14%.
- **Cleaning:** Grain after threshing will be contaminated with dust, chaff, short straws, leaves, shriveled and broken seeds, the grains of other cereals and weeds, insects, soil, stones, etc. It is important to remove all theses from the seed to improve the keeping quality and to avoid the damage to handling machinery. Cleaning the seeds, besides making them fit for long storage, will also add to the quality of the seed. It is important to keep in mind that while cleaning the seeds, care should also be taken not to damage the seeds, especially if the seed is intended to be used for future sowing or malting, where any damage reduces the suitability of the seed for that.
- **Storage:** Bad storage allows decrease in bulk, and in quality; seed and malting grain become animal feed; animal feed stuffs are reduced in nutritional value or become unusable. Contamination of grain, for example with insect, pests or rodent faeces, may cause the rejection of samples of otherwise high quality. Grain which is segregated by type and quality, roughly cleaned, dried, very thoroughly cleaned and stored under rigorously controlled conditions may be kept for several years without deterioration. Grain stored under cooler condition ($<15^0$C) with moisture content of 12 % does not require ventilation and will retain its viability. Grain for animal feed may be stored damp if the temperature is reduced quickly and is maintained at a low value. High quality barley is not normally stored for more than two years (Briggs, 1978).

22. CONSTRAINTS IN PRODUCTION

Many constraints have been observed in barley production despite a lot of progress in barley improvement work. Singh (1979) has listed the following constraints in barley production in India.

- Barley has to compete with fertiliser responsive high yielding dwarf varieties of wheat in the same season.
- Want of suitable fertiliser responsive high yielding dwarf plant types in barley.
- Hulled grains of barley is inferior to wheat grain and also less consumer demand and low market price.
- Barley is an industrial crop but less developed industries in the country so little or no demand for industrial uses.
- Common domestic uses of barley are not much explored.

- Alternative uses of barley (feed and fodder) not much explored
- Production and availability of wheat and rice in abundance in the country making less demand for barley.
- Lodging tendency in tall types discourages higher input use and thus barley is not preferred in high intensity agriculture.

In addition to the above points Dr. Mahabal Ram, the former project coordinator of All India Co-ordinated Barley Improvement Project, opines that

- Non-availability of high yielding varieties of barley with better built in resistance to major diseases and pests and also with improved grain quality characteristics.
- Cultivation of barley on the marginal land.
- Poor price-support.
- Poor procurement and utilization system

These are some of the other major constraints in barley production in India.

23. REFERENCES

Anonymous, 1973. National commission on agriculture -Report of the study team on wheat, barley, oats, rye and triticales, New Delhi.

Berrie, A.M.M. 1977. An Introduction to the botany of the major crop plants. Heyden and Son Ltd. London. pp.31-32.

Bose, R.D. and Dixit, D.D. 1931. *Indian Journal of Agricultural Sciences* **1**: 90-108.

Briggs, D.E. 1978. Barley. Chapman and Hall, London. pp. 526.

Delorit, R.J., Greub, L.J. and Ahlgren, H.L. 1984. Crop production. Vth Edition. Prentice Hall Inc. New Jersey. pp.170-171.

FAI, 1998. Fertlizer statistics 1997-98. Fertiliser Association of India, New Delhi. pp. III 46-50.

FAO, 1998. FAO Production year book 1997. **51**:69-70.

Fedek, G. 1994. Barley. *In*: Encyclopedia of agricultural science Vol 1. (eds. Arntzen, C.J. and Ritter, E.M.) Academic press, San Diego, California. pp.234-235.

Foster, E. and Prentice, N. 1987. Barley. *In*: Nutritive value of cereal grains: Genetic and Agronomic improvement. (eds. Olsen, R.A. and Frey, K.J.) 28 Agronomy series. ASA inc., CSSA inc, SSSA inc. Madison, Wisconsin, USA. pp.338-396.

Harlan, J.R. 1968. On the origin of barley. *USDA Agricultural Handbook* **338**:9-31.

Harlan, J.R. 1976. Barley. *In*: Evolution of crop plants (ed. Simmonds, N.W.) Longman, London. pp.93-98.

Harlan, J.R. and Zohary, D. 1966. Distribution of wild wheat and barley. *Science N.Y.* **153**:1074-1080.

Harris, G. 1962. The structural chemistry of barley and malt. *In*: Barley and malt: Biology, Biochemistry, Technology.(ed. A.H. Cook). Academic Press, New York. pp.431-582.

Misra, B.N., Verma, B.S. and Singh, K.N. 1979. Effect of major nutrients (NPK) in relation to grain yield and malting quality of 2 row barley (*Hordeum distichum* L.) *Proceedings of first national symposium on barley*, Karnal. 24-27, February 1979. pp.207-211.

Pederson, A. 1950. *Landbrugets Plantekultur, II* Kandrup og Wunsch Bogtrykkeri, København, **31**:34.

Ram, M. 1982. Genetic revolution in barley. All India Co-ordianted Barley Improvement Project (ICAR), IARI Regional Station, Karnal. pp. 12 & 38.

Singh, K.N. 1979. Barley agronomic research - today and tomarrow. *Proceedings of first national symposium on barley*, Karnal. 24-27, February 1979. pp.202.

Singh, K.N. and Sandhu, H.S. 1981. Barley agronomic research- An overview. *In*: Quarter century of agronomic research in India (1955-1980),(eds. Rajendra Prasad, Parashar, K.S. and Singh, R.P.). The Indian Society of Agronomy, New Delhi. pp.107-118.

Suryanarayana, D. 1979. Plant disease problems of barley crop in India and prospects of their control. *Proceedings of first national symposium on barley*, Karnal. 24-27, February 1979. pp.155-163.

Vavilov, N.I. 1951. Origin, variation, immunity and breeding of cultivated plants. *Chronica Botanica* **13:**(No.1-6).

Verma, M.M. 1979. Recent concept in the breeding methodology of barley. *Proceedings of first national symposium on barley*, Karnal. 24-27, February 1979. pp.105-107.

Wendrof, F., Schild, R., Hadidi, N.E., Close, A.K., Kobusiewicz, M., Wieckowska, H., Issawi, B. and Haes, H. 1979. Use of barley in the Egyptian late paleothic. *Science* **205** (4413):1341-1347.

9. Oats

N. P. Shukla, J. B. Singh and Shiva Dhar

Oats are an extremely versatile *rabi* cereal fodder crop of India. They can be grown in wide range of environments on diverse soil types for forages and as a companion crop. It is a quick growing, palatable, succulent and nutritious crop. It is rich in energy, protein, vitamin B, Phosphorus and Iron (Mehra, 1978) and it is used as a green fodder, hay, silage and its grain make good concentrate for poulty, cattle, sheep and other livestock. In addition to that it forms an excellent conmbination when fed along with winter season legumes like, berseem, lucerne, senji (Indian clover) and saftal (Persian clover).

1. ORIGIN

There are various school of thoughts regarding the exact timing and location for the domestication of oats. According to Thellung (1912) oats was domesticated after wheat and barley and perhaps it originated in Asia Minor. The ancient Greeks called it *bromos*, and latins called it as *Avena*, but these names were commonly applied to species which are not cultivated and which were weed mixed with cereals (De Candolle, 1890), Danaiffe and Sirodot (1901) stated the derivation of Latin word *Avena*, remains some what obscure. It seems probable that it is from the Latin word *Aveo* (to desire), that it is say, forage desired by all animals. The syllable *Av*, in its composition found in a number of languages. The syllable *Se* is found in the Sanskrit in the sense of supply food. *Ava*, nourishment; *Avasa*, pasturage, a word assumed to be equivalent to the Russian world *Osevu*, the Polish word *Owies*,: the Romanian word *Ovesia*, and Servian word *Ovas*. According to vernacular names De Candolle (1890) stated "an ancient existence of oats in north and west of the Alps and the borders of the Europe towards Tartary and Caucasus". In India, it is not known exactly about the timing of introduction of cultivated oats. However, in medieval India (1590 A.D.) there is reference of green oats as a horse fodder in 'Ain-e-Akbari' authored by Abul Fazal. Shaw and Bose (1918) stated that the cultivation of oat in India was introduced in the early part of the 19th century specially for growing on military farms to feed the horses.

2. DISTRIBUTION

Total world area under oat cultivation is approximately 17166 thousand hectare with production of 32840 thousand million tonnes (Table 9.1). The major oat producing countries are USSR, USA, Canada, Poland, Ukrine and Oceania. In India it is grown in Punjab, Haryana, Jammu and Kashmir, Himachal Pradesh, Uttar Pradesh, Rajasthan and Maharashtra.

TABLE 9.1: Area and production of oats in world (1997)

Country	Area 000 ha	Production 000 Mt	Productivity kg/ ha.
Canada	1499	3485	2325
USA	1178	2556	2170
Russia Fed.	6904	11000	1593
Oceania	860	1738	1469
Ukraine	524	731	2197
Total	17166	32840	1913

3. CLASSIFICATION

Oats belongs to genus *Avena*. Some economically important species of genus *Avena* are as follows:

3.1. Hexalpoid (2n=42)

1. *Avena sativa*: It is the most common and important cultivated species of oats and grown in several parts of the world for grain, fodder and other purposes.
2. *Avena fatua*: This is a wild species and troublesome weed in cultivated fields. It is also cultivated for grain, green fodder and other purposes in India and parts of south-west Asia.
3. *Avena sterillis*: Commonly known as a fly oats, is grown mainly for fodder in parts of Mediterranean region and south-west Asia. The spikelets of this species are larger than *Avena sativa*.
4. *Avena byzantina*: The red oats, is grown around Mediterranean regions, Europe and north Asia and warmer parts of subtropical world for grain and fodder. It is also cultivated in India.
5. *Avena ludoviciana*: The wild red oats, is cultivated for grain in Mediterranean and south-west Asian regions.
6. *Avena chinensis*: Cultivation of this species is not much popular. It is mainly grown on hilly parts of China for its grain. The lemma and palea of this species incompletely fused with the grain, yielding naked seed easily on threshing.

3.2. Tetraploid Species (2n=28)

1. *Avena abyssinica*: An abysinnia oat, grows wild and is also cultivated in desert area of north Africa.
2. *Avena barbata*: It is tall, weak and slender range grass of western United States of America.
3. Diploid species (2n = 14)

4. *Avena strigosa*: The sand oats, is cultivated for fodder in the Mediterranean region.
5. *Avena brevis*: The short oats, is grown in southern Europe, mostly for green fodder or hay.

4. CLIMATE

Climatic factors are very important in determining where oats are adopted, and among these, temperature and moisture are the most important. Cool and moist climate are considered best for oat because they require more moisture to produce a green unit of dry matter than any other cereal except rice. Oat requires about 15-25 °C temperature for its optimum growth. Frost and severe cold retards its growth, whereas hot and dry conditions hasten its maturity resulting in poor yield and inferior quality of fodder. Oats can be successfully grown where the average annual rainfall ranges from 380 to 1140 mm however, a well distributed rainfall of 400 mm during crop duration is sufficient to meet its requirement as a fodder crop.

5. SOIL AND SOIL PREPARATION

Oats are grown on different soil types around the world except sodic soil and waterlogged ones. Next to rye (Secale cereal) oats are the most versatile of the cereals in regards to soil types. Almost any reasonably fertile, well drained soil is suited to oats if temperature and moisture conditions are favourable, however fertile and well drained loam to clay loam soils are the best. The crop suffers badly due to lodging when heavily fertilized with nitrogenous fertilisers on poorly drained clay soils. Nutrient (N, P, K) requirements for the oats is less than those of wheat and maize, and can be tailored to the yield level sought. Oats are substantially less salt tolerant than barley, slightly less tolerant than wheat or rye, but slightly more tolerant than sorghum. It can tolerate slight acid and saline conditions but not above 8.5 pH. Under non saline conditions oats are moderately tolerant to Na, but less tolerant than wheat and barley.

For field preparation one deep ploughing followed by three or four harrowing and planking will serve the purpose to get friable and well levelled seed beds. If a tractor is used 2 or 3 operations by disc harrow or cultivator would suffice the purpose to destroy the weeds, stubbles or previous crop and create the desired tilth.

6. CROPPING SYSTEM

The suitable mixtures with early sown oat crop is Japanese rape or Chinese cabbage because they become ready in 60-70 days and could be harvested along with oat. Oat can also be grown in association with berseem, persian clover or lucerne. The seed rates (kg/ha) for different mixtures are as follows:

Oat + berseem (40+25), oat + persian clover (40+15), oat + lucerne (40+15), oat + senji (70 + 25), oat + pea (70+40), oat + Chinese cabbage (100+3.5), oat + Japanese rape (100+3.5), oat + turnip (100+2.5) and oat + vicia (100 + 10).

Some of the common rotations for oat are cowpea-oat, guar-oat, jowar+cowpea-oat and maize + cowpea-oat.

7. MANURES AND FERTILISERS

To obtain maximum yields adequate nutrients must be present throughout the growing season. Oat plants will be less vigorous, have smaller leaves and panicles and will have less developed root system if nutrients are short in the first half of the growing season. Later deficiencies limits seed set. Proper fertiliser management is influenced by the previous crop, soil, moisture and rainfall pattern, capability of soil to supply nutrients, cultivars used and time of applications. In case of oat grown after legume less nitrogen and more phosphorus will be needed. Similarly if the oat is preceded by an oilseed crop it will require more sulphur in case of sulphur deficient soil.

Production potential of oats to a great extent depends on application of nitrogen. Oat being a graminaceous crop responds favourably to nitrogen fertilization. For obtaining maximum forage yield of good quality from every kg of nitrogen it is essential to apply the optimum quantity of nitrogen at proper stage. Generally 2-4 cuts are taken in oat and part of the nitrogen has to be applied after each cut. In general, the nitrogen requirement of multicut oat varieties are higher as compared to single cut varieties and unless nitrogen is appropriately applied at various growth stages and more particularly after each cut a high yield of fodder may not be possible. Nitrogen not only increases the production of succulent and leafy fodder but is also know to influence the quality of fodder as well.

In general, 60 kg of nitrogen per hectare should be applied and properly mixed in soil before sowing. Additional dose of 30 kg nitrogen per hectare should be broadcasted in two splits i.e. (10 kg and 20 kg) after first irrigation (at tilliring stage) and at first cut, respectively. In light textured soils 40 kg of phosphorus per hectare should be applied at the time of sowing as it shows a favourable effect on shoot height and tillering during the earlier growth stages.

If manure is available the crop may be manured with 25 tonnes of Farm yard manure 10-15 days before sowing. Besides that a mixture of 65 kg Urea, 200 kg single super phosphate and 50 kg Muriate of potash should be applied at sowing. The second dose of urea at the same rate may be applied after six weeks from sowing or immediately after first cutting. If the foliage appears to be pale green at boot stage 2 per cent solution may be sprayed. Zinc sulphate may be aplied at the rate of 10 kg per hectare in areas prone to zinc deficiency.

8. VARIETIES

There are large number of varieties varying in duration, height, growing conditions, number of cuts, etc. As a result of introduction, inter varietal hybridization and selection programme a large number of new improved stains were developed and recommended for cultivation. The following material/varieties has been found promising for cultivation:

1. **Kent:** Plant type of this variety is erect, very good in tillering capacity, synchronous in flowering. It is a tall (115-150 cm) culm is straight, slightly red, leaves are droopy, seeds are medium plump and bold. It is resistant to lodging, shattering and free from rust, leaf blight and insect. It flowers in 90-110 days and matures in 145-150 days. Yield potential of this variety is 500

quintals green and 105 quintals per hectare dry fodder. Seed yield ranges from 18-22 quintals per hectare. It is widely grown in India.

2. **UPO-94:** This variety was released in 1982 for northern India, Rajasthan, Gujarat and M.P. states. This variety is medium late and erect type selected at Pantnagar. Plat height is 135-140 cm at flowering, stem light green in colour with 8-12 tillers bearing dark green foliage. Panicle is long having 60-65 spikelets per panicle. Lemma is yellowish white and medium in length. It can produce upto 600 quintal green fodder, 130 qu intal dry fodder and 16-18 quintal seed yield per hectare. It is a multi-cut variety and 3 cuts may be taken easily.
3. **OS-6:** It is a single cut variety and found suitable for entire oat growing areas of the India. It has an erect growth habit, early vigour, medium tall in height (109 cm) possessing long and broad green leaves. Flag leaves are open and long remains erect at the time of panicle emergence. It flowers in 80-85 days and matures in 130-140 days. The production potential of this variety is 525 quintals green fodder, 110 quintals dry fodder and 20-30 quintal seed yield per hectare.
4. **OS-7:** This variety is most suitable to Haryana state. It is erect and fast growing variety with height of 110-125 cm. Leaves are long, broad and light green in colour. Flag leaf is erect at emergence of panicle and this variety is suitable for single cut with yield potential of 540 quintals green fodder and 115 quintals dry fodder per hectare.

Besides the above, other promising varieties are listed in Table 9.2.

TABLE 9.2: Important varieties of oat along with the area of cultivation and yield potentiality

Variety	Area of cultivation	Green forage yield (q/ha)
JHO-810	Kashmir Valley	500-600
JHO-829	Hill zone of India (Multi-cut)	500-550
JHO-822	Central India (Multi-cut)	
UPO-212	Whole country (Multi-cut)	400-575
OL-9	North, North- west and South hills (Single/Double cut)	400-525
OL-125	Central and North-west regions (Single cut)	400-550
OS-96	North India, A.P. and South hills	375-500
OS-108	North-east and Central India	425-500

Source: (Hazra 1989 and Hazra 1995).

9. SOWING

9.1. Sowing Time

The normal sowing time starts from October to November. However, middle of October to middle of November is the optimum sowing time for higher production. Mutli-cut varieties if sown earlier will help to tide over the scarcity period from middle of December to middle of January. Early seeding helps to ensure efficient use of available moisture.

9.2. Seed Rate

There is seldom one optimum seeding rate because oat plants adjust their tillering and growth depending upon plant density. Seed should be drilled at the rate of 80-90 kg/ha for small varieties, whereas it ranges 100-125 kg/ha. for bold seeded varieties e.g., Kent. A moderate reduction in seeding rate seldom results in a yield reduction. This is due to more vigorous plants and the production of more tillers. Very high seeding rate result in increased competition, weaker culms and frequently increased lodging. For maximum yields, only good quality seed of recommended cultivars should be sown.

9.3. Spacing

In case of low tillering varieties spacing between rows should be kept 20-25 cm apart. However, high tillering varieties should be sown at 25-30 cm apart.

9.4. Method of Sowing

Seed should preferably be drilled. If seed drill is not available the sowing can be done either by *Pora* or *Kera* method behind the plough.

10. IRRIGATION

Water requirement of crop depends on soil texture, prevailing climatic conditions, etc. The oat crop needs copious irrigation for succulent and quality herbage production, nevertheless, it can be grown with acceptable level of yield even under limited irrigation. Productivity of oat greatly depends on availability of water during growing season and accordingly scheduling of effective irrigation technique is thus an essential factor, keeping in view its limited availability, cost and efficient utilization. A pre sowing irrigation is required for assured moisture in seed bed to ensure good germination and early establishment. A total number of 4-5 irrigations are required for oats. Generally, the first irrigation should be given 20-25 days after sowing. Subsequent irrigations may be provided at tillering, jointing, panicle primordia initiation, flowering and at milk stage. Judicious and timely application of irrigation water could check the leaching losses of nutrients and could also save the deterioration of soil by avoiding over irrigation. Further, during boot stage irrigation is essential due to high water requirement at flowering stage.

Development of suitable irrigation schedules offers an important approach to water saving because timely supply of water to crops in adequate quantity is the crux of the efficient water management for ensuring optimum yield and quality of the oat. Forage crops needs suitable soil moisture environment (Kumar and Verma, 1981) to achieve luxuriant vegetative growth, higher dry matter yield and better forage quality. Several studies (Verma and Kumar 1984, Menhi Lal and Shukla 1988b, Reddy and Hussain 1990a) on oat reflects that optimum soil moisture regime for oat is 75% available soil moisture (ASM). During the recent past, sincere efforts have been made to visualise the facts and facets of irrigation scheduling base upon ratio of irrigation water to cumulative pan evaporation (IW/CPE ratio) because of its rational approach to integrate the moisture characteristics of soil and evaporative demand of atmosphere (Parihar *et. al.*, 1974, Khera *et. al.*, 1988). Shukla *et. al.*, 1988 studied the response of three moisture regimes i.e. 0.5, 0.8 and 1.1 IW/CPE for single cut variety of oat i.e. Cv. Kent and found that 1.1 IW/CPE ratio produced highest green and dry forage yield. Irrigation in oats needs to be reconciled to avoid moisture stress at consecutive growth stages both during vegetative and reproductive stages. The jointing stage is more sensitive to moisture stress than tillering (Shukla *et. al.* 1982).

Planned study of quantify evapo-transpiration with help of lysimeter revealed that the crop utilizes 1/5 to 1/3 of the total water requirement from sowing to jointing stage, whereas, the remianing 4/5 to 2/3 is required from jointing to milk/dough stage, emphasizing the need of rationalizing irrigation water during later phases of crop growth (Menhi Lal and Shukla, 1988b). In an experiment conducted at IGFRI, Jhansi it was observed that ET need of oat crop increased progressively with advancement of growth, attaining a peak rate of 5.7 mm per day during both stage (Pradeep Behari and Hazra, 1989). Similarly, Menhi Lal and Shukla (1988a) also observed a peak ET rate at boot stage of oat crop. In water scarcity areas moisture conservation practices must be done. Some hydrophilic polymer such as Jalshakti and agrolite-400 has the characteristics to absorb, retain and release moisture as per crop need. Menhi Lal and Shukla (1996) found that the use of 5.0 kg/ha of Jalshakti is optimum for producing higher forage yield of oat with improved protein content and better water use efficiency.

11. WEED CONTROL

Weed emerges with the emerging crop seedling and if not controlled in early stage of crop growth, these may cause reduction in yield depending upon the intensity and the nature of weeds. The most prominent weeds are wild oats (*Avena fatua*), Krishnaneel (*Anagalis arvensis*), Satyanashi (*Argemone maxicana*), Bathua (*Chenopodium album*), Khar bathua (*Chenopodium murale*), Canada thistle (*Circium arvense*) Gajri (*Fumaria parviflora*) munmuna (*Spergula arvensis*), Senji (*Melilotus alba*), Akri (*Vicia sativa*), Chatri-matri (*Vicia spp.*), *Lathyrus* spp. etc.

Herbicides are often used as a control measure. As a pre-emergence herbicide linuron at the rate of 0.5 kg/ha may be spread to control associated weeds and this will cause 60-70 per cent reduction in weeds.

The most widely used herbicide is 2,4-D, which is available in the market under different trade name. The major reasons for its preference are lower cost and the fact that it controls several broad leaved weeds which are common in oats. Sodium salt of 2,4-D may be applied as post emergence herbicide @ 0.75 kg/ha.

Other herbicides commonly recommended for oats are MCPA and a mixture of bromoxynil (3,5 dibromo-4-hydroxy benzonitriles) and MCPA. MCPA at the rate of 1.0 kg/ha after three weeks would reduce weed intensity to about 70-80 per cent. Post emergence application of herbicides considerably control *Convolvulus arvensis* which otherwise resistant to pre-emergence herbicides.

12. PLANT PROTECTION

Effect of pest and disease in reducing forage yields, unless they are in epidemic proportions, are not fully appreciated or rarely considered in the formation of animal production system based on forage crop production. It is only recently that the significance of pest and diseases have began to be recognised. Pest and diseases can be managed through chemical agents, cultivation practices and use of resistant varieties, however an integrated approach involving all the three strategies as appropriate probably offers the best solution. The best long term prospects lies in developing disease and pest resistant cutlivars. Insect pests do not cause much damage to the oat crop. However, termites and white ants are the serious pests which adversely affect the stand of crop. Termites are a serious problem in unirrigated sandy loam soils. Application of 4% endosulfan @25 kg/ha is an effective control measure.

Some of the important diseases of the oat crop are leaf blotch (*Pyrenophora avenae*), covered smut (*Ustilago hordei*), loose smut (*Ustilago avenae*), seed and seedling blight. (*Fusarium* spp., *Pythium* spp. and *Helminthosporium* spp.), crown rust (*Puccinia coronata-avenae*), stem rust (*Puccinia graminis-avenae*) and sclerotial wilt (*Sclerotia rolfsii*) which may cause considerable damage to the crop. For the control of leaf blotch disease treatment with organomercurial fungicides like ceresan or agrosan @ 0.25 per cent completely inhibit the pathogen from seed. In field infection spraying of mencozeb @ 0.25 per cent can provide good control. Rest of the diseases can be controlled with the use of zineb or mencozeb @ 0.25 as foliar spray at 15 days interval.

13. HARVESTING

For oats, number of cutting depends upon the type of varieties and requirement of fodder for specific time and purpose. The single cut crop may be harvested at 50 per cent flowering to milk stage for better quality and quantity. In two cut management system, first cutting should be taken at flag leaf stage (about 60 days) and last cutting at 50 per cent flowering to milk stage. For multi-cut crop the first cutting should be done at 55 days, second at 105 days and last at 50% flowering stage. Delay in harvesting beyond milk stage may result in sharp decline in digestible protein.

14. YIELD

Generally, the oat forage yield varies from 400-500 quintals per hectare, but with proper management practices the forage yield of 500-600 q/ha can be obtained. Contrary to general belief it has been observed that the oat crop harvested once provides higher yield compared to harvesting in 3-4 cuts. However, the advantage of multi-cut system is that the fodder supply could be maintained for longer period and the quality of the forage will be better because in case of single cut, delay in harvesting deteriorate the crude protein and increases the lignin and hemicellulose content.

15. QUALITY

Brickman and Forsberg (1988) noted that levels of crude protein, neutral detergent fiber (NDF) and acid detergent fiber (ADF) constitute measures of oat forage quality. The nutritive value of fodder crops decreases with advancement of maturity of plant (Pal *et. al.*, 1971, Sharma and Mudgal 1966). Mc Elroy and Gervais 1983 observed sharp reduction in the protein fraction of oat forage with advancement in maturity, attaining the plateau at heading stage. The oat fodder harvested at flowering stage was superior in terms of chemical composition and nutritive value over other stages of growth (Om Prakash and Verma 1995).

16. SEED PRODUCTION

Seed is an essential component in forage production programme because crop is harvested in its vegetative stage for forages. The cultivation practices for seed crop of oat is quite similar to, its forage production upto the flowering stage. The seed crop needs irrigation at milk and dough stage for healthy seeds. The roughing must be done before grain filling stage. The seed crop should be free from any fungal disease. The research done at Indian Grassland and Fodder Research Institute, Jhansi revealed that spray of 2.0 Kg/ha active ingredient of cycocel (CCC) before flowering increases the seed yield by 35 to 50 per cent.

17. REFERENCES

Brickman, M.A. and Forsberg, R.A. 1988. Oats for Silage.

Proceedings of the 3rd International Oat Conference, (Eds. B Mattsson and R Lyhagen), Lund, Sweden, Svalott AB, Sweden, pp. 190-195.

DeCandolle, Alphonse. 1890. Origin of cultivated plants. The Inst. Sci. Series, New York. D. Appleton and Co.

Denaiffe et Sirodot. 1901. L' Avoine. J.B. Ballie're et fils, Paris. 848 pp. Supplement 1, au Traite and Sur L' Avoine. 39 p.

Hazra, C.R. 1989. Technology for increasing forage production in India. Bull. AICRPFC, IGFRI, Jhansi pp. 1-25.

Hazara, C.R. 1995. Advances in forage production technology. Bull. AICRPFC, IGRFI, Jhansi pp.10-27.

Khera, K. L., Parihar, S.S., Sandhu, B.S., Jalota, S.K. and Baldev Singh. 1988. Response of berseem to irrigation and nutrients. J. Res. PAU 25: 9-14.

Kumar, V and Verma U. K. 1981. Effect of moisture regimes and nitrogen levels on yields of oats. Indian J. Agron. 25: 87-88.

Mc. Elary, A.R. and Gervais P. 1983. Yield quality relationship in barley and oats grown for forage. Naturalistic Canadian 110: 32-33.

Mehra, K. L. 1978. Oats. ICAR Pub., New Delhi.

Menhi Lal and Shukla, N.P. 1988a. Evapotranspiration and forage yield of oat in relation to growth phases and weather conditions. Indian J. Agron. 30: 117-120.

Menhi Lal and Shukla, N.P. 1988b. Water management in forage crops; in Pasture and forage crop research – A State of Knowledge report PP. 211-220. (Eds. Panjab Singh) Range Mgmt. Soc. of India, Jhansi, India.

Menhi Lal and Shukla, N.P. 1996. Studies on irrigation schedules for oats varieties in relation to Jalshakti. Haryana J. Agron. 12 (1): 4-6.

Om Prakash and Verma, D.N. 1995. Chemical coposition, nutritive value and yield of oat fodder at different stages of maturity. Indian J. Anim. Nutr. 12(1): 41-43.

Pal, R.N., Gill, R.S., Rana, N.D and Negi, S.S. 1971. Composition and yield of forage oat varieties at different stages of growth. Indian J. Agric. Sci. 41: 1088.

Parihar, S.S., Gajri, P.R. and Narang, R.S. 1974. Scheduling irrigation to wheat, using pan evapotranspiration. Indian J. Agric. Sci. 44: 567-571.

Pradeep Behari and Hazra, C.R. 1989. Forage Productivity of Sorghum and Oat evaporational Studies. Forage Res. 15: 162-166.

Reddy, M.G. and Hussain, M.M. 1990. A study on application of nitrogen and cutting management on production of oat. J. Res. APAU 18 (1):60-62.

Sharma, C.B. and Mudgal, V.D. 1966. Studies on the lignin and Cellulose content of fodder crops and effect of lignification on Cellose digestion. Indian J. Dairy Sci. 19:100.

Shaw, F. J. F. and Bose, R. D. 1933. Studies in Indian oats, Indian J. Agri. Sci. 3(5): 754-770, 771-807.

Shukla, N.P., Hukkeri, S.B. and Menhi Lal 1982. Effect of irrigation schedules of different stages of growth on seed production of oat. Indian J. Agri. Sci. 52 (12): 837-840.
Shukla, N. P., Menhi Lal and Ramji Lal 1988. Effect of moisture regimes and nitrogen management techniques on forage yield of oat. Haryana J. Agron. 4:9-13.
Thellung, A. 1928. Die Ubergangsformen Vom Wildhafer-Typus (*Avena agrestes*) Aum Saathafer-Typus (*Avena sativae*). Rec. Trav. Bot. Neerl. 25a:416-44.
Verma, U. K. and Kumar, V. 1984. Effect of moisture and nitrogen levels on the yield of forage oats. Agric. Sci. Digest India. 4:82.

OIL SEED CROPS

10. Groundnut

P. K. Ghosh, Rinku Dey, K.K. Pal and R. K. Mathur

Groundnut is an important oilseed crop. Besides being an important source of vegetable oil, it is also used as an important source of food, feed, nutrition and fodder. The present contribution of groundnut to total oilseed production is about 40 per cent. Assuming that the groundnut will account for a similar proportion to the total oilseeds production by the year 2020 AD, the total demand for groundnut will be about 14 million tonnes. The present production level is around 8.2 million tonnes. Hence to meet the gap of 5.8 million tonnes, a growth rate of about 2.2 per cent is required. This growth has to come mainly from the increase in productivity. This chapter deals in details about the improved package of practices to meet the future challenges ahead.

Groundnut haulms make a very palatable fodder to cattle when fed in green state. It may be fed in the form of hay after drying or after converting it into silage. The haulms are rich in nutrition containing about 8-11% protein. Groundnut kernels have about 25 % protein which is about 1.3 times higher than meat, 2.5 times higher than eggs and 8 times higher than fruits. The oil content of kernels ranges from 40-50 % and is extensively used for cooking purposes. Oil is a rich source of vitamin A, B and E. Groundnut kernels are used in the roasted form for culinary purposes. Peanut milk, butter, curd and *chatani* is prepared from raw peanut kernels.

Groundnut has got numerous industrial uses (Asiedu, 1994). The oil is also used for preparing vegetable oil (*vanaspati*). Low quality oil is used for making soaps. Groundnut oil is used in beauty-aids, shaving cream, cold cream, etc. It is used for medical aids such as plasters, ointments, etc. The groundnut cake powder is used as laundry starch and in the manufacture of paper, gummed tapes and plastics. Activated carbon is prepared from groundnut shells. It is a good foreign exchange earner. Groundnut cake is a very good organic manure (8% N, 1.4% P_2O_5 and 1.2% K_2O). It can be used as a nutritive cattle feed. The cake flour blends easily with wheat and other flours and consequently used by bakers, confectioners, candy makers and ice cream manufacturers.

1. ORIGIN

The groundnut (*Arachis hypogaea* L.), is an important oil, food and forage crop and is grown widely throughout the tropical and subtropical zones of the world. It is a native new world crop. Early explorers found it cultivated extensively in both Meso-America and South America. Remnant pericarp tissue recovered from archaeological sites in Peru dates its purposeful agricultural use there at approximately 3900-3750 years before the present. It's domestication probably first took place in the valleys of the Parana and Paraguay river systems in the Gran Chaco area of South America. All wild species in the genus *Arachis* are found only in South America. The centre of diversity of *Arachis* could be in the Mato Grosso or Planalto profile from Corumba to Joazeiro (Gregory *et. al.*, 1980) in Brazil.

2. DISTRIBUTION

The area, production and productivity of groundnut in India and the world is given below:

TABLE 10.1: Area and production of groundnut of the world

Country	Area (mha)	Production (mt)	Yield (Kg/ha)
India	8.50	8.40	988
China	3.62	9.72	2688
Sudan	1.09	0.81	742
Nigeria	1.48	1.20	1145
Senegal	0.89	0.68	760
Zaire	0.66	0.55	833
USA	0.65	1.94	2995
Total	22.89	31.59	-

TABLE 10.2: Area and production of groundnut in India

State	Area (mha)	Production (mt)	Yield (Kg/ha)
Gujarat	1.99	2.38	1196
Andhra Pradesh	2.17	1.77	816
Tamil Nadu	1.15	1.84	1604
Karnataka	1.21	0.95	791
Maharashtra	0.60	0.63	1043
India	8.50	8.40	988

3. CLASSIFICATION

Cultivated groundnuts are allotetraploids (2n=40) in contrast to wild diploid species (2n=20). The wild species of this section can be hybridized well and through ploidy and genomic manipulations desirable genes can be introgressed into the cultivated form. Gregory *et. al.* (1951) had given the following classification of groundnut on the basis of branching pattern, presence or absence of reproductive axes on main axis and the arrangements of reproductive and vegetative axes on the primary branches.

3.1. *Arachis hypogaea* sub-species *fastigiata*

Erect and bunch type: branching sequential, inflorescence usually present on the main axis and first branching on the cotyledonary laterals are reproductive, seed dormancy usually absent, foliage light green. It includes two varieties:

1. **Var. *fastigiata* called Valencia:** Vegetative branches on primary branches absent or regularly placed at the distal end nodes; inflorescence usually simple.
2. **Var. vulgaris called Spanish:** vegetative branches occasional and irregularly placed; inflorescence compound. B. *Arachis hypogaea* sub-species hypogaea
3. **Semi-spreading or spreading type:** Branching alternate, inflorescence simple and absent on main axis, seed dormancy usually present, foliage dark green, first branch on cotyledonary lateral always vegetative. It includes two varieties:
   a) **Var. hypogaea called Virginia:** some completely prostrate called Virginia runner and some semi-spreading called Virginia bunch; main axis in procumbent forms usually short; stem not very hairy; medium-late in maturity.
   b) **Var. hirsuta called peruvian runner:** habit procumbent, main axis may exceed 1 m, stem fairly hairy, very late in maturity.

All the varieties are cross-compatible. The crosses within a subspecies are called inter-varietal crosses and between varieties in different sub-species are called infra-specific crosses.

4. GROWTH STAGES

Stage Abbreviated stage title	Description
Vegetative stages	
VE	Emergence Cotyledons near the soil surface with the seedlings showing some part of the plants visible
V0	Cotyledons are flat and open at or below the soil surface
V-1	First tetrafoliate One to N developed nodes on the V-(N) to Nth tetrafoliate main axis, a node is counted when its tetrafoliate is unfolded and its leaflets are flat
Reproductive stages	
R1	Beginning bloom one opening flower at any node on the plant
R2	Beginning peg One elongated peg (gynophore)
R3	Beginning pod One peg in the soil which turned swollen, overy at least twice the width of the peg
R4	Full pod One fully-expanded pod to the dimensions characteristic of the cultivar
R5	Beginning seed One fully-expanded pod in which seed cotyledon growth is visible when the fruit is cut in cross- section with a razor blade (Past the liquid endosperm phase)
R6	Full seed One pod with cavity apparently filled by the seeds when fresh
R7	Beginning maturity One pod showing visible natural colouration or blotching of inner pericarp or testa

Stage Abbreviated stage title	Description
R8	Harvest maturity Two-thirds to three-fourths of all developed pods have testa or pericarp colouration. Fraction is cultivar dependent, lower for virginia types.
R9	Over-matured pod One undamaged pod showing orange-tan colouration of the testa and/ or natural peg deterioration

5. CLIMATE

5.1. Growing Seasons

In India groundnut is generally sown in four seasons:

1. ***Kharif*** **(monsoon):** About 85% of groundnut in India is sown in the *kharif* season under rainfed conditions.
2. ***Rabi:*** Groundnut is grown in *rabi* season on a limited scale in areas where winter is not severe and night temperatures do not go below 15°C. This crop is usually cultivated in rice follows to utilize the residual moisture after the harvest of rice or with minimal irrigation (5-9). *Rabi* groundnut is raised in coastal regions, river deltas of Krishna, Godavari and Cauvery and in other irrigated areas in parts of Tamil Nadu, Andhra Pradesh and Karnataka.
3. **Summer:** Grown in 6 states namely Orissa, Tamil Nadu, Andhra Pradesh, Karnataka, Gujarat and Maharashtra under irrigated condition (9-12 irrigations).
4. **Spring:** Uttar Pradesh, Punjab and West Bengal with irrigation

Soil temperature is an important factor and its effects are critical on seed germination, emergence of the seedling, early plant growth, rate of flowering and pod development. When soil temperature goes below 18°C, emergence of seedling is low. The optimum temperature for vegetative growth of groundnut is between 27°C and 30°C depending on the cultivar. Reproductive growth is maximum at 24-27°C. Maximum rate of growth of pods is between 30°C and 34°C.

Groundnut has been found to grow and yield well in areas receiving 50-125 cm of well distributed rainfall. Adequate rains at sowing are necessary for proper germination and good plant stands, and well distributed rainfall during the crop period ensures normal vegetative growth, increased flowering and proper development of pods.

Light influences both photosynthesis and respiration in groundnut. The canopy responds to increased light intensity upto full sunlight. The opening of flower buds and total number of flowers formed depend highly on light.

6. SOIL AND SOIL PREPARATION

The best soils suitable for groundnut cultivation are well drained sandy loam/light soil, with sufficient calcium and moderate organic matter having EC 4 dsm-1 and p^H of 6.0-7.5. Generally country plough and harrows are used for the preparation of soil. In Maharashtra and Karnataka two ploughings and in Tamil Nadu 4-6 ploughings are sufficient for preparing the optimum soil tilth followed by 2 harrowings while in Madhya Pradesh, Gujarat and Andhra Pradesh ploughing is done several times till fine tilth is obtained.

7. CROPPING SYSTEMS

About 60% of *kharif* groundnut is grown as a sole crop, year after year, on the same piece of land in which no other crop is grown. Continuous groundnut cultivation on the same piece of land affects the long time sustainability of the system by exhausting soil fertility, increasing the build up of insect- pests, diseases and weeds and ultimately lowering the yield. A sustainable system will involve the well planned alternate uses of the land through sequential cropping, intercropping, relay cropping etc. keeping groundnut as the central commodity of the system.

7.1. Sequential Cropping

The promising crop sequences being practised in different parts of India are given in Table 1. In general, groundnut does well if planted after grain crops and groundnut-wheat sequence is common mostly in western (Gujarat and Maharashtra) and northern (Punjab, Haryana and Uttar Pradesh) parts of India. In recent years due to frequent shortage of water supply in the command areas, rice-groundnut sequence has gained importance. This system is found to be better than rice-rice sequence from monetory, load of insect pest populations and total water requirement points of view. This system is productive also in residual moisture conditions of the coastal areas of Karnataka, Orissa and Andhra Pradesh. Per unit water used in rice-rice sequence is higher than rice-groundnut sequence. About 60 Kg N and 90 Kg P_2O_5/ha instead of 30 Kg N and 40 kg P_2O_5/ha in case of groundnut grown in fallow fields could be applied in high and medium texture soils to overcome the adverse soil physical environment in the rhizosphere (Singa Rao and Prabhu Prasadini, 1994).

7.2. Intercropping

Area under *kharif* groundnut is not likely to expand any further in India. The most potent way of increasing the area is by finding a place for groundnut as an inter crop. This system gives some safety to farmers against the natural calamities and helps in better utilization of farm resources. As a result of concerted research efforts stable and remunerative intercropping systems with cereals, pulses and other oilseeds have been identified for dry land areas (Table 10.3).

TABLE 10.3: Groundnut-based cropping systems in different regions of country

Soil zone and regions	Water availability period (days)	Double cropping system	Intercropping (with row ratio)
Vertisols and related soils			
Saurashtra (Gujarat)	130-140	Groundnut- wheat	Groundnut + castorbean/ pigeon pea (3:1)
North/central Karnataka	130-150	Groundnut-sunflower	Groundnut + pigeon pea (3 or 7:1)
Southern Maharashtra	160-180	Groundnut-wheat	Groundnut + sunflower (2 or 3:1)

Soil zone and regions	Water availability period (days)	Double cropping system	Intercropping (with row ratio)
Southern Rajasthan	160-180	- do-	Groundnut + pigeonpea (2:2)
Vidarbha (Maharashtra)	190-210	Groundnut – safflower	-
Akola (Maharashtra)	180-190	-	Groundnut+ sunflower (3:1)
Bundelkhand (Madhya Pradesh)	210-230	Groundnut- chick pea	-
Inceptisols and related soils			
Plains of Uttar Pradesh	200-230	-	groundnut + pigeon pea (2:1)
Oxisols			
Subhumid region of Orissa	200-220	Groundnut – horsegram	Groundnut + pigeon pea (5:1)
Subhumid region of southern Bihar	160-180	-	Groundnut + pigeon pea (2 or 3:1)
Humid region of West Bengal	230-260	Rice- groundnut	Rice + groundnut
Humid and Coastal belt of Andhra Pradesh, Tamil Nadu and Karnataka	140-190	Rice-rice-groundnut	-
Alfisols and related soils			
Southern karnataka	190-220	-	Groundnut + pigeon pea (4:1)
Rayalseema (Andhra Pradesh)	110-130		Groundnut +pearl millet (6-8:1-2) Groundnut + castor bean (2:1)

The most promising intercropping systems identified in recent years are, intercropping of groundnut and pigeonpea followed by inter-cropping of groundnut with sunflower and groundnut with cereals.

8. MANURES AND FERTILISERS

Groundnut, being a legume and an oilseed crop, has greater requirement for sulphur and phosphorous. P is a constituent of enzymes, phospholipids and proteins while S is a constituent of S containing amino acids and helps in fatty acid synthesis. Calcium is another important mineral for groundnut production. Groundnut has unique characertistics of absorbing about 75% of S and Ca through the gynophores and developing pods. These important nutrients can be applied either through manures and/ or fertilisers.

8.1. Manures

Organic farming is the most suitable eco-friendly approach in crop production. Farm Yard Manure (FYM) and compost are mainly used as organic manures in India and are known to improve the fertility and productivity of the soils. Well decomposed FYM or compost at 5-10 t/ha, depending upon the types of soil should be applied one month before sowing and be incorporated into the soil. Application of 10 t FYM along with 25 Kg N and 50 Kg P_2O_5 could improve the yields of summer groundnut in vertisols i.e., medium black soil (Ghosh, 1997). But, in alfisols of Andhra Pradesh substantial improvement in pod yield of groundnut has been recorded with FYM at 5 t/ha (Anonymous, 1993) and could save 10-15 Kg N, 20 Kg P_2O_5 and 25 Kg K_2O/ ha. In acid soil use of rock phosphate charged with FYM is beneficial compared to FYM alone.

8.2. Fertilisers

Response/Nutrient removal: Groundnut removes on an average 170, 30, 110 and 39 Kg of N, P, K and Ca per ha, respectively, but the quantities vary with soil type, fertility status, location and yield level. Groundnut showed response of 5.6 to 16.9 Kg pods/ Kg N, 3.9 to 11 Kg pods/Kg P and 3.0 to 9.2 Kg pods/Kg NPK under different situations and soils (Pasricha and Tandon, 1990). The highest response of P was observed upto 60 Kg P_2O_5/ha. There are number of reports about the non-responsiveness of groundnut to P application even in apparently low P level. Response of groundnut to K application are not consistent but marked increase in response to K has been reported because K deficiency or depletion of reserve K increases with intensive cropping. The highest response of K was upto 67.5 Kg/ ha (Pasricha *et. al.*, 1988). Out of the total nutrients absorbed by the groundnut crop, the kernel contained 41% N, 52% P, 28% K, 11% Mg and 1% Ca. The leaves and stalks contain 45-55% of the total NPK absorbed (Kanwar *et. al.*, 1983; Ghosh and Mathur, 1996).

8.3. Macro Nutrients

TABLE 10.4: The recommended doses, sources, methods of fertiliser application

Nutrient	Doses	Sources	Methods	Stages/ time
N	Requirement is not high due to Biological Nitrogen Fixation. Only starter dose of 10-20 Kg (rainfed)/ 10-30 Kg/ ha (irrigated)	Ammonium sulphate is preferred to urea due to S content	Furrow or Broadcast	Two splits (Basal & top dressing at 25 DAS
P	i.) 20-60 Kg P_2O_5 depends on soil type (large variation due to differences in soil properties, moisture regimes, source and rotation ii) higher level on high P fixing soils	SSP is preferred to DAP and TSP due to 12% S and 19.5% Ca contents	Drilling below the seed	Basal
K	60-67.5 Kg K_2O/ ha	Muriate of potash	Broadcast	Basal
Ca	40 Kg/ ha (to provide atleast 1 meq exchangeable Ca/ 100 g soil in the root zone & 3 meq/ 100 g soil in pod formation zone	Gypsum (@ 125-500 Kg/ ha to meet required quantity of S & Ca)	To be applied to 1/2 basal the base of the 1/2 at pegging plant/ pod zone to facilitate Ca diffusion from soil solution to developing pods	

Nutrient	Doses	Sources	Methods	Stages/ time
S	30-40 Kg/ ha	Ammonium sulphate, Gypsum, elemental S Pyrite, SSP	Pyrites should be Sowing, flower applied 2-3 weeks initiation & in advance	pod formation

8.4. Micro Nutrient

Soils of Saurashtra region of Gujarat and Rayalseema region of Andhra Pradesh, the two most important groundnut growing areas in India, generally suffer from zinc deficiency. The deficiency of Zn is also common in Gujarat, Andhra Pradesh, Tamil Nadu and Punjab. Molybdenum and Boron deficiencies are acute in acid soil. Soils of Tamil Nadu, Maharashtra and Punjab have also been found deficient in boron. Iron chlorosis in groundnut has been reported from all the major groundnut growing areas especially in calcareous soils having free carbonate and bicarbonate. The details of rectification of the Micronutrients deficiency are given below:

TABLE 10.5: Micro nutrient forms and their application rates

Micro nutrient	Forms	Application rate	
		Soil	Spraying
Iron	Ferrous sulphate	10 Kg/ ha	0.15% $FeSO_4$ + 0.9% Citric acid
Boron	Borax	5-20 Kg/ha	0.2% Borax
Zinc	$ZnSO_4$	15-20 Kg/ha	0.5% $ZnSO_4$ + 0.25% lime
Molybdenum	Sodium molybdate	-	0.4 g/ Kg seed as seed dressing

8.5. Balanced Fertilization

Application of nutrients in balanced proportion is essential for obtaining higher yields considering the antagonistic and synergistic effect among the nutrients. The effect of S × Zn in groundnut was positive and a combined application of 15 Kg S/ha and 10 Kg Zn/ha gave maximum yield of 26 q/ha (Pasricha and Aulakh, 1986; Pasricha *et. al.*, 1988). S and P has positive interaction in groundnut. A basal application of 40 or 60 Kg P_2O_5 and 45 or 50 Kg S gave maximum pod and oil yield and protein contents in groundnut (Baldev Singh *et. al.*, 1993; Dimree and Dwivedi, 1994). P and K had antagonistic effect but Ca and K had synergistic effects in groundnut. Application of K beyond 60 Kg K_2O/ha significantly reduces uptake of N, P, K, Ca and Mg (Yakadri and Satyanarayana, 1995). Interaction of Zn and Mo in groundnut was synergistic. The maximum pod yield of 4960 Kg/ha was obtained at Zn 5.6 Kg/ha + sodium molybdate as seed dressing at 2 g/ Kg seed (Gowda *et. al.*, 1995).

8.6. Biofertiliser

With ever increasing cost of chemical fertilisers, greater emphasis has been given on the use of alternative and renewable sources of fertiliser. Rhizobium inoculation is beneficial in areas in which groundnut is introduced for the first time. An increase of 15-30% of groundnut yields in summer and 13-53% in *kharif* groundnut has been obtained with Rhizobium inoculation (Wani and Lee, 1992). Two Bradyrhizobium strains IGR 6 and IGR 40 increased the pod yield of bunch groundnut by 9-18% in *Rabi*/summer (Joshi *et. al.*, 1989). Mycorrhizae especially vesicular arbuscular mycorrhizae (VAM) have been found to increase groundnut yields by 20-30% (Subha Rao, 1994). *Glomus fasciculatum* was reported to be the best among the VAM fungi for groundnut.

9. VARIETY

Prior to 1980, breeding efforts were directed mainly towards the improving the yield per se. With the identification of resistance sources to major diseases and insect pests, resistance breeding received a strong stimulus resulting in identification of tolerant varieties. The special features of the recently released important groundnut varieties at national and zonal levels are listed in Table 10.6.

TABLE 10.6: Important groundnut varieties of India

Name of Variety	Habit group	Pod yield (Kg/ha)	Shelling (%)	100 kernel Weight	Oil (%)	Salient features
National release						
M 13	Virginia Runner	2750	68.0	58.1	49	Dark green waxy leaves, large size pod (HPS type), prominent reticulation and medium constriction, leaf spot tolerant
JL 24	Spanish	1800	75.0	53.7	50.7	Large dark green leaves, Early (95 days), Bunch smooth pod, compact bearing widely adopted
ZONE I (NORTHERN ZONE): Uttar Pradesh, Punjab, Haryana, Bihar and Northern Rajasthan (Jaipur, Ajmer, Bharatpur, Sawai Madhopur and Tonk districts).						
Bihar						
BG 3	Virginia Bunch	2500	72.0	52.0	49.0	Early maturity
BAU 13	Virginia Bunch	2191	63.0	71.0	48.0	Bold seeded HPS for export trade
Punjab						
M 335	Virginia Runner	2300	67.0	71.0	49.0	Plant has less lateral spread, leaflets large and dark green colour, pod bold with 1-2 seeded & rarely 3- seeded. Reticulation prominent, constriction medium, kernels bold with light brown testa
SG 84	Spanish Bunch	2062	64.0	49.0	50.0	Plant erect with medium size, dark green leaves, 2-seeded pod with medium constrict-ion and slight beak, testalight brown

Name of Variety	Habit group	Pod yield (Kg/ha)	Shelling (%)	100 kernel Weight	Oil (%)	Salient features
Rajasthan						
Somnath	Virginia Runner	1900	72.3	62.0	50.1	High Yielding, early maturing, bold seeded of HPS grade for rainy season
CSMG 84-1	Virginia Runner	2704	64.0	48.0	50.4	Tolerant to thrips, termites, pod borer and foliar diseases
RSB-87	Semi-spreading	1400	66	-	51.7	Long pods with three dark red kernels, recommended for zone IIA and III A of Rajasthan
RG-141	Spanish Bunch	2000	72-74	-	50.4	Medium size pods with shallow constriction and less prominent back. Two kernel. Good for heavy soils.
ZONE II (WESTERN ZONE): Gujarat, S. Rajasthan (Chittorgarh, Udaipur & Jhalawar districts).						
Gujarat						
Somnath	Virginia Runner	1900	72.3	62.0	50.1	Early maturing, bold Seeded of HPS grade for rainy season
GG 3	Spanish Bunch	1283	68.7	32.0	51.0	Early maturity, green foliage at harvest stage
ZONE III (CENTRAL ZONE): Madhya Pradesh (excluding Mandsour and Ratlam district) and Maharashtra (excluding Satara, Sangli, Solapur, Kolhapur, Osmanabad and Nanded districts)						
Maharashtra						
TKG 19A	Virginia Bunch	2260	63.0	60.5	46.0	Bold and attractive kernels of HPS grade
B 95	Virginia Bunch	3345	68.0	90.0	48.0	Suitable for Summer cultivation, extra bold seeds of HPS grade
TG 26	Spanish Bunch	1596(R) 2425(I)	65.0	35.0	49.0	Early maturity, fresh seed dormancy
ZONE IV (SOUTH-EASTERN ZONE): Orissa and N. Andhra Pradesh (Srikakulam, Vishakapatnam, and Vijayanagaram districts)						
Orissa						
Ak 12-24	Spanish Bunch	1250	75.0	29.4	48.5	Wider adaptability, suited for medium to heavy soils
Jawan	Spanish Bunch	1230 (R) 2000 (I)	66.3	36.5	47.6	Elongated pod, slightly beaked, rose kernel
ZONE V (PENINSULAR ZONE): Karnataka, Andhra Pradesh (excluding Srikakulam and Vishakapatanam, districts) and S. Maharastra (Satara, Sangli, Solapur, Kolhapur, Osmanabad and Nanded districts)						
Andhra Pradesh						
DRG 17	Virginia Bunch	2100	68.0	49.0	49.0	Suitable for rainfed conditions, tolerant to moisture stress
DRG 12	Spanish Bunch	2604	69.0	39.0	47.0	Suitable for summer cultivation, tolerant to foliar diseases and BND

Name of Variety	Habit group	Pod yield (Kg/ha)	Shelling (%)	100 kernel Weight	Oil (%)	Salient features
Karnataka						
KRG 1	Spanish Bunch	1200 (R) 2230 (I)	73.0	42.0	49.0	Suitable for *kharif* and summer seasons
Dh 8	Spanish Bunch	3200	71.7	33.0	46.0	Dark green leaves, compact plant, tolerant to late leafspot, smooth seeded rose round at one end sharply pointed at the other
ZONE VI (SOUTHERN ZONE): Tamil Nadu and Kerala						
Tamil Nadu						
VRI 2	Spanish Bunch	1500	74.0	50.0	48.0	Bold seeded, slightly constictions suitable to both rainfed and irrigated
VRI 3	Spanish Bunch	1668	75.0	33.0	48.7	Pods and seeds are small. Pods have slight to moderate constriction withlittleor no beak

10. DORMANCY

Varietal differences have been found in groundnut for the presence or absence of fresh seed dormancy. In groundnut it is present in Virginia types while in Spanish and Valencia types, it is absent. Even different lots of the same variety may show differing dormancy characteristics depending on production conditions. Dormant seeds may lose dormancy if digging is delayed and the pods are allowed to remain in the soil after maturity. Usually, the presence of dormancy in groundnut is not much of a problem as it will usually wear off in storage before the next planting season. But, the lack of dormancy in Spanish and Valencia types can cause vivipary in some areas when rain delay harvest. Dormancy in seeds may be attributed to one or the other of the following reasons, viz., impermeability of the seed-coat to water, mechanically resistant seed-coats, incompletely developed or rudimentary embryos, dormant embryos, etc. The seedcoat imosed component of groundnut seed dormancy involves water soluble germination inhibitors and, in some cases, resistance to initial water uptake by the dry seed (Bhapkar *et. al.*, 1986). Removal of seed coat improves germinability of many maturity classes. The evolution of ethylene by the embryo is an essential pre-requisite for germination. It has been estimatedt hat groundnut seeds must be able to produce 2-3 nl of ethylene/g fresh weight in order to break dormancy (Ketring and Morgan, 1972). Lower level of hydrolases like amylase and protease during seed development may be responsible for seed dormancy. A nondormant groundnut line possessed higher level of amylase and protease during seed development than a dormant cultivars (Sharma and Sengupta,1988).

11. SOWING

11.1. Sowing Time

1. ***Kharif:***
   a) **Normal:** Second fortnight of June to first fortnight of July depending upon the onset of monsoon.
   b) **Pre-monsoon:** Pre-monsoon sowing with 1-2 irrigations was found good. In Gujarat, advancement of sowing in the first week of June increased pod yield by 46% and in Punjab, shifting of sowing date from normal time of first week of July to 20th June increased the pod yield by 19% (Devi Dayal *et. al.*, 1989).
2. ***Rabi:*** Crop is sown from September to December depending upon the harvest of rice.
   a) **Summer:** The crop is sown from the second fortnight of December to first week of February depending upon the temperature.
   b) **Spring:** Crop is sown from second fortnight of February to first week of March after harvest of *toria* and potato.

Groundnut suffers from yield reduction due to delayed sowing and it is more pronounced in *rabi* season as compared to *kharif* season. In Rajasthan, Uttar Pradesh, Orissa and West Bengal low soil temperature (°C) during *rabi* season delays germination and high temperature at pod filling stage interferes with pod development. Research conducted at the National Research Centre for Groundnut, Junagadh, showed that application of chopped wheat straw at 5 t/ha on the soil surface immediately after sowing of groundnut raised soil temperature by 2-3°C at seedling emergence and lowered the soil temperature by 3-5°C during pod development phase. Groundnut crop under wheat straw mulch, thus, maintained good vigour and growth and ultimately gave 20-24% higher yield than control plot (Ghosh *et. al.*, 1997b).

11.2. Sowing Implements

Bullock drawn seed drill and sowing behind the country plough are the traditional implements generally used by the farmers. These implements can maintain only row to row spacings. Improved tractor drawn groundnut seed drill developed by Tamil Nadu Agricultural University can maintain both row and plant spacing.

11.3. Sowing Techniques

11.3.1. Conventional

Groundnut is generally sown in flat beds with 30 x 10 cm spacing. In Saurashtra region of Gujarat, set-furrow system of groundnut cultivation is still a practice by the farmers. In set-furrow system farmers use same furrow (90 cm) year after year for groundnut cultivation. Putting of murram and sand in the set-furrows is also in practice.

11.4. Improved Techniques

Some of the improved techniques are as follows:

1. **Criss-cross Pattern:** In this method, total seed lot is divided into two equal halves, first half of the seed is sown in one direction adopting recommended row to row spacing and then remaining half in perpendicular to first direction by adopting same row spacing. It maintains optimum plant population because of uniform distribution of seed. About 18% yield improvement was realized (Devi Dayal *et. al.*, 1994) and may be useful mostly where groundnut succeeds rice.
2. **Paired row:** In this system, two pairs of rows are spaced at 45/60 cm with a spacing of 22.5/30 cm within the pair. About 20% yield improvement was realized (Devi Dayal and Reddy, 1987).
3. **Broad bed and furrow method:** This method is useful for high rainfall areas having deep vertisol where drainage of excess water is a problem. The moisture, thus stored in the furrows, could be utilized effectively. On an average 15% higher yield of groundnut has been recorded over flat bed method (ICRISAT, 1993).

11.5. Depth of Sowing

Generally 2.5 to 5 cm is found optimum (Ghosh *et. al.*, 1997a). If the seed is sown in deep (5 cm), germination will be badly affected.

11.6. Seeds

1. Seeds should be pure, viable and free from diseases.
2. Seeds should have minimum germination of 70 per cent.
3. Seeds should be sown after shelling the pods, not the pods as such since it results in delayed and partial germination.
4. After shelling, all the damaged seeds should be rejected before sowing.

In general, pod yield, germination and seedling vigour index are higher for larger seed (10 mm) than medium (8-10 mm) and small seed (7-8 mm) of grounnut (Dorate *et. al.*, 1995).

11.7. Seed Treatment

For the control of seed borne diseases, treatment of groundnut seeds with thiram (3 g/ Kg seed), Mancozeb (3 g/ Kg seed) and Carbendazim (2 g/ Kg seed) is recommended.

11.8. Seed Rate and Spacing

Use of optimum seed rate is the key factor for maintaining recommended plant population. In summer season, mostly the bunch varieties are grown. The exact seed requirement varies with seed size and row spacing. The seed rates of varieties of different habit groups and the corresponding plant density and spacing are given below:

TABLE 10.7: Recommended seed rate, spacing and plant density in groundnut

Types	Seed rate (Kg/ha)	Plant density (Lakhs/ha)	Spacings (cm) between	
			Plants	Rows
Bunch	100	3.33/4.44	10	30/22.5
Semi-spreading and spreading	100-110	2.22	10/15	45/60

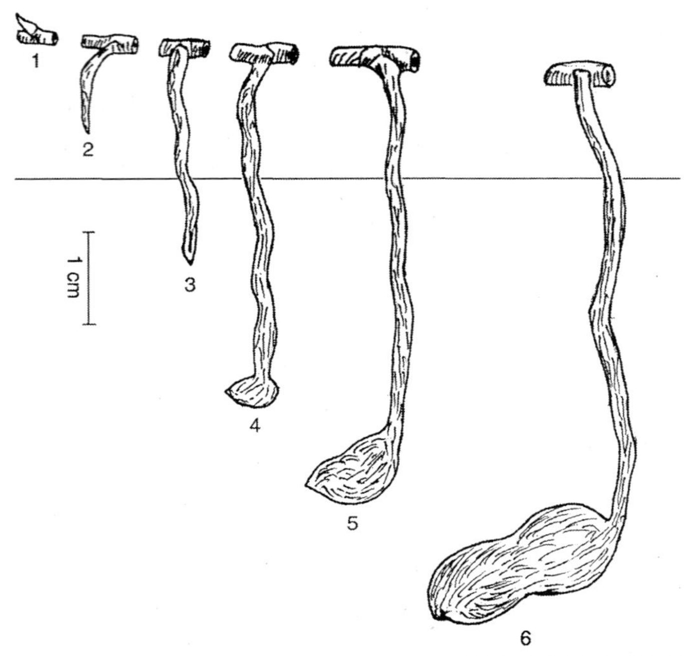

FIGURE 10.1: Process of pegging and pod development. (1) ovary after fertilization, (2) aerial pegs 5-7 days, (3) peg penetration soil 18-22 days, (4) pod enlargement 14-21 days), (5) early stage in pod development, (6) immature pods (after Smith, 1950).

12. PROCESS OF PEGGING

After fertilization is complete, the flower withers rapidly and over a period of several days the fruit develops as a positively geotropic stalk like structure, the `peg' which eventually penetrates the soil. The cells which elongate comprise the basal tissue of the ovary itself and thus a more appropriate term for the peg is a carpophore. At first the peg extension is slow, taking 5-6 days to penetrate the bracts. When the fruit reaches 3-4 mm in length it becomes positively geotropic and starts to grow towards the soil, the rate of elongation increasing rapidly between 5 to 10 days after fertilization. Penetration of the soil occurs after 8-14 days of fertilization and growth usually continues till then. The mechanical stimulus of the peg entering the soil is important for successful pod development, stimulating ovary wall thickening and the diageotropic orientation of the pod due to rapid cell enlargement on the dorsal side of the carpel near the basal ovule (Zamski and Ziv, 1976). The basal ovule develops first. On reaching the maximum depth, the peg become diageotropic and almost horizontal. The pod development starts when the peg ceases to elongate. Pod enlargement proceeds from base to apex

12.1. Factors Affecting Pegging

Genetical: The number of pegs developed are higher in the bunch types than the spreading and semi-spreading types. The cultivars of sub sp. *fastigiata* (valencia) has thicker pegs than those of sub sp. *hypogaea* (virginia runner)

12.1.1. Environmental

1. Daily minimum mean temperature is positively correlated with peg formation but maximum temperature has a negative effect
2. High light intensity affects pegging
3. The process of pegging may stop and the already formed pegs may dry if the soil moisture drops to wilting point

12.1.2. Others

1. Specific advantages of early formed flowers: Early formed flowers may inhibit the development of other flowers into pegs
2. Reduction in embryo growth during peg elongation may simply be a function of the rapidly growing carpophore. Hence, initial proembryo development is an essential prerequisite for peg elongation. Auxin and GA produced by the proembryo promotes its growth.
3. Length of pegs is related to the differences in branching pattern. In wild species, the length of the peg is highly variable and may be up to a metre in some species
4. The strength of the peg is an agronomic attribute governing the case of harvesting of the crop

13. IRRIGATION

13.1. *Kharif* Season

Kharif groundnut is often subjected to various weather aberrations viz., (i) early-season drought associated with late sowing (ii) mid-season drought due to failure of rain during peak growing period and iii) late-season drought due to early cessation of monsoon. If groundnut receives total rainfall of 400-600 mm, well distributed over the entire growth period it is sufficient to harvest a good crop. But very often non-availability of sufficient moisture at pod development stage (in September) is common in many places. Providing life saving irrigation at this stage may result in increased yield to the tune of 33-63 % (Devi Dayal *et. al.*, 1989).

13.2. Scheduling of Irrigations

Based on critical stages: Maintenance of optimum soil moisture at critical growth stages is the key factor for releasing higher yields. Early vegetative growth, flowering, pegging and pod formation are considered as critical stages, of which peak flowering and pod formation stages are the most crucial. Imposing drought of 20 days, after 15-20 days of sowing (with pre-sowing irrigation) followed by releasing water stress providing two irrigations at five days interval helps in the development of deeper root system, synchronized flowering, higher biomass production and higher pod yield. Irrigation should be stopped one week before harvest so that the quality of seeds does not deteriorate.

Based on irrigation intervals and calculation: Irrigations are scheduled at fixed intervals according to the soil and climatic conditions. Total amount of water and number of irrigations required for groundnut vary with locations and soil types (Table 10.8. In general *rabi*-summer groundnut requires 8-18 irrigations. First irrigation should be given at 15-20 days after sowing and second at 20-25 days after first irrigation. Subsequent irrigations should be given at 7-10 days interval.

Based on depletion of soil moisture: Irrigation should be given when 25 per cent depletion of available soil moisture (DASM) occurs but when water supply is crucial, irrigation should be scheduled at 50 % DASM in 0-15 cm of soil surface through sprinkler system (Devi Dayal *et. al.*, 1989).

Based on IW/CPE: Irrigation water/cumulative pan evaporation (IW/CPE) ratio for groundnut ranges from 0.6 to 1.0. Ramachandrappa *et. al.*, (1993) reported that irrigations should be scheduled at 0.5 IW/CPE during 10-40 DAS and later on at 0.75 IW/CPE ratio to realize higher pod yield but further increase in irrigation efficiency up to a ratio of 1.0 did not show significant yield advantage (Tiwari and Dhakar, 1997).

TABLE 10.8: Water requirement of groundnut for different regions and soils types in India

State/Place	Soil types	Growing season	No. of irrigation	Water requirement (mm)
Bhavanisagar (Tamil Nadu)	Red sandy loam	Rabi	9-13	500-700
Madurai (Tamil Nadu)	Sandy loam	Rabi	8	480
Hyderabad (Andhra Pradesh)	Sandy loam	Rabi	8	655
Tirupati (Andhra Pradesh)	Sandy loam	Rabi	17	420
Chiplima (Orissa)	Sandy loam	Rabi	11	620
Chaturi (Orissa)	Loamy sand	Rabi	10	690
Akola (Maharashtra)	Medium black	Summer	18	950
Parbhani (Maharashtra)	Clay loam	Summer	8	550
Mysore (Karnataka)	Sandy loam	Rabi	-	498
Junagadh (Gujarat)	Medium black	Summer	11	270
Ludhiana (Punjab)	Sandy loam	Kharif	2	140
Kanpur (U.P)	Sandy loam	kharif	2	150

13.3. Irrigation Methods

Following methods of irrigation are used for groundnut:

1. **Border strip method:** Strips should be of 3-5 m width and water from main channel is diverted into border strips. This is one of the most suitable surface irrigation method and labour cost for forming bund is less than check basin method.

2. **Check basin method:** When fields are undulated this method is followed. Water is diverted into the sub-channels and then finally to sub-plots.
3. **Furrow method:** This method is the most effective with maximum water use efficiency of 3.71 Kg/ha-mm and also saves 2-3 irrigations compared to border strip and check basin methods (Kathmale and Chavan, 1996).
4. **Sprinkler and drip method:** Use of sprinkler and drip irrigation methods are becoming popular since water requirement in these methods is about half and the water use efficiency is also quite high but the only drawback is the high initial cost involved. An yield advantage of 32 % over the check basin method was realized with the sprinkler irrigation system (Devi Dayal *et. al.*, 1989).

13.4. Drainage

Good drainage should be provided for realising good yields of groundnut. Present practice of flood irrigation aggravates problem and also makes it difficult for the penetration of pegs into the soil. Groundnut needs aeration in the rooting and pegging zones. Inadequate aeration results in abnormal respiration, which in turn inhibits root growth and retards metabolic functions. Besides, the plants become chlorotic due to deficiency of N and or S in coarse textured soil or Fe in calcareous soil in oxygen deficient environment.

14. WEED CONTROL

Weed infestation in groundnut is one of the main factors for loss in yields to the tune of 13-80 per cent (Yaduraju *et. al.*, 1980; Kalaiselvan *et. al.*, 1989). Erect and bunch types suffer more than spreading types. Similarly, *Kharif* groundnut is affected more than *rabi*-summer groundnut (Singh *et. al.*, 1987). The weed competition at early stage of groundnut is maximum because of initial slow growth but it is relatively low at the end of expansion of phenophases where complete ground cover is attained by groundnut canopy. The critical period of crop-weed competition was reported to be between 28 and 45 days after sowing (Singh *et. al.*, 1987; Hamada, 1988). The major weeds generally found in groundnut fields are: *Cynodon dactylon*, *Echinochloa colonum*, *Dactyloctenium aegyptium*, *Digitaria sanguinalis*, *Sorghum halpense*, *Echinochloa crusgalli* (among monocots) and *Amaranthus viridis*, *Celosia argentea*, *Euphorbia hirta*, *Tridex procumbens*, *Abutilon indicum*, *Phylanthus niruri* and *Chenopodium album* (among dicots). The secretions of roots of the weeds like *Cirsium arvense*, *Abutilon indicum*, *Chenopodium album* etc. could be a serious impediment in the early establishment of groundnut and thereby, could limit the absolute realisation of groundnut yields if not controlled timely. Cultural method of weed control is a common practice in groundnut. Generally, one hand weeding at 30 days after sowing (DAS) or two weedings at 15 DAS and 35 DAS control(s) weed but it is time consuming, expensive and tedious. Herbicides found effective against weeds and tested in all parts of the country are given below:

Name	Doses	Mode of application	Remarks
Alachlor	1.5 Kg/ha	pre-emergence (shallow incorporation i.e. 5 cm if no rain within one week after application)	controls mostly monocots
Nitrofen	1.5 Kg/ha	pre-emergence (apply just after sowing)	controls mostly monocots
Fluchloralin	1.5 Kg/ha	pre-plant incorporation (5 cm deep immediately after application)	controls both monocots and dicots
Trifluralin	1.5 Kg/ha	pre-plant incorporation (apply one day prior to sowing)	controls both monocots and dicots
Pendimethalin	1.0 Kg/ha	pre-emergence (apply within 24 hours)	Mostly grassy weeds
Dual	1.5 Kg/ha	post-emergence (Apply at 20-25 DAS)	Mostly grassy weeds
Fluzy-flop-p-Butyl	0.5 Kg/ha	post-emergence (Apply at 20-25 DAS)	Mostly grassy weeds

Use of proper doses of these herbicides generally do not hamper nodulation in groundnut. But continuous use of herbicides might result in soil and ground water pollution in long run. The integrated weed management thus offers a suitable alternative. When the above herbicides are combined with one or two hand weedings at 30 and 45 DAS controlled weeds efficiently and recorded higher pod yield of groundnut (Kori *et. al.*, 1997; Tiwari *et. al.*, 1997)

15. PLANT PROTECTION

The major insect pests and diseases, symptoms and their control measures are presented below:

Insects/diseases	Symptoms	Control measures
Insects		
Leaf miner (*Aproaerema modicella* Deventor)	The young larvae mine into the leaves as soon as they hatch from egg. If mines are opened, the minute caterpillar can be seen inside. When they become too large to occupy the mine, the larvae emerge and web adjacent leaflets together, and continue to feed on leaf tissue from inside the webbed leaves. A severely attacked field looks `burnt' from a distance. Leaf miners are favoured by the hot dry conditions of the post-rainy season.	Dimethoate at 200-250 ml./ha or monocrotophos at 150-200 ml at. per ha. Chemical control is recommended if five or more active larvae per plant are found at 75 DAS or later.
Aphids (*Aphis craccivora* Koch)	Nymph and adults suck the sap from the tender growing shoots, flowers, and pegs, causing stunting and distortion of the foliage and stems. Under drought conditions, when the crop is less than 30 days old, all terminal buds are infested with aphids.	Dimethoate at 200-250 ml/ha
Thrips (*Scirtothrips dorsalis* Hood) (*Caliothrips indicus* Bagnall) (*Frankliniella schultzei* Trybom)	Thrips live in the flowers and folded leaflets of groundnut. Nymph and adults suck sap from the surface of the leaflets. This results in white patches on the upper and necrotic patches on the lower surface of the leaves. Injury is normally seen in seedlings. In severe infestations, particularly in the *rabi* crop, leaf distortion causes stunted plants.	Dimethoate at 200-250 ml/ha if there are more than five thrips/terminal leaf, only up to 30 DAE in the post-rainy season

Insects/diseases	Symptoms	Control measures
White grub (*Holotrichia* sp)	The young grubs feed on rootlets and nodules, the older grubs in third instar devour the entire tap-root and such plants wither and dry. The maximum damage occurs during August and September, in patches in field and the crop nearer to the hosts of adult beetles suffers heavy damage.	Spray their feeding tress with carbaryl 50 WP (2 g/L) 3-4 times until mid-july.
Red hairy caterpillar (*Amsacta albistriga* Walk *Amsacta moori* Butler)	Young larvae feed gregariously on the under surface of leaves by scraping them. This feeding continues for 4-5 days and the dried scraped-patch of the leaves can be easily detected. As larvae grow they disperse and feed individually by devouring leaves, flowers and growing points. When a large number of caterpillars infest the crop, only the bare stems of plants remain resulting in heavy and occasionally a total yield loss.	During the first 50 DAE, if defoliation exceeds 25% or if one or more larvae per plant is found then apply Endosulphan at 350 ml/ha or fenvalerate at 100 ml/ha.
Diseases		
Early leaf spot (*Cercospora arachidicola*)	Lesions are sub-circular, dark brown on the upper leaflet surface and also on petioles, stems, stipules and pegs. When the disease incidence is severe, affected leaflets become chlorotic, then necrotic and lesions often coalesce resulting in premature senescence ad shedding of leaflets.	1. Bavistin at 0.05% + Dithane M-45 0.2% sprayed at 2-3 weeks interval 2. Groundnut + pigeonpea/bajra intercropping with spray of Carbendazim (0.05%) and Mancozeb 0.02% at 55 and 70 DAS 3. Carbendazim 0.05% + Mancozeb 0.2% + neem leaf extract (2%) + Penicillium islandicum at 40,55 and 70 DAS
Late leaf spot (*Cercosporidium personatum* Berk & Curt. Deighton)	Leaf spots are nearly circular and darker than early leaf spot. Occurs on the lower leaflet surface. When the disease incidence is severe, the affected leaflets become chlorotic, then necrotic and lesions offen coalesce resulting in premature senescence and shedding of leaflets. Late leaf spot attack usually coincides with that of rust.	1. Bavistin at 0.0-5% + Dithane M-45 0.2% sprayed at 2-3 weeks interval 2. Groundnut + Pigeonpea/ bajra intercropping with spray at 55 and 70 days after sowing 3. Carbendazim 0.05% + Mancozeb 0.2% + neem leaf
Collar rot (*Aspergillus niger* van *Tieghem*)	Affected tissue is covered by the sporulated fungous at the soil surface. Infection commonly occur in cotyledonary hypocotyle region shortly after emergence. The infected area become water soaked and light brown and soon covered with black fungal spores. This disease occur mainly in a moist soil environment.	Seed treatment with 5 g Thiram or 3 g Dithane M 45 or 2 g Bavistin/ kg seed

Insects/diseases	Symptoms	Control measures
Stem Rot (*Sclerotium rolfssisacc.*)	Yellowing and wilting of branches near the base of the plant, and environment of the seed of white mycelium at or near soil line around the affected area of the stem are the symptoms. Crop wither independently or together with stem infection. Brown coloured lesions may be found on the pods.	Seed treatment with 5 g Thiram or 3 g Dithane M 45 or 2 g Bavistin per kg seed

16. HARVESTING

Groundnut being an indeterminate crop, plants do not show much visual symptoms of maturity on the foliage. The best way to judge the maturity of groundnut crop is to assess the pod maturity by the inside colouration of shell. Care should be taken to harvest the crop when it is fully mature. As a thumb rule dig plants when 75 % of the pods are fully mature. The important symptoms of maturity are:

1. Yellowing of foliage
2. Spotting of leaves
3. Dropping of old leaves
4. Hardening and toughness of pods
5. Dark tannin discolouration inside the shell
6. Unwrinkled kernels and colouration of testa

Apart from pod loss during harvest, pre-mature or delayed harvest affects the quality and viability of the seeds. In general, spreading varieties are harvested between 140-145 days, the semi spreading between 120-135 days and bunch types between 90-110 days after sowing. Irrigations, if given to the crop, must be stopped at least 8-10 days before harvest.

16.1. Methods of Harvest

The bunch and semi-spreading types are generally harvested by uprooting either by hand, bullock drawn blade harrows or by tractor-drawn diggers, while the spreading types are harvested by digging out the plants with the help of khurpi or spade or bullock drawn harrows. Stripping of pods is done manually, or by using comb type hand stripper or pedal-operated stripper.

17. YIELD

The yield of groundnut is controlled by several factors like the distribution of rainfall during the crop period, soil type, management and plant protection practices followed and the genotype. The yield of an irrigated crop in summer is almost double of the rainfed *kharif* crop. The higher yields in summer irrigated crops are due to controlled moisture, bright sunshine, better utilization of nutrients and less pests and disease incidence. In India, the average productivity of groundnut is about 1042 Kg/ ha. The yield of haulms is usually 2 to 2.5 times to that of pod yield.

18. QUALITY

The important factors that determine the quality of groundnut are volume weight of pods, shelling out-turn, 100 kernel weight, oil content, free fatty acid content, colour of pods, colour of testa, composition of fatty acids and amino acid composition of protein.

18.1. Volume Weight of Pods

It is the weight per unit volume of the pods and indicates the maturity and development of kernels. Varieties with small pods have higher volume weight than those with bigger pods.

18.2. Shelling Out-turn

It is the percentage of kernels to pods by weight. It is dependent upon the thickness of the shell, and development of the kernels.

18.3. 100 Kernel Weight

It gives an indication of proper filling of kernels. It is a varietal character and is also dependent on nutrition management of the crop.

18.4. Oil Content

The oil content is dependent on temperature during the first three weeks of pod development, maturity of the kernels and S content in the pod zones. High temperature favours high oil content. It also varies with the genotypes as well as cultivars.

18.5. Free Fatty Acid Content

It determines the keeping quality of the oil. The higher limit of the free fatty acid allowed is 2.5% and if it is higher the oil deteriorates faster. Harvesting immature pods and incomplete drying before storage leads to higher fatty acid content in oil.

18.6. Colour of Pods and Kernels

Light golden-yellow colour of pods and rose testa of kernels are preferred. The colour of the testa is a genotypic character but the pods must be harvested at proper maturity for the testa to attain the natural colour.

18.7. Aflatoxin Content

Aflatoxin production in groundnut caused by the *Aspergillus* flavus group of fungi takes place mainly during post harvest processing and storage. In India 81 % of the kernels are immediately crushed for oil and thus, aflatoxin is not a problem in the domestic market. However, aflatoxin is a concern for exportable commodities i.e., kernel, groundnut cake etc. The permissible limit of aflatoxin content is six ppb. Recently, bold- to medium bold-seeded genotypes resistant to invasion, colonization and production of aflatoxin by *A. flavus* have been identified for export purposes (Basu and Ghosh, 1996).

19. POST HARVEST TECHNOLOGY

Drying: After harvesting, the plants with intact pods should be left for 1-2 days either in heaps within the field or in any shaded areas or in different drying structures. Pods should be dried in sunlight for 3-4 days by spreading in a thin layer on a threshing floor. The proper drying of the produce may be judged from the following tests:

1. The pods should give rattling sound when shaken.
2. When a kernel is pressed between the thumb and index finger, it should easily split into 2 cotyledons.
3. When the surface of the kernel is rubbed hard a portion of the testa should come off.

Processing: Groundnut is processed into many kinds of products like:

- **Peanut butter:** It is a paste prepared from dry-roasted, clean and sound groundnut kernels from which the coats and germs are removed and salt, hydrogenated fat and sugar, anti-oxidants and flavours are added to improve smoothness, spreadability, flavour and shelf life.
- **Roasted and salted groundnuts:** Groundnuts are consumed in India mostly in the roasted forms. Whole nuts may be boiled in salt solution. Kernels may be dry roasted or fried in oil; blanching precedes oil roasting. For extending the shelf life of roasted or fried nuts, they may be coated with acetylated monoglycerides or butyrated hydroxy anisole.
- **Peanut candies and confections:** Roasted nuts are used in making different types of candies and confections. Peanut brittles is prepared by boiling a mixture of corn syrup, sugar and water (20:30:7) to which groundnut is added. Sodium bicarbonate is added to aerate the brittle.
- **Groundnut oil:** Groundnut oil may be obtained from kernels through mechanical pressing or by the use of a solvent. Mechanical pressing uses various types of equipments such as mortar and pestle, grinding stones, hydraulic presses, expellers or screw presses, crushing rolls, rotary ghanis, through solvent extraction etc. 17.3 Storage

Groundnuts for seed purpose are generally stored in earthen pots, mud bins or in bamboo baskets which are often plastered with mud or cowdung. All the produce intended for storage should be well dried and should not have more than 5-8% moisture. A stack should not be higher than 10 bags, depending on the height of the godown and the space available. Each stack should be separated from the wall for ventilation. For storing groundnut pods, use of polythene lined gunny bags are found to be better. However, storing in polythene-lined gunny bags along with calcium chloride (250 g for a 30 Kg bag) is an improvement upon the polythene lined gunny bags alone. It helps in increasing the viability of the pods because of low moisture content and high dehydrogenase activity (Padmanabhan *et. al.*, 1995). At about 8°C temperature and 65-70% relative humidity, insect damage is less. The important storage pests causing substantial damage to the stored produce are listed below:

19.1. Groundnut bruchid: *Caryedon serratus* Olivier

The larva burrows straight through the egg shell and the pod wall, and starts eating the seed. The first sign of attack is the appearance of 'windows' cut into the pod wall by the larva to allow the adult to leave the pod after emerging from the pupal cocoon.

19.2. Red flour beetle: *Tribolium castaneum* Herbst

The larvae cause severe damage to the kernels. The sign of damage is powdery remants along with the damaged seeds.

19.3. Rice moth: *Corcyra cephalonica* (Stainton)

The creamy white larvae start feeding on the seed immediately after they hatch. The sign of damage is the presence of webbing in the damaged kernel.

These storage pests can be controlled through the following means:

1. Groundnut should be stored unshelled,
2. Fumigation with celphos - 3 g tablet per sack of groundnut (40 Kg) and covering the sacks with a polythene sheet for 5 days can control bruchids,
3. Broken seeds should not be stored for long periods, and
4. Dusting with an inert substance such as attapulgite-based clay dust (ABCD) can help to minimise storage insect problems

19.4. Marketing

In India, about 81% of the groundnut produced is used for oil extraction, 12% as seed, 6% for edible use and 1% for export. Groundnut has a well-organised marketing system. Groundnut is marketed mostly as nuts in shell and very little as kernel. The period of marketing is mainly November to December when about 70-80% of the marketable surplus is sold in the market directly by farmers. The vegetable oil wing of the National Dairy Development Board (NDDB) is authorised by the Government of India to undertake the Market Intervention Operation (MIO) to ensure a fair price to producers and a fair deal to consumers.

20. CONSTRAINTS

Following biotic and abiotic factors limit the production of groundnut.

20.1. *Kharif*

1. Drought stress at various phenophases,
2. Use of low levels of inputs by small holders and marginal farmers in dryland areas, and
3. The high incidence of foliar fungal diseases (rust and leaf spots), peanut bud necrosis disease, attack by white grub (North India) and insect pests (Spodoptera, Helicoverpa, and leaf miner)

20.2. Post-rainy Season

1. Poor plant stands resulting from loss of seed viability,
2. Lack of fresh seed dormancy in bunch varieties,
3. Lack of early-maturing varieties for residual moisture situations,
4. Poor soil tilth in rice follows,
5. Lack of cold-tolerant and high temperature tolerant varieties, and

6. Nutrient deficiencies in acid soils.

The estimated losses of yield resulting from these factors either singly or in combinations, have been found vary from 10-70% depending upon severity.

21. REFERENCES

Agasimani, C.A., Balelad, H.B. and Hosmani, M.M. 1992. Response to time, form and level of calcium application in groundnut. Indian Journal of Agronomy 37(3); 493-495

Anonymous. 1986. Recent advances in groundnut productivity research. National Research Centre for Groundnut, Junagadh. pp. 108-109

Anonymous.1993. Annual Report. Krishna Godavari Zone, APAU, Hyderabad

Asiedu, J.J.K.B. 1994. Industrial utilization and processing. In: The groundnut crop: A scientific basis for improvement, Smartt, J. (ed), Chapman and Hall, London, pp. 480-508

Baldev Singh, Banani Singh and Khandelwal, R.B. 1993. Effect of phosphatic fertilisers, gypsum and their mode of application on yield, oil and protein content of groundnut. Indian Journal of Agronomy 38(1);56-59 Bandyopadhyay, A., Desai, S., Radhakrishnan, T., Misra, J.B. and Pal, K.K. 1998. Status of biotechnological research in groundnut. Two day pre-project formulation workshop on: Biotechnological options for crop improvement in groundnut and pigeon pea. 9-10th March, 1998, Hyderabad, pp. GN 3.1-GN 3.22

Basu, M.S. and Ghosh, P.K. 1996. The status of technologies used to achieve high groundnut yields in India. In: Achieving high groundnut yields: Proceeding of an International Workshop, 25-29th August, 1995, Laixi City, Shandong, China, pp. 27-40

Boote, K.J. 1982. Growth stages of peanut (Arachis hypogaea L.). Peanut Science 9;35-39

Devi Dayal, Basu, M.S. and Reddy, P.S. 1989. Water management in groundnut. Technical bulletin No. 6, pp. 10

Devi Dayal, Naik, P.R. and Dongre, B.N. 1994. Criss-cross method of sowing for *rabi*-summer groundnut cultivation. Groundnut News 6(1);6-7

Devi Dayal and Reddy, P.S. 1987. Effect of paired row planting on the yield of groundnut. Indian Journal of Agronomy 32(2); 183-184

Dimree, S and Dwivedi, K.N. 1994. Response of S and P on groundnut. Journal of Oilseeds Research 11(2);193-195

Dimree, S., Dwivedi, K.N. and Hasi Ram. 1993. Effect of S and P nutrition on yield and yield attributes of groundnut. Indian Journal of Agronomy 38(2);327-328

Directorate of Economics and Statistics. 1996. Department of Agriculture and Cooperation, Ministry of Agriculture, Goverment of India, New Delhi.

Dorate, D.N., Dumbre, A.D. and Bhingarde, M.T. 1995. Effect of seed size on growth, yield and seed quality of groundnut Aarachis hypogaea L.) under summer conditions. Seed Research 21(2);107-109

FAO. 1994. Yearbook Production, Vol. 48, pp. 108-109

Ghewande, M.P., Desai, S., Prem Narayan and Ingle, A.P.1993. Integrated management of foliar diseases of groundnut in India. International Journal of Pest Management 39;375-378

Ghosh, P.K. 1995. Development of suitable agronomic practices in groundnut. Annual Report. National Research Centre for groundnut pp. 21-25

Ghosh, P.K. 1996. Development of suitable agronomic practices in groundnut. Annual Report. National Research Centre for groundnut pp. 53-58

Ghosh, P.K. 1997. An integrated approach of nitrogen management in groundnut (*Arachis hypogaea* L.). Crop Research 13;565-567

Ghosh, P.K. and Mathur, R.K. 1996. Nitrogen partitioning pattern and its nature of association with biomass harvest index and nitrogen harvest index in spanish groundnut. Journal of Oilseeds Reserach 13;275-277

Ghosh, P.K., Devidayal and Naik, P.K. 1997a. Effect of seed maturity, sowing depth and intra-row spacing on plant stand, growth and yield of groundnut (Arachis hypogaea L.). Indian Journal of Agricultural Sciences 67;322-325

Ghosh, P.K., Devidayal, Singh, V. and Naik, P.R. 1997b. Improvement of yield of summer groundnut through mulching and criss- cross sowing in Gujarat, India. International Arachis Newsletter 17;61-62

Gowda, G., Shivaraj, B. and Gowda, A. 1995. Effect of Zn and Mo on the yield and quality of groundnut. Journal of Oilseeds Research 12(1);133-136

Gregory, W.C., Krapovickas, A., and Gregory, M.P. 1980. Structure variation and evaluation and classification in Arachis, In Advances in legume Science (Eds. R.J. Summerfield and A.H. Bunting), Royal Botanical Gardens, Kew, London, pp. 469-481

Gregory, W.C., Smith, B.W. and Yarbrough, J.A. 1951. Morphology, Genetics and breeding in the peanut- the unpredictable legume. National Fertiliser Association, Washington, D.C. pp. 28-85

Hamada, A.A. 1988. Weed competition in irrigated groundnut, variety Ashford, in the Rahad Scheme, Sudan, Beitrage Zur. Tropisehen Land Wirt Schaft Urd. Veterinarmedia. 26;25-31 ICRISAT. 1993. On-farm research on groundnut, pigeonpea, chickpea and transfer of technology to semi-arid tropics farmers in India. Patancheru, A.P., 502 324, India: International Crops Research Institute for the Semi-Arid Tropics

Joshi, P.K., Kulkarni, J.H. and Reddy, P.S. 1989. IGR 6 and IGR 40. Two effective Bradyrhizobium strains for *rabi*/summer groundnut production National Research Centre for Groundnut, Technical Bulletin, pp. 1-10

Kalaiselvan, P.A., Ramalingam, S. and Bhat, M.Y. 1989. Influence of weed competition in different stages on growth and yield of groundnut. Andhra Agricultural Journal 32;197-198

Kanwar, S., Nijhawan, H.L. and Raheja, S.K. 1983. Groundnut nutrition and fertiliser responses in India, ICAR, New Delhi

Kathmale, D.K. and Chavan, N.R. 1996. Effect of irrigation frequencies and methods on yield of summer groundnut. Journal of Oilseeds Research 13; 263-264

Koppalkar, B.G. and Sheela Vantar, M.N. 1990. Studies on the oilseed production potential of groundnut and sunflower intercropping systems during summer condition. Journal of Farming 6;1-7

Kori, R.N., Goudareddy, P.S., Hunshal, C.S., Patil, S.L. and Nadagouda, B.T. 1997. Effect of weed control on nutrient uptake, weed height and yield of groundnut. Journal of Oilseeds Research 14;51-52

Mishra, C.M. 1994. Response of groundnut varieties to phosphorous under rainfed conditions. Indian Journal of Agronomy 39(2);326-327

Padmanabhan, G., Nagarajan, M., Manian, K and Annamalainathan, K. 1995. Prologing seed viability in groundnut (Arachis hypogaea L.) with calcium chloride. Journal of Agronomy and Crop Science 174;87-90

Pasricha, N.S. and Aulakh, M.S. 1986. Role of sulphur in the nutrition of groundnut. Fertiliser News 31(9);17-21

Pasricha, N.S., Aulakh, M.S., Bahl, G.S. and Badehesha, H.S. 1987. Nutritional requirements of oilseed and pulse crops in Punjab. Deptt. of Soils, PAU. Ludhiana. pp. 13-55

Pasricha, N.S., Aulakh, M.S., Sahota, N.S. and Baddesha, H.S. 1988. Fertiliser use research in Oilseed crops. Fertiliser News 33(9); 15-22

Pasricha, N.S. and Tandon, H.L.S. 1990. Fertiliser management on oilseeds. In: Fertiliser Management in commercial crops. Tandon, H.L.S (ed.). Fertiliser Development and Consultation Organization, New Delhi, pp. 28-56

Pradhan, L. Dixit, L. Rout, D. and Mishra, A. 1992. Effect of source and level of nitrogen on growth and yield of groundnut in acid lateritic soils of Orissa. Indian Journal of Agronomy 37(2);273-275

Ramachandrappa, B.K., Kulkarni, K.R. and Nanjappa, H.V. 1993. Effect of irrigation schedules and land managment practices on dry matter production and growth attributes in summer groundnut. Mysore Journal of Agricultural Sciences 27(2);135-140

Ravi Kumar, A., Raghavulu, P., Reddy, G.V. and Subbaiah. 1994. Effect of doses and time of nitrogen and gypsum application on dry matter production and yield of groundnut. Indian Journal of Agronomy 39(2);323-325

Sankara Reddi, G.H. 1988. Cultivation, storage and markrting. In: Groundnut, Reddy, P.S.(ed.), Publication and Information Division, ICAR, New Delhi, pp. 318-371

Singa Rao, M. and Prabhu Prasadini, P. 1994. Improved production technologies for augmenting yield of groundnut in rice- fallows. In: Prasad. M.V.R. et al (Ed.) Sustainability in Oilseeds. Indian Society of Oilseeds Research, Hyderabad, pp. 317-321

Singh, A., Panwar, R.S., Bhan, V.B. and Malik, R.K. 1987. Effect of different methods of weed free maintenance on crop weed competition in groundnut. Indian Journal of Weed Science 17;9-14

Singh, P., Verma, B.S. and Sahu, M.P. 1994. Effect of level and source of Phosphprous and bio-regulators on groundnut. Indian Journal of Agronomy 39(1);66-70

Singh, R.P. 1994. Sustainability of oilseeds productions systems. In: Sustainability in oilseeds, Prasad *et. al.* (ed), Indian Society of oilseeds Research, Hyderabad, pp. 228-240

Subba Rao, I.V. 1994. Integrated nutrient management for sustainable productivity of Oilseeds in India. In: Sustainability in Oilseeds, Prasad *et. al.* (ed), Indian Society of Oilseeds Research, Hyderabad, pp. 264-270 Subbaiah, H., Nanjappa, H.V. and Ramachandrappa, B.K. 1997. Nutrient uptake by crop and weed as influenced by intercropping systems in groundnut (Arachis hypogaea). Indian Journal of Agronomy 42;59-62 Subrahmanyam, P., Wongkaew, S., Reddy, D.V.R., Demski, J.W., McDonald, D., Sharma, S.B. and Smith, D.H. 1992. Field diagnosis of groundnut diseases. Information Bulletin No. 36. Patancheru, A.P., 502 324, India: International Crops Research Institute for the Semi-Arid Tropics, Patencheru, Hyderabad

Tandon, H.L.S. 1990. Plants nutrients and fertilisers. In: Fertiliser recommendations for oilseed crops. Tandon, H.L.S. (ed), Fertiliser Development and Consultation Organization, New Delhi, pp. 17-26

Tiwari, R.B. and Dhakar, L.L. 1997. Productivity and economics of summer groundnut as affected by irrigation, fertilisers and weed control. Indian Journal of Agronomy 42;490-494

Tiwari, R.B., Dhakar, L.L. and Singhi, S.M. 1997. Growth and water use by summer groundnut (Arachis hypogaea) as affected by irrigation schedule, fertiliser and weed control methods. Indian Journal of Agronomy 42;338-341

Wani, S.P. and Lee, K.K. 1992. Role of bio-fertilisers in upland crop production. In: Tandon, H.L.S. (Ed). Fertilisers, organic manures, recyclable workers and biofertilisers, Fertiliser Development and Consultation Organization, New Delhi

Wightman, J.A. and Ranga Rao, G.V. 1993. A groundnut insect identification handbook for India. Information Bulletin No. 39, International Crops Research Institute for the Semi-Arid Tropics, Patencheru, Hyderabad

Yakadri, M. and Satyanarayana, V. 1995. Dry matter production and uptake of nitrogen, phosphorus and pottassium in rainfed groundnut (Arachis hypogaea). Indian Journal of Agronomy 40(2);325-327

Yaduraju, N.T., Kulshrestha, Gita. and Mani, V.S. 1980. Herbicides studies in groundnut. Indian Journal of Agronomy 25;447-452

Bhapkar, D.G., Patil, P.S. and Patil, V.A. 1986 Dormancy in groundnut- a review. Journal of Maharashtra Agricultural Universities 11;68-71

Ketring, D.L. and Morgan, P.W. 1972. Physiology of oilseeds IV. Role of endogenous ethylene and inhibitory regulators during natural and induced after ripening of dormant Virginia type peanut seeds. Plant Physiology 50;382-387

Sharma, A. and Sengupta, U.K. 1988. Changes in L-amylase and protease activity during seed development indormant and non-dormant groundnut (Arachis hypogaea) cultivars. Indian Journal of Experimental Biology 26:732-733

Zamski, E. and Ziv, M. 1976. Pod formation and its' geotrophic orientation in the peanut, Arachis hypogaea L., inrelationto light and mechanical stimulus. Annals of Botany 40;631-636

11. Sesame

T. N. Balasubramanian and SP Palaniappan

Sesame is regarded as *queen of oilseeds* by users because of the quality (fatty acid composition) of its oil and for its resistance to oxidation and rancidity even when stored at ambient air temperature.

In India sesame seeds are used for oil extraction (78%) for edible purposes (20%) and for seed purpose (2%). Sesame oil is mainly used for edible purposes (70%) as salad and as cooking oil and for non edible purposes like domestic and toilet (14%) and for manufacture of margarine (16%). Sesame oil forms the basis of most of the fragrant and scented oils and in the perfumeries it is used as laxative. Sesame cake is mainly used as cattle feed.

1. ORIGIN

Sesame, *Sesamum indicum* L.(Syn. *S. orientale* L.) belongs to family Pedaliaceae (order: Tubiflorae). Archaeological evidences indicate that sesame was cultivated in Palestine and Syria by 3000 BC. In the Inidian sub continent, charred sesame seeds were recorded in Harapa, the oldest archaeological site in the region, in addition largest sesame acreage and diverse forms of sesame are seen in India. On the other side, 25 species of sesame are found in Africa though Africa has limited cultivated varieties. Two wild species of sesame (*S. capense* and *S. schenkii*) which have 2n=26 with widest distribution are found in Africa, India and far East. Based on antiquity, diverse forms of cultivated sesame and occurrence of wild sesame, *S. indicum* sp. *malbaricum* in India. India is now considered as the basic centre of origin.

2. DISTRIBUTION

Sesame (*Sesamum indicum* L.) is basically a crop of the tropics and subtropics. Its main distribution is between 25° S and 25° N latitude, but it can be grown well upto 40° N latitude in china, Russia and the USA, 30° S latitude in Australia and 35° S latitude in South America. It is a crop of both tropics and subtropics. India is the world's major producer with a third of the world's acreage and approximately a quarter of the total production.

The important countries in the world which grow sesame are India, China, Philippines, Burma, Korea, Thailand, Sri Lanka, Pakistan, Egypt, East, Africa, Ethiopia, Sudan, Kenya, Tanzania, Somalia, Uganda Nigeria, Mexico, Venezuela, USA and Australia. The area production and productivity of top five countries are given in Table 11.1.

TABLE 11.1: Area, production and productivity of top five countries in the world (mean value)

Country	Area ('000 ha)	Production ('000 tonnes)	Productivity (kg/ha)
India	2134	503	236
Sudan	1048	251	246
China	993	559	617
Burma	866	212	245
Nigeria	250	75	300
World	6572	2272	346

The major sesame growing states are Rajasthan, Uttar Pradesh, Madhya Pradesh, Maharashtra, Andhra Pradesh, Orissa, Karnataka, Gujarat and Tamil Nadu. The top most five states in India, which grow sesame are given in Table 11.2.

TABLE 11.2: Area production and productivity of five top most states in India (average values)

States	Area (000 ha)	Production (000 tonnes)	Productivity (kg/ha)
Rajasthan	292	25	80
Maharashtra	282	64	228
Uttar Pradesh	267	11	55
Madhya Pradesh	218	40	187
West Bengal	209	64	481
India	2239	615	245

3. CLASSIFICATION

Many classification have been given in literature.(Table 11.3) comprises 16 genera and about 60 species, of which several can be crossed with *S. indicum* and a few also cultivated for their seeds.

TABLE 11.3: Species and characteristic of genus sesame

Species	Wild/ cultivated	Author	Distribution	Characteristics
Sesamum schenckii	wild	Abraham (1945)	South West Africa, India (Travancore) East Indies	Erect, inter media leaves 5-7 partite, Purple corolla, seeds Free from wings
Sesamum malabaricum	Partly cultivated	John *et. al.*,(1950)	Africa, India (Malabar)	Fertile in cross with *S. indicum*

Species	Wild/ cultivated	Author	Distribution	Characteristics
Sesamum indicum (S. orientale)	Cultivated	Nohara (1934); Richharia (1936); Raghavan (1947) and Kobayashi (1947)	Tropical to Temperate zone	Erect, Simple or branched variable leaves, whitish to pink colour corolla and companulate, nectar sessile, nectar sessile, coloured to white seeds, smooth, edible.
Sesamum indicum (S. orientale)	Cultivated	Richharia and parrai,(1940) Langham,1945, Kobayashi and Shimamura (1947), Mazzani (1962)	India, USA, Japan and Venezuela	Autotetra ploid
Sesamum Prostratum	Partly cultivated	Ramanujam (1941)	India (Tamil Nadu)	Prostrate
Sesamum laciniatum	wild	Ramanujam and Joshi (1948)	India (Deccan hills)	
Sesamum angolanse	wild	Kedarnath (1950)	Angola	Erect, simple or branched leaves entirely violet to purple corolla, nectar sessile, long capsule blackish seed, rugose
Sesamum radiatum	Partly cultivated	John and Rao, 1941	Upper Guinea Sri Lanka	Erect, simple or branched leaves entil purple corolla, nects sessile, Short capsule smooth and blackish seeds
Sesamum occidentale	Partly cultivated	Sampath and Ramanathan	Africa, Sri Lanka, India	
Sesamum indicatum	Partly cultivated	Ramanujam (1942) Raghavan and krishnamurthy (1947)	India	Experimental amphidiploid of *S. orientale* (n=13) x *S. Prostratum* (n=16)
Sesamum lacinientale	Partly cultivated	Aiyadurai, Srinivasalu and Amirtha Devarathanam (1963)	India	Experimental amphidiploid of *S. indicum* (n=13) x *S. Laciniatum* (n=16)
Sesamum angustifolium	Partly-cultivated		Congo, Mozambique	Erect, simple or branched leaves entire; rosed corolla nectary sessile, long capsule
Sesamum calycium	Partly cultivated	unknown	Angola Mozambique	Errect, branched, lanceolate entire leaves, rosette corolla nectary sessile, small capsule
Sesamum baumii	Partly cultivated	Unknown	Angola	Erect, slender, leaves entix palerosed corolla, nectary subsessile, rugose seed
Sesamum marlothil	Wild	-	South Africa, East Indies, Australia,	Erect, lower leaves 3 foliate, nectar sessile, blackish seed

Species	Wild/ cultivated	Author	Distribution	Characteristics
Sesamum rigidum	Wild	-	Angola	Erect, branched, leaves entire, pale rosettes corolla, seed oblique flat or convex

4. GROWTH STAGES

Similar to other annual crop sesame also has different growth phases. In general, only three stages are indicated

1. Pre flowering stage (establishment to 30 DAS)
2. Flowering stage (31-45 DAS)
3. post flowering stage (46-85 DAS)

In reality four different stages could be identified for sesame of 85 days duration and they are:

Establishment stage	0-12 days (12 days)
Vegetative stage	13-36 days (24 days)
Reproductive stage	37-60 days (24 days)
Ripening stage	61-84 days (25 days)

In respect of sesame crop, once flowering starts flowering, flower setting and capsule development are continuous and over lapping and hence it is very difficult to observe the changes in phenological stages.

The Physiological maturity occurs 40 to 45 days from flowering (30-35 days) In a study on dry matter production in sesame at Hyderabad, it was observed that the rate of dry matter production was slow upto flowering. Only 16 Percent of total seasonal dry matter was produced at the time of flowering, while most of the dry matter (80%) was produced during reproductive phase (50 to 92 DAS). The growth rate during 50 to 92 days period was more than 200 kg\ha\day. Total dry matter ranged from 11291 to 11485 kg\ha with a seed yield ranging from 109 to 156 kg\ha. The dry matter production in matured plants was 11 per cent in roots, 18.5 Per cent in leaf, 51 Per cent in stem and 19 Per cent in reproductive parts.

5. CLIMATE

Sesame also known as benni seed, gingelly, simsim and tie. It is highly sensitive to physical factors especially to light period and its intensity. Among the Factors of production, climate is important for its productivity. Sesame is well grown upto 1250 m, although some varieties may be locally adopted upto 1500 m.

Sesame is fairly drought resistant crop, Once established, sesame is capable of withstanding a higher degree of water stress, The seedling stage, However, is extremely susceptible to moisture shortage.

Sesame produces an excellent crop with a rainfall of 500 to 650 mm, but a low of 300 mm and a high or 1000 mm, will also produce a crop with a good distribution over different pheno stages.

For maximum yields, precipitation should be distributed over the period of plant growth as follows:

- Germination to first bud formation 35 Per cent
- First bud to main flowering 45 Per cent
- Main flowering to maturity 20 Per cent

Heavy rain at flowering will drastically reduce yield and if cloudy weather persists for longer period at this time, yield can be exiguous. Similarly, rainfall during harvest also reduces the yield by increasing susceptibility to diseases and prolonging the period required for capsule to dry out. sesame is extremely susceptible to water logging continuous rain during growth will greatly increase the incidence of fungal diseases.

Sesame requires fairly hot condition during growth to produce higher yields and the total heat units required during the critical 3 to 4 months growth period are 2700°. A temperature range of 25 to 27° C encourages rapid germination, initial growth and flower formation. If the temperature falls below 20° C germination may be inhibited and a temperature of less than 18° C after emergence will severely retard growth. Low temperature at flowering can result in the production of sterile pollen or premature flower fall, conversely periods of high temperature (40° C) at flowering will seriously affect fertilization and the number of capsules set. If the night temperature is either 33° C or 15° C, flower initiation is delayed, but this effect could be counteracted by a low and higher day time temperatures respectively. Sesame is susceptible to damage from strong winds after the main stem is elongated and the tall stemmed combinable types are not recommended for areas, where high winds at harvest occur. The more sturdy but dehiscent types are also unsuitable for high wind localities, as high proportion of the seeds will be lost through shattering.

Sesame is a short day plant and it requires a minimum of 10 hours light per day. The effect of different light period was assessed and it was found that there was reduction in the two main growth periods (time to bud initiation and time for flower initiation) but the structure of the plant was radically altered (Table 11.4).

TABLE 11.4: Effect of photoperiod on growth of sesame

Treatment (Light hours)	Time to bud initiation (days)	Time of flower initiation (days)	Height at flowering (cm)
12.5-13.5	68	92	122
10	26	43	50

6. SOIL AND SOIL PREPARATION

Sesame comes up in varieties of soils but drainage is of paramount importance, since sesame is highly susceptible to water logging. High clay soils, sandy soils, saline and alkali soils are mostly unsuitable for sesame. Medium fertile soil, with high drainage and a mixture of sand and clay texture are highly suitable. Generally sandy clay or sandy loam soils are preferred for sesame. The soil should have a pH range between 5.3 to 8.0.

Land preparation is very important for sesame since sesame seeds are very small and sensitive to tilth of the soil. Sesame seed germination is positively correlated with soil tilth. finer the tilth, maximum is the germination and establishment. Upto seven ploughing with country plough may be adequate to obtain the required tilth. The field should be weed free and well leveled for sowing.

Generally, for *kharif* sesame, soil is ploughed deep in summer and with the onset of south west monsoon, two to three shallow ploughing, or harrowings in a crosswise are done either with country plough or blade harrow. for semi *rabi* crop, two to three harrowings are enough followed by leveling to bring the soil to fine tilth. In mixed cropping system, where sesame is grown with a main crop, the soil is prepared based on the tilth need of the main crop.

7. CROPPING SYSTEM

Cropping system in a given area is governed by a complex of agroclimatic, socioeconomic and trade related factors and very from region to region at macro level and from farm to farm at micro level.

Sesame is cultivated both under rainfed situation in uplands as well as under irrigated situation. Sesame is one among the component crops in different cropping systems across India and it is one of the crops in the intercropping system also. There is a general belief that the crop succeeding sesame in a rotation would suffer for nutrients and would give lower yield. Since sesame is an exhaustive crop when it is grown without any nutrient application, it removes the available nutrients from the soil As a result, the soil is exhausted and this had a deleterious effect on the succeeding crop. If sesame is grown under optimum nutrient management, then the succeeding crop is not affected.

7.1. Crop Rotations

Cropping systems involving sesame as one among the component crops in different agro ecological regions are presented in table 11.5.

TABLE 11.5: Important cropping systems (sesame) in agro-ecological

Agro ecological region	Important Cropping systems	
	Irrigated	Rainfed
Northern plain and central high lands including aravali hot semi arid eco region with alluvium derived soils (parts of Gujarat, Haryana, Madhya Pradesh, Punjab, Delhi, Uttar Pradesh and Rajasthan)	Cotton/pearl millet/ groundnut/	sesame-fallow
Deccan plateau and Eastern Ghats hot Semi arid eco region with red & black soils (parts of A.P.)	rice sesame	-
Eastern Ghats (T.N. upland and Deccan), Karnataka plateau, hot semi arid with red loamy soils (parts of A.P., Karnataka, T.N.)	-	Sesame-Castor horsegram

Agro ecological region	Important Cropping systems	
	Irrigated	Rainfed
Eastern (Chhota Nagpur) plateau and Eastern Ghats hot sub humid with red and lateritic soils (parts of A.P., Bihar, M.P., Maharashtra, Orissa and West Bengal)	-	Rice/Pigeonpea/Sesame/ Millets-fallow
Eastern plain, hot sub humid with Alluvium derived soils (parts of Bihar, U.P)	-	Sesame-maize/wheat
Bengal and Assam plains, hot sub humid to humid with alluvium derived soil (parts of West Bengal Assam and Tripura)	Rice-potato-sesame	-

State wise crop rotations	
State	**Crop rotations**
Tamil Nadu	Rice-Rice-Sesame (sandy loam soil Rice-sesame (sandy loam soil) Sesame-horsegram/Bengalgram (rainfed, red soils)
Karnataka	Sesame-horsegram (rainfed, red soils)
Andhra Pradesh	Paddy-Sesame (*Rabi*)
Madhya Pradesh	Cotton and sesame-wheat cotton and sesame-Jowar-groundnut
Uttar Pradesh and Punjab	Early sesame-Gram, lentil, pea or barley
Gujarat	Sesame-wheat
Rajasthan	Sesame-mustard
Orissa	Potato-sesame
Maharashtra	Sesame-Safflower

7.2. Intercropping/Mixed Cropping

Indian farmers conventionally intercrop sesame with pulses, maize and cotton and in Africa with maize and sorghum. Sesame is sown as mixture with pulses in certain parts of India. Sesame and pulse crops interact in such way that when sesame yields are low, pulse yields are higher and vice versa. Another significant point is that in 10 out of 16 combinations of sesame and pulses, the yields of sesame have increased over the control.

The common mixtures followed in different states:	
Tamil Nadu	Groundnut+sesame (rainfed), Bajra+sesame (rainfed), Cotton+sesame (rainfed and irrigated) and Redgram+sesame (rainfed)
Bihar	Maize+sesame and Redgram+sesame
Gujarat	Kodomillet+sesame, Redgram+sesame, Cotton+sesame and castor+sesame
Karnataka	Redgram+sesame, castor+sesame and Bajra+sesame
Madhya Pradesh	Maize+sesame and Redgram+sesame
Punjab	Groundnut+sesame, cotton+sesame, Maize+sesame and Bajra+sesame
Rajasthan	Clusterbean+sesame and Pulses+sesame
Uttar Pradesh	Maize+sesame, Bajra+sesame, Sorghum+sesame and Redgram+sesame
Maharashtra	Redgram+sesame

As compared to intercropping of sesame in sorghum, bajra and groundnut, intercropping sesame in redgram was found remunerative and sesame yield was not affected.

8. MANURE AND FERTILISERS

Sesame like other annual crops requires macro, secondary and micronutrients. But in reality the Crop is raised by small and marginal farmers on relatively poor soils without any proper fertiliser management. However, for satisfactory yield, adequate nutrients are to be provided. In Tamil Nadu, one study indicated that sesame removes 39 kg N, 10 kg P and 23 kg K from one hectare. Thus varieties capable of giving high yields require additional plant nutrients. As compared to nitrogen and potassium, the removal of phosphorus by sesame crop is low. In matured sesame plants, the average accumulation of N was the highest in the leaves (35%) followed by capsules (31%), whereas the accumulation of P and K was the highest in the stem with 55 and 53 percent respectively. During peak periods of absorption, daily uptake per hectare was 2.9 kg N; 0.6 kg p; and 3.0 kg K. The amount of dry matter needed to produce 100 units of seed contained 23 units of N; 6 units of P and 26 units of K.

Nutrient concentration in all the vegetative parts increases until six weeks after sowing followed by gradual decline towards maturity. Remobilisation of NPK occurs from leaf, stem and roots to pods during reproductive phase.

NPK recommendation for different States

States	N:P:K (kg/ha)
Gujarat	50:25:25 (irrigated) and 25:25:0 (rainfed)
Madhya Pradesh	30:30:0 (irrigated) and 15:15:15 (rainfed)
Orissa	30:30:20
Andhra Pradesh	40:20:20
Maharashtra	50:0:20 (N in two splits)
Rajasthan	15:20:0 (rainfed) and 50:50:25 (irrigated, N in two splits)
Karnataka	40:20:20
Tamil Nadu	23:23:13 (Rainfed) or 17:13:13+seed treatment with *Azospirillum* at 600 g/ha 35:23:23 (irrigated) or 21:23:23+seed treatment with *Azospirillum* at 600 g/ha 40:40:60+Boralx 10 kg/ha+ Micro nutrient 12.5 kg/ha for sesame crops raised for seed purposes
Uttar Pradesh	30:20:30

Nitrogen recommendation for different states is given earlier. Split application of nitrogen is essential. Basal application of 50 Per cent recommended nitrogen followed by top dressing of remaining 50 Per cent between 25th to 30 DAS would resulted in higher yield. Response to applied phosphorous is very much limited. However a maintenance dose of 40 kg/ha is generally recommended. Similar to nitrogen, sesame responds to potassium up to 67.2 kg/ha and it influenced dry matter as well as leaf area index in sesame. With enough moisture in the soil, anticipating dry spell, spraying of KCl 1 per cent at 30 and 40 DAS could be done against any possible moisture stress.

8.1. Micronutrients

Sesame growth is limited by three micronutrients. They are manganese, zinc and boron.

Manganese deficiency: Leaves develop intervenal chlorosis and chlorotic tissues later develop light brown or husk coloured necrotic lesions. Mixing 5 kg/ ha of $MnSO_4$ with 45 kg of soil and broadcasting evenly in the beds after sowing is recommended wherever deficiency is at chronic problem.

Zinc deficiency: Middle leaves develop chlorosis in the intervenal areas followed by necrosis along the apical leaf margin. Application of 12.5 kg of $ZnSO_4$/ha at the time of sowing is to be done wherever deficiency is common enough sand (50 kg/ha) may be mixed with $ZnSO_4$ for easy application.

8.2. Secondary Nutrients

Basal application of gypsum at 200 kg/ha had increases sesame yield by over 27 per cent when sulphur is not adequate.

Plant growth regulators: Foliar application of succinic acid, indolacetic acid and maleic hydrazide had significantly increased sesame seed yield Dormancy is completely broken by soaking seeds in GA_3 solution (100 ppm) for 48 hours these treatments also hastened emergence.

Since India is nearest to equator, organic manuring seems to be important, not only to improve the physical and biological properties of the soil, but also to supply certain needed microelements. In this direction the result from West Bengal state indicated the superiority of combined application of 4 tonnes of FYM/ha over 45 kg N plus 22.5 kg P_2O_5/ha for sesame crop. Similarly in Tamil Nadu, application of 12.5 tonnes of FYM/ha is recommended for sesame as organic manuring, while for rainfed sesame 750 kg enriched farm yard manures is recommended per hectare

In order to reduce the bill of cost on fertilisers and also to maintain soil health integrated nutrient management concept has to be adopted for sesame cultivation. it is judicial combination of organic manures, biofertilisers and chemical fertilisers applied to meet the crops demand.

9. VARIETIES

The yield improvement achieved through selection/hybridization followed by selection has been only marginal in sesame. Hence attempts are being made worldwide to produce sesame hybrids. In this aspect, China is said to have produced one sesame hybrid with yield potential upto 3000 kg/ha under irrigated conditions. In India, each major sesame growing stage releases its own sesame varieties suited to their agroclimatic conditions. Important varieties suitable for different states are given in Table 11.6.

TABLE 11.6: Varieties recommended in different States of India

Andhra Pradesh	Gowri (*kharif* and *rabi*), T85 (*Kharif*)
Gujarat	Mrug 1 (*kharif*), Purva 1 (semi *rabi*), Patan-64 (*Kharif*) and Patan 65 (*Kharif*)
Madhya Pradesh	G5 (*kharif*), T4 (*kharif*), G35 (*kharif*), No 128 (*kharif*) and No. 32 (*kharif*)
Maharashtra	D.7-11-1 (*kharif*), N.58-2(*kharif*), No.128 (*kharif*) and Chanda 8(Semi *rabi*)

Orissa	Vinayak (*kharif/rabi*)
Rajasthan	Pratap (C50)(*kharif*), RT 125 (*kharif*) and Type 13(*kharif*)
Tamil Nadu	RT- 125 (*kharif*), TMV3 (*kharif*), TMV4 (summer), TMV6 (summer), SVPR 1 (Summer), VRI.1 (summer) and TMV5 (*Rabi*)
West Bengal	B.67

10. SOWING

10.1. Times of Sowing

The relation between crop yield and environment condition is an important aspect for intensive investigation in agricultural research. Determining of the optimum time of sowing appears to be a pre-requisite for exploring the full yield potential of sesame. Sesame growing period in India below:

In India, sesame yield is higher in summer irrigated crop than the monsoon crop. Reduction in monsoon crop yield is mainly due to heavy rain received during the crop growth period. In most of the states in India sesame is being grown during *Kharif* season as monsoon crop. In Madhya Pradesh, Maharashtra, Andhra Pradesh, Tamil Nadu, Orissa and Karnataka it is grown both in *kharif* and *Rabi* season. It is also grown in summer in parts of Orissa and Tamil Nadu under irrigation. Optimum times of sowing have been identified for different states of India and they are as follows:

State	Sowing season
Tamil Nadu	June 15th to July 7th, November 15th to November end and February 15th to March 15th
Kanpur (UP)	End of July
Punjab and Haryana	Before 15th July
Rajasthan	Before 15th June
Bihar	15th February (irrigated)
West Bengal	March (irrigated)
Orissa	First fortnight of March (irrigated)

10.2. Method of Sowing

Broadcasting by hand is the rule in all states of India where sesame is being grown except in few pockets, where farmers sow the seeds in line either by using specific local *gorru* or by employing light country plough. For better establishment, an optimum depth of 3 cm sowing is adequate. Normally about 5 kg/ha seed rate is used.

Normally, for broadcasting, seeds are mixed with dry sand (4 times the quality of seeds used) to make it as bulk and broadcasted uniformly by hand both lengthwise and widthwise. After sowing, the seeds are covered either with country plough or by leveler. In some pockets farmers use brunches of *Prosopis juliflora* to cover the sown seeds. If the seeds are not covered properly, ants will eat sesame seeds. Rainfed crop should be sown under optimum soil moisture, while in irrigated condition it is sown in the prepared dry soil followed by irrigation. Tamil Nadu Agricultural University has developed one centrifugal seed broad castor for broadcasting sesame seeds which is very convenient and makes sowing easy. This tool completes one hectare sowing within 100 minutes. The cost of the tool is Rs.1500/-. One man is enough to operate the tool.

10.3. Seed Treatment

Two seed treatments are to be done before sowing, one after another. They are Fungicidal and *Azospirillum* seed treatment. Fungicidal seed treatment is used against seed born fungus and *Azospirillum* seed treatment for enhanced nitrogen fixation. Method of seed treatment is given below:

10.3.1. Dry Seed Treatment

Required seeds are placed in seed treating drum or in polythene bag and mixed with carbondazim/captan/thiram at the rate of 2 gram per kilogram of seeds and thoroughly mixed. This should be done 24 hours before sowing.

10.3.2. Wet Seed Treatment (1 Kg Seed Treatment)

This seed treatment not only provides protection to seeds against seed born diseases, but also hardens the seeds for early vigour and drought tolerance. 750 ml of water is taken for 1 kg seed three grams of carbandazim is dissolved in 750 ml of water and the seeds are added to this solution and kept for soaking for five hours. Once in 30 minutes the seeds should be stirred. Caution should be taken not to press the seeds by hand during periodical stirring. The container should not be covered during soaking and it should be done in a shady place.

After five hours of soaking the water is drained and the seeds are kept in a gunny bag under shade till they reach their original weight of one kg. Normally 15 hours will be taken to dry the seeds under shade and during drying periodical stirring may be done. This type of treatment can be done 7-10 days earlier to sowing.

10.3.3. Bio Seed Treatment (One kg Seed Treatment)

For each kg of seeds, four grams of *Trichoderma viride* powder, which contains active fungi is be mixed with the seeds and kept under shade for an hour before sowing.

10.3.4. Azospirillum Seed Treatment: (1 ha Seed Treatment)

This is done to fix atmospheric nitrogen. Five packets of *Azospirillum* (1 kg) is added to 625 ml of cold rice gruel in an open container. the five kilogram of fungicide treated seeds as in 1 or 2 are added in this paste a little by little and thoroughly coated seeds are shade-dried for 30 minutes. This can be done 30-40 minutes before sowing.

10.4. Transplanting

This type of establishment is done in certain states in India. In Tamil Nadu it is being done in new delta area of Thanjavur district. This method facilitates maintenance of required population per unit area, through line plantings in addition, formation of capsules starts from the base of the main stem. This method of establishment offers an extra sesame yield of 100 kg/ha over broadcasting method under irrigated condition. for transplanting sesame, the pre-requisites are as follows:

1. Soil should be sandy loam
2. Suited for summer sesame under irrigated condition
3. The age of the seedlings should be from 15 to 20 days
4. Irrigation water should not be a limiting factor
5. Seedlings may be pulled after 3 P.M. and should be planted immediately on the same day
6. The land should be irrigated in the main field before transplanting.

Both basin and ridges and furrows methods could be used for transplanting sesame seedlings. It is always safe to have ridges and furrows for early and better establishment of transplanted sesame seedlings.

10.5. Nursery Preparation

To plant one hectare, 300 to 500 m^2 raised bed nursery is required. The soil of the nursery should be dominated by sand. Plough or hoe with showel to bring the soil to fine filth. Form 6 x 1.5 m raised beds with 30 cm channels all around. Sprinkle 750 g seeds treated with fungicide and *Azospirillum* after mixing with 3.250 kg of seed on the raised bed followed by covering the seeds by the fingers or small wooden rake. Sprinkle little malathion dust (1.5 kg) over the seed bad against ants. Allow water to stand in the channels formed all around the raised seed bed and make the water to seep into the bed. Give another life irrigation on 3rd day after sowing.

For nursery bed do not apply any organic manures or fertilisers. It causes root injury to the seedlingsa at the time of pulling out. It is better to apply carbofuran at 375 g per 300 m^2 area on 10th day against any sucking pests in the nursery.

10.6. Planting

Bring the main field to fine tilth by periodical ploughing with the application of 12.5 tonnes of FYM. In the last ploughing apply 125 kg of complex fertiliser (17:17:17) per hectare and incorporate. Form furrows of 22.5 cm depth with an interval of 30 cm. Before transplanting, irrigate the furrow to soak to 50 per cent of its capacity.

Recommended spacing for different States.	
Bihar	15 x 30 cm and 30 x 30 cm
Kerala	15 x 30 cm and 30 x 30 cm
U.P.	45 x 15 cm
Gujarat (Sourastra)	45 x 15 cm
Madhya Pradesh	15 x 45 cm (for semi *rabi*) and 30 x 10 cm (for other seasons)
Southern Karnataka	30 x 10 cm and 30 x 20 cm
Andhra Pradesh	30 x 15 cm
Maharashtra	30 x 15 cm
West Bengal	30 x 10 cm
Tamil Nadu	30 x 30 cm (irrigated), 22.5 x 22.5 cm (rainfed) and 25 x 15 cm
Rajasthan	35 x 15 (Branched varieties) and 30 x 10 (unbranched)

Transplant one seedling in the soaked area of the furrow on its sides with an interval of 30 cm at 2 cm depth. Pulling of seedlings and transplanting should be done in the evening. Nursery may be wetted in the evening one day prior to transplanting. After 3 rd day irrigation may be given in the furrows where seedlings are transplanted before sunrise. Irrigate thereafter, once in 10 days as per the requirement. Ten days after transplanting. 40 kg urea/ha may be applied as top dressing after necessary weeding and hoeing. After 25 day hand hoeing followed by earthing up should be done. Before and after this operation, irrigation should be given. Necessary plant protection measures may be adopted as per recommendation

The sesame plants should be properly spaced to ensure satisfactory stand and growth. It depends upon branching nature of the cultivar, soil type, soil fertility and season of planting. For monsoon crop it is recommended to have closer spacing as compared to summer irrigated crop.

10.7. Thinning

Thinning is an important operation to be done in sesame. Single seedling had to be maintained per hill, otherwise, growth and yield will be affected. Two thinning are to be done for population maintenance. First thinning is done by keeping two plants per hill on 15th day followed by second thinning on 30th day before flowering by keeping single well grown plant per hill.

11. IRRIGATION

Irrigation requirement of sesame varies with cultivar, soil type and season of cultivation. The water requirement of sesame is from 250 mm to 350 mm. Research evidences indicates the critical nature of reproductive stage for irrigation. Moisture stress at vegetative stage had reduced yield upto 50 per cent, while if moisture stress was given at reproductive stage the yield loss was upto 55 per cent. Irrigation in sesame may be done in the following pattern.

- I^{st} Irrigation : at sowing
- IInd Irrigation : 7 DAS
- IIIrd Irrigation : 25 DAS (pre flowering)
- IVth Irrigation : 35-40 DAS (flowering)
- V^{th} Irrigation : 55 DAS (capsule setting)
- VIth Irrigation : 65 DAS (grain formation)

Scientifically irrigation may be scheduled at 25 per cent available soil moisture (ASM) at vegetative stage followed by 50 per cent ASM irrigation at reproductive phase. Irrigation may also be scheduled at 0.6 IW/CPE ration in vegetative stage followed by 0.4 IW/CPE ratio at reproductive stage.

12. WEED CONTROL

In general weeds are more in *kharif* season as compared to *rabi* and summer seasons. The yield loss in sesame is computed to 49 to 70 per cent due to weeds. Because of slow growth of sesame in the early stage, weeds if not controlled properly exert greater influence on the growth and yield. Experimental results indicated that the critical weed free period for sesame is from 15 to 30 days after sowing

Under normal condition, hand weeding and hoeing twice on 15 and 35 DAS is recommended. However the method of hand weeding and hoeing is very costly considering the human labour availability for this operation. Integrated weed Management (IWM) practices recommended for sesame is given below.

1. Pre-emergence application of alachlor 1.5 kg /ha + one hand weeding and hoeing on 30 DAS
2. Pre-emergence application of fluchloralin at 1 kg /ha + one hand weeding and hoeing on 30 DAS

13. PLANT PROTECTION

It is reported that sesame is affected by 20 species of insects and mites. Similarly, sesame is infested by 19 pathogens, of which are fungi (13), bacteria (2), virus (1), mycoplasma like organism (1), and nematodes (2).

The reported yield loss in sesame is around 40 per cent due to pests and diseases.

General recommendation for controlling sesame pests:

Pest control schedule	
15 Days after sowing	Dimethoate 250 ml/ha
25 Days after sowing	Endosulfan 625 mk/ha
35 Days after sowing	Manocrotophos 625 ml/ha
50 Days after sowing	Phasolone 750 ml/ha
60 Days after sowing	Malathion dust 5% 50 kg/ha (on the soil application)

13.1. Diseases

The following are the diseases

Phyllody: Possibly the most destructive sesame disease is phyllody or green flowers particularly in India and Burma; where incidence has been reported as high as 100 per cent and 90 per cent respectively. The disease is characterized by the transformation of the floral parts into green leafy structures associated with abundant vegetative growth. Phyllody in sesame has been shown to be associated with presence in the phloem of a Mycoplasma Like Organisms (MLO), which is transmitted in India by leaf hoppers in the genus *Orosisus.*

Some of the possible solutions suggested for the control of phyllody are:

1. Use of insecticide to control vector green leaf hopper.
2. Growing resistance varieties both to vector and Mycoplasma.
3. Cultural practices, which minimise the infestation by vector.

13.2. Powdery Mildew

White masses of mildew spores cover the leaves of the plants, which later turn yellow and wither. This disease appears on 45-60 day old crops. This is controlled by dusting of sulphur at 25 kg/ha or spraying of wettable sulphur at 2.5 kg/ha.

13.3. Black Stem Wilt

This is serious disease in sesame, when sesame is grown during *kharif* season and as well as under irrigated condition in summer season. This disease is more severe under drought and appears at later stages of crop growth. Yield losses up to 57 per cent has been reported. The stem is attacked mostly at the collar region. The charcoal rot appears in the form of light brown discolouration at the collar region or at the base of stem near the soil surface. The discoloration slowly extends above the stem to several centimeters in height. This discoloration also can be seen in the branches also as extension from stem. The infected region turns dark and necrotic, which is more pronounced in the stem near soil surface. Leaves exhibit yellowing and drying, resulting the hollowness of the stem. Root also affected at later stages. The plant dies with in a week time of attack. On the stem of dried plants, abundant pycnidia are produced as erumpent bodies. This disease is controlled by the following methods.

1. Seed treatment with carbendazim or chlorothalonil or captan at 3 g/kg of seed.
2. Seed palleting with spore suspension of *Trichoderma viride* (10^6/ml) at 50 mL/kg of seed.
3. Spot drenching with 0.1% carbendazim.
4. In the chronic area, 150% application of phosphorus plus application of alachlor and *neem cake* application at 250 kg/ha have to be included as packages since these application reduce the disease incidence.

13.4. General Recommendation to Control Sesame Diseases

1. Dry seed treatment with captan/thiram/carbendazim at 2 g/kg seed or wet seed treatment with carbendazim at 3 g/kg seed or bio-treatment with *Trichoderma viride* at 4 gram/kg of seed.
2. Spraying 500 g carbendazim or 1000 g mancozeb/ha at 35 DAS along with insecticide.
3. Spraying of wettable sulphur at 2.5 kg/ha along with insecticide at 50 DAS along with insecticide

14. HARVESTING

Scientifically harvesting of sesame can be done within two weeks from the physiological maturity. Normally dry weigh at of 1000 grain weight is increased rapidly from 25th to 35th day after flowering and reaches its peak on 40th day after flowering in upper part of capsule and 45th day after flowering in lower and middle part capsules. This is considered to be period of physiological maturity. At this period, 1000 grain weight of sesame would be around 2.14 g with 90 per cent germinability and 26 per cent of grain water content.

Harvesting should be done when the capsule have turned yellow and the leaves have dropped as the highest oil content in seed is obtained at this stage further delay in harvesting causes shattering of grain without any increase in oil content. Harvesting at early stages results in low oil content which increases quickly up from 10th to 25th day after flowering from 25 to 35 percent and reaches its maximum at 35th day after flowering. Grains of upper capsule are much lower in oil content than those of middle or lower capsules. The optimum time of harvesting is when the moisture content in the capsule drops below 70 per cent, leaf senescence reaches upper nodes and 50% of capsule loose green colour. General recommendation for harvesting and curing of sesame are as follows:

14.1. Decision for Harvest

1. observe the crop, considering the average duration of the crop.
2. 25 per cent of bottom leaves might have been shed and the top leaves might have lost their green colour and turned to yellow.
3. The colour of the stems would be yellow
4. Upto middle height of the plant, the capsules might have changed to yellow colour.
5. Harvest is to be done before the bottom capsules turn brown and ready to dehisce with light drizzle or moisture addition.
6. Examine the 10th capsules from bottom by opening. If the seeds turn black (varies with varieties) harvest may be done.
7. If harvest is delayed the capsules will dehisce and there might be yield loss. The estimated yield loss due to delayed harvest is around 30 per cent

14.2. Cutting Stalk, Curing Threshing and Drying

1. Pull out the plants from the bottom or cut at ground level.
2. Stack in the open, one over the other in a circle form with the stems pointing out and the top portion pointing inside.
3. Cover the top with straw, so that humidity and temperature increase.
4. Cure for 3 to 5 days so that the seeds which are at different stages of maturity, mature uniformly.
5. After keeping for 3 to 5 days, spread the plants in the thrashing floor and alone the capsules to dry and dehisce.
6. After drying for 3 days, shake the plants. About 75 per cent of the seeds will fall off.
7. Dry the plants for one more day and again shake the plants. All the mature seeds will fall off.
8. Winnow the seeds and dry in the sun for 3 days to five percent seed moisture. Stir once in 3 hours to give uniform drying.
9. Collect the seeds and store in gunny begs with necessary plant protection.

15. YIELD

Under well managed condition a normal sesame components for a variety crop yield 5-6 q/ha.

16. QUALITY

From sesame plants the economic products obtained are - sesame seed, sesame oil, sesame oil cake and sesame flour.

16.1. Sesame Seeds

Sesame seeds are small in size (1000 seed weight varies from 2.0 to 3.5 gram) and vary in colour from white through brown to black. The pericarp forms 9.8 to 10.2 per cent of the weight and the dehulled kernel, irrespective of the colour of the pericarp is white in colour. The composition of white and black varieties of seeds are given in Table 11.7.

TABLE 11.7: Chemical composition of sesame seed and hull (g/100 g material)

Constituent	White seeded variety			Black seeded variety		
	Whole seed	Dehulled seed	Hull	Whole seed	Dehulled seed	Hull
Moisture	5.4	6.2	4.2	5.2	6.1	4.1
Fat	50.2	54.1	10.2	49.8	53.5	9.9
Protein	19.8	21.1	8.0	20.0	21.5	8.1
Crude fibre	3,2	1.4	18,5	3.3	1.3	19.5
Carbohydrate	14.9	14.7	22.0	14.7	14.9	15.9
Mineral matter	4.8	2.3	22.8	5.2	2.6	24.2
Oxalic acid	1.72	0.2	14.3	1.8	0.09	18.3
Calcium	1.06	0.19	9.75	1.21	0.06	12.1
Phosphorus	0.47	0.48	0.51	0.62	0.62	0.66

Hulls form 9.8% of the weight of the whole seed; Sesame seeds losses viability relatively quickly unless it is stored with enough care. Activities of enzymes like, catalase, peroxidase, amylase and acid phosphatase are decreased in dry seeds. Total dehydrogenase activity has shown positive correlation with the loss of viability.

Electro conductance, free sugars, orcinol and folin positive compounds and inorganic phosphate are increased in the seed leachate with the loss of viability. Seeds stored under high relative humidity had more leakage as compared to those stored under lower relative humidity.

It was found that at the end of eleven months there was reduction in viability and oil content and increase in acidity, when stored in mud pots and gunnies. Under 50 per cent relative humidity and 18° C temperature the seeds could be stored for a maximum of one year without any loss of germination. Well dried sesame seeds (4% moisture) retain their full viability at a storage temperature of -10 to 32° C for two years. Seeds with 10 per cent moisture did not retain their viability, while those with 7 per cent moisture remain viable at low temperature (upto 10° C). If sesame seeds are stored in air tight container the viability of seeds can be maintained upto three years.

The seeds which are to be used for sowing should be stored after using following practices:

After winnowing, the seeds are dried continuously for three days both in the morning and evening to achieve 4 to 5 Per cent seed moisture. Then they are mixed with malathion dust (2 kg/100 kg seed) and kept in polythene/gunny. On every new moon day these seeds are dried under sun and kept again in the gunny, if necessary again adding malathion dust. This method has kept the seeds viable for two years.

16.2. Sesame Oil

The characteristics of the oil obtained from different varieties of seed lie with in the following ranges:

Specific gravity 20°	0.917-0.922
N 40°	1.4660-1.4671
Acid value	1.0-4.0
Saponification value	186.9-195.0
Iodine value	104.0-118.0
Thiocyanogon value	75.0-76.2
Acetate value	1.1-9.8
Unsaponifiable matter	1.5-2.3%

The keeping quality of sesame oil is related to its acidity. Generally oils containing high levels of free fatty acids do not store well.

16.3. Oil Cake

Sesame cake or meal obtained as a by product of the oil milling industry is highly valued as a livestock feed because of its high methionine content.

16.4. Sesame Flour

Sesame flour from dehulled and defatted seeds has been used as an exclusive protein source in nitrogen balance studies on children (3-7 Years) at levels of biological value and net protein utilised value of the flour are 88, 62 and 54 respectively. These values compare favourably with those reported for groundnut and cotton seed flours.

17. POST HARVEST TECHNOLOGY

One of the efficient ways to increase oilseed availability is to control the losses occurring between the field and the consumer or in other words it is much economical to save what is already produced. The estimated loss of sesame seed is around 10 per cent from harvest to threshing.

17.1. Drying

The drying operation (sun drying or mechanical drying) reduces the level of moisture in the harvested crop, enhances its storability, transportability and flavours. By reduction of moisture content of sesame to 4 to 5 per cent pest infestation and microbial deterioration are eliminated

Drying can be done in two ways.

1. **Sun drying:** Even though sun drying is a traditional one with little cost, it always results in non-uniform drying if done carelessly.
2. **Use of mechanical dryers:** Mechanical dryers can be used even during rainy period however its high cost limits its broad use by farmers.

The cheap drying method for sesame is as follows:

1. After winnowing of the harvested sesame seed,

2. Drying under sun for three consecutive days to reduce the seed moisture to 4 to 5 Per cent is recommended and then it may be stored in polythene sacks. Drying may be repeated on every new moon day.
3. For seed purposes, after drying to seed moisture content of 4 to 5 per cent, mix malathion dust 5D @ 2 kg per 100 kg of seed and store it in polythene sacks or mix with silica-gel at 350 g/kg of seed.

17.2. Cleaning and Grading

Cleaning and grading are important steps in the production of high grade seed as well as production of quality oil and cake. Sesame seeds obtained after harvesting and threshing may contain a lot of foreign materials. Impurities higher than the seed size may be removed by a blower or aspirator. Large and small impurities can be removed by reciprocating or rotary sieves, while those with same size, but with weight differences can be separated by specific gravity separators. Based on these principles many types of winnowers and graders are available for cleaning and grading sesame.

17.3. Storage

Both sesame seeds and its cake are attacked by the following pests during its storage. This is mainly due imbibing of moisture from the surroundings by the seeds and cakes, hereby initiating biochemical activities.

17.3.1. Pest on Sesame Seeds

1. Rice moth (*Orcyra cephalonica* stainton)
2. Redflour beetle (*Tribolium castaneum* herst)
3. Khapra beetles (*Trogoderma granerium* Everts)
4. Almond moth (*Cadra cautella*)

Among *T. castaneum* and *C. cephalonica* have been found feeding both sesame seeds and cakes. These pests feed on germ portion and damage the viability of the seeds. The damage rate increases with increase in storage time.

The Susceptibility of sesame seeds to storage pests increases with increase in oil content and white seeded sesame is highly susceptible to tribolium than black and brown seeded sesame.

17.4. Control

The weight loss in stored seed is low (6.6%) when the seeds are stored in urea bag (polythene sack even after four months of storage. But gunny bag and mud pot storage results in a weight loss of 15.5 and 16.7 per cent. The dusts endosulfan 4D, and ekalux 4D at 1 per cent are good in controlling tribolium pests followed by malathion 5D (1.8%). *Neem* oil and *neem* seed kernel powder and castor oil can also used.

18. PRODUCTION CONSTRAINTS

The attributed main reason for low sesame yield in many countries including India is poor crop management rather than the low genetic potential of the cultivars grown. The extreme sensitivity of sesame crops to climatic factors and susceptibility to diseases and pests are some of the other reasons of low productivity.

The production constraints of sesame in India are as follows:

18.1. Varietal Constraints of Sesame

- Mostly present day varieties are indeterminate with shattering capsules. They have restricted adaptability and are thermo and photosensitive. They lack compact dwarf structure, early maturity, good branching and non-shattering capsules with high harvest index and so are unable to compete with other crops
- Varieties with wide adaptation are required, which can stabilise and increase the production, but such varieties are not available at present.
- The induction of characters like drought tolerance, disease and pest resistance and deep root system is required.
- Sesame is self pollinated crop, nature restricts the exploitation of hybrid. Sesame hybrid has to be developed for high technical management.
- Photoinsensitive types are not available.

1. Mainly sesame is cultivated in marginal and sub marginal soils under dry farming situation. In India only 10 Per cent of total sesame area is under irrigation.
2. Scientific seed production programme is not existing.
3. Poor technology adoption.
   a) Required population per unit area is not maintained.
   b) Mainly seeds are sown by broadcasting.
   c) Grown as mixed or inter-crop with little attention.
   d) Poor plant protection given.
   e) Subnormal fertiliser dose to sesame crop.
   f) Poor weed management.
   g) Because of higher oil content, it requires attention for proper storage for profitable marketing.
4. Weak extension activity
   a) Neglected crop to farmer, scientists and farmers
   b) Traditionally treated as minor and marginal crop.
   c) Competition of oils of other major oilseed crops like groundnut, mustard, rapeseed
5. Non availability of organised marketing system.
6. Sub optimal research programmes.

19. FUTUROLOGY OF SESAME

V-shaped cultivation: In China, in respect of branching sesame type, nipping of sesame plants after leaving 2 pairs of leaves from the cotyledon node between 15-20 days after sowing is done this operation is preceded by thinning. By clipping, sesame plant gives more number of primary, Secondary and tertiary branches, result in higher yield. They also apply 150 per cent of recommended fertiliser under irrigated condition. This technology was tested in India and found encouraging.

1. Whatever, might be the procedure adopted to create male sterility in sesame, release of hybrids in sesame is in the near future. This area needs immediate attention.
2. Efficient cropping zone for sesame may be identified both for irrigated and rainfed situations.
3. Nematode appears to be becoming a severe problem. In depth research is needed.
4. Intensive research is needed to control black stem disease.
5. Pelleting of sesame seeds like groundnut seed size with inert material is required to sow the seeds through seed drill. This will help to maintain population.

12. Castor

D. D. Malaviya, V. K. Poshiya and H. L. Dhaduk

Castor is remained as poor man's crop with recipient of minimum inputs hitherto. However, with changing scenario of modern agriculture with high-tech production systems accompanied by free market through globalization. It has now emerged out as one of the most commercial non-edible oilseed crops in the country. Primarily it contains 25-58 per cent oil in different varieties with an average range of 40-55 per cent. Its stem, leaves, seeds, oil and cake are useful in various ways. Castor oil exhibits a wide variety of uses, including industrial, lubrication, medicine both human, veterinary and domestics. Oil cake is used as manure and plant stalks are used as fuel or thatching material and pulp for paper industries. The castor leaves are used for rearing silk worms in silk producing areas. In recent past, the castor oil has gained an important place as commercially high valued non-edible oil in market including use in aero-engines.

Castor oil is also used as purgative. It can be used medicinally as cathartic and obstetrics. It is also applied externally as a bland emollient to the skin as a skin ointment for skin diseases. Castor oil can also be used externally as emollient as a soothing medium when applied into eye after removal of foreign bodies and also laxative for many animals.

Castor oil is used in the manufacture of resins. It is mainly used for surface coating on household equipment, furniture, refrigerators and kitchen equipment. Besides the resins it also form the base materials for several paints, enamels and varnishes of superior quality.

Castor oil is used in the preparation of leathers, adhesives, synthetic perfumes and flavours, paints, varnishes making, Heveacrumb rubber and nylon. The oil is also used in the preparation of synthetic detergents, wax and a variety of rubber goods, hair oils etc. Castor oil is used in obtaining clear bright colours in dying fabrics and for soaking raw skins in fur trade.

1. ORIGIN

Castor is considered as one of the ancient non-edible oilseed crops. It is indigenous to eastern Africa and most probably originated in Ethiopia (Weiss, 1971). Russian plant breeders consider that there are four large centre of variability, each with its own specific plant characteristics to support a polyphylletic origin for the crop. The most important is the Irano-Afghanistan-USSR region. The other three located in Palestine south-west Asia, India-China and Arabian Peninsula (Moshkin, 1977).

2. DISTRIBUTION

Castor is grown under varied conditions of climate including tropical, subtropical and temperate. However cultivation is mostly confined to countries lying between 52° N to 40° S latitudes. Major countries growing the castor are India, China, Brazil and Thailand. India ranks first in respect of acreage and production among these countries (Table 12.1). Major castor growing states in India are Gujarat, Andhra Pradesh, Tamil Nadu, Orissa and Rajasthan. It is also grown in states of Uttar Pradesh, Maharashtra, Karnataka and Madhya Pradesh, Bihar too. The Gujarat state ranks first position with respect to area, production and productivity among all major castor growing states in the country (Table 12.2).

TABLE 12.1: Area, production and yield of castor in major countries of the world

Country		1992	1993	1994	1995	1996
India	Area	659	688	821	980	1050
	Production	617	643	883	930	1000
	Yield	936	935	1076	949	952
China	Area	240	240	240	240	240
	Production	290	280	260	250	280
	Yield	1208	1167	1083	1042	1167
Brazil	Area	175	141	106	75	123
	Production	102	43	53	32	46
	Yield	583	305	500	427	374
Thailand	Area	35	35	22	20	19
	Production	30	32	18	16	15
	Yield	857	914	818	800	789
World	Area	1257	1251	1342	1465	1582
	Production	1131	1094	1314	1324	1418
	Yield	900	875	979	904	896

Area = 1000 ha; Production = 1000 MT; Yield = kg/ha

TABLE 12.2: State-wise estimates of area, production and yield of castor

State		1992-93	1993-94	1994-95	1995-96	1996-97	Average
Gujarat	Area	306.1	312.3	383	388.1	371.2	352.1
	Production	496.4	509	687	613.4	737.1	608.6
	Yield	1622	1630	1794	1581	1986	1722
Andhra Pradesh	Area	245.0	259.6	269.6	264.1	240.9	255.8
	Production	67.0	75	90.4	71	61.2	72.9
	Yield	273	289	335	269	254	284
Tamil Nadu	Area	27.4	33.1	41.6	32.5	40.3	35.0
	Production	8.5	10.3	13	10	12.7	10.9
	Yield	310	311	313	308	315	311

State		1992-93	1993-94	1994-95	1995-96	1996-97	Average
Orissa	Area	25.6	25.6	26.3	28.9	26	26.5
	Production	13.1	13.2	14.1	15.6	8	12.8
	Yield	512	516	536	540	308	482
Rajasthan	Area	12.9	15	25.1	38.6	35.8	25.5
	Production	14.5	8.8	20.5	43.8	48.5	27.2
	Yield	1124	587	817	1135	1355	1003
Karnataka	Area	24.2	22.3	22.8	22.4	18	21.9
	Production	17.4	13.2	18.6	18.6	18	17.2
	Yield	719	592	816	830	1000	791
All India	Area	659.7	684.5	787.2	789.0	747.5	733.5
	Production	623.1	634.8	849.2	777.3	890.7	755.1
	Yield	945	927	1079	985	1192	1029

Area = '000 ha, Production = '000 tonnes, Yield = kg/ha

3. CLASSIFICATION

The name *Ricinus* is a Latin term meaning *Dog tick* and was given to this plant by Linnaeus because of the resemblance of the seed to that canine pest. Linnaeus grouped all the varieties of the plant under one species *Ricinus communis* Linn., family euphorbiaceae Popova (1930) a Russian worker, has tried to divide the species communis into the following subspecies.

This classification is based on eco-geographical groupings. But the last three so resemble each other that it is very difficult to differentiate. Hence Hilderbrandt suggests that the last three be placed under one group namely 'Indo-African'.

3.1. The Persian subspecies: Percius G. popova

It is found in Iran, India, Middle Asia and parts of Russia. This subspecies has two outstanding morphological characteristics. The inflorescence is elongated, narrow, conical and the fruits may cover the flower stalk to the base. Its seed is oval, small and has no caruncle when ripe. It is one of the most productive of all the known forms. It is medium duration variety, which is divided into four types possessing shattering capsules.

3.2. The Chinese subspecies: Chinensis G. Popova

The forms of Chinese subspecies are found in West China and Manchuria. This subspecies is early takes 75 to 97 days for maturation. The inflorescence is short and loose with a broad, conical shape. Seed is oval and large and grey coloured with a small caruncle. Plants bear up to 13 branches. This subspecies is less productive and the seed is distributed over a large number of small inflorescence. The important features of this subspecies are its earliness and non-shattering characters of the capsules. This subspecies is further divided into three types.

3.3. The Indo-African subspecies: Africanus G. popova

This subspecies is found in four continents. The ecological diversity of this subspecies is very high. It possesses 88 per cent of all the known characters of castor plant and therefore it is very valuable for breeding purposes. The important morphological feature of this subspecies are as shown below.

1. Seed elongated, narrow at the caruncle which is well developed.
2. The shape of inflorescence varies.
3. The lower part of the flower stalk is usually free of pistillate flowers.
4. Many anthocyanin forms are found.

3.4. The Zanzibar subspecies: Zanzibarinus G. popova

This subspecies is found in the tropical parts of Africa and America and in Europe as an ornamental plant. Most forms of this subspecies are late maturing. The subspecies has almost square shaped seeds, with well developed caruncle. The inflorescence is cylindrical or ellipsoid in shape. The capsules are large and spherical. The plants are strongly developed with thick branches and large sized leaves.

There exists a remarkable multiplicity of forms in castor. However the all cross freely with each other and thus they constitute a single group namely *Ricinus communis*, again which is the mainly cultivated species.

Castor is highly sensitive to environmental changes. The characters like plant height, height at which main spike appears, the length and compactness of spike, branching habit are greatly affected by environmental changes. Plants when grown in iso-competition produce 2 to 3 lateral spikes. However, it produces as many as 30 lateral spikes when grown in absence of iso-competition on highly fertile soil. Therefore any classification based on such characters will be temporary and will only be applicable to a specific set of condition. Such a classification does not comply with the requirements of botanical classification.

4. GROWTH STAGES

Three periods of growth of castor have been identified such as germination, vegetative growth and capsules formation. During the first period (i.e. from sowing to sprouting) swelling, growth of the first rootlet and hypocotyl and emergence of cotyledons on the surface takes place. Generally its duration is 10 to 15 days, depending upon the variety and the temperature conditions.

During the period of vegetative growth (i.e. sprout to flowering of central spike), the formation of secondary roots growth in height of main stem with the formation of nodes and leaves takes place and the central spike flowers. Its duration is 35 to 70 days.

During the fruiting period seeds of the central spike ripen, lateral shoots grow and the seeds of lateral spikes occur. Flowering and ripening of the seeds of lateral spikes of the first and second orders belong to intermediate phases of development. The main intermediate phase of development is the ripening of seeds on the central spike, which gives up to 80 to 90 per cent of total yield. The duration is 33 to 90 days.

5. CLIMATE

Basically a semi-tropical perennial, castor now grows throughout the warm-temperate and tropical regions and flourishes under such a variety of climatic conditions. It has been commercially cultivated from 40° S to 52° N, from sea-level to 3000 m but will grow up to 3000 m with an optimum 300 and 1500 m, the limiting factors being frost during growth. A hard frost will normally kill dwarf or annual castor at any stage of growth, young plants being most susceptible and a frost free growing period of 140-190 days necessary. It grows in tropical and subtropical regions as a perennial plant and in temperate climate as an annual plant. Long, clear, sunny days are most suitable and cloudy or humid days irrespective of temperature will reduce yields.

Castor requires a moderately high temperature 20-26° C with low humidity throughout the growing season to produce maximum yields. Castor is sensitive to high temperature especially at flowering. The interaction between temperature and light intensity can also affect seed size and composition. Very high temperature at flowering 40° C or above even for short periods results in blasting of flowers and poor seed set. High temperature can also adversely affect seed composition; above 35° C will reduce oil and protein content; below 15° C reduce oil content and alter oil characteristics.

High soil temperature decreases time span between sowing to emergence and low soil temperature extends it. Thus temperature and time of sowing are closely related.

Castor is a deep rooted crop and shows fair degree of resistance to drought. In India castor is widely grown as a rainfed crop with sowing in May and July. It requires a well distributed rainfall of 500 to 600 mm during growing period for good yields. It can thrive well on sandy soils under the conditions receiving a rainfall of 300 to 500 mm, when the varieties showing extremely quick root growth and establishment are used for commercial planting.

Castor is a long day plant. However it grows well without marked reduction in yield under wide range in day length. No significant differences with respect to plant growth and development are noticed when grown under different photo periods of 12, 14, 16 and 18 hours.

6. SOIL AND SOIL PREPARATION

Any type of soil, except heavy clays and poorly drained is suitable for castor growing. It grows well on the light sandy soils. Deep, moderately fertile, with slight acidic condition (pH 5.0 to 6.5) are more suitable. It can also tolerate a pH of 8.0 but at this condition, the soil structure and physical properties of soil, will become the major limiting factor for castor growing. Soils with high fertility are not desirable as they favour excessive vegetative growth at the expense of seed yield. The castor grows well in soils, which is not fit for other valuable commercial food crops.

Deep ploughing is beneficial especially for rainfed castor because this practice will help in absorption and conservation of rainwater. Harrowing once or twice with the receipt of pre-monsoon rains is advantageous. In lateritic soils deep ploughing is beneficial as it helps in breaking up the hard layers in sub-soils. One or two ploughing followed by one or two harrowing with disc or blade harrows to pulverise the clods for obtaining optimum tilth is a normal practice of land preparation.

7. CROPPING SYSTEM

Crop rotation is useful in castor as it controls wilt disease. Therefore the crop rotation with major crops of a particular area like pearl millet, sorghum, other oilseed crops and pulses should be followed for good harvest. As soon as the castor crop is over it is beneficial to grow the crops like pearl millet or groundnut or green gram during summer season. Castor gives higher yield when shown after pulses. Winter wheat happens to be the best preceding crop for the castor. Thus wheat-castor, pulses-castor, castor-pulses and castor-pearl millet appears to be most common crop rotation system.

7.1. Inter-cropping System

Castor is also grown as a mixed crop. Though inter-cropping system varies from state to state, following systems are widely accepted.

States	Suggested Inter-cropping	Ratio
Gujarat	Castor + Groundnut	1 : 3
	Castor + Black gram	1 : 3
	Castor + Pigeonpea	1 : 1
Andhra Pradesh	Castor + Groundnut	1 : 3
	Castor + Sorghum	1 : 2
	Castor + Cowpea	1 : 2
	Castor + pearl millet	1 : 2
Karnataka	Castor + Chilli	1 : 1
	Castor + Rai	1 : 1
	Castor + Bean	1 : 4
Tamil Nadu	Castor + Groundnut	1 : 3
	Castor + Green gram	1 : 2
Madhya Pradesh	Castor + Soybean	1 : 3
	Castor + Gram	1 : 3
Orissa	Castor + Groundnut	1 : 3
	Castor + Horse gram	mixed cropping

8. MANURES AND FERTILISERS

Castor is a long duration crop standing in fields up to 120-150 days and 180-210 days for short and medium duration varieties/hybrids, respectively. The nutrient requirement specifically depends on variety besides the type of soil and moisture availability. In general seed yield of one tons ha^{-1} removes 30 kg N, 12 kg P_2O_5 and 10 kg K_2O ha^{-1}. Castor being an oilseed crop its sulphur requirement should not be overlooked.

Soil nutrient depletion is a common problem in India. It has been reported for the country as a whole (Tandon and Pratap Narayan, 1990); for each of the 15 agro-climatic regions (Biswas and Tewatia, 1991) and within a region (Yadav and Prasad, 1992).

8.1. Manures

Application of farm yard manure or compost at the rate of 5-6 tons ha^{-1} during the last ploughing is beneficial to improve the soil properties and it enhances the availability of micro-nutrients. Green manuring with glyricidia and lencaena was found beneficial for crop production.

8.2. Fertilisers

Nitrogen: Nitrogen recommendation vary from region to region and system of cultivation. As castor is grown as rainfed and irrigated crop in the country, the response to applied 'N' vary under rainfed and irrigated conditions.

Nitrogen recommendations for rainfed castor for different states of the country

State	N (kg ha^{-1})	Remarks
Gujarat	40	Half dose at sowing and remaining at flowering
Andhra Pradesh	40	Half dose at sowing and remaining at flowering
Karnataka	37	All as basal dose
Tamil Nadu	30	All as basal dose
Madhya Pradesh	40	Half dose at sowing and remaining at flowering
Orissa	50	Half dose at sowing and remaining at first inter culturing
Rajasthan	40	Half dose at sowing and remaining at first inter culturing

Nitrogen recommendations for irrigated castor for different states of the country

State	N (kg ha^{-1})	Remarks
Gujarat	75	½ 'N' as basal, ¼ 'N' at 30-40 DAS ¼ 'N' at 70 DAS
Andhra Pradesh	80	½ 'N' as basal, ¼ 'N' at 30-40 DAS ¼ 'N' at 70 DAS
Karnataka	62	½ 'N' as basal and remaining at 40-50 DAS
Tamil Nadu	60	½ 'N' as basal and ½ 'N' 40-50 DAS
Madhya Pradesh	60	½ 'N' as basal and ½ 'N' 40-50 DAS
Rajasthan	40	All as basal dose

DAS: Days After Sowing

8.3. Phosphorus

TABLE 12.3: Phosphorous recommendations for rainfed castor for different states of the country

State	P_2O_5 (kg/ha)	Remarks
Gujarat	40	All as basal dose
Andhra Pradesh	20	As basal, preferably by placement method
Karnataka	37	At sowing by incorporation
Tamil Nadu	15	Incorporate into soil during last ploughing
Madhya Pradesh	40	All as basal dose
Orissa	30	As basal by incorporation
Rajasthan	20	All as basal dose

TABLE 12.4: Phosphorus recommendations for irrigated castor

State	P_2O_5 (kg/ha)	Remarks
Gujarat	50	Basal placement at 5-7 cm depth, 5-7 cm away from crop
Andhra Pradesh	30	Basal placement at 5-7 cm depth, 5-7 cm away from crop
Karnataka	60	Basal placement, 4-6 cm away from crop in a ring
Tamil Nadu	30	Basal placement at 5-7 cm depth, 5-7 cm away from crop
Madhya Pradesh	30	Basal placement at 5-7 cm depth, 5-7 cm away from crop

State	P_2O_5 (kg/ha)	Remarks
Rajasthan	40	Basal placement at 5-7 cm depth, 5-7 cm away from crop

8.4. Source of Phosphorus

The water soluble single super phosphate is generally preferred for castor. In addition to phosphorus this fertiliser also supplies sulphur and calcium.

TABLE 12.5: Potassium recommendations for castor

State	RF / I	K_2O (kg ha^{-1})	Remarks
Gujarat	-		Under 'K; deficient soil as basal application
Andhra Pradesh	RF	40	As basal by incorporation
Karnataka	RF and I	25	--do--
Tamil Nadu	RF	15	--do--
Madhya Pradesh	RF	40	--do--
Orissa	I	50	As basal by placement method

Note: RF = Rainfed and I = Irrigated

8.5. Scope of Potassium

The most widely used potassium source is muriate of potash. It is water soluble and also less costlier.

9. MICRO NUTRIENTS

- Zinc: Common symptoms of Zn deficiency are stunted growth, poor branching, development of light green, yellow, bleached spots, chlorotic bands on either side of the midrib. Little leaf condition is also a common symptom. Inter-nodes are short; flowering, fruiting and maturity are delayed; shoots may die off and leaves fall prematurely.
- Iron: Iron deficiency begins on younger leaves. Otherwise deficiency symptoms are somewhat similar to Mn as both lead to chlorosis. Yellowing of the interveinal area of leaves commonly referred to as iron chlorosis. In case of severe deficiency yellow and green strips along the length of the leaf blade may be observed.
- Sulphur: Sulphur is a part of amino acids cysteine and methionine. This is essential for protein production. Sulphur deficiency symptoms in many ways resemble to those of 'N' in that they produce pale yellow or light green leaves. Unlike N, S deficiency symptoms usually appear first on the younger leaves and persist even after N application. Plants deficient in 'S' are small and spindly with short and slower stalks, their growth is retarded.

Sulphur deficiency is most likely to occur in coarse textured soils low in organic matter in high rainfall areas and the in the systems of crop rotations which include pulses and oilseeds.

10. INTEGRATED NUTRIENT MANAGEMENT

No single nutrient source in India is in a position to meet the entire nutrient demand of the order of about 22 million tonnes which is likely to reach 28 million tonnes by 2000 A.D. Besides increasing use of chemical fertilisers, there is an urgent need to exploit potential alternative sources of plant nutrients fully. The complementary use of chemical fertilisers, organic manure and bio-fertiliser referred to as 'Integrated Nutrient Supply System' (INSS) has a major role to play to solve the problem of nutrient mining. The basic concept of this system is the maintenance of soil fertility to an optimum level of sustaining the desired crop productivity through all possible sources of plant nutrients in an integrated manner. The main components of INSS are chemical fertilisers, organic manure, green manure and bio-fertilisers.

The research work carried out at various research centres under All India Co-ordinated Research Project on Castor under rainfed and irrigated conditions clearly indicated the beneficial effect of integrated nutrient management not to chemical fertilisers.

10.1. Rainfed Situation

The application of 50 per cent of recommended dose of inorganic fertiliser over a basal application of 50 % N through farm yard manure at 6 t ha^{-1} being comparable with 100 % RDF, 50 % RDF + *Azospirillium* seed treatment + 25 % N (FYM) and 50 % RDF + 25 % N (FYM) + 25 % N (Neem cake) resulted in significantly higher seed yield of castor over 50 % RDF + green manure and absolute control.

10.2. Irrigated Situation

Studies carried out at Sardar Krushinagar (Gujarat) on the efficacy of different sources of fertiliser under irrigated conditions revealed that 100 % recommended dose (80-50-0) + castor cake 1 t ha^{-1} + PSB (Phosphate Solubilizing Bacteria) gave highest yield (3757 kg ha^{-1}) which was closely followed by 50 % RDF + castor cake 1 t ha^{-1} + PSB (3560 kg ha^{-1}).

11. VARIETIES

TABLE 12.6: Popular varieties of castor with important characteristics

Variety	Characteristics	Recommended area
GAUC-1	Early well branched with non-shattering types of capsules, with oil content 47.1 per cent. Released in 1973	Gujarat
GAUCH-1	Medium tall (100-200 cm) hybrid released in 1973; for which VP1 and GAUC-1 formed female and male parents respectively. It is an early maturing hybrid with triple bloom, green stem spiny capsules and drought escape mechanisms but it is susceptible to wilt and root rot complex. Oil content 49.5 per cent	Gujarat
GCH-4	Hybrid (VP-1 x 48-1) is resistant to wilt. It is medium tall, triple bloom, with semi spiny capsules, maturing in 140-150 days. It gives 20-25 % higher yields than GAUCH-1 and GCH-2. Definite contrasting characters of male and female parents of GCH-4 facilitates easy detection of off types in seed production. Oil content 49.0 per cent.	All regions

Variety	Characteristics	Recommended area
GCH-5	Hybrid (Geeta x SH-72) is resistant to wilt and tolerant to root rot. Stem colour mahogany, double bloom, possesses spiny capsules and loose spikes. It has flat leaf shape. Oil content 48.6 per cent.	Gujarat and Rajasthan
Bhagya	It is a dwarf and early maturing variety (120-150 days). Tolerant to jassids and white flies. Oil content 54.0 per cent.	Andhra Pradesh
Saubhagya	Dark green leaves, short petioles covering lateral shorts; released in 1974. Developed from a double cross involving Aruna and a short mutant Mauthners dwarf. The convergent plant types of slightly longer duration (180-210 days) made it more suitable for inter-cropping systems. Oil content 51.0 per cent.	Andhra Pradesh
Aruna	It is a mutant variety obtained by radiation and semi-dwarf and early maturing (150-160 days) with double bloom and non-shattering capsules. Suitable for double cropping, high female tendency in the spikes, released in 1970 from Andhra Pradesh. Oil content 50.8 per cent	Andhra Pradesh, Karnataka, Orissa and Rajasthan
TMV-5	Stem rose coloured, triple bloom, spiny capsules; non-dehiscent, small seeded; 12-14 nodes, tolerant to jassid, released in 1985. Oil content 50.8 per cent.	Tamil Nadu
DCS-9	Stem colour red, double bloom, spiny capsules flat leaf and compact spike (mostly female). Oil content 47.42 per cent.	Rainfed areas of peninsular India
GC-2 (SKI-73)	Parents (1-21 x VI-9) stem colour mahogany, triple bloom spiny capsules, spike loose and contains mostly female flowers, leaf shape flat.	All irrigated areas in the country

12. SOWING

12.1. Seasons

Usually castor is a *kharif* season crop. With introduction of new, early and high yielding varieties and hybrids, its cultivation is extended to *rabi* and summer seasons, also. Though summer season is not ideal for commercial cultivation of castor, except for the seed production of hybrids.

12.2. Time of Sowing

The rainfed castor is sown with the onset of monsoon. Generally it is sown in the month of June and earlier half of July. Whereas under irrigated condition the sowing period can be extended up to 15th August without any drastic reduction in yield.

12.3. Method of Sowing

Castor is sown by three methods viz., sowing in furrow, sowing with mechanical or conventional seed drills and by dibbling method. Dibbling method is particularly preferred for seed production and hybrid cultivation.

12.4. Seed Rate and Spacing

Seed rate depends upon size of seed. Generally a seed rate of 12-15 kg is sufficient for one hectare. Spacing varies from region to region depending upon the fertility of soil, season and condition under which the crop is sown. Nevertheless under rainfed condition spacing is maintained at 90 x 45 cm, for poor soils and 120 x 60 cm for fertile soils. Sowing can also be done at 45 x 30 cm or 60 x 30 cm under particular set of conditions.

12.5. Depth of Sowing

The seeds should be sown at a depth of 8 to 10 cm for rainfed crop and 6 to 7 cm for irrigated crop and should be properly covered with soil to ensure quick and better germination.

13. IRRIGATION

Crop in *kharif* season is grown as rainfed. Hence if the onset of monsoon and amount and distribution of rains over the crop period is normal, the crop growth and yields would be satisfactory. However 2 or 3 supplementary irrigation helps in enhancing the yield if the crop is exposed to prolonged dry spell. The *rabi* and summer season crops require 6 to 8 irrigations depending upon the soil type and prevailing weather conditions. Less number of irrigations are required if soils are heavy, on the other hand light soil requires frequent irrigation.

Kharif season crop requires first irrigation after 70 to 75 days of sowing, if monsoon is regular. However under dry spell early irrigation is necessary to maintain the yield level. The remaining irrigations should be given at an interval of 15 to 20 days after withdrawal of monsoon. The interval depends on types of soil besides the prevailing climate. In general frequent irrigations should be applied in sandy or light textured soils as compared to heavy textured. On the basis of depletion of available soil moisture, the castor crop should be irrigated at 40-50 per cent available soil moisture for economic production. The total water requirement of castor varies from 250 to 400 mm.

The castor crop requires 8 irrigations each of 50 mm depth in light soils. The first four irrigations should be applied at an interval of 15 days after cessation of monsoon (September-November). Remaining four should be applied at an interval of 20 days (December-February). Similarly, after cessation of monsoon six irrigations each of 50 mm depth are sufficient for medium black soils at an interval of 20 days, while only four irrigations each of 70 mm depth are adequate at an interval of 24-28 days for heavy soils.

13.1. Method of Irrigation

This is very simple and traditional method. The water is allowed to flow on surface either in flat bed or ridge and furrow. The length of beds depends on type of soil, Short beds are preferred for sandy and light textured soils in order to minimize the percolation losses as compared to heavy textured soils. In general water use efficiency (WUE) hardly exceeds 50 per cent under this method, which indirectly restrict the fertiliser use efficiency (FUE).

13.2. Micro Irrigation

It is modern technique of irrigation system, which includes sprinkler, mini sprinkler and drip system.

The drip irrigation is one of the best methods of irrigation especially in the arid and semi-arid regions, where the water is very costly input. Castor is one of the most dominating long duration wide spaced commercial crop. Therefore, drip system suits well as compared to other crops. The WUE under this system of irrigation is the highest (> 90 %).

The research work carried out at Junagadh and Sardar Krushi Nagar under Gujarat Agricultural University, clearly indicated its superiority over traditional method with respect to yield, WUE and FUE (Malavia, *et. al.*, 1995) and, therefore, this system has been recommended for the farmer's use. This system should be laid out at a dripper spacing of 60 cm and be operated at a pressure of 1.2 kg cm^{-2} (120 kPa) with 4 lph dripper. From September to November the system should be operated for about 40 minutes and from December to February for 30 minutes on an alternate day. With the equal quantity of water used for surface method of irrigation one can increase the area of crop by 1.5 times. It gives higher yield (36 %), saves 24 per cent water and 20-25 $kgNha^{-1}$ through fertigation.

14. WEED CONTROL

There are no specific crop bound weed species for castor as Phaliris minor and Chil in wheat fields in Northern wheat belt of India. Different types of weed flora are seen in castor fields under different agro-climatic regions. Similarly the weed intensity may vary with the types of soil, sowing time, variety, cultural practices, moisture availability etc.

Castor yields are highly affected by the weed competitions in early stage i.e. up to 45 to 60 days after sowing. However the severity of effect varies from variety to variety. Frequent interculturing commencing from 15 to 20 days from sowing with an interval of 12 to 15 days is recommended. The crop requires commonly three intercultural operations.

Research findings suggest that two inte-rculturing and two hand weeding operations at 30 and 60 days after sowing in castor hybrid (GCH-4) were found effective and economical for weed control. In castor weeds can also be controlled by using the following herbicides.

Common weeds of castor

Common name / English name	Botanical name	Family
Sheshmul	*Brachiaria spp.*	Gramineae
Bhumsi	*Eragrostis major*	Gramineae
Samo grass	*Echinochloa crusgalli*	Gramineae
Kanajaro	*Digera arvensis*	Amaranthaceae
Quail grass	*Celosia argentea*	Amaranthaceae
Land calthrops	*Tribulus terrestris*	Zygophyllaceae
Mexican poppy	*Argemone maxicana*	Papaveraceae
Kubi	*Leucas aspera*	Labiatae
Bermuda grass	*Cynodon dactylon*	Gramineae
Nut grass	*Cyperus rotundus*	Cyperaceae

Herbicides for castor		
Herbicide	**Rate (kg/ha)**	**Time of application**
Fluchloralin	1.0	Pre-sowing, soil application
Pendimethalin	1.5	Pre emergence to crop and weed (1-2 days after sowing)
Alachlor	1.3	Pre emergence to crop and weed (1-2 days after sowing)
Metolachlor	1.0	Pre emergence to crop and weed (1-2 days after sowing)

15. NIPPING

On emergence of primary spike, the axillary buds of main shoot are removed. This leads to increase in yield and crop matures earlier.

16. PLANT PROTECTION

Major insects-pest and nature of damage

16.1. Castor Semi-looper (*Achoea janata* Linn)

The catterpillar feed voraciously on castor leaves, starting from the edges inwards and leaving behind midribs and stalks only. Damage is found to be the highest in August-September, and with the excessive loss of foliage, the seed yield is reduced considerably.

Control measures: (1) Collection and destruction of larvae in early stages., (2) Deep ploughing of land to expose pupae, (3) Castor crop should be sown after 15th August to escape the pest infestation. (4) Chemical control measures.

Spraying the crop with any one of the following insecticides.	
Insecticide	**Dose ml per litre of water**
Endosulfan 35 EC	2.0
Monocrotophos 36 SL	1.0
Quinalphos 25 EC	2.0
Dichlorvos 76 EC	0.5
Cypermethrin 10 EC	0.9

16.2. Prodenia (*Spodoptera litura* Fab.)

The larvae feed on leaves and fresh growth. They are mostly active during night and cause extensive damage.

Control measures: (1) Collection and destruction of egg masses. (2) Use of light trap to collect adult insect and destroy them. (3) Use of pheromone trap to collect male insects and destroy them. (4) Deep ploghing after harvesting for destruction of pupae. (5) Application of NPV (Nuclear Polyhedral Virus) suspension at the rate of 250 larval unit/ha/week. (6) Making heaps of grasses in the field to collect the larvae.

Spraying the crop with any one of following insecticides.	
Insecticide	**Dose (ml per litre of water)**
Chlorpyriphos 20 EC	2.5
Endosulfan 35 EC	2.0
Dichlorvos 76 EC	0.5
Fenvalrate 20 EC	0.5
Decamethrin 2.8 EC	1.0

16.3. Castor Capsule Borer (*Dichocrocis punctiferalis* Guenee.)

The caterpillar bores into the capsules and eats the seed causing excessive damage to the crop. They bore into shoot and inflorescence when capsules is not found as a result affected parts of the plants withers and dries up. The pest is active from September to February.

Control measures: Spraying the crop with any one of the following insecticides.

Insecticide	Dose (ml per litre of water)
Monocrotophos 36 SL	1.0
Dichlorvos 76 EC	0.5

Dusting the crop with Quinalphos 1.5 % dust or Methyl Parathion 2.0 % dust at 25 kg/ha.

16.4. Castor Hairy Caterpillar (*Euproctis lunata* Walker)

Caterpillars feed on leaves of plants and in case of severe infestation, they may cause complete defoliation. The attacked plants remain stunted and produce poorly developed seeds.

Control measures: (1) Quinalphos 1.5 % dust should be applied on field boundaries to kill migrated larvae. (2) Dusting the crop with Methyl Parathion 2.0 % dust or Endosulfan 4.0 % dust at the rate of 20 to 25 kg/ha.

16.5. Jassid (*Empoasca devestans* F.)

Both nymphs and adults suck the sap from the leaves. As a result a characteristic "hopper burn" symptom is noticed wherein, margins turn yellowish initially, and later on turn reddish and turns downwards. In case of heavy infestation, the leaves show brown necrotic patches and the growth of the plants remained stunted, which adversely affects the yield.

Control measures: Spraying the crop with any one of the following insecticides.

Insecticide	Dose (ml per litre of water)
Methyl-o-demeton 25 EC	1.0
Phosphamidon 85 SL.	0.3
Dimethoate 30 EC	1.0
Monocrotophos 36 SL.	1.0

16.6. Thrips (*Scirtothrips dorsalis* Hood.)

The insects scrap the epidermal tissues of leaves and suck the oozing cell sap from the leaves, as a result whitish patches noticed on the leaves, later on, it turns into brownish in colour. Excessive feeding on leaves leads to their curling and the growth of the plants is stunted.

Control measure: As per jassids.

17. DISEASES

17.1. Seedling Blight (*Phytophthora parasitica* Dustur)

The first indication of the disease is the appearance of a roundest patch of unhealthy dull green colour on both the surfaces of cotyledons of seedling of 15-20 cm high. The cotyledons soon droop down. The disease spread from the leaf to the petiole, then to the stem and the growing point, killing the seedlings. In older plants, the spots are localized on the lamina or leaf blade. As the spots get older they turn yellowish and then become concentrically zoned. The disease is soil borne and is favoured by low lying, ill drained field.

Control measures: (1) Use of disease resistant varieties if possible. (2) For protecting the seedlings up to 20-25 days, seed treatment with thiram at 3 g / kg seed is recommended. Other fungicides like mancozeb, captafol can also be used successfully. (3) Well drained fields should be selected for sowing the crop. (4) The older plants, showing the true-leaf infection, should be sprayed with bordeaus mixture or with any other protectant fungicide like mancozeb 75 per cent (dithane M-45) at 2.5 g/Litre of water is recommended.

17.2. Alternaria leaf spot [*Alternaria ricini* (Yoshii) Hansford]

All the aerial parts including inflorescence and capsules are attacked by the pathogen under humid and cloudy conditions. Brown irregular spots with concentric rings are developed on the leaves. Concentric rings may not be clear on stems and capsules. Premature defoliation occurs in severely infected plants. Immature capsules wilt suddenly turning brown. Pathogen can survive on seed. Damping off symptoms are produced when infected seeds are sown.

Control measures: (1) Seed treatment with Thiram at 2 g / kg seed or Captafol is recommended to minimize the seed borne inoculum. (2) Three sprays of Mancozeb at an interval of 15 days starting from 90 days of crop growth have proved effective.

17.3. Cercospora Leaf Spot [*Cercospora ricinella* (Sacc. and Berlese) Speyazzini]

Symptoms of the disease appear first as minute, circular to angular pale yellow lesions, which later enlarge to turn brown with deep brown margin and greyish-white centres. The fructifications of the fungus appear as tiny black dots in the white centres. The infected leaf tissue may break off causing holes in the leaf blade. Sometimes the circular spots may coalesce to cover a much larger surface of the leaf causing blighting of the older leaves, while the tender young leaves remain green.

Control measures: (1) Spraying the crop twice with Copper Oxychloride 50 per cent w.p. (Blitox, Flytolan) @ 5.0 gm per Litre of water or mencozeb 75 per cent wp (dithane M-45) @ 2.5 gm per litre of water. (2) Resistant varieties, if available, should be used.

17.4. Fusarium Wilt (*Fusarium oxysporum* f.sp. *ricini* Nanda and Prasad)

The first symptom appeared in young plants during the period from sprouting to formation of two to three real leaves. Yellow sickly appearance of plants and marginal necrosis of leaves which later advances to interveinal areas and covers the leaves completely. In some plants wilting is preceded by appearance of spots or blight and dark strip formed on the stem. In later stage, up to harvesting individual branches or the entire plants acquire a dark violet ting and during moist weather, luxuriant mycelium of the fungus develop near the root neck. In the field, the first symptom of the disease appears on individual plant or in a small circle of plants, which increase and spread all over the field. Sick plants either do not bear capsules or give puny seeds.

Control measures: (1) Follow crop rotation. (2) Use wilt resistant varieties.

17.5. Root Rot [*Macrophomina phaseolina* (Tassi) Goid]

The root rot fungus *M. Phaseolina* cause typical symptoms at collar region as small brown depressed lesion, which later on sunken and spread upward and downward or the stem and plants break at this point. Shredding of tissues are the conspicuous symptoms in dried plants. Roots of the affected plant shows shredding of bark along with brown discolouration. Numerous black sclerotial bodies are also observed on the infected stem and roots. Many infected plants are dried without producing capsules.

Control measures: (1) Apply neem cake at 20 qt / ha before sowing. (2) Application of *Trichoderma harzianum* in soil before sowing. (3) Give the irrigation to crop in water stress condition. (4) Use of disease resistant varieties. (5) Follow crop rotation.

17.6. Bacterial Leaf Spot [*Xanthomonas campestris* pv. Ricini (Yoshii and Takimoto) Young]

Tiny angular dark water-soaked lesions appear on cotyledons and leaves of affected plants. They enlarge 2 to 5 mm in diameter and later become dark brown to jet black and coalesce to cause extensive brown foliar necrosis that can result in serious defoliation. Elongated dark lesions may develop on petioles and young branches. Bacterial gummy, shining ooze is seen in the form of small beads or fine scales on both sides of affected leaf.

Control measures: (1) The affected crop residue should be destroyed which help in the reduction of the spread of the disease in the field. (2) Seed treatment with streptocycline and 'Sandoz seed dressing 6334' have been found effective in control of the disease (Singh *et. al.*, 1976). (3) Hot water treatment of seeds at 50-60° C for 10 minutes eliminates the bacteria from the seed. (4) Spraying the crop with Streptomycine (agrimycin) @ 1 gm per litre of water is found effective in controlling the disease.

17.7. Fruit Rot or Blossom Blight (*Botrytis cinerea* Pers ex Fr.)

A dense wooly growth on flowers and capsules varying in colour from pale to olive grey are noticed. Though the disease is mainly restricted to flowers and capsules, the infection may spread to leaves and stems from affected racemes. First symptoms are small blackish spots from which drops of yellow liquid may exude. Fungal threads, which grow from these spots spread the infection and produce characteristic appearance of the affected raceme.

Control measures: Application of carbendazim (0.05 %) or thiophanate methyl (0.05 %) spray twice at 15 days interval after emergence of racemes.

18. HARVESTING

Castor produces spikes in the main shoot as well as on the primary, secondary and tertiary branches. The plant produces varying number of spikes, which matures at different intervals hence the harvesting is done in two or four stages as and when the spikes are fully dry. The capsules are picked out when the spike turn from yellow to black, starting at 120, 140 and 160 days after planting in dwarf varieties. The number of picking for other varieties is about six.

19. YIELD

- Capsule Yield: (1) Rainfed: 1200 to 1500 kg ha^{-1} (2) Irrigated: 2500 to 4000 kg ha^{-1}
- Stover yield: (1) Rainfed: 1200 to 1500 kg ha^{-1} (2) Irrigated: 2500 to 4000 kg ha^{-1}
- Grain to stover ratio: (1) Rainfed: 1:1 (2) Irrigated: 1:1

20. POST HARVEST TECHNOLOGY

20.1. Drying

Castor seeds have peculiarities, which differentiate them from seeds of other agricultural crops in drying. Under their comparatively thick and compact seed hull, there is a uniform air cavity; it is three times bigger in the embryo than in the remaining part. The seed hull consist of a hard coat and accounts for 23 to 26 per cent of the seed mass, but it contains more than half of the total moisture of the seed.

There can be open or artificial drying of seeds and capsules of castor. For open drying, capsules, spikes or threshed seeds are mixed on floors and exposed to the action of air and sun. For this purpose, the most suitable are asphalt floors with a 30 to 35 cm elevation in the middle. The material is spread in rows of 0.5 to 1.5 m wide, with a thickness up to 15 cm and it is shoveled every 3 to 4 hours. Drying is effective during the days when the relative humidity of the air does not exceed 60 per cent. Drying continuous for 3 to 12 days, depending on initial moisture and weather. This method is simple and practical but has some limitations like requirement of floors of quite large dimensions up to 10 per cent of the harvest area, greater investment in manual labour, a continuous period of drying and the drying process is completely dependent on weather conditions.

Artificial drying provides effective dehumidification and is not dependent on weather conditions. In this process seeds are exposed to the action of heated air for a gas-air mixture.

20.2. Processing

Extraction and Refining of oil: The seed are separated from the spiny husk mechanically and dehulled by an impact or disk mill decorticator to obtain the kernels. The oil is extracted from the kernels by a hydraulic press or continuous screw press followed by solvent extraction. The most satisfactory approach is hot pressing using a hydraulic press followed by solvent extraction to remove the bulk of oil remaining in the press cake. Hot pressing through a hydraulic press extracts about 75 to 85 per cent kernel oil. Subsequent solvent extraction yields the bulk of remaining oil and a meal with 1 to 2 per cent oil.

Medicinal castor oil is extracted from the decorticated seeds by cold pressing or pressing at 32 to 38° C, depending upon the techniques employed. This yields oils of pale straw colour, which can be bleached by ultra violet light. Filteration of this oil provides oil suitable for medicinal uses.

Storage: Normally there is no pest and disease, which can damage the seed in storage. The seed is stored in gunny begs in warehouses with good / proper aeration. There should not be high humidity and moisture inside the store house.

21. CONSTRAINTS OF PRODUCTION

1. Lack of availability of certified seed improved varieties / hybrids of castor.
2. Castor is generally taken as rainfed crop.
3. Poor crop management results in low yield of castor.
4. Heavy damages by the pests like castor semi-looper and castor capsule borer.
5. Castor is taken as a mixed crop with some other main crops.
6. Sex-reversion in castor reduces the yield.
7. Large area covered by local varieties of castor.
8. Lack of availability of disease resistant varieties.

22. REFERENCES

Biswas, B.C. and Tewabia, R.K. 1991. Nutrient balance in Agro-climatic regions of India-An overview. Fert. News, 22(8):1-7.

Das, P.C. 1997. Oilseed crops of India. Kalyani Publishers, Noida (U.P.).

Kulkarni, L.G. 1959. Indian Central Oilseeds Committee, Gandhi Bhavan, Hyderabad-1.

Malavia, D.D., Khanpara, V.D., Shobhana, H.K. and Golakiya, B.A. 1995. A comparison of irrigation methods in arid and semi-arid western Gujarat, India. In the 5th International Micro Irrigation Congress held at Orlando, Florida (U.S.A.) during April 206, 1995.

Moshkin, V.A. 1977. On the origin and distribution of castor as a crop. Byull 69, Inst. Rast. N.I. Vavilova, Rosyov-na-Don, USSR.

Moshkin, V.A. 1986. Oxonian Press Pvt. Ltd., New Delhi.

Popova, G.M. 1930. Castor oil plant. Lenin Acad. Agril. Sci. Inst. Appl. Bot. And New Culture Leningrad. P.63.

Schropp, W. 1938. Contribution to the knowledge of potash deficiency symptoms in some oil and fibre-yielding plants. Ernahr Pfl, 34(10):165.

Singh, J., Banerjee, A.K. and Swarup, J. 1976. Efficacy of fungicides in controlling bacterial leaf spot and blight of castor and varieties resistant to it. Indian J. Farm Sci., 4, 125.

Tandon, H.L.S. and Pratap Narayan. 1990. Fertilisers in Indian Agriculture – Past, Present and Future (1950-2000), FDCO, New Delhi 160 +viii.

Tandon, H.L.S. 1992. Fertiliser and their integration with organics and biofertiliser. In Fertilisers, Organic Manures, Recyclable wasters and Fertilizsers (ed. H.L.S. Tandon). FDCO, New Delhi pp. 12-35.

Weiss, E.A. 1971. Castor, sesame and safflower. Leonard Hill, London: 1-901.

Weiss, E.A. 1983. Oilseed crops. Longman, London.

Yadav, D.S. and Prasad, N. 1992. Fertiliser use pattern in the 1980's in different agro-climatic zones of northern region. Fert. Mktg. News, 23(7):1-19.

13. Sunflower

Gajendra Giri

Sunflower (*Helianthus annuus* Linn) is grown in India as an ornamental plant since ancient times. But as an oilseed crop, it was introduced in 1969 only. Commercial cultivation of this crop in India was started in 1972 in few hectares land in Southern states. Owing to the certain qualities, sunflower has made in roads in northern states also. Today, sunflower is grown in about 2.0 million hectare area and considered as a potential oilseed crop in India.

Among the vegetable oils, sunflower oil is considered as most suitable for human consumption because of high level of linoleic acid and absence of linolenic acid. Oleic and linoleic-unsaturated fatty acids constitute about 90% of the total fatty acids in its oils. This characteristic has made sunflower oil nutritionally better than groundnut or mustard oil (Table 13.1).

TABLE 13.1: Fatty acid composition (%) of some important edible oils

Crop	Saturated	Mono	Polyunsaturated	Total
		Unsaturated	Linoleic	
Groundnut	18	52	30	30
Rapeseed	6	67	17	27
Soybean	16	24	53	60
Sunflower	12	21	67	67
Safflower	10	15	75	75
Palm oil	50	39	10	11

Because of this, it has been accepted that higher level of unsaturated fatty acids in a diet reduces the level of blood cholesterol, a factor which is responsible for the incidence of coronary heart disease. Thus, the major use of sunflower oil is as a cooking oil. This is also a major ingredient in some margarine and shortening products.

Almost the entire volume of sunflower seed produced in India is the oil bearing type, which contains 40-44 per cent oil. The United States, however, produces and exports a more lucrative type sunflower seeds which are confectionery grade. The Oil content of this seed is less than 35 per cent. The kernel of this seed are also consumed in view of their high nutritional value and flavour. Major markets for confectionery grade sunflower seed are in countries like Germany, Canada, Spain, Mexico, Denmark and U.K.

Use of sunflower oil for industrial purposes is not very common. It is used to some extent in certain paints, varnishes and plastics because of good semi-drying properties. This is also used in the manufacture of soaps and detergents in the Russia. Sunflower meal is used as an animal food protein concentrate. Since protein content varies with the hull content of the meal, it ranges from 28 per cent in hulled seeds to 42 per cent in completely dehulled seeds. Sunflower hulls are used as roughage in animal food and as fuel. Other uses are making insulation board and litre for livestock.

FIGURE 13.1: Sunflower plant with seeds.

1. ORIGIN

Sunflower probably originated in the south-west America (Heiser, 1976; Vranceanu, 1974) since it grows naturally throughout this region. Sunflower was introduced to Europe in the Sixteenth century and the first published description appeared in a herbal by the Belgian in 1568. It reached Europe from Mexico via Spain. It became very popular as an ornamental and was established as an oilseed crop in eastern Europe. It was imported into Russia from Holland in the eighteenth century and the first commercial production of oil occurred in 1830-40. After second world war, the introduction of Russian varieties such as Peredovik, Mennonite and Sunrise has an immense impact on the development of this crop for commercial purpose in Europe and Americas. In India, the sunflower was introduced with the import of peredovic and Armveski varieties from Russia in 1969. Today, sunflower is one of the most important oilseed crop not only in India but at Global level also.

2. DISTRIBUTION

Sunflower stands third among annual oil seeds grown in the world. Globally, sunflower occupied about 21.0 mha during 1995-96 which accounts about 11 per cent of total area under oilseeds. During this period, production of sunflower seed (25. 9 m tonnes) contributed about 10.1 per cent of world's oilseed production. Among six crop producing continents, Europe is the main producer of sunflower seed accounting for about 60 per cent of the global production, Whereas, Russia is the largest producer as an individual country. Argentina stands next to Russia. Other main sunflower growing countries are China, U.S.A., Spain, France, Romania, Turkey, Hungry, Bulgaria, India, Australia, South Africa and Yugoslavia.

TABLE 13.2: World production of sunflower

Year	Area in mha	Production m tonnes	Yield kg/ha
1979-81	12.35	14.41	1167
1991	17.03	22.95	1345
1992	18.19	21.98	1210
1993	18.15	20.49	1268
1996	21.01	25.94	1320

In India, sunflower as an oilseed crop was introduced in 1969 in Karnataka. However, large scale cultivation of sunflower was started in 1972 with the introduction of open-pollinated varieties of Russian (U.S.S.R.). Origin on 0.22 million hectare in southern states, since then, area under sunflower is on ascent touching almost 2.0 million hectare during 1991-92. Until mid 80's, the cultivation of sunflower remained confined to southern states of Karnataka, Tamil Nadu, Maharashtra and Andhra Pradesh. Owing to the short duration nature of the crop and development of hybrids, sunflower gained popularity in northern states and could be fitted in various cropping systems. This made sunflower an important spring season crop in the States of Punjab, Haryana, Uttar Pradesh, Madhya Pradesh and Rajasthan.

TABLE 13.3: State-wise area, production and yield of sunflower

State	Area (M. Ha.)	% of total Area	Production (M. Tonnes)	% of Total Production	Yield (Kgs/ Hect.)
Karnataka	0.88	44.0	0.37	28.0	420
Maharashtra	0.54	27.0	0.34	25.8	623
Andhra Pradesh	0.30	15.0	0.21	15.9	718
Punjab	0.12	6.0	0.19	14.4	1542
Haryana	0.06	3.0	0.11	8.3	1758
Uttar Pradesh	0.03	1.5	0.05	3.8	1404
Tamil Nadu	0.04	2.0	0.04	3.0	1040
Others	0.03	1.5	0.01	0.8	
All India	2.00	100.00	1.32	100.00	659

3. CLASSIFICATION

The sunflower (*Helianthus annuus* L.) having somatic number 2n = 34, belongs to the family compositae. The genus *Helianthus* is derived from the Greek *heliouas* meaning sun and *anthos* meaning 'flower'. According to Heiser *et. al.* (1969) the genus is comprised of sixty-seven species. The genus is divided into four major groups viz. Annui-thirteen species mostly annuals of North-America, Ciliares-six perennial species of Western-North America, Divaricati- Thirty perennial species mostly of eastern North America and Fruticosi shrubby species of South America. Commercial varieties grown for seed are considered to be *H. annuus* var. *macrocarpus*.

4. CLIMATE

Sunflower is a plant of temperate zone but varieties adapted to a wide range of environments have been developed. Tall, strong growing varieties are planted in Russia as snow breaks to protect spring wheat in U.S.A. as wind breaks for soybeans and to trap snow in Canada. Sunflower is grown from 40° S to 55° N but major production is between latitudes 20 and 60° N and 20-40° S. It can be grown to an attitude of 2500 m above mean sea level, but gives highest oil yield below 1500 m. Sunflower grows well within a temperature range of 20-25° C but under controlled conditions, tests indicated 27-28° C as optimum. Tolerance of 8-34° C by sunflower indicated its adaptation to regions with warm days and cold nights. Temperatures affect the rate of development as prolonged high temperatures reduces the time of maturity. Temperature is known to affect seed oil content, seed and oil characteristics. In general, temperatures above 25° C at flowering reduces seed yield and its oil content. Low temperatures below 16° C can also reduce seed set and oil content.

Sunflower is considered to be drought resistant but oil yield is substantially reduced if plants are stressed at peak growth period and flowering. Number and size of leaves are reduced if crop faces moisture stress at peak vegetative stage. If the sub soil moisture is adequate, a good crop can be harvested with 500 mm of irrigation, Sunflower will give a moderate yield on rainfall of 300 mm. The relationship between rainfall and yield is almost linear from 200 to 500 mm rain. Evenly distributed rainfall of 500 to 700 mm over the growing period and ceasing just before flowering and seed filling is ideal for good harvest. When rainfall is above 1000 mm, losses from diseases and lodging can be severe, especially if there are heavy rains at fully grown crop. Sunflower plant is susceptible to damage by lodging due to high winds particularly when irrigated because of its height. Sunflower at seedling stage is highly susceptible to hail. If terminal shoots are damaged, these seedlings seldom recover.

5. SOIL AND SOIL PREPARATION

Soils with an appreciable sand content is considered better than more clayey. Irrespective of type, soils with good drainage are best suited. Sunflower grows well on neutral to moderately alkline soils, with a pH range of 6.5 to 8.0 but acidic soils are not suitable. Poor drainage conditions are unsuited because it increases susceptibility to fungal disease and lodging. Soil salinity affects plant growth and development and also reduces disease resistance. An exchangeable sodium percentage more than 16 delayed germination and development of flower heads (Chhabra *et. al.*, 1979).

6. CROPPING SYSTEM

No other crop in our country has gained as much popularity as sunflower because of its wide adaptability, short duration, high yield potential, besides high quality oil. Introduced in southern states in 1969, sunflower has made in roads in North-Western parts also. In the assured irrigation areas, there is a feasibility of its horizontal expansion through cropping system approach either as intercrop or as sequential crop.

7. MANURES AND FERTILISER

Sunflower is considered as a heavy feeder of nutrients. Hence, for obtaining higher yields adequate supply of plant nutrients for sunflower is most essential. A crop producing one tonnes of seed removes about 63 kg N, 8.5 kg P_2O_5 and 100 kg K_2O. However, response of sunflower to various nutrients varies from soil to soil, season to season depending upon the variety, cropping systems and climatic conditions.

Most of the experimental evidences strongly pointed towards positive response of sunflower to N. However, the level of N for the highest yield fluctuated to a wide range. Pal and Yadav (1974) reported a significant increase in the seed yield of sunflower upto 120 kg N/ha. Whereas, Bindra and Kharwara (1992) working at Palampur reported response upto 60 kg N/ha. only. Experiments conducted at IARI, New Delhi during *kharif* by Rao and Saran (1991) and during Spring by Legha and Giri (1999) recorded significant increase in sunflower yield upto 80 kg N/ha. On the basis of these findings, the optimum dose of N was worked out to be 115 kg/ha for spring season sunflower crop in Delhi. During monsoon season, loss of N is quite obvious hence its rational application at right stage of crop growth is desired for higher productivity. Reddy and Giri (1997) have recommended the application of 60 kg N/ha to *kharif* season sunflower in three splits e.g. at sowing and 25 and 50 DAS. This application coincides with buttoning and flowering stage of sunflower crop.

Sunflower, being a short duration crop can be easily fitted in cropping systems. Under such conditions, application of phosphorus is inevitable to maintain the soil fertility for higher productivity. However, the actual dose of P depends upon the availability status of soil. In rice-fallow gangetic alluvial soil of W. B. Appreciable improvement in yield attributes and yield of sunflower with the application of 60 kg P_2O_5 has been reported by Sarkar *et. al.* (1995). Whereas, in the alluvial plain zones of Western U.P., sunflower grown during spring season has shown response upto 80 kg P_2O_5/ha. (Tomar *et. al.*, 1997). Increase in seed yield to the tune of 58.7 per cent with the appliction of 80 kg N and 40 kg P_2O_5/ha has been reported from Hissar by Singh and Singh (1997). However, application of 40 kg N and 20 kg P_2O_5/ha in combination with 10 tonnes FYM /ha gave the seed yield at par with 80 kg N and 40 kg P_2O_5/ha. They have worked out economic optimum dose as 90.6 kg N and 45.3 kg P_2O_5/ha. Improvement in the root length and root: shoot ratio due to P application (Tomar *et. al.*, 1997) might be responsible for higher growth and yield of sunflower.

There is good scope of increasing seed yield and oil content of oilseed crop by proper nutrition particularly potassium. Appreciable increase in yield and oil content of oilseeds with the application of 60 kg K_2O/ha has been reported from Nadia, W.B. (Mondal *et. al.* 1997). Major nutrients could be made available by applying 10 t FYM/ha is it has given the yield equivalent to the 50 per cent of the recommended dose of NPK (Ujjinaia *et. al.*, 1994).

The response of sulphur application in sunflower is spectacular as increase in yield to the tune of 6 to 30 per cent and in oil content to the tune of 3.8 per cent has been reported by Tandon (1991). Srimanarayana and Raju (1994) recorded highest yield of sunflower with 40 kg S/ha and found ammonium sulphate superior to gypsum in black soils of Hyderabad. But in alfisols highest yield was recorded at 60 kg S/ha and gypsum was found superior. Experiments conducted at IARI reported response of sunflower to 25-30 kg S/ha (Kameshwar Rao and Gangasaran, 1991 and Legha, 1996). Response of sunflower to sulphur is more pronounced in the presence of N. Hence 30 kg S/ha in combination with 80 kg N/ha has given a significant boost to seed yield of sunflower in comparison to their individual application. (Legha, 1996).

Among micronutrients, Boron has shown its favourable effect on the seed yield of sunflower. Its exudate on the stigma influences germination of pollen grain and growth of pollen tube through its association with enzymatic activities associated with plasmalemma (Shorrocks, 1984). Experiments conducted on crop improvement project on sunflower revealed increase in seed yield of sunflower to the tune of 32 per cent with spraying 0.2% borax to capitulum on ray floret stage. Dusting of 2 to 5 kg borax/ha has also resulted 22 per cent increase in seed yield (Venugopal *et. al.* 1994).

8. VARIETIES

Sunflower cultivation in India was started in 1972 with the introduction of open pollinated varieties from U.S.S.R. Though, sunflower production in the country did not increase as anticipated in late 70's, identification and release of early maturing variety Morden in 1978 and release of the first hybrid BSH-1 in 1979 helped in reviving interest in sunflower (Seetharam, 1978 and 1979). During the last 20 years, many open pollinated varieties have been developed which have broadened the varietal base. Besides, modern, several open pollinated varieties such as Surya, CO 1, CO 2, S 5-56 and TANU-SUF-9 have been released in many states. Some of the improtant varieties/hybrids released in India are given in Table 13.4.

TABLE 13.4: Characteristic features of varieties/hybrids of sunflower released in India.

Varieties/ Hybrid	Recommended for	Duration (Days)	Head diameter (cm)	Seed yield Under rainfed condition	Oil content (%)
Varieties					
Morden,	All states	75-80	12-15	600-800	30-38
EC-68414	All States	100-110	15-20	800-1000	40-42
EC-68415	Karnataka	100-110	15-20	800-1000	40-42

Varieties/ Hybrid	Recommended for	Duration (Days)	Head diameter (cm)	Seed yield Under rainfed condition	Oil content (%)
Surya	Maharashtra	90-95	15-20	800-1000	32-35
CO 1	Tamilnadu	65-70	8-10	500-700	38-39
Co 2	Tamilnadu	85-90	15-20	800-10	38-40
Hybrids					
BSH-1	All States	85-90	12-15	1000-1500	40-42
KBSH-1	All States	90-95	15-20	1200-1500	42-44
APSH-11	All States	90-95	15-20	1000-1500	42-44
MSFH-1	All States	90-95	15-20	1000-1500	38-42
MSFH-17	All States	90-95	15-20	1000-1500	38-42
MSFH-8	All States	90-95	15-20	1000-1500	35-37
LSH-1	Maharashtra	85-90	12-15	900-1200	37-39
LSH-3	Maharashtra	95-100	15-20	1000-1500	38-40

Recently developed varieties/hybrids of sunflower are:

- **Varieties:** TANUSUF-7 AND GAUSUF-15
- **Hybrids:** PSFH-67, PKVSH-27, Jwalamukhi, PAC-36, Arun and Paras.

9. SOWING

9.1. Time of Sowing

Sunflower, being a photo and thermo insensitive could be grown throughout the year in India. However, it would not produce uniform yield as it requires certain bioclimatic requirements congenial for its normal growth and development. Soil temperature of 25-30° C is preferable, although higher temperature is tolerated in the presence of adequate moisture. Temperatures below 15° C generally retards germination. The variation in temperature during the growing season modify the plant activities and growth processes to a large extent. Besides, genetical characteristics, photoperiodic variations also cause difference in days of flowering and maturity (Chandra, 1974). Sowing at optimum time ensures better harmony between the crop and weather. There is sufficient evidence to show that spring season is the ideal and promising in north-western India. However, in southern states, sunflower can be grown throughout the year. Experiments conducted at Jodhpur (CAZRI, 1972) on sunflower variety sunrise indicated that the crop sown after 15 March did not come up well because of high temperatures and hot winds prevailing during summer months. Crops sown during December and January took more days to germination, flowering and maturity. Sowing in the first week of February recorded highest seed yield. Experiments conducted at IARI. also suggested 2nd week of February as optimum time of sowing for spring season sunflower. Sowing in the first week of March resulted in poor seed setting because of higher temperature prevailed during flowering and seed filling stages. However, it took only 95 days to mature as against 105 days in February sown and 125 days in January sown sunflower hybrid MSFH-8 (Legha, 1996). For a *kharif* crop of sunflower, the time of sowing should be adjusted such that flowering and pollination period do not coincide with peak rainfall. Because prevalence of excessive and continuous wet weather or high temperatures during flowering periods leads to poor seed set. Highest yield from the crop sown in the first week of July has been reported by Reddy and Giri (1997). They have reported reduction in seed yield by 28.1 and 75.8 per cent when crop was sown in August and September, respectively.

9.2. Spacing and Seed Rate

Maximum yield is the function of plant competition between rows and within rows. When sunflower is grown under different seasons and times within a season, vegetative phase varies depending upon the prevailing temperatures. This causes variation in the size of the plant and dry matter accumulation (Chandra, 1974). Under such situations planting geometry i.e. adjustment between and within rows play a significant role in different energy for higher productivity. Experiments conducted in different parts of country have suggested 45 cm spacing for short duration and dwarf genotypes and 60 cm for hybrids.

Plant population in the range of 55 to 65 thousand/ha for hybrids and 75 to 100 thousands/ha for short duration dwarf genotypes is desirable for optimum productivity from an irrigated crop of sunflower. Under dryland conditions, the population which can be sustained must be 20 per cent lower than that in normal conditions. To maintain the desired population a seed rate of 7 to 8 kg/ha for hybrids and 8 to 10 kg for dwarf genotypes is suggested. Sunflower seed looses its viability at a faster rate, hence one should not use older seed for sowing unless it is supported by germination test. Seed should normally be planted at 3-8 cm depth, depending on the moisture status of the seed bed. Soil in the vicinity of the seed must remain moist throughout germination. Sunflower is generally sown on flat beds but to ensure germination sowing on ridges prepared at desired distance followed by irrigations recommended. On flat beds sowing can be done either behind *deshi* plough or with seed drill adjusted for desired distance. However, dibbling of 2-3 seeds per hill is recommended for maintaining optimum population, particularly for ridge sowing.

10. IRRIGATION

Sunflower is considered to be drought resistant, however, it responds well to irrigation. Moisture stress in the vegetative phase causes reduction in the size and number of leaves. If it persists further, lower leaves shed and plant height is reduced. Under moisture stress conditions, size of capitulum and filling of seeds are also adversely affected. Hence availability of sufficient moisture at critical stages is equally important. For obtaining higher yields from a sunflower crop, most critical stages for irrigation are - bud formation, buttoning, flowering, seed setting and seed development. However, scheduling irrigation is equally important which may vary, depending upon the climate, season, soil and cultivar. In normal conditions, water use by sunflower is mainly confined to the upper 150 cm depth of the soil. In the areas where there is naturally high sub-soil moisture, sunflower can produce good yield with 300-500 mm irrigation water. When crop is fully dependent on irrigation, water required by sunflower varies between 600-750 mm.

A pre-sowing irrigation to ensure germination is essential in the fields where spring sunflower is preceded by another crop in winter season. It is more important to maintain a irrigation schedule at stated interval rather than irrigating when plants show stress. Experiments, conducted in red chalka soils of Andhra Pradesh recorded highest yield of sunflower when it was irrigated at 1.0 I.W. / CPE ratio (Reddy and Kumar (1996). However, experiments conducted in sandy loam soils of western U.P. recorded highest yields at IW/CE ratio of 0.8 during spring season (Tomar *et. al.*, 1997).

Irrigation should be avoided at the time of high wind velocity as plants are easily uprooted in wet soils. Plants are most vulenrable to lodging at seed filling stage because of heavy capitulum. Excess moisture is equally detrimental at any stage and the extent of damage is dependent on the duration of water logding. Young plants are more susceptible to excess moisture even for 2-3 days. Hence leveling of fields and maintenance of good drainage is essential for good harvest.

Saline water is not considered suitable for irrigating sunflower. Experiments conducted at U.A.S. Bangalore indicated reduction in germination of sunflower genotypes due to sodium chloride salinity (Chandru *et. al.*, 1994). The critical salinity level, at which 50 per cent reduction in vigour index occurs was found to be 12 dsm^{-1}.

11. WEED CONTROL

Sunflower is sown in wider rows (45-60 cm) which provides enough opportunity to weeds to establish. They cause enormous loss to sunflower crop as they compete for nutrients, moisture and light. Weeds when allowed to compete with crop till harvesting they remove 84.3 kg N, 21 kg P_2O_5 and 31.4 kg K_2O/ha (Girjesh and Pati, 1989). Reduction in seed yield of sunflower yield to the tune of 42 to 67 per cent has been reported (Balyan, 1993 and Legha, *et. al.*, 1992) due to weeds.

11.1. Weeds Flora

The prominent weed flora encounters with sunflower crop during *kharif* and spring season comprises of *Trianthma portulacastrum* L., *Digera arvensis* Forsk., *Phyllanthus niruri* under dicot weeds and *Cyprus rotundus* L., *Cynodon dactylon* L. and *Echinochloa colonum* L. under monocot weeds. However, weed flora may vary from region to region depending upon climate, soil type and cropping pattern.

11.2. Critical Period of Weed Competition

This is the period of development in the life cycle of a crop when it is most sensitive to competition and responsive to weed control measures. This period varies from cultivar to cultivar and season to season. The most critical period for weed competition in sunflower has been reported as 40-45 days after sowing. Sunflower smoother the weeds at later stage due to rapid growth. Hence, weeds emerging after this period do not pose serious problem.

11.3. Control Measures

Weeds in sunflower crop can be effectively controlled by mechanical means e.g., *khurpi, kasola*, hand hoe etc. This gives assured control of weeds particularly in wet season where weeds emerges early and grows fast. Depending upon the intensity and frequency of weed infestation, two hand hoeings at 20 and 40 days after sowing controls the weeds effectively as increase in the seed yield of sunflower to the tune of 68.7 per cent over control has been reported (Jayakumar *et. al.* 1998).

Experiments conducted in different parts of country have revealed that herbicides can also be used successfully in sunflower for effective control of weeds.

Fluchloralin is recommended for pre-planting application to control weeds. Suppression in weeds and increase in seed yield of sunflower due to pre-planting incorporation of flunchloralin @ 1.25 kg/ha has been reported by Jaykumar *et. al.* (1998).

Pendimethalin is also recommended as pre-emergence application. Weed control efficiency to the tune of 80-92 per cent due to pre-emergence application of pendimethalin @ 1.25 kg/ha has been reported by Durigan and Motta (1989). Increase in sunflower yield by 122 per cent due to weed control by pendimethalin application has been reported by Legha *et. al.* (1992).

Integration of various weed control measures is always advisable for increasing efficiency and reducing input cost. Findings of Solunkhe *et. al.* (1990) revealed efficient weed control with fluchloralin @ 0.9 kg/ha followed by one hand weeding at three weeks after sowing and increase in seed yield of sunflower by 28 per cent. Application of pendimethalin @ 1 kg/ha + one hand weeding at 40 days after sowing recorded higher seed yield of sunflower over pendimethalin alone (Nalayini and Sankaran, 1992 and Legha *et. al.*, 1992).

11.4. Pollination

Poor seed setting is yet another constraint to limit the productivity of sunflower and this problem is two dimensional; occurrence of hollow achenes and partial filling of seeds resulting in reduced specific weight. The probable factors affecting filling of seed in sunflower are inadequate population of pollen vectors, presence of axillary heads, too large heads, micro nutrient deficiency, self incompability, heavy rainfall during flowering. Leolopes (1972) observed death of pollen grains when anthesis phase of sunflower coincided with hot weather and scorching sun. Singh (1977) was of the opinion that any weather factor such as rainfall or high temperatures may result in washing off pollen or desiccation leading to high achene sterility.

The protandrous and self incompatible nature of sunflower favours cross pollination and is predominantly brought about by insects. The pollination is largely done by honey bees and although other insects like bumble bees also play a role. As sunflower is highly cross pollinated crop, sufficient number of pollinators are necessary for better pollination of all the florets in capitulum to ensure higher seed set and yield. The introduction of bee-hives into the field at an appropriate time can help in seed setting in sunflower. Panchabhavi, *et. al.* (1976) have recorded 27 per cent increase in seed setting due to the introduction of honey bee colonies. Supplementing natural pollination with hand pollination has also resulted in higher seed set, more number of seeds per head, higher seed weight per head and significantly higher yield in sunflower (Devendrappa and Hosmani, 1990). Experiment conducted at IARI., New Delhi also indicated increase in seed filling, seed weight per plant, and seed yield of sunflower (Table 13.5), due to hand pollination.

TABLE 13.5: **TABLE 13.5:** Effect of hand pollination on seed set and yield of sunflower

Parameters	Pollination	
	Natural	Natural + Hand
No. filled seeds per head	426	470
Seed filling (%)	53.8	64.4
Seed weight per plant (g)	12.5	13.4
Seed yield (q/ha)	7.95	9.10

The practice of hand pollination should be done for ten days at anthesis stage early in the morning on alternate days. For this purpose flowers should be rubbed gently with hand covered with soft cloth. This would help pollens to reach all the female flowers and result in higher fertilization.

12. PLANT PROTECTION

Mention is often made of a number of constraints which limit the production and productivity of oilseeds. Lack of plant protection is one of the major constraints. Various insect-pests and biotic pathogen which infest sunflower crop any time from seed germination to harvest are responsible for its low productivity. In India yield loss of 25 per cent due to capitulum borer (AICORPO, 1981) and 20 to 45% due to blight and rust (Mayee, 1988) have been reported. Hence, adoption of plant protection measures at optimum stage is of paramount importance to augment the productivity of sunflower.

12.1. Insect-Pests

Bihar hairy caterpillar (*Spilosoma obliqua* Walker): This is a polyphagus pest which appears in all sunflower growing areas. The crop is damaged by young larvae which feed gregariously on the undersurface of the leaves. In later stage, caterpillar defoliates the complete plants. The insects can be controlled by spraying endosulfan 35 E.C. @ 2 mL or dimethoate 30 E.C. 1 mL per litre of water.

Capitulum borer (*Heliocoverpa armigera* Hubner): The capitulum borer, an important pest of sunflower in all major sunflower growing areas of India, the insect directly damages to the developing seeds and causes enormous loss to yield. Loss in yield is more if infestation occurs in bud stage since it enables larvae to damage more ovaries than it could damage seed if it occurs during and after seed set. Two sprays of endosulfan (0.05 per cent) at 25 and 40 DAS is advisable for the effective management of capitulum borer on sunflower. However, in the absence of its incidence at an early stage, single spray at 40 DAS is essential.

Root Weevil (*Myllocenus discolor* Fabr.): The pest appears in all the sunflower growing areas of the country. Damage is caused by both grubs and adults. The adults of the insect feed on foliage whereas grubs feed on roots which causes mortality of the plants. Soil application of the insecticides at the time of last ploughing help to control the pest. Any of the chlorinated hydrocarbons e.g. lindane 3.5 kg a.i., heptachlor 1.5 kg a.i., dieldrin 1.0 kg a.i. or toxaphane 250 g a.i./ha may be used to control this pest.

12.2. Diseases

In India, sunflower in its early period of adoption was mainly a rainy season crop and disease problems were restricted to Alternaria blight, rust, head rot and occasionally charcoal rot. With changes in cultivation practices, cultivars and area under cultivation, disease scenario is undergoing detrimental alterations. It is feared that some diseases may limit the cultivation of the crop. Hence, controlling of diseases at optimum stage is an important aspect for recording higher yield.

Alternaria leaf blight [*Alternaria helianthi* (Hansf) Tabuki and Nishihara]: It is a destructive disease causes reduction in seed and oil yield to the tune of 80 and 33% respectively (Balesubrahmanyam and Kolte, 1980; Reddy and Gupta, 1977). The pathogen is seed-borne and survives in host debris, weed host plants. Secondly, it is disseminated by wind, rain splash. The disease produces characteristic black, circular spots on foliage. These spots coalesce and cause blight. Rotting of heads may also occur. The disease is widely distributed in India. Mancozeb at the rate of 0.30% sprayed at 7-10 days interval has been found effective in AICORPO trials.

Rust (*Puccinia hellianthi* Schw.): It is a potentially destructive pathogen of sunflower in all sunflower growing areas in India. Losses in yield of EC 68414 ranging from 20 to 45% have been reported by AICORPO. Reddish brown pustules appear on the lower surface of the young leaves and later spread over entire vegetative surface covering stems, petioles, floral bracts and petals. Dusting of elemental sulphur at 15 kg/ha of spraying of any organosulphurs (Zineb 2 kg/ha, mancozeb 1.5 kg/ha, ziram 1.5 kg/ha) or oxathiins (Carboxin 0.1% and oxycarboxin 400 ppm) are reported to be highly effective in managing the disease. Sunflower hybrids viz. MSFH-1, 6, 8, 9, 10, BSH 1, VSF 1 and 2 are fairly resistant to rust in India.

Downy mildew [*Plasmopara halestedii* (Farl): Berl & de Toni]: This is a seed, soil and air borne pathogen. It producers variety of symptoms from seed emergence to ear head formation. Chlorosis, downy growth, excess stunting, based gall and damping off are the main characteristics of the disease. The young seedlings are highly prone to infection which declines with advancement in age from 15 to 20 days. The potential losses due to the disease range from 80-85 per cent in seed yield. Seed dressing with apron 35 reduces the disease infection.

Charcoal rot [*Macrophomina phaseolina* (Tassi)]: The fungus survives mainly in soil and plant debris as sclerotia. Damping off, seedling blight, basal stem rot and charcoal rot symptoms are produced in sunflower. The damaging stage is at flowering to seed development when photosynthates are translocated to reproductive organs and sinks are produced in root and stem portions. The disease incidence, increases due to low soil moisture at flowering accompanied by high temperatures above 30° C for prolong period. Seed treatment with thiram at 2-4 g/kg seed helps in minimizing seed borne inoculum.

Head rot (*Rhizopus* spp): This is very damaging disease in sunflower. *Kharif* crop is more susceptible to the infection. At first brown irregular water soaked spots appear on the surface of the head near the stalk and gradually spread to the head causing rot of the head. The disease adversely affects the seed development. This can be controlled by spraying mancozeb @ 0.2 per cent during head emergence.

12.3. Birds

Sunflower heads attract the birds especially parrots due to the attractive colour. They start damaging the heads just after seed setting. They feed on seeds and make the heads empty, if not taken care. Hence, the crop should be protected for about 20 to 30 days depending upon the duration of seed setting and maturity. Manual scaring of the birds of covering the crop by nets are most effective methods. However, using coloured shining taps as light reflectant and application of 5 per cent malathion on individual flowery after seed setting are also suggested.

13. HARVESTING

Harvesting of sunflower crop at its optimum stage of maturity is very important post production operation to minimize field losses and to get best quality seed with higher viability. Delay in harvesting causes field losses as sunflower is very prone to bird damage. Rainfall at maturity causes rotting of capitulum which results in fungi attack and discoloration of seed. Premature harvesting is a very important endeavour. Generally, maturity of sunflower is judged by unscientific practices like visual observation of colour of capitulum and seed as well. Removal of seed from capitulum and hardness of seed coat are also indication of sunflower maturity. Sunflower is harvested when colour of the capitulum turns to lemon yellow and colour of the seeds turns to darker. When seed turns to darker in colour, its coat also becomes hard and seed is easily removed from capitulum. At this stage, if seed is pressed with fingers, kernel remains intact whereas in case of immature seed, kernel is smashed on pressing. Measurement of moisture content in seed is a scientific way to judging sunflower maturity. At the time of harvesting seed moisture content, should not exceed 15 per cent. Harvesting of sunflower is done mostly by sickles. At the time of harvesting, capitulum is separated from stem and left for sun-drying.

14. YIELD

Sunflower is becoming popular due to its high yield potential and short duration. However, our national average of 565 kg/ha is the lowest in the world. In fact, the productivity of sunflower varies from region to region, season to season and also depends upon the selection of cultivar and level of management. Experiments conducted in different parts of country have reported wide variability in yield of sunflower. Morden a short duration variety gives 6 to 8 q/ha yield under rainfed conditions where as same cultivar with assured irrigation may give upto 12 q/ha. Experiments conducted at IARI during 1991 and 1992 reported seed yield in the range of 11 to 15 q/ha with harvest index of 46.6 per cent. From July sown morden variety of sunflower (Reddy and Giri, 1997). Whereas sunflower hybrid MSFH-8 sown in February recorded as high yield as 23.7 to 25.0 q/ha with harvest index of 30 to 40% from same station (Legha, 1996) short duration variety morden has wide seed: straw ratio where hybrids has narrow ratio.

15. QUALITY

Sunflower, an important oilseed crop has gained popularity because of high quality and digestible oil. The seeds generally contains 30 per cent hull. A thorough screening of germplasm indicated the content of oil in the range of 41.2 to 55.1 per cent (AICORPO, 1986).

TABLE 13.6: Characteristics of sunflower oil

Specific gravity	0.9109-0.9187
Refractive index 40°C	1.4636-1.4681
Colour (Y + R 5)	1.0 to 12.7
Acid value	0.7 to 2.5
Iodine value	101.4 to 135.0
Sponificaton value	1.93 to 2.00
Unsaponifiable matter	0.20 to 1.93

However, oil content and its quality varies from variety to variety, season to season depending upon the sowing time and temperatures prevailed during seed filling period. However, in general higher oil is obtained in *rabi* season and higher iodine value in oil in spring season. Crop sown in the period between third week of April to third week of May recorded high oleic acid content in oil. But crop sown in June yielded lowest oil content (Bhattacharya *et. al.* 1982). Under Indian conditions, a sunflower seed oil has the characteristics given in Table 13.6 (Kamath *et. al.*, 1979).

The oleic acid content of 42-57% and a linoleic acid content of 33-48 per cent has been reported by Chakraborty *et. al.* (1979). Balasaraswathi and Raj (1983) have reported better keeping quality of oil from a rainfed crop than that of irrigated one. However, the oil from the dehulled seeds could be stored for longer periods (Thayumanavan, 1981). The defatted seed contained 48.5 to 19.3 protein (Rahma and Narsing Rao, 1981).

Phenols are undesirable as they affect discolouration of proteins hence considered as a nutritional constituents. The phenol content ranges between 2.6 to 3.8%. Haemagglutinin activity ranged from 50.6 to 132.8 units/mg of protein.

16. POST HARVEST TECHNOLOGY

16.1. Cleaning

After threshing generally oilseed mixture contain impurities, which are to be removed prior to storage or drying or oil expression. Normally, cleaning/ separation of foreign materials from oil seeds is done using a vibrating sieve with a blower for winnowing.

16.2. Decorticating

The cleaned sunflower seeds need to be shelled for further processing. The shelling or decorticating is the process of removal of outer husk. Normally this process is done by the principle of impacting the seeds with a harder surface.

A sunflower seed sheller has been developed by Tamil Nadu Agricultural University. The graded seeds are fed into hopper by means of a bucket elevator and then into the rotor. The rotor throws the seeds into stator and creates an impact on the seeds. This impact force induced on the seeds due to high revolution of rotor resulted in a rapid velocity of seeds. The force due to rapid movement of seeds shelled the sunflower. The shelled materials are winnowed to separate the lighter husk and foreign material from the seeds. Since shell contain mainly cellulose, hemicellulose and lignin and removal of these materials increased the oil content by 20% in seeds.

16.3. Oil Extraction

After shelling and cleaning, seeds are ready for extraction of oil. Oil within sunflower seed is stored in the form of globules surrounded by tough membrane. Oil from oilseeds are expelled using methods like mechanical screw press, extrusion-expression and solvent extraction. Ninety per cent of oilseeds are crushed mechanically to expel oil using screw press. In the mechanical screw press expeller due to dynamic compression process, cell wall ruptures, tough membrane broke out and oil ooze out by disintegration.

16.4. Solvent Extraction Process

Oil cakes come out of mechanical screw expeller normally contain 10-12% oil. For extraction of this oil as well as direct extraction from seeds, solvent extraction method is used. The process of extraction of oil is performed using n-hexane solvent. This solvent penetrates and diffuses into cells of seeds and picks up oil. This oil is separated from solvent by condensation and the solvent in the oil cakes were extracted by evaporation.

16.5. Extrusion Expression

In this extrusion process, oil seed is heated well beyond boiling point of water, which results in buffing and expansion. The main objectives of heating is to denature the available amino acid and make cell wall sufficiently brittle so that oil is freed from globules. By this process semi-fluid extruded is obtained. Normally, it is done at high pressure for a short period of time. Thus extruded sunflower oilseed products are expelled to yield oil, cakes and other products. With in a single pass of extruded sunflower oilseed products into the expeller resulted in 70% extraction of oil.

Among all these methods of oil expression from oilseeds, extrusion-expression has been found better in terms of energy requirement, cost and quality of oil.

17. REFERENCES

AICORPO (1981), Annual Progress Report, Sunflower, 1980-81, D.O.R. Hyderabad.

AICORPO (1984), Annual Progress Report of *Kharif* Oilseeds Workshop, D.O.R. Hyderabad.

AICORPO (1986), Annual Progress Report of *Kharif* oilseeds workshop. D.O.R. Hyderabad.

Balasaraswathi, R and Raj, W. (1983). Studies on sunflower oil with reference to its keeping qualities. J. Food Sci. Tchnol. India 20(2): 21-24.

Balasubrahmanyam, N. and Kolte, S.J. (1980). Indian J. Agril. Sci. 50: 701-705.

Balyan, R.S. (1993). Integrated weed management in oilseed crops in India. Proc. Int. Symp. Indian Soc. of seed Sci. 1: 317-323.

Bhattacharya, P., Samui, R.C., Dasgupta, S.K., Chandrasekhar Rao, T. and Laxminarayana G. (1982). Effect of date of sowing and harvesting on characteristics and fatty acid composition of sunflower seeds. J. Food Sci. Tech. (India) 9 (6): 246-248.

Bindra, A. and Karwara, P.C. (1992). Response of spring sunflower (Helianthus annuus) to nitrogen application and irrigation. Indian J. Agron. 37 (2): 283-284.

C.A.Z.R.I. (1972). A report of Central Arid Zone Research Institute, Jodhpur (ICAR), (*Rabi* 1971-72).

Chakraborty, M.M. Saha, S. and Datta, R. (1979). Cultivation of sunflower (Helianthus annuus L.) in West Bengal. Studeies on changes in lipid composition according to date of sowing in *rabi* season. J. Oil submitted to P.G. School, IARI., New Delhi.

Chandra, Vinod (1974). Effect of sowing dates, plant population, levels of nitrogen, phosphorus and irrigation on yield and oil content of sunflower. Ph. D. Thesis, submitted to P.G. school, IARI., New Delhi.

Chandru, R. Chikkadevfaiah and Marwade, M.H. (1994). Salt tolerance sutdies on sunflower. Sustainability in D.O.R., Hyderabad, pp. 340-343.

Chhabra, R., Singh S.B. and Abrol, I.P. (1979). Effect of exchangeable sodium per centage on the growth, yield and chemical composition of sunflower. Soil Science 127 (4): 242-247.

Devendrappa, K.J. and Hesmani, M.M. (1990). Studies on hand pollination and growth regulators on seed setting of sunflower varieties. Karnataka J.Agric.Sci. 3(1 and 2): 113-116.

Durigan, J.C. and Motta, M. (1989). Weed control with pre-emergence herbicide in sunflower. Pesquisa Agropecuaria Brasileira 24 (9): 703-710 (C.F. F.C.A. 44(5): 3314).

Girjesh, G.K. and Patel, V.C. (1989). Effect of weed control in groundnut and sunflower intercropping system with reference to nutrient uptake and oil yield. J. Oilseed Res. 6(2): 334-340.

Heiser, C.B. (1976). Sunflower in Evolution of crop plants, ed. N.S. Simmonds, Lohgman Green. London p. 36.

Jayakumar, R., Sekar, M.P., Kempuchetty, N. and Subramanian, S. (1988) effect of integrated weed mangament on yield and quality of sunflower. Madras Agric.J. 75(3-4): 85-88.

Khargakharate, V.G. and Nirwal, B.G. (1991). Growth and yield of sunflower as influenced by inter and intra row spacing and nitrogen. J. Maharashtra Agricultural University 16 (2): 291-292.

Kene, H.K., Thosar, V.R. and Vlemale, R.B. (1992). Optimum sowing time of sunflower varieties is summer season. J. Maharashtra Agricultural Universities. 17 (3): 411-412.

Legha, P.K. (1996). Influence of date of sowing, crop geometry, and doses of N and S on spring sown sunflower in Semi-Arid condition of N-W India. Ph.D. Thesis submitted to P.G. School, IARI.

Legha, P.K. and Giri, Gajendra (1999). Influence of nitrogen and sulphur on growth yield and oil content o sunflower grown in spring season. Indian J. Agron 44:

Legha, P.K., Malik, R.K. and Faroda, A.S. (1992). Weed management in *Kharif* sunflower (Helianhus annuus L.) Crop Res. 5 (2): 376-378.

Leolopes, M.D. (1972). Effect of date of planting and row spacing on sunflower drop in Anolaluoia (South Spain). Proc. 5th Int. Nat. Conference on Sunflower, 27-29th July, 1972, at Clermont, Ferrand, France.

Mondal, S.S., Dasmahapatra, A.N., Chatterjee, B.N. and Maiti, P.K. (1997). Effect of potassium application on the yield and oil content of sesame, mustard and groundnut on a k-deficient soil. J. of Potassium Research 13 (2): 153-158.

Mayee, C.D. (1988). Important diseass of sunflower and their management in India. Proc. Nat. Seminar on Strategies for making India self reliant in vegetable oils. 5-9 Sept. 1980 pp. 202-213.

Nalayini, P. and Sankaran, S. (1992). Effect of moisture regimes and weed control treatments on sunflower (Helianthus annuus L.) Indian J. Weed Sci. 24(3-4) 1-5.

Pal, M. and Yadav, S.C. (1974). Response of sunflower varieties to soil moisture regimes and nitrogen levels. Indian J.Agric. Res. 8: 157-161.

Panchabhavi, K.S., Devaiah, M.A. and Channa basavanna, G.P. (1976). The effect of keeping *Apis cernua* (*indica*) colonies on the seed set of sunflower (*Helianthus annuus* L.) Mysore J. Agric. Sci. 10(4): 631-636.

Rahma, E.H. and Narasinga Rao, M.S. (1981) Removal of polyphenols from sunflower meal by various solvents. Effect of functional properties. J. food Sci. 46 (5): 1521-1522.

Rao, J.V. and Reddy, B.N. (1991). Intercropping of sunflower in groundnut through adjustment in plant density. J. Oilseeds Res. 8: 271-274.

Rao, S.V.C.K. and Saran, G. (1991). Response of sunflower cultivars to planting density and nurient application. Indian J. Agron. 36 (1): 95-98.

Reddy, B.N.. Sudhakar Babu, S.N. and Prasad, M.V.R. (1994) Sunflower based cropping system approach for enhancing productivity and sustained production. Sustainability in oilseeds. Indian socitey of oilseeds Research, D.O.R. Hyderabad, India pp. 410-412.

Reddy, G.P. and Giri, Gajendra (1997). Influence of time of seeding. pollination and nitrogen on yield of sunflower. Indian J. Agron. 42 (3): 506-511.

Reddy, M.D. and Kumar, K.A. (1996) Performance of sunflower in different irrigation and nitrogen levels during *rabi* in red chalka soils. J. Oilseeds Res. 13(2): 187-192.

Reddy, P.C. and Gupta, B.M. (1997). Indian phytopath 30 (4): 569-570.

Sarkar, R.K., Chakraborty, A. and Bala, B. (1995). Response of sunflower (*Helianthu annuus*) to crop geometry and nitrogen and phosphorus application rice-fallow gangetic soil. Indian J. Agron. 40(4): 657-659.

Seetharam, A. (1980). Hybrid sunflower. Indian farming 24: 15-16.

Seetharam, A., Giriraj, K. and Kusumakumari, P. (1978). Cerianka 66-A dwarf early variety of sunflower. Curr. Res., 7: 164-166.

Shorrocks, V.M. (1984). Boron deficiency its prevention and cure. Borax holdings Ltd., London pp 5-7.

Singh, Jagdev and Singh, K.P. (1997). Integrated nutrient management in sunflower (Helianthus annuus). Indian J.Agron. 42(2): 370-374.

Sidhu, M.S.; Sawhney, J.S., Sahota, P.S. and Sharma, B.D. (1991). Agronomic management of sunflower in different cropping systems. Proc. Nat. Seminar on strategy for making India self-reliant in vegetable oils. Sept 5-9, 1988 pp. 193-201.

Solunkhe, U.V., Shelke, D.K. and Bhosle, R.H. (1990). Weed management in sunflower during the *rabi* season in Maharashtra. J. Maharashtra Agric. Univ. 12(3): 367.

Singh, V. (1977). Problem of Empty seeds in sunflower. Paper presented at Xth Annual workshop (*Kharif* crops), A.I.C.R.P. on oilseeds at A.P.A.U. Hyderabad in May, 1977.

Tandon, H.L.S., (1991). Sulphur research and agricultural production in India. 3rd Ed., The Sulphur Institute, Washington, D.C.U.S.A. pp. 140.

Thayamanavan,. (1981). Keeping quality of oils from de hulled and whole seeds: Oilseeds J. 12: 54-56.

Tomar, H.P.S. Dadhwal, K.S. and Singh, H.P. (1997). Root Characteristics and moisture use pattern of spring sunflower (*Helianthus annuus*) as influenced by irrigation, nitrogen and phosphorus. Indian J. Agro. 42(3): 515-519.

Tomar, H.P.S., Singh, H.P. and Dadhwal, K.S. (1997). Effect of irrigation, nitrogen and phosphorus on growth and yield of spirng sunflower (*Helianthus annuus*). Indian J. Agron. 42 (1): 169-172.

Ujjinaiah, U.S., Balakrishn, P., Munegowda, M.K. and Seenappa, K. (1994). Effect of organic manure and inorganic fertilisers on seed yield of sunflower (*Helianthus annuus* L.) Sustainability in Oilseeds I.S.O.R., D.O.R. Hyderabad pp. 309.311.

Veenugopal, N., Virupakshappa, K. and Seenappa, K. (1994). Directed spray/dusting of borax to capitulum at sunflower. Sustainability in oilseeds, I.S.O.R.; D.O.R. Hyderabad, pp.: 201-202.

Verma, N.S. Dumka, P.K. and Tyagi, S.M. (1977). Biochemical studies in Research on sunflower. Expet. Stn. Tech. Bull. G.B.P.U. Agric. & Tech., Pantnagar pp. 44-46.

Vranceanu, A.V. (1974). Floarea-Soarelui. Acad. Rep. Soc., Romania, P. 332.

Wang, D.S. and Fan. F.Y. (1987). Study on the response of sunflower to rotation. Oil crops of China No. 3: 33-39, F.C.A. 1988, 41 (10): 857: Abst. No. 7031.

14. Rapeseed and Mustard

P. S. Rathore

India is one of the leading oilseeds producing country. Oilseeds form the second largest agricultural commodities after cereals. Rapeseed and mustard occupy second position after groundnut in area and production. Sarson and toria are commonly known as rapeseed while *rai* or *raya* or *laha* or *raiya* as mustard.

Rapeseed and mustard are the main oilseed crops grown in *rabi* in India. India occupies third position with regards to production of rapeseed and mustard in the world. The green tender plant is used for preparing vegetable commonly called *sarson ka saag*. The whole seed is used in preparing pickles and flavouring vegetables and curries. The oil is mainly used for cooking, frying and in pickles. Oil is also used in preparing vegetable *ghee*, hair oil, medicines, soap, lubricating oil and in tanning industries. The seed and oil of *Brassica* species have a peculiar pungency. This pungency in oil is due to pressure of a glucoside sinigrin ($C_{10}H_{16}O_9NS_2K$). The oil cake left after extraction is utilized as cattle feed and manure. The oil cake contains 25-30 percent crude protein, 5% nitrogen, 1.8-2% phosphorus and 1.0-1.2% potassium. However due to its pungency and less palatability animals can consume only small amounts.

1. ORIGIN

Ancient sanskrit records (2000-1500 B. C.) have a mention of rapeseed and mustard. Prain (1898) and Bailey 1922, 1930 and Sinsky (1928) mentioned China as origin of Rai. Origin place of other crops of the group are mentioned below:

1. Banarsi rai (*Brassica nigra*) - Uracia
2. Toria (*Brassica campestris* var. *toria*) - East Afghanistan
3. Brown sarson (*Brassica campestris* var *brown sarson*) - East Afghanistan and adjoining parts of Indian subcontinent.
4. Yellow sarson (*Brassica campestris* var yellow sarson) - North East India.

2. DISTRIBUTION

The leading countries which produces rapeseed and mustard are India, Canada, China, Pakistan, Poland and Sweeden. India ranks first with regard to area and third in production. Of the total world area (22 mha) India comprises 6.9 mha with production of 7.0 mt (1996-97). The leading states in India, producing mustard are Rajasthan, Uttar Pradesh, Madhya Pradesh, Haryana, Punjab and Gujarat. Among the states Rajasthan ranks first in area and production with 34% share in production of India. The area and production in world is given in Table 14.1.

TABLE 14.1: Area and production of rapeseed-mustard in world

Countries	Area (000 ha)		Production (000 t)	
	1996	1997	1996	1997
India	6659	6813	6071	6964
China	6734	6790	9201	9540
Canada	3451	4813	5062	6190
France	875	988	2912	3486
Germany	854	918	1970	2843
Total	22013	24198	30636	35154

3. CLASSIFICATION

Classified Indian rapeseed and mustard (Singh, 1958)

Group & international name	Local or common name	Botanical name	Seed colour	Seed size	Maturity	Oil %
Sarson (Indian colza, rape colza)	Yellow sarson. (Turnip rape)	*Brassica cempestris* var. *yellow sarson*	Yellow	Medium	Late medium	45
		Brassica campestris var. *brown sarson*	Reddish brown	Medium	Medium late	43
Toria (Rape)	Yellow toria. (Indian rape)	*B. campestris* var. *yellow toria*	Yellow	Small	Early	35
		B. competris var. *black toria*	Black	Small	Early	35
Rai (Mustard)	Rai, Raya, Laha. (Indian mustard)	*Brassica juncea* (L.) Czern & Coss	Dark brown	Medium	Late	35
	True mustard	*B. nigra*. Kook	Brown	Very small	Late	29
Wild mustard	-	*B. teuniforti* Kook	Brownish grey	Very small	Medium	18
Hill mustard	-	B. juncea rugosa	Brown	Small	Very late	18
White mustard (Ujli sarson)	-	*B. alba/hitra* Moench	White	Small	Medium	20
Taramira (Rocket)	Taramira	*Eruca sativa* Mill	Reddish brown	Medium	Medium late	28

4. RAPESEED V/S MUSTARD

The main distinguishing characteristics of rapeseed and mustard are given below:

Character	Rapeseed (Sarson/toria/lahi)	Mustard (Rai, Raya, Laha)
Plant height	45-150 cm.	90-200 cm.
Leaves	Sessile, leaf lamina clasps the stalk partially or completely	Leaves stalked and do not clasp the stalk.
Siliquae (Pod)	Short and thicker	Long and slender
Pollination	Cross pollinated	Self pollinated
Seed coat	Smooth	Rough

5. BROWN SARSON V/S YELLOW SARSON

The main distinguishing characteristics of brown sarson and yellow sarson are given below:

Character	Brown sarson	Yellow sarson
Leaves	Pale, thin	Dark green and fleshy.
Branching	Erect - spreading	Erect
Siliquae (pod)	Thin, narrow	Thick and broad
Seed	Dark brown to reddish brown and mucilaginous	Yellow and non mucilaginous.

6. CLIMATE

Rapeseed and mustard are cultivated in *rabi* season from September-October to March-April. Crop requires cool, dry and clear weather with temperature of 18-25° C for its growth and development. Growth of crop ceases if temperature falls below 3° C and above 35-40° C. Long day with adequate sunshine are ideal for its good production. Crop need fairly good soil moisture and can be grown in areas receiving 30-40 cm annual rainfall. High rainfall, high humidity and cloudy atmosphere at flowering increases incidence of aphids and other diseases and at the same time pollination is adversely affected. As movement of honey bees is restricted because of rains. The crops of the group are also susceptible to frost and excess water.

7. SOIL AND SOIL PREPARATION

Crop can be grown in soils varying from sandy loam to clay loam. It thrives well on light loam soils. Saline-alkaline soils are not suitable for its cultivation. Soil having neutral pH (6.5-7.5) are ideal but can be grown on moderately saline conditions.

Seeds of the crop of this group is very small, hence cultivation requires a clean and fine seed bed. For ensuring good germination, proper support and moisture supply to seedlings, one deep ploughing with soil turning plough immediately after harvest of *kharif* crop followed by 2 cross ploughing with harrow or country plough. Clods should be broken by planking the field after ploughing. Under rainfed conditions moisture may be conserved by harrowing after every rain along with contour bunding and other water conservation measures including removal of weeds.

8. CROPPING SYSTEM

Crop is generally grown as pure crop in irrigated condition and in conserved soil moisture after *kharif* fellow. The crop can also be grown after *kharif* crop if adequate moisture is available. The cropping sequences generally followed in different states are mentioned below:

Cropping sequences commonly practices in different states	
States	**Cropping sequence**
Rajasthan	Fallow-mustard
	Sorghum-mustard
	Maize-mustard
	Pearlmillet-mustard
	Green gram/Cow pea/ Black gram-mustard
	Cluster bean-mustard
Uttar Pradesh	Green gram/Black gram-mustard
	Maize-mustard
	Fallow-Brown sarson
	Paddy-mustard
	Paddy-toria
	Fallow-mustard
	Fallow-toria
	Sorghum-sarson
Madhya Pradesh	Fallow-mustard
	Black gram-mustard
	Soybean-mustard
	Rice-mustard
Haryana	Fellow-mustard
	Pearlmillet-mustard
Punjab	Maize-mustard
	Pearl millet-mustard
	Maize-Brown sarson
Bihar	Maize-mustard
Assam	Jute-rapeseed
	Paddy-rapeseed
West Bengal	Jute-mustard
	Maize-mustard
	Paddy-mustard

8.1. Intercropping

System involves growing of two or more crops. It is based on complementary nature of associated crops and adjustment of plant population for getting high cropping intensity without affecting population of main crop. Intercropping under dry land situation gives insurance against unfavourable. Weather conditions and disease-pest incidence. In Uttar Pradesh sarson was grown with wheat but due to its high susceptibility to aphids now mustard is grown with wheat in almost 80% area. Based on number of research findings scientists identified profitable intercropping system for different places as given in Table 14. 2.

TABLE 14.2: Profitable intercropping systems

States/Places	Crop combination	Ratio
Haryana (Hissar)	Mustard + chickpea	1:5
	Mustard + sugarcane (Autumn)	-
Uttar Pradesh (Pantnagar)	Mustard + barley	-
	Mustard + potato	-
	Mustard + lentil	1:5
Dholi	Mustard + chickpea	1:5
Kanpur	Mustard + chickpea	1:3
Bhubneshwar (Orissa)	Mustard + wheat	1:4
Bhilwara and Udaipur (Rajasthan)	Mustard + chickpea	1:7
Madhya Pradesh	Mustard + chickpea	-
	Mustard + barley	-
	Mustard + wheat	-
Punjab	Potato + mustard	3:1

9. MANURES AND FERTILISERS

Rapeseed and mustard respond favourably to the applied manures and fertilisers. It has been estimated that mustard crop yielding 40 q removes 134 kg N, 54 kg P_2O_5 and 12 kg K_2O Crop responds to both organic and inorganic fertilisers. Well rotten FYM at 10-15 t/ha at 15-20 days prior to sowing enhance growth and yield of mustard and other crops of rapeseed group. Response to inorganic fertiliser largely varied with crop, soil type, climatic conditions and availability of moisture. Nitrogen imparts dark green colour to leaves and is component of protein, involved in photosynthesis, respiration and thus increases growth of plant. Mustard needs more amount of nitrogen compared to sarson, toria need further less. Rathore (1989) obtained linear increase in yield of mustard with increase in levels of nitrogen from 30 to 120 kg N/ha. Singh et. al. (1997) obtained linear increase upto 180 kg N/ha. In general a dose of 60-90 kg N/ha may be applied. However a negative correlation was found between N and oil %. Phosphorus help in root development and enhance flowering and seed formation. Potash is important for protein, fat and carbohydrate metabolism. Phosphorus and potash must be applied, based on soil test values. Jaggi and Sharma (1997) recommended 60 kg P_2O_5/ha and observed 7.3 q more yield over no phosphorus. In absence of soil test values 40-60 kg P_2O_5/ha may be applied to crop. Information on potassium fertilization of mustard and *toria* is limited. Though it is to be applied on the basis of soil test, a general dose of 20-40 kg K_2O was recommended by many workers.

Sulphur also play an important role in oilseed crops. With intensive cultivation, use of high yielding varieties and use of high analysis S free fertilisers deficiency of sulphur is also spreading. Rathore (1989) obtained 15% higher yield due to application of 160 kg S/ha over its no application, besides. 4% increase in oil content. Similarly, Singh et. al (1997) obtained 3.5-5 q more seed yield by application of 80 kg S/ha over control, and oil content increased from 38.29 to 40.97% (Table 14.3).

Soil having less than 10 ppm available sulphur may be regarded as S deficient soils. Application of 20-40 kg S through elemental sulphur 21 days before sowing or with gypsum at the time of sowing. If no fertiliser S source is available phosphorus must be applied through single super phosphate serve the purpose. Crop also respond to zinc. If deficiency 5-10 kg Zn be applied through zinc sulphate.

TABLE 14.3: Effect of S application on seed yield and oil content in mustard

S levels (kg/ha)	Seed yield q/ha		Oil content %	Oil yield (kg/ha)
	1987-88	1988-89		
0	16.19	16.79	38.29	631.86
40	18.01	19.21	39.75	739.52
80	19.08	20.20	40.97	804.68
CD (P = 0.05)	0.18	0.23	0.11	0.83

Crop do not yield high if there is imbalance of N, P_2O_5 and K_2O. Balanced application of nutrients is essential. Thus nutrients should be applied in 4:2:1 proportion. The recommended doses of nitrogen, phosphorous and potassium for different parts of India are given below in Table 14.4.

TABLE 14.4: Fertiliser recommendations* for rapeseed and mustard in different states of India under irrigated condition

States	Crop											
	Mustard			Toria			Brown Sarson			Yellow Sarson		
	N	P_2O_5	K_2O	N	P_2O_5	K_2O	N	P_2O_5	K_2O	N	P_2O_5	K_2O
Rajasthan	60	40	-	40	20	-	60	40	-	-	-	-
Uttar Pradesh	80	40	40	60	30	20	50	30	30	50	30	30
Haryana	60	30	-	60	-	-	60	20	-	60	20	-
Punjab	80	50	-	40	20	-	50	20	-	50	20	-
M P	60	30	20	60	30	20	-	-	-	-	-	-
Gujarat	40	40	-	-	-	-	-	-	-	40	40	-
Bihar	60	40	40	50	30	30	-	-	-	60	40	40
West Bengal	80	40	40	-	-	-	-	-	-	80	40	40

*Doses for unirrigated conditions are half of the irrigated conditions.

Nutrient requirement of crop may also be decided based on critical concentration. Rathore (1989 and 1990) reported 6.07-6.62% N in top 2-3rd leaf at 60 days stage and 0.408-0.412% S in 4-5th leaf from top as critical level for mustard.

9.1. Integrated Nutrient Management

Combined use of chemical fertilisers, organic manures and bio fertilisers with an objective of obtaining sustainable and environmentally safe crop production. Application of 10 t FYM/ha along with recommended levels of fertiliser i.e. 60 kg N + 40 kg P_2O_5 and 40 kg K_2O and seed treatment with *Azotobacter* form the integrated nutrient management package for mustard. Impressive results have also been obtained from experiments conducted at Pantnagar, Navgaon, S K Nagar, Ludhiana and Hissar.

9.2. Time of Fertiliser Application

Time of fertiliser application is important for nitrogen. Depending on soil type N should be applied in splits. Half quantity of total nitrogen and full amount of phosphatic potassic, sulphur and zinc be applied at the time of sowing. Remaining 1/2 quantity of N should be applied at 30 days after sowing just after first irrigation. Under rainfed situations all amount of all fertilisers can be applied at the time of sowing. In case of deficiency of Zn in standing crop, zinc sulphate may be applied as foliar spray along with 1/2 quantity of lime. Fertiliser as far as possible should be applied through drilling. N splits may be top dressed just when water recede.

10. VARIETIES

TABLE 14.5: Important varieties of rapeseed and mustard

Variety	Recommended for	Maturity (days)	Oil (%)	Av. yield (q/ha)	Suitable for	Remarks
Toria						
Agarni (B-54)	West Bengal, Assam	70-80	41	8-10	UI	
BR-23	West Bengal, Assam	100	43	9-10	UI	
T9	Uttar Pradesh, Madhey Pradesh, Rajasthan	100	40	12-15	I/UI	Aphid resistant
T36	Uttar Pradesh, Madyey Pradesh, Rajasthan	100-105	43	12-15	I	
Sangam	Haryana	95-105	42	15	I	
ITSA	Punjab, Haryana	100-105	42	8-10	I	
TL 15	Punjab	85-90	44	10	I	
M 27	Assam and Orissa	90-95	42	12-14	U	
DK 1	Himachal Pradesh	75	44	8-10	I	
Ts 29	Assam	85-90	44	10-12	U	
BR-29	Bihar	110	42	17-19	-	
PT 507	Uttar Pradesh	85-90	42	14-16	-	
Bhagirathi	Orissa	120-125	40	20-22		
Krishna	U.P.	125-130	40	25		
RCC 4	Himachal Pradesh	150	40	10-75		
Mustard						
Laha-101	Uttar Pradesh	150	41	15	I	
Varuna (T 59)	Rajasthan, Uttar Pradesh, Bihar, U.P. & W.B.	125-130	40	20	I/U	
Shekhar	Uttar Pradesh, Rajasthan	135	40	18	I/U	
Kranti	Whole country	135	40	15-18	I	
Durgamani	Rajasthan	135-140	39	10-12	I/U	Orobanche resistant

Variety	Recommended for	Maturity (days)	Oil (%)	Av. yield (q/ha)	Suitable for	Remarks
Pusa bold	Delhi, Rajasthan, East India, Orissa	140	40	18	I	Bold seed, shattering resistant
Prakash	Haryana	130-135	39	15-20	U/I	Blight resistant
RH 30	Haryana	-	40	16	U/I	Shattering resistant, late sown
RLM-198	Punjab	152-166	38	17-18	I	Aphid resistant
RLM-514	Punjab	152	40	15-20	U/I	
Vabhav	M.P.	120-125	33	13-15	-	Blight & pest
Vardan	M.P.	120-125	40	10-16	-	Blight & pest
RL 18	Punjab, Haryana	140-150	40	15-20		
BR-40	Bihar	115	40	12	-	
Datan 67	Gujarat	115-120	38	18-20		
Seeta	West Bengal	95-100	38	U/I		
NDR 8501	East Uttar Pradesh	125	38	14-16	-	Alkaline and saline
Brown sarson						
BSH-1	Punjab, Haryana	130-140	42	15-20	-	Blight
Pusa kalyani	Rajasthan, Punjab & Uttar Pradesh	125-135	40	15-20	-	
BS-70	-	130	45	10-12	I	
KOS-1	Hills of north	-	44	10-12	U	
Yellow sarson						
YS-24	Punjab, Haryana & Rajasthan	140-150	42	12-15	I	
K-88	Uttar Pradesh	125-130	45	12-18	-	
Patan sarson 16	North Gujarat	110	42	10-12	I	
Bonoy	West Bengal	90	46	10-12	U	
M18	Assam	-	45	-	-	
T151	Uttar Pradesh	120	46	14-16	-	
Pusa gold	Bihar, West Bengal, Parts of Orissa	100-105	45	20-25	I	

*U = unirrigated, I = irrigated

10.2.1. Other Recent Releases

Indian mustard	Laxmi, Rajat, Rohini, TM-2, Sanjukta, Asesh.
Toria	Jowahar Toria-1, Panchali, RH 68, TS29
Brown sarson	KBS 3
Yellow sarson	NDYS 921, Rajendra sarson 1, Subinoy, YS 923

11. IDEAL PLANT TYPE

Mustard plant is photosynthetically inefficient. Bhargava *et.al.*(1984) described an ideal plant model for *Brassica* sp. as: plant height must be 1-1.25 m with 5-6 primary branches emerging at 30-45° angle from main stem. There should be only 40 siliquae on main stem with 20 siliquae in top two branches and 15 siliquae each on lower three branches. Each siliqua must have 20 seeds weighing 5 g. There must be 7-10 leaves with deep root system.

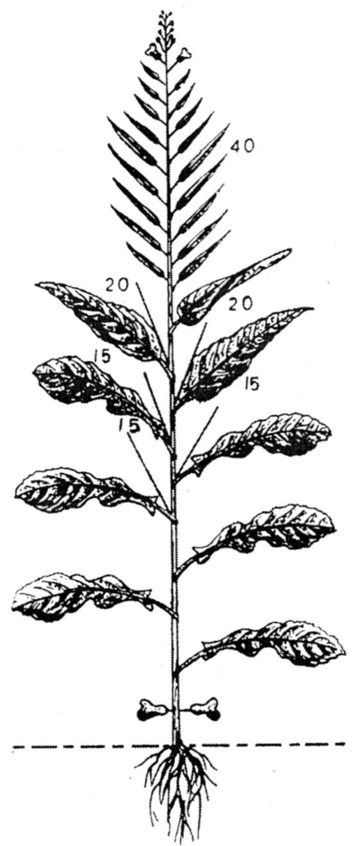

FIGURE 14.1: Ideal plant type of mustard.

12. SOWING

Sowing time is mainly dependent on temperature and it is the single important factor which determines the yield of crop. Sowing too early or too late result in not only decrease in yield due to reduced growth, poor flowering and less test weight but also reduce oil content also. Under too early sowing, crop may face high temperature at germinating stage and may be attacked by low temperature at most susceptible flowering stage. Too late sowing, germination may be delayed because of low temperature and at maturity stage crop may face high velocity hot winds. Looking to these factors mustard should be sown in first fortnight of October and *toria* in second fortnight of September. The specific sowing time for different states/places are given below:

TABLE 14.6: Sowing time of rapeseed and mustard in different states

Crop	Uttar Pradesh	Rajasthan	Madhya Pradesh	Punjab	Gujarat	Haryana	Bihar
Mustard	1st fortnight October	Middle of September to middle of October	Beginning of October	2nd fortnight of October	2nd week of October	Middle of October	Middle of October
Toria	1st week of September	Middle of Sept.-middle of October	1-3rd week of September	1st fortnight of September	-	Middle of September	Middle of Sept. to middle October
Brown sarson	-	Middle Sept. to middle of October	-	End of Sept. to 1st week of October	-	End of Sept. to middle October	-
Yellow sarson	-	1-15th Sept.	-	-	2nd week of October	End of Sept. to 15th Oct.	Middle October

12.1. Method and Depth of Sowing

Crop of this group should be sown in lines. Under dryland agriculture, moisture in seeding zone is limiting factor. The crops should be sown in such a way that there is no much soil cover on seeds and seed could be sown at some what deep in soil. Shallow sowing, result in poor germination because of drying of top soil. To achieve this, crop should be sown by ridge seeder developed at Hissar. In irrigated conditions sowing should be done with seed-cum-fertiliser drill. Fertiliser should be drilled at 7-10 cm deep and seeds are placed at 3-4 cm. The crop can also be sown by country plough by pora method (putting seeds behind the plough through funnel tube).

12.2. Spacing

Maintaining optimum plant population is important to realize higher yields. Spacing fluctuates with growth habit of variety, date of sowing, manuring and irrigation. In general crops of the group may be sown at row to row spacing of 30 cm. and plant to plant space of 10-15 cm. Thinning (removal of extra plants) may be done to adjust plant to plant spacing. Choudhary *et.al.* (1985) obtained 1.41 q more yield when mustard was sown at plant to plant space of 10 cm. as compared to 20 cm. spacing at Tabiji (Ajmer).

12.3. Seedrate

Pure crop of rapeseed and mustard can be sown with 4-6 kg seeds/ha. If crop are sown as mixed or intercropped with other crops seed rate of 1.5-2.0 kg/ha is sufficient depending upon row ratio. Selection of seed is very important to get more yields. As far as possible certified seeds should be used for sowing.

12.4. Seed Treatment

Seed must be treated with seed dressing fungicides to check seed borne and to some extent soil borne fungal diseases. Seeds may be treated with Benlate/ Vitavex at 2.0-2.5 g/kg seeds.

13. IRRIGATION

Generally, crops of rapeseed and mustard are raised on conserved soil moisture. Efforts should be made to conserve as much water as possible by bunding the field and ploughing before onset of monsoon. There should not be deficiency of moisture at pre-flowering and pod filling stage. The total water requirement of these crops is 400 mm. First irrigation should be given late to promote root development. First irrigation be given at 30-35 days after sowing, second at initiation of flowers and third at pod filling stage. About 3-4 irrigations are required in light soils and 2-3 irrigations in heavy soils. Chaudhary (1985) harvested 10 q more yield by applying 3 irrigation over control at Tabiji (Ajmer) (Table 14.7). Scientifically irrigation in rapeseed and mustard be given at 50% deplition of available soil moisture. Based on IW/CPE ratio crop may be irrigated at ratio of 0.6-0.8.

TABLE 14.7: Effect of irrigation on mustard yield

Treatment	Yield (q/ha)
No irrigation	7.94
One irrigation (30 DAS)	12.09
Two irrigation 30 & 60 DAS	15.75
Three irrigation 30, 60 and 90 DAS	17.31

14. WEED CONTROL

Weeds are negative factor in crop production. To avoid nutrient, moisture and space competition, crops should be kept weed free upto 40-50 days stage. Thereafter weed growth is automatically suppressed by smoothering effect of crop. About 25-30% loss is caused by weeds in rapeseed and mustard. The weeds which predominates in these crops are *Chenopodium album, Chenopodium murale, Asphodelus teunifolius, Anagallis arvensis, Convolvulus arrensis, Melilotus alba/Parisflora, Cynodon dactylon, Physalis minima, Orobanche cernua*, parasitic weed.

TABLE 14.8: Herbicidal weed control in rapeseed and mustard

Herbicide	Common name	Rate kg/ha.	Mode of application
Fluchloralin	Basalin	0.75-1.0	Pre plant incorporation
Pendimethalin	Stomp	0.5-1.5	Pre emergence
Oxyflurofen	Goal	0.2-0.5	Pre emergence
Oxadiazone	Ronstar	1.0-1.5	Pre emergence
Isoproturon	Tolkcan	1.0-1.3	Pre emergence
Isoproturon	Tolkcan	0.5-0.75	Post emergence 30 DAS

Weeds may be controlled by one intercultural operation with khurpi or hand hoe 5-10 days after first irrigation. Hoeing also facilitates aeration and conserve soil moisture. Weeds can also be controlled through use of herbicides efficiently and timely. The herbicide along with their rates and mode of application is given in Table 14.8.

When rapeseed or mustard is grown mixed with gram Fluchloralin at 0.75-1.0 kg/ha may be used as per-plant incorporation to control weeds.

15. PLANT PROTECTION

15.1. Frost

Rapeseeds and mustard are very sensitive to low temperature, specially if it occur at flowering. Very cool temperatures (below 0° C), calm atmosphere and clear sky on any day indicate occurrence of radiation frost. Apply irrigation if temperature drop below 0° C, sprinkle water. To reduce extent of damage spray 78 g DMSO in 1000 litre of water if crop is at flowering stage. Repeat spray if low temperature prevails. Effect may also be minimised by spraying 0.1% H_2SO_4.

15.2. Insect-Pests

Mustard saw fly: The larvae of mustard saw fly (*Athalia proxima*) feed on leaves and make holes. In case of severe infestation entire leaf lamina is eastern. The female of this insect have saw like ovipositor thus called saw fly. Damage of insect is severe 7-10 days after germination of crop. Infestation may be controlled by dusting endosulfan dust 4% or quinolphos 1.5% or malathion 5% or methyl parthon 2% at 20-25 kg/ha in morning. Spray of monocrotophos 36 WSC at 1.0 L/ha also takes care of this insect.

Aphid: Aphid (*Lipaphis erysimi*) is a green small 2 mm tiny insect. Both nymph and adult infest the crop by sucking the cellsap from leaves, stalk and inflorescence and siliquae. The affected plant appear wilted and sticky. Infestation is more severe under cloudy conditions and cause severe damage if occur at flowering or pod formation. Economic threshold level for aphid is 50-60 aphids/10 cm terminal portion of central shoot (Singh, 1999). It is controlled by dusting the crop with 2% methyl parathion or 5% malathion at 20-25 kg/ha under rainfed conditions while under irrigated conditions. Spray phosphomiden 85 wsc at 250-300 ml or endosulfan 35 EC or malathion 50 EC at 1.25 L/ha in 500-600 Lit water. It is also controlled by spraying oxydemeton methyl (0.025%) or monocrotophos (0.05%).

Painted bug: The nymph and adult of painted bug (*Bagrada cruciferanum*) suck the cell sap from leaves and top stem. At pod formation stage, causing reduction in yield and quality of produce. Dusting the crop with 4% endosulfan or 1.5% quinalphos or 5% malathion at 20-25 kg/ha or spray with monocrotophos 36 WSC at 1.0 L/ha checks the infestation.

15.3. Diseases

Blight: The disease is identified by appearance of concentric black spots on leaves, stalk and pods. The infected pods have small and shriveled grains. It is a fungal disease spreading through seed and causal organism is *Alternaria brassicae*. To take care the seeds be sown after treating with apron. SD 35, 6 g/kg seeds or spray ridomil 0.2%, (ALCORPO, 1967-1992). Spray of 1.5-2.0 kg Mencozeb or captafal or 2.5 kg zineb/ha in water just after seeing first symptom. Repeat the spray 20 days after first spray, if needed.

Downey mildew: This fungal disease is caused by *Peronospora brassicae*. The leaves of infected plant have yellow, irregular spots on upper side and whitish spots on lower side. Use of clean and healthy seeds for sowing, prevent infestation. Spraying 0.2% apron SD 35 followed by spray of dithane M-45 (0.2%) decreases spread of disease. Spray of metalaxyl 0.2% followed by mancozeb 0.2% spray at flowering checks the spread of this disease.

Powdery mildew: Disease is caused by fungi *Erysiphe polygoni*. Symptoms of disease are deposition of greyish white powder in leaves. On increase infestation white powdery growth changes to dark grey and black and leaves develop small-big lesions resulting to death of leaf. Infestation of disease is more under cloudy conditions. It is controlled by applying 20 kg sulphur dust or 2.5 kg wettable sulphur or 750 mL dynocap in water.

White blister: Plant leaves stalk and floral parts have white raised blisters resulting into deformation of inflorescence. It is a fungal disease caused by *Albugo candida*. To prevent infection, use of healthy seed is necessary. Use of crop rotation to break continuity of disease checks its spread. Two spraying with 0.2% zineb at 10 days interval check its further infestation and spread.

16. HARVESTING

When crop attained maturity, the colour of leaves, stem and siliquae is pale green to yellow. The lower siliquae have a dried appearance while the upper siliquae may look still green. Seeds are dark and produce rattling sound on shaking. Rapeseed and mustard should be harvested at physiological maturity. When 75% of siliquae mature, crop can be harvested. Too early harvesting produce wrinkled and small seeds of low viability with low oil content. Late harvesting, result in loss of grains by shattering. Crop should be harvested in the morning to avoid shattering. Crop is harvested leaving 5-10 cm. bottom plant with the help of serrated sickle. Crop should be dried for 5-7 days.

17. THRESHING AND WINNOWING

Plants are threshed by beating the pods with wooden sticks or by running bullocks or tractor. Crop can also be threshed by threshers. Winnowing is done in slow blowing wind or by winnovers.

18. YIELD

Irrigated rapeseed	15-20 q/ha
Irrigated mustard	20-25 q/ha
Unirrigated rapeseed	10-15 q/ha
Unirrigated mustard	15-20 q/ha

With use of all improved technology one may harvest these potentials. Thakur and Singh (1997) studied contribution of agronomic inputs in Indian mustard. The magnitude of each/combind factor is given below in Table 14.9.

TABLE 14.9: Contribution of agronomic inputs in Indian mustard

Treatments	Seed yield (q/ha)	Net returns Rs./ha
Recommended package	19.64	15,370
RP-Improved variety	14.23	9983
RP-Fertiliser	8.18	4588
RP-Irrigation	14.88	11016
RP-Disease control	15.86	12194
RP-Pest control	14.63	11025
RP-Disease & pest control	14.53	11269
RP-Fertiliser & irrigation	6.85	3732
RP-Fertiliser & plant protection	6.58	4027
RP-Irrigation and plant protection	11.93	9022
RP-Weed control	13.64	9791
CD (P = 0.05)	1.46	-

* RP = recommended package

19. POST HARVEST TECHNOLOGY

Cleaned grains should be dried well to 8-10% moisture before it is stored. Seeds are not subjected to the stored grain pests.

19.1. Milling

Oil content in rapeseed mustard ranged from 30-40%. Extraction is done in local ghanis on small scale or oil meals on large scale. Specific property of oil of different cropsis given below:

Crops	Characteristics			
	Iodine value	Refractive index at 40° C	Acid value	Saponification value
Toria	75-100	1.461-1.466	-	-
Sarson	100-102	1.477-1.480	-	-
Mustard	100-102	1.460-1.463	0.5-40	169-177

In recent years, problem of adulteration in mustard oil is becoming very serious. *Argemone maxicana* commonly called *satyanashi* or prickly poppy is cheap adulterant used to improve pungency of mustard oil. High level of adulteration of *Argemone* causes dropsy in persons using adulterated mustard oil. Seeds of *Argemone* resembles to rapeseed mustard. Toxic alkaloid dihydro-sanguinarine and sanguinarine are responsible for dropsy. Intake of mustard oil with more than 0.1% *Argemone* oil causes dropsy. The alkaloid separates water from the blood and cause death of consumer. Swelling of limbs, vomiting, loose motions, nausea, breathlessness, renal failure and cardiac failure are the main symptoms. Epidemics of dropsy was also reported in 1935 in Nanded (M.P.), Mumbai in 1966, West Bengal in 1977 and in Delhi 1983, 1989 and 1998.

20. CONSTRAINTS OF PRODUCTION

Production varies widely from year to year. Frost and aphids attack is the main cause. Moisture stress is another factor responsible for its low production. The existing genotypes because of long duration does not fit in dry conditions. High yielding genotypes with short duration are required. Genotypes which could tolerate frost and insect-pests and diseases should be identified to harvest high yields. Production may still be increased by adopting available improved varieties, with proper agronomic management. Poor plant stand is a major problem in almost all mustard growing areas. Use of recommended levels of fertilisers, weed control and plant protection could boost the yield to great extent. Govt. may initiate crop insurance particularly to unpredictable frost damage.

21. REFERENCES

ALCRPO 1967-92. Research high lights, 1967-92. All India coordinated research project on oil seeds. Directorate of oilseeds research, Hyderabad.

Reddy, P.S. and D. Pati (1998). Indian oilseeds: Present status and future needs. Indian J. agric. sciences. 68(8): 453-9.

Rathore, P.S. and Manohar, S.S. (1989). Response of mustard to levels of nitrogen and sulphur (II). Effect of N and S on aneillary character and yield of mustard. Indian J. agric. 34 (3): 333-336.

Rathore, P.S. and Manohar, S.S. (1989). Plant analysis as a guide to sulphur fertilization of mustard. Indian J. agric. Sci. 59(12): 820-821.

Rathore, P.S. and Manohar, S.S. (1990). Plant analysis as an aid in N fertilization in mustard. Indian J. agric. Sci. 35(1 & 2): 151-152.

Anonymus (1996). Annual report on rapeseed & mustard. Directorate of oilseeds research, Hyderabad.

Bhargava, S.C., Tomar D.P.S. and Sinha S.K. (1984). Research and development strategies for oilseed production in India ICAR P 103.

Yadav, R.L., Singh, Punjab, Prasad, Rajendra and Ahlawat, I.P.S. (1998). Fifty years of agronomic research in India. Indian society of agronomy. New Delhi.

Singh, R.A. and Singh, M.P. (1999). Performance of Indian mustard variety, NDR 8501 in Eastern Uttar Pradesh. Indian farming 48(10): 7.

Bhatia, P.C., Yaduraju, N.T. and B. Daury (1999). Argemone-the killer weed. Indian farming. 48(10): 16.

Thakur, K.S. and Singh C.M. (1997). Contribution of agronomic inputs in Indian mustard (Brassica juncea) under mid-hill conditions of Himachal Pradesh. Indian J. agronomy 42(4): 665-669.

Choudhary, A.K., Doval, S.L. and D.S. Oberio (1985). Raising production of mustard in Rajasthan. Indian farming 35(5): 19-21.

Sinha, S.K. (1980). Plant physiology research. Indian farming 30(8): 41-46.

FAO, 1997. Production year book.

Singh, D.P. (1958). Rape and mustard. The Indian central oilseed committee, Hyderabad.

Singh, T., Singh, S., Kumar, S. and Tomar, S. (1997). Response of Indian mustard (Brossica juncea) to nitrogen, phosphorus and sulphur fertilization. Indian J. agronomy 42(1): 148-151.

Jaggi, R.C. and Sharma, D.K. (1997). Effect of sulphur and phosphorus in yield and their uptake by Indian mustard (*Brassica juncea*). Indian J. agronomy. 42(2): 352-336.

Anonymus (1999). 4th Agricultural science congress. Sustainable agric. export. National academy of agric. sciences, New Delhi, RAU, Bikaner and Govt. of Rajasthan.

Thakur, H.L., Bhateria, S., Sagwal, J.C., Sood, O.P., Thakur K.S., Kumar, A. and Srivastava, A. (1997). New Indian mustard variety for Himachal Pradesh. Indian farming 47(8): 15-16.

Katiyar, R.K. (1997). Pusa-gold: A new high yielding early maturing yellow sarson. Indian farming 47(8): 25-27.

15. Linseed

P. S. Rathore

Linseed is a dual purpose crop. It is grown for both oil and fibre (flax). The oil (35-45%) is used as cooking medium. Because of quick drying, its oil is used in paints, varnish and printing ink industries. The oil is also used in preparation of lubricants, polishes and greases. The seeds contains a mucilagious substances called spermoderm, which is used as laxatice and in cosmetics and pharmaceutical industries. The oil cake too is nutritious and palatable cattle feed or can be used as organic manure to improve soil physical condition and fertility. Cold press linseed cake may cause hydrocyanic acid poisoning but hot pressed linseed cake is safe. The fibre commonly known as lignin. Fibre is used for manufacturing linen, canvas, carpet and thick cloths. Out of total production 5% is used for sowing, 88% for extraction of oil and 7% fibre.

1. ORIGIN

Linseed is being cultivated from bronze age. Mention of linseed is available in old Egyptian records. Vavilov (1926) reported South West Asia namely India and Afghanistan as centre of origin of oilseed linseed, while of flax type, Mediterranean region.

2. DISTRIBUTION

Linseed is grown in all the climatic conditions, tropical, sub-tropical and temperate from 11-30 °N latitude. It is widely grown in India, China, Canada, Germany and Russia. The area, production and productivity of linseed is given in Table 15.1.

TABLE 15.1: Area, production and productivity of linseed in world (1996).

Countries	Area (000 ha)	Production (000 mt)	Productivity (kg/ha)
India	940	308	328
China	600	440	733
Canada	575	843	1466
Russia	153	-	327

Countries	Area (000 ha)	Production (000 mt)	Productivity (kg/ha)
Germany	82	78	851
Total	3109	2221	715

Indian share in world's linseed area and production is around 28 and 14%, respectively. Linseed in India is mainly grown in Madhya Pradesh and Uttar Pradesh, accounting 36 and 30% area of country, respectively. The other states which produce linseed are Bihar and Maharashtra. During 1994 area under linseed was 920 thousand ha with production of 340 th tonnes and productivity of 370 kg/ha.

3. CLASSIFICATION

Linseed (*Linum usitatissimum* L.) belongs to family linaceae with chromosome no. 30 and 32. Riccharia (1962) classified the 26 varieties of Indian linseed on the basis of seed colour and size.

1. Based on seed colour.
   a) Yellow coloured
   b) Brown coloured
   c) Fawn coloured
2. Based on seed size.
   a) Small seeded
   b) Medium seeded
   c) Bold seeded.

4. CROPPING SYSTEM

Linseed is grown after harvest of *kharif* crops. The common rotation are, sorghum - linseed, maize - linseed, groundnut - linseed, soybean - linseed. Based on requirements linseed is also sown mixed with other crops like wheat or gram. Experimental results also showed that intercropping of linseed with gram or lentil is economical. The most popular inter cropping system practiced in different parts of India are mentioned below:

TABLE 15.2: Popular inter-cropping system

System	States
Safflower + linseed	Karnataka, Madhya Pradesh and Maharashtra.
Linseed + gram	Punjab, Uttar Pradesh.
Linseed + lentil	Punjab, Uttar Pradesh.
Linseed + wheat (2:1)	Uttar Pradesh, West Bengal.
Linseed + potato (1:3)	Uttar Pradesh.

Crop is mostly grown as *paira* crop in Bihar or *utera* crop in Uttar Pradesh on residual moisture. Linseed is broad casted in standing *kharif* paddy at flowering or dough stage. Moisture stress is the major cause of poor yield under these systems.

5. SOIL AND SOIL PREPARATION

Crop can be grown on any soil type, however, black cotton soils or alluvial loam soils having good water retaining capacity and adequate drainage are best. Crop is tolerant to both acid and salts, though perform better in neutral to mild acid conditions.

Crop needs fine seed bed. One deep ploughing followed by 2-3 harrowing and planting are adequate. To save the crop from termites and an early attack of bud fly, mix 20-25 kg/ha endosulfan dust 4 per cent at the time of last ploughing.

6. MANURES AND FERTILISERS

With the introduction of high yielding and hybrid varieties of cereals, linseed is cultivated on poor marginal soils. Water stress and poor fertility are the main constraints in its production. Application of 5-7 t FYM/ha not only improve nutrient availability, but also soil physical condition, particularly water retaining capacity. Crop is grown both under rainfed and irrigated conditions. The requirement of fertiliser and mode of application is given below.

Fertiliser requirement of linseed.

Crop condition	N	P_2O_5	K_2O	Mode of application
Irrigated	40-60	20-30	-	1/2 quantity of N + full amount of phosphorus placed at sowing. Remaining 1/2 N at 35 DAS.
Rainfed	20-30	10-15	-	Full amount at sowing

Dutta et al (1995) recommended 60 kg N/ha under sandy loam soils at Hissar having 138 kg N/ha. Similarly Samui (1995) recommended 80 kg N/ha for Kalyani, (West Bengal). While Singh and Verma (1997) recommended 90 kg N, 45 kg $P_2O_5$and 30 kg K_2O for light textured soils of Ghazipur (U.P.). Sulphur, nutrition also found to increase crop yield under Nagpur conditions. Application of 30 kg S as gypsum gave significantly higher yield (Anonymous 1994-95). Crop, responded to zinc also. Foliar spray of 0.5% $ZnSO_4$ at 50 DAS improved seed yield. Foliar spray of 0.5% $ZnSO_4$ was found better compared to 5 kg $ZnSO_4$ soil applied zinc at Kota, Rajasthan.

7. VARIETIES

Selection of variety is an important factor in catching high yield. Improved varieties of linseed with characteristics are given below in Table 15.2.

TABLE 15.3: Improved varieties of linseed and area of adaptability.

Variety	Duration (Days)	Oil (%)	Av. yield q/ha	Area of adaptability
Chambal	130	-	11.50	Rajasthan
K-2	175	46	11.00	Haryana, Punjab, Himachal Pradesh
L C 54*	160	42	13.00	Haryana, Punjab, Himachal Pradesh
Hira	135	43	12.00	Uttar Pradesh
Mukta*	130	45	12.00	Uttar Pradesh

Variety	Duration (Days)	Oil (%)	Av. yield q/ha	Area of adaptability
Jawahar 17	120	43	8.00	Madhya Pradesh
Neelam*	145	43	13.00	Uttar Pradesh and West Bengal
Himalini*	165	42	13.00	Himachal Pradesh, Punjab, Rajasthan and Haryana
B-67	125	42	13.00	West Bengal
R-17	140	47	13.00	Madhya Pradesh.
S 36	110	-	12.00	Karnataka
Garima-	125	42	15.00	Uttar Pradesh
Sweta	130	44	12.00	Uttar Pradesh
Shubhra	130	45	12.00	Uttar Pradesh
Pusa-2*	150	43	13.00	Major linseed growing areas.
Kiran	130	43	8.00	-
Gaurav+	137	43	10.0	Good for fibre
Jeevan+	175	-	10.5	Good for fibre
R L 102-71+	122	45	16 q (I) 10 q (R)	25% fibre, South and S.E. Rajasthan.

* Seed purpose, + Seed & fibre, Utera-Jawahar-7, L C 185 & Surabhi.

8. SOWING

Time of sowing depend on climatic conditions and availability of moisture. Irrigated crop of linseed could be sown in October till first fort night of November. Delay in sowing beyond first fort night of November give poor yields. In Jammu and Kashmir crop is sown in February-March. Rainfed crop should be sown early with cessation of monsoon, to utilise residual soil moisture.

Crop may be sown by broad cast or in lines. Line sowing is better. Crop is sown in rows 30 cm apart. Plant to plant distance within row is kept 5 cm. To sow crop in lines 25 kg seeds are required, while for broadcasting 30 kg seeds are required. In paira or utera sowing 10 kg seeds are sufficient. Crop is generally sown behind the plough by pora method, but could be sown by seed drill. Depth of sowing should be 5 cm.

9. IRRIGATION

Crop is mainly cultivated, rainfed, however it responds well to irrigation. Singh and Verma (1997) reported seed yield with no irrigation, one irrigation at branching and one irrigation at capsule initiation as 6.79, 8.10 and 7.79 q/ha respectively. Thus two irrigations first at 35 days after sowing (seedling development) and second 65 day after sowing (flower initiation) augmented the yield to great extent. Richharia (1943) reported seedling and pre flowering as critical stages. Yadav et al (1998) based on results of various experiments reported 0.5-0.8 IW/CPE ratio for scheduling irrigation in linseed and estimated water requirement as 80-200 mm. Dutta et al (1995) obtained 11.68 q/ha at 0.8 IW/CPE as compared to 10.75 q/ha at 0.4 IW/CPE ratio, while under no irrigation yield was 9.88 q/ha.

10. WEED CONTROL

Linseed plant is short statured, grows slowly in the beginning and have less number and area of leaves. Therefore, it cannot compete with weeds. Mishra (1997) reported 30-40 % reduction in grain yield due to weeds. Study indicated that 20-45 days of growth is regarded as critical period from crop-weed competition view. Weeds not only decrease crop yield, but also impair quality of oil. The common weeds infesting linseed are given below-

Common weeds of linseed	
Broad leaf	**Grassy**
– *Chenopodium album*	– *Avena ludoviciana*
– *Carthamus oxycantha*	– *Phalaris minor*
– *Asphodelus tenuifolius*	– *Cyperus rotundus*
– *Melilotus alba*	– *Cynodon dactylon*
– *Convolvnulus arvensis*	– *Lolium temulentum*
– *Anagallis arvensis*	

Hand weeding is an important cultural method. Two hoeing at 3rd and 6th week after sowing control most of the weeds. Plant to plant space should also be adjusted during first hoeing. Research carried out in different parts of India indicated that weeds can be controlled efficiently by using herbicides. Pali et. al. (1997) obtained 5.25, 10.27 and 10.04 q/ha yield under un-weeded check, two hand weeding (30 & 60 DAS) and under isoproturom 1 kg post emergence treatment respectively at Madhya Pradesh. The important herbicides and their mode of application is given in Table 15.3.

TABLE 15.4: Herbicidal weed control in linseed

Herbicide	Dose kg/ha	Mode of application
Fluchloralin	1.0	Pre plant incorporation
Tnfluralin	1.0	Pre plant incorporation
Fluchloralin	1.0	Pre emergence
Terbutryn	0.75	Pre emergence
Iso proturom	0.75-1.0	Post emergence, one month after sowing.
MCPA	0.2-0.5	Post emergence, pre branching stage of crop.

11. PLANT PROTECTION

Important insect pests and diseases infesting linseed are given below-

Insect pests and diseases of linseed.	
Particular	**Control measures**
Insect pests	
Linseed bud fly (*Dasyneuralini*)	Early sowing, sowing of resistant varieties. (Neelum, Jawahar-1, Kiran, Himalini and Surabhi) spray phosphomidon 0.03% at 15 days interval or 0.07% endosulfan. Bio control using *systesis dasyneurae*, *Elasmus* sp. *Eurytoma* sp. bio agent is beneficial.

Particular	Control measures
Leaf minor	Dusting with 10% methyl parathion 25 kg or spray of 0.04% monocrotophus 36 EC.
Caterpillar	Spray 1.25 Lit. thiodan 35 EC in 500-750 Lit. water.
Diseases	
Blight (*Alternaria* sp.)	Early sowing, crop rotation, seed treatment with bavistin help in limiting infestation. Spray 0.251 %. dithane M-45 or calaxine 0.2%.
Rust (*Melampsora lini*)	Grow resistant varieties (Hira, mukta). Spray dithane Z-78 @ 2 kg/ha in 500-750 Lit. water or dusting 20 kg sulphur dust.
Powdery mildew (*Oidium lini*)	Spray wettable sulphur compounds as sulfex @ 2-3 kg/ha in 500-750 Lit water. Repeat spray after 15 days.

12. HARVESTING

Stage at which plant almost ceases its physiological processes and formation of the grain or synthesis of produce of economic value is complete, is termed as maturity. The maturity in linseed can be judged by colour of leaves, stem and grain. The lower leaves wither away, leaves and stem are yellow and bolls are ripe. Stem at base is yellow and bolls dries up and turn brownish. Crop is harvested by cutting close to ground by sickle or by uprooting the whole plant. For oil crop, cutting is preferred and for fibre, uprooting is better because of more fibre length. After harvesting, produce is kept for sun drying. Grains are separated after 2-3 days by beating plant with sticks or by feet of bullock. Winnowing is done to remove seed from straw.

13. YIELD

- Pure crop 12-15 q/ha.
- Mixed crop 5-8 q/ha.

14. POST HARVEST TECHNOLOGY

Seeds are dried in sun till moisture is 10-12%. After drying seed may be stored or extracted for oil by village ghanis or power driven mills. Fibre crop requires retting. Retting in general is done by two ways (i) Dew retting (ii) Water retting. Out of the two, water retting is good. In this method, plant is kept in running water or tank. At temperature of 80 °C retting is completed in 7-10 days. After 7-10 days, bundles are placed upright for drying and fibre is separated by passing retted plant between fluted rotters and then beaten with wheel to loose and separate fibre. The average length of fibre is 500 mm.

15. QUALITY

Linseed is cultivated for both oil and fibre. Linseed oil is yellowish brown in colour and contains 30-60 per cent lenolenic acid, 6-16 per cent stearic and palmitic acid, 13-36 per cent oleic acid, 10-25 per cent linoleic acid and fraction myristic acid. The keeping quality of its oil is good, if protected from light and air. The composition of linseed is given below:

Constituents	Composition (%)
Water	6.60
Protein	20.30
Fat	36.00
Carbohydrates	29.00
Fibre	4.80
Ash	3.30

16. REFERENCES

Anonymus (1994-95) Annual report, ICAR. New Delhi:

Anonymus (1995) *Rabi* phaslo ki unnat kheti Directorate of Ext. N.D. Univ. Agric and Tech., Faizabad.

Anonymus (1998) A decade of achievements 1987-96. Raj. Agric. Univ. Bikaner.

Dutta, H.K., Ram, D.S., Rao. Mana Singh, H (1995). Response of linseed (*Linum usitatissimum*) to irrigation and nitrogen. Ind. J. Agron.

FAO Production Year Book Vol. 51.

Malik, Y.P. (1999) Integrated management of bud fly in linseed. Indian farming. 48 (10): 29-30.

Mishra, J.S. (1997). Critical period of weed competition and losses due to weeds in major field crops. Farmer and parliament 33 (6): 19-20.

Pali, G.P., Patel, S.R., Sarkar, C. and Tripathi, R.S.(1997). Integrated weed management in irrigated linseed (*Linum usitatissimum*). Indian J. Agron 42 (1): 177-180.

Pundir, P and Dahama (1994) Crop varieties of RAU, Directorate of Research, RAU Bikaner.

Reddy, P.S. and Pati, D. (1998) Indian linseeds: Present status and future needs. 68 (8): 453-459.

Riccharia, R.H. (1962) Linseed: Indian central oilseed committee, Hyderabad. pp 155.

Richharia, R.H. (1950). Wealth from waste. The Indian linseed plants from fibre point of view. United press ltd. Bhagalpur, India.

Samui, R.C., Roy, B., Tripathy, S.K. and Patra, A.K.(1995). Response of linseed (*Linum usitatissimum*) varieties to nitrogen in alluvial soils of West Bengal. Indian J. Agron. 40(1): 132-133.

Singh, N.B., Verma, K.K. (1997). Response of linseed (*Linum usitatissimum*) to varying irrigation and fertility levels. Indian J. Agron. 42(4): 696-98.

Vavilov, N.I (1926). Studies on the origin of cultivated plants. Bulletin of applied botany and plant breeding. 16(2).

Yadav, R.L., Singh, P., Prasad, R and Ahlawat IPS (1998). Fifty years of agronomic research in India. Indian society of Agron.

16. Safflower

P. S. Rathore

Safflower popularly known as *kusum* is an oil seed crop. The oil (24-36%) consists of unsaturated fatty acids (94 %). Because of its high linoleic acid (78%) it is good for heart patients. The oil is also used in paints, and varnish industry. Oil cake is a good cattle and sheep feed (45% protein). Carthamin and safflower yellow (20%) are extracted from florets for making dyes. Being deep rooted and drought tolerant it is a remunerative crop under dryland situations. Crop is also sown on borders to guard other crops against animals.

1. ORIGIN

Vavilow (1926) reported that it might have originated from either India, Afghanistan or Ethiopia. De Candolle (1886) however, reported Arabia as probable centre of origin and *Carthamus palaestinis* is considered its wild progenitor. Safflower might have originated from *Carthamus oxycantha* (wild safflower).

2. DISTRIBUTION

Crop is mainly grown between 14° and 45° N and 15° and 35° S in India, Mexico, USA and Ethiopia. These countries accounts nearly 95 per cent of world's area and production (Table 16.1). India ranks first with respect to area and second in production. It is grown between 14°-22° N and 73.5°-79° E. About 99 per cent of safflower in India is cultivated in Maharashtra, Karnataka and Andhra Pradesh. It is also grown in Orissa and Madhya Pradesh in small scale. The state-wise area and production is given in Table 16.2.

TABLE 16.1: World's area and production of safflower (1997)

Country	Area th.ha.	Production 000 t.	Productivity kg/ha.
India	704	430	671
Canada	97	164	1688
USA	95	195	2051
Ethiopia	69	35	507
China	11	20	1905

Country	Area th.ha.	Production 000 t.	Productivity kg/ha.
Australia	34	20	588
Total	1039	933	899

TABLE 16.2: State-wise area and production of safflower (1992-93).

States	Area lakh/ha	Production lakh t.	Yield kg/ha
Maharashtra	4.85	2.33	480
Karnataka	1.92	0.81	420
Andhra Pradesh	0.21	0.08	380
Gujarat	0.11	0.07	600
Madhya Pradesh	0.11	0.06	580
Total	7.30	3.39	465

3. CLASSIFICATION

Safflower (*Carthamus oxycantha* L.) belongs to the family compositae and 2n chromosome number are 24. The related wild species are *Carthamus oxycantha* (wild safflower) and *C. lanatus* (Thistle safflower). Based on flower's colour it is classified in to white, yellow, light orange and reddish orange.

4. CLIMATE

Safflower is a *rabi* crop with deep root system and xeromorphic characters. It is a day neutral, but thermo-sensitive plant generally grown rainfed. Germination takes place at 15-16 ° C. Temperature of 25-30° C during flowering is conducive for its high yield. Crop is tolerant to low temperature at seedling and negative phase. Low temperature at flowering cause an adverse effect on pollination and seed set. Excessive water or humidity is not at all desirable. Crop is well grown in areas receiving 600-900 mm precipitation. High rainfall or prolonged rains at flowering reduce pollination, seed formation and impair quality of produce.

5. SOIL AND SOIL PREPARATION

Deep, well aerated, water retentive soils having neutral reaction are excellent for raising safflower. Heavy soils of poor permeability are not fit for its cultivation. There may be severe problem of wilt and root rot.

Soil is prepared by 2-3 harrowing and planking to crush clods. Seed bed must be well pulverised and free from weeds. Mix well 20-25 kg endosulfan dust 4% at last ploughing to protect the crop from termites and other soil insects.

6. CROPPING SYSTEM

Safflower is raised after preceding *kharif* crop. The common crop sequences are given below:

Crop sequence	State/s where practiced
Green gram – safflower	Maharashtra, Karnataka, Uttar Pradesh
Sorghum - safflower	Andhra Pradesh, Karnataka, Tamil Nadu

Crop sequence	State/s where practiced
Maize - safflower	Andhra Pradesh, Madhya Pradesh, Rajasthan, Bihar
Ground nut safflower	Karnataka
Pearlmillet - safflower	Uttar Pradesh, Andhra Pradesh
Fallow - safflower	Karnataka, Uttar Pradesh
Soybean - safflower	Madhya Pradesh
Cowpea - safflower	Rajasthan

Continuous cropping of safflower leads to build up of root rot, wilt and other diseases. Intercropping of safflower with other *rabi* crops have been found remunerative. Scientists at different places identified some profitable intercropping system (Table 16.3).

TABLE 16.3: Remunerative intercropping system with safflower in different states

States	System
Karnataka, Andhra Pradesh, Maharashtra and Tamil Nadu	Safflower + gram (1:3).
Andhra Pradesh, Maharashtra, Tamil Nadu	Safflower + coriander (1:3)
Karnataka, Maharashtra, Madhya Pradesh	Safflower + wheat
Maharashtra, Madhya Pradesh	Safflower + linseed (1:2)

7. MANURE AND FERTILISERS

Light soils, need adequate dressing with FYM at 8-10 t/ha. It should be well mixed at least 15-20 days prior to sowing. Crop responds well to applied nitrogen. Rajput (1992) recommended 60 kg N/ha for sandy loam soils of Morena, Madhya Pradesh. He obtained 20 q more yield over 40 kg N/ha. Based on number of experiments, response upto 60 kg N/ha was noted with optimum around 30-40 kg N/ha. Nitrogen and oil content are negatively correlated. Singh (1992) reported 32.5% oil under control as compared to 27.1% in 60 kg N/ha but total oil yield/ha was high. Response to phosphorus, depends on inherent status and phosphorus fixing capacity of soil. Soils deficient in phosphorus, respond well to basal dressing of 20-25 kg P_2O_5/ha. Patel et. al. (1994) obtained response of 25 kg P_2O_5/ha over control at Navsari Gujarat. In areas with adequate moisture 20 kg K_2O may be applied, however phosphorus and potassium should be applied based in soil test. Response to micronutrients have been noted only in specific pockets.

8. VARIETIES

Group of individuals within a species with certain specific characteristics is called variety. Selection of variety according to soil and climatic conditions is important to harvest sizable yield. A list of improved varieties of safflower, suitable for different parts of country is given in Table 16.3.

Improved varieties of safflower				
Variety	**Oil (%)**	**Maturity (Days)**	**Yield q/ha**	**Adaptability**
Bhima	32.5	-	12-14	Maharashtra
JSF 1	30.0	-	10-15	Rajasthan, Madhya Pradesh

Variety	Oil (%)	Maturity (Days)	Yield q/ha	Adaptability
Co 1	-	120-125	7-8	Tamil Nadu.
N 62-8	30	140-150	-10	Parts of Maharashtra
A 1	31.0	125-130	8-9	Andhra Pradesh and Karnataka
Manjira	32.5	105-110	8-10	Andhra Pradesh
T 65	30.0	-	12-13	Uttar Pradesh
K 1	-	120-125	6-8	Bihar, Uttar Pradesh, Tamil Nadu
Nira	30.0-	-	14-18	Maharashtra, Tamil Nadu
HUS 305	35.0	-	11-13	Uttar Pradesh and Peninsular India
JSF-2	33.0-	-	12-14	Madhya Pradesh
J L 28-1	31.4	114-167	18-20	Maharashtra

9. SOWING

Crop is generally sown from last week of September to second fort night of October in different parts of India. Early sowing are found beneficial particularly with regards to less infestation of aphids and other insects. Late sowing after 1st week of November, markedly decreases crop yield. Based on research findings, specific sowing time in different parts is given below:

States	Sowing time
Maharashtra (Rainfed areas)	15th Sept.- 30th Sept.
Maharashtra (Adequate moisture) North East Karnataka, Andhra Pradesh, South Tamil Nadu and South-East Rajasthan	1st week Oct.- 15th Oct.
Uttar Pradesh, parts of Maharashtra	3rd-4th week of October.

9.1. Seed Rate

Crop is generally sown using 15-20 kg/ha. Row to row spacing is kept at 45-60 cm depending upon climate and soils conditions. For heavy rainfall zone of Gujarat, Gujarat Agricultural University recommended sowing of Bhima variety at 60 cm spacing using 15 kg seeds/ha (Anon., 1995). Plant to plant spacing may be kept at 30 cm and depth of sowing is 5-6 cm. Under mixed cropping a seed rate of 4-5 kg is sufficient depending upon proportion of crops.

10. WEED CONTROL

Mishra (1997) reported 35-60 per cent loss due to weeds in safflower. Weeds grow luxuriantly, because of wide row spacing, initial slow growth and little leaf area. Thus 15-45 days of growth is considered as most critical from crop weed interference view point. The most common weeds found in crop are *Chenopodium album*, *Chenopodium murale, Angallis arvensis, Spergula arvensis, Melitotus indica* as broad leaf weeds and *Phalaris minor, Cynodon dactylon* as grassy weeds.

Hand weeding at 20 and 35 days after sowing, keep most of the weeds under control. Weeds could also be controlled by using herbicides in less time and economically. Singh and Sandhu (1996) reported, pre emergence pendimethalin at 0.75 Lit/ha equally effective to two hand weeding. The herbicides which could be used in crop are given below in Table 16.4.

TABLE 16.4: Herbicidal weed control in safflower

Herbicide	Mode of application	Recommended level kg/ha
Fluchloralin	Pre plant incorporation	0.75-1.0
Trifluralin	Pre plant incorporation	1.0-1.25
Oxadiazone	Pre emergence	0.75-1.0
Pendimethalin	Pre emergence	0.75
Terbutryn	Pre emergence	0.75

11. IRRIGATION

Safflower is deep rooted, xerophytic plant, can survive under scarce soil moisture. Under stress conditions, one or two irrigations at 25 days and 75 days after sowing enhances yield. Rosette and flowering are the most critical stages for irrigation. In water retentive soils, one life saving irrigation may yield good crop of safflower. Patel and Patel (1993) reported 22, 51 and 68 per cent increase in yield with one irrigation at grand growth, at flowering, and at seed formation respectively over no irrigation.

11.1. Nipping

Subbiah and Swaram (1965) recommended nipping of central shoot to induce branching and to promote yield.

12. HARVESTING

Crop should be harvested when lower leaves turn yellow and start drying. Generally crop attend maturity within 125-150 days. The plants are either uprooted or cutoff from ground level with sickle. Crop should be harvested in morning to avoid effect of spines. Harvested plants be dried and threshed. Threshing may be done by beating manually with sticks. Produce is cleaned by winnowing in moderate wind. Mechanical thresher may be used for quick threshing and winnowing.

13. YIELD

Irrigated crop - 15-16 q/ha and rainfed crop - 6-8 q/ha.

14. POST HARVEST TECHNOLOGY

Cleaned grains are dried till 8 per cent moisture. Storage is safe, if seed have about 6 per cent moisture. Seeds of crop are crushed as per demand, because keeping quality of its oil is poor.

15. QUALITY

Safflower oil is recommended for heart patients as it keeps blood cholesterol level low because of its high linoleic acid. Oil is pale yellow with specific gravity of 0.92, iodine value 130-150 and saponification value 186-194. Composition of safflower seed (Apple white 1966) is hull (40%), oil (37%) and meal (23%). The oil includes linoleic acid (78%), oleic (11%), stearic (3%) and palmitic (6%).

16. PLANT PROTECTION

The major insect pests and diseases of safflower with their control measures are given below-

16.1. Insect Pests/Diseases

Insect pests: Aphids (*Uroleucon compositae theobold*): Early sowing check its infestation, spray of dimethoate or monocrotophous or endosulfan 0.05%. Cut worm and safflower caterpillar. Spray 0.05% endosulfan or monocrotophous. Safflower bud fly spray dimethoate or phosaline or endosulfan 0.05%.

16.2. Diseases

16.2.1. Alternaria Leaf Spot (*Alternaria carthami*)

Seed treatment with carbendazim 2 g/kg seed prevent infestation. Spraying dithane M-45 at 0.25% control leaf spot.

16.2.2. Rust (*Puccinia carthami*)

Grow resistant variety APRR, SF 438, seed treatment with daconil at 3 g/kg seed prevent infestation. Spray calixin at 0.05% if occur in field.

16.2.3. Ramularia Leaf Spot

Spray 0.1% bavistin.

16.2.4. Wilt (*Sclerotinia sclerotiorum*)

Sow resistant variety and follow crop rotation.

17. REFERENCES

Anonymus (1993). Salient aspects of research achievements and recommendations GAU, Directorate of research, Ahmedabad.

Apple white, T.H. (1996). The composition of safflower seed J. oil chem. soc. 43 (6): 406-08.

Patel, Z.G. and Patel N.C. (1993). Effect of irrigation scheduling on yield of safflower (*Carthamus tinctorius*). Indian J. Agron 38 (3): 509-510.

Patel, Z.G., Mehta, S.C. and Raj, V.C. (1994). Response of safflower (*Carthamus tinctorius*) to row spacing and nitrogen and phosphorus fertilisers in vertisols of South Gujarat. Indian J. Agron 39 (4): 699-700.

Rajput, R.L., Verma, O.P. and Gautam, D.S. (1992) Relative performance of safflower (*Carthamus tinctorius*) varieties with different levels of nitrogen under rainfed conditions. Indian J. Agron. 37 (2); 290-292.

Singh, J. and Singh, B.S. (1996). Chemical, weedless try in safflower (*Carthamus tinetorius*). Indian J. Agron 40 (4): 510-574.

Singh, S.B., Chauhan, Y.S. and Verma, G.S. (1992). Effect of row spacing and nitrogen levels on yield of safflower (*Carthamus tinctorius*) in salt affected soils. Indian J. Agron. 37 (1): 90-92.

Subbiah, N and Swaram, M.K. (1965). Increasing yield in safflower. Indian oil seed J. 9: 155

Umrani, N.K., Deokar, AB and Nimbaear, V.S. (1984) Bhima - A high yielding variety of safflower.

PULSE CROPS

17. Soybean

P. S. Rathore

Soybean is considered highly nutritive crop. It contains 40 per cent high biological value protein and 20 per cent oil. Soybean protein has 5 per cent lycine, which is deficient in most of the cereals. Enriching cereal flour with soybean improves nutritive quality. Soybean contains less starch, thus, it is good for diabetic patient. Its oil is used as cooking medium and also for making *vanaspathi ghee*. The industrial uses include soymilk, soya flour, soya cake, biscuits, varnish and paints. 1 kg of bean may yield 5-6 kg of soymilk. Soylecithen – a biproduct of oil industry is used as emulsifier in cosmetics and pharmaceuticals. Soybean plant is used as fodder and cake as an excellent concentrate for livestock. It also enriches soil by fixing atomospheric nitrogen. Approximately 85% of soybean produced is utilized for oil extraction, 10% for seed and 5% for food.

1. ORIGIN

Based on historical and geographical evidences North eastern china is generally regarded as the origin place of soybean. It has been cultivated from pre-historic time in China. It was introduced in USA 1804 and in India during 1880.

2. DISTRIBUTION

Soybean is an important crop of the world. It is cultivated on 67.62 mha with a production of 147.62 mt. The important soybean producing countries are listed in Table 17.1. Soybean is getting attention in India too. It was grown on 0.9 mha in 1982-83 but now it is grown on 4.2 mha. The important states producing soybean are given in Table 17.2.

TABLE 17.1: Area and Production of soybean in world

Countries	Area 000 /ha.	Production (000 mt)	Productivity kg/ha
USA	28281	74224	2625
Brazil	11552	26546	2298
China	8385	13508	1611
Argentina	6600	145000	2197

Countries	Area 000 /ha.	Production (000 mt)	Productivity kg/ha
India	5600	5350	953
Total	67619	147025	2174

TABLE 17.2: Area and Production of soybean in India

States	Area (000 ha.)	Production (000 t.)
Madhya Pradesh	3703.8	3781.6
Uttar Pradesh	44.8	35.0
Rajasthan	427.7	400.8
Maharastra	635.4	718.8
Total	4887.2	4992.7

3. CLASSIFICATION

Soybean belongs to family leguminosae, subfamily papilionceaceae belongs to genus *Glycine*. It has chromosome number 2n = 40. Based on seed size, colour shape of grain soybean may be classified into three groups.

3.1. Manchurian Classification

1. Yellow group - Seeds of this group are yellow with-
   a) light hilum
   b) Golden hilum
   c) Dark brown hilum
2. (ii) Black group - Seeds are of black colour with
   a) Small seed
   b) Large seed
   c) Flat seed.
3. Green group- Epidermis of seed is green with
   a) Yellow embryo
   b) Green embryo
4. Martin classified the soybean, on the basis of shape and size of seeds.
   a) Soja elliptica with elliptical seed
   b) Soja compressa with compressed seed
   c) Soja spherica with rounded seed
5. Hertz on the basis of pods classified soybean as:
   a) Soja tumida with swollen pods
   b) Soja platycarpa with flattened pods

4. CLIMATE

Soybean is cultivated in warm and moist climate and grown in monsoon season from June-October. Temperature of 15-32° C is required for its germination. Optimum temperature required for its growth and yield is 30-33° C. If temperature is below 10° C, crop growth is retarded. Similarly, if temperatures are above 38° C, crop growth is also retarded. Day temperatures of 25° C are good for its flowering. Crop can be grown in areas receiving 600-650 mm rainfall. Rainfall during maturity deteriorates the grain quality. Cloudy weather prolongs vegetative phase. The crop is generally cultivated at an attitude of 1200-2000 meter.

5. SOIL AND SOIL PREPARATION

Soybean is grown on variety of soils ranging from light to black cotton soils in different pockets of India. Well drained fertile loams soils are ideal for its cultivation. Soils should be loose and well aerated. Highly compacted soils are harmful for root nodule development. Crop is sensitive to both saline and acidic conditions and can be grown with in pH range of 6.0-7.5. Water logging is also harmful for the crop.

Soil preparation for soybean consists of one deep ploughing with mould board plough followed by two harrowings and planking. To protect the crop from soil borne insects specially termites 25 kg endosolfan dust 4 per cent should be mixed into the soil just before last harrowing.

6. CROPPING SYSTEM

The following crop sequences are generally observed with soybean:

- Soybean – Wheat
- Soybean - Potato
- Soybean - Bengal gram
- Soybean – Potato - Wheat
- Soybean - Sunflower

Soybean is remunerative ideal inter-crop. The most important intercropping system are given below:

- Maize + soybean
- Pigeopea + soybean
- Cotton + soybean
- Sorghum + soybean
- Groundnut + soybean
- Sesame + soybean

7. MANNERS AND FERTILISERS

Like other leguminous crops, requirement of nitrogen is substantially fulfilled from symbiotic nitrogen fixation through *Rhizobium*. Seed treatment with *Rhizobium* not only enhances yield but enriches the soil also. Integrated nutrient management involving use of organic manures, crop residues and inorganic fertiliser is beneficial. FYM at 15-20 tonnes/ha alongwith 20 kg N/ha as an starter dose enhances yield of soybean. *Rhizobium* inoculation can fix 150 kg N/ha in soybean. A crop fertilized with nitrogen have more protein per cent as compare to unfertilized crop. If *Rhizobium* culture is not available, soil @1 t/ha from nearby field in which soybean was previously cultivated for 2-3 years may be collected from 15 cm depth and broadcasted in fields (autoculture) and mixed into the soil.

Soybean need large quantity of phosphorus, particularly at beginning of pod formation till ten days prior to seed development. phosphorus requirement of crop should be based on soil test value. In absence a soiltest a dose of 50-60 kg. P_2O_5/ha is generally recommended for the crop. This not only increases test weight but yield also (Rajput et. al., 1991).

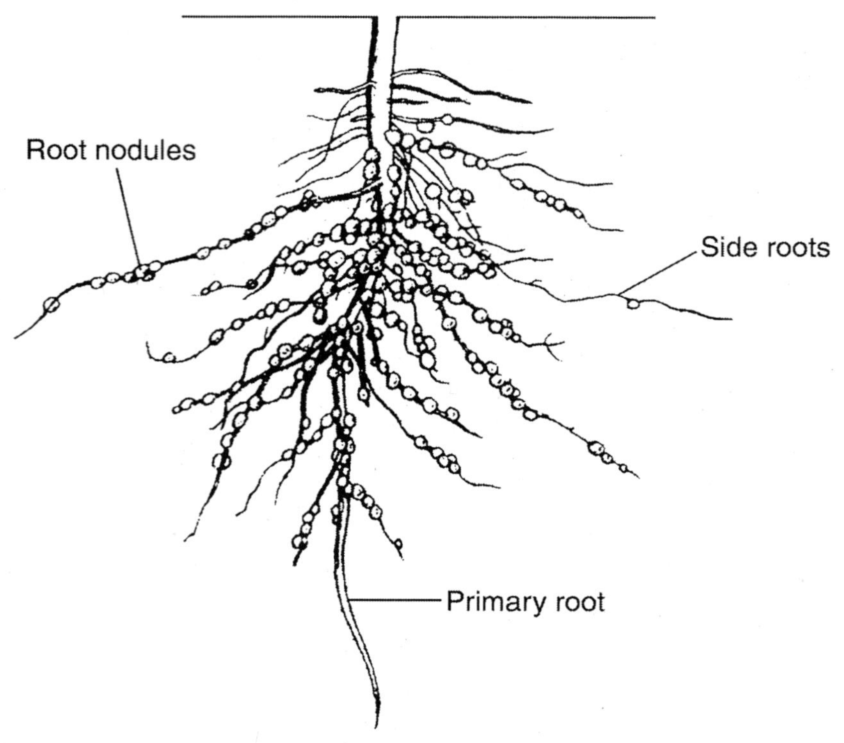

FIGURE 17.1: Nodulation in soybean.

8. VARIETIES

Yield potential of varieties differ from one agro-climatic region to another. The varieties found suitable for particular region are given below:

Varieties of soybean suitable for different regions	
Regions/zones	**Variety recomonded**
North Hill zone	Bragg, shilajeet
North plain zone	Bragg, Alankar, PK 327, PK 262, PK 308, PK 416
Central zone	Jawahar, Gaurav (JS-72-44) Durga (JS-72-280), Bragg, Ankur, MACS-13
Southern zone	KHSB-2, Hardee, Improved pelicon.

The other varieties of soybean recommended for cultivation are: ADT 1, Ankur, Birsa soy 1, Bragg, Co 1 Durga, Gaurav, Gujarat soy, Hardee, J S 2, JS 75-46.

KHBS 2, MACS 13, Monetta, NrC 2, PK 308, J 362, J 564, Pusa 16, R/RC 7, Shivalik, SL4, VL soy.

Characteristics of important soybean varieties

Variety	Maturity (Day)	Seed	Oil (%)	Yield (q/ha)	Place
Lee	110-115	Light yellow black hilum	23-25	15-20	Srinagar valley, HP, Hills of U.P.
Clark	90-95	Yellow and black hilum	18-20	25-30	
Improved Pelicon	110-120	Light yellow	20-22	20	Southern zone
Punjab-1	120-130	Yellow with blackish hilum	20-22	20-25	Punjab, Delhi Haryana, U.P.
Shilajeet	105	Bold	20-22	20-25	MP Gurjarat Rajasthan, U.P. and M.P.
PK 262	130	Bold and yellow	22	30-35	U.P., M.P.
K-30	125-130	Black	20	12-18	

Potassium requirement of soybean is also high. The maximum potassium is required for vegetative development of plant. In absence of soil test a general dose 20 kg k_2O is recommended for the crop (Mishra and Vyas, 1992).

In zinc deficient soils one foliar spray of zinc sulphate at 5 kg zinc + 2.5 kg lime in 1000 L is necessary to over come zinc deficiency.

9. SOWING

Sowing of soybean soon after onset of monsoon rains is best as compared to its earlier or later sowings. The optimum time of sowing for different zones are given below:

TABLE 17.3: Optimum sowing time of soybean in different zones

Zone	Sowing time
Northern hill zone	Beginning of June - mid July.
Northern plain zone	Last week of June - Mid July.
Central zone	Mid June - mid July
Southern zones	Mid June - mid July
Eastern zones late June	Early July

Seed rate depends upon size of the seed, test weight, spacing and method of sowing. An optimum plant population of 3 lac plants ha must be maintained for higher yield. Generally 60-70 kg seeds are required/ha. If sowing is delayed due to any reason, the seed rate must be increased by 25%.

Crop is sown at row to row spacing of 45-60 cm in north and 30-45 cm in central and south (Rajput et. al., 1991). At the time of thinning (removal of extra plants). Plant to plant spacing be kept at 5 cm. The seed should be sown at 2-3 cm in depth in heavy and 4-5 cm in light soils. Seeds may be sown behind country plough.

9.1. Seed Treatment

Besides the *Rhizobium* culture seeds should also be treated with seed dressing fungicides. Many fungi may attack, directly on seed as well as seedling. Most important disease causing damage in soybean are *Rhizoctonia*, *Fusarium* and *Phythium* sp. The damage of these fungi may be reduced by treating the seeds with thiram/bavistin @ 3 g/kg seed.

10. IRRIGATION

Usually the crop is raised as rainfed. However, in situation of early withdrawal of monsoon rains irrigations are found beneficial. Flowering and pod filling stage are the most critical stages at which moisture stress severely damages crop yield. Moisture stress at flowering and pod formation result in flower and pod dropping. Thus in case of moisture stress 1-2 irrigation at flowering and pod filling stages is beneficial in realizing good yield. The last irrigation should be scheduled at least 30 days before harvest. The total water requirement of the crop is 450 mm Scientifically, irrigation in soybean may be scheduled at IW/CPE ratio of 0.6 (Rajput et al 1991). Check basin or border strip method is most popularly used for irrigating soybean.

11. WEED CONTROL

Soybean, being a rainy season crop, is heavily infested with many grassy as well as broad leaf weeds. Yield loss in soybean may range from 25-70 per cent depending upon intensity and infestation of weeds. Besides yield loss, quality is also adversely affected. The major weeds of soybean are given below. The most critical period for crop weed interference is initial 15-45 days (Mishra, 1997). Weeds emerging after this period are suppressed by smothering action of crop.

Weeds in soybean can be controlled by using *khurpi* or hand-hoe. Generally two weedings, one at 20-25 days after sowing and other at 35-40 days after sowing keep most of the weeds under control. Timely availability of labour and survival of weeds within the row are main difficulties of hand weeding.

Herbicidial weed control has been found efficient and economical in soybean. The most common herbicide with their rate and time of application are given below.

Crop is sensitive to weed competition, particularly in early growth phases. The most common weeds infesting the crop are: grassy weeds (*Echinochloa colonum, Sorghum halepense, Cyperus rotandus, Cynodon dactylon*) and broad leaf (*Celosia argentia, Digitaria* sp., *Trienthema partulactrum, Vernonia* sp., *Tribulus terristris* and *Phyllanthus niruri*)

Weeds of soybean can be controlled effectively by using herbicides as mention below:

Herbicide	Rate (kg/ha)	Mode
Fluchloralin	0.5-1.0	Pre-plant
Trifluralin	0.5-1.0	Pre-plant
Alachlor	1.0-2.0	Pre-emergence
Oxadiazon	0.5-1.0	Pre-emergence

Herbicide	Rate (kg/ha)	Mode
Butachlor	2.0	Pre-emergence
Metolachlor	1.5-2.0	Pre-emergence
Pendimethalin	1.0	Pre-emergence
Methabenzuron	0.2-0.3	Pre-emergence
Fluazitop- p	0.15-0.25	Post-emergence 10-15 DAs.
Sethoxydim	0.25-0.50	Post-emergence 20-25 DAs

Joshi and Billore (1998) recommended 1.0 kg./ha fluchloralin or 1.5 kg clomozen for controlling weeds with an weed control efficiency (WCE) of 94.45 and 96.63 (Table 17.4).

TABLE 17.4: Effect of herbicides on weeds and yield of soybean

Treatment	Rate (kg /ha)	Weed density/m^2	WCE (%)	Seed yield (kg/ha)
Fluchloralin	1.0	28.67	94.45	2475
Alachlor	2.0	155.67	86.07	2314
Clomozen	1.5	63.67	96.63	2483
Sethoxydin	1.5	641.67	65.21	2256
Alachlor	2.0	201.33	84.45	2403
2-HW	-	163.00	93.43	2370
Unweeded check	-	676.33	-	2031
CD 5%		158		

12. PLANT PROTECTION

12.1. Insects

- **Stemfly** (*Melanagromyza phaseoli*) cause 15-20% reduction in yield. The larvae generally bore in to the stem of soybean plant.
- **Gridle Beetle** (*Obserea brevis*) may cause 50-60% loss in crop yield. The beetle make tunnels in stem. It can be controlled by applying 0.07% endosulfan (35 EC) or monocrotophos 0.4% (36EC).
- **Bihar Hairy caterpillar:** Caterpillar feed on leaf epidermis. It causes heavy damage in soybean. This insect can be controlled by dusting 4% endosulfan dust @ 25 kg/ha at early stages and by spraying 0.07% endosulfan or monocrotophs 0.4%.

12.2. Diseases

Seedling rot: This is a fungal disease caused by *Aspergillus flavus, A. niger* and *Fusarium semetum.* The disease results in poor seedling emergence because of seed rotting. The seed treatment with thiram at 3.0 g/kg of seeds is beneficial.

Pod blight: This fungal disease is caused by *Colletotrichum truncatum.* The pods of attected plant turns yellowish green and later on dry up. The grain becomes shriveled and mouldy. Use of resistant varieties, sowing of crop in rotation and spray of 2.0-2.5 kg/ha zineb in 500-600 of water can control the disease.

Yellow mosaic: It is a viral disease and transmitted by white fly (*Bemisia tabaci*). The disease is dangerous and widely spreaded. The affected plant looks yellowish. The yellowness of leaf increases with the intensity of infection. Damage in crop is severe when it is infested at early stages (70-75 days of sowing). Growing of resistant varieties as *Shilajit* reduces the damage. Inspect the crop regularly, remove the infected plants and burn them. Spray metasystox 1 L/ha in 500-600 water at 10-15 days interval to control the disease.

13. HARVESTING

At maturity plant shed the leaves. The leaves turn yellow. The seed in pod is hard and dough. The moisture content in grain should not be more than 12-14%. Preferably the crop should be harvested in the morning to avoid losses of shattering. Harvesting is done by uprooting the plants or by cutting with sickle. Crop is threshed by beating with sticks or by bullock or tractor thresher or by mechanical soybean thresher. The grain should be dried in to sun to 10-12% moisture.

14. YIELD

20 to 25 quintals

15. STORAGE

Soybean to be used for seed in next season should be stored under low temperature and low humidity conditions. The grains should be dried in sun and stored in ventilated rooms. Because of oil, bitterness may develop if stored for long time.

16. QUALITY

Soybean is an excellent protein source and called *wonder bean.* Composition of soybean grain is given in Table 17.5. It has been reported that black seeded soybean varieties are rich in protein but low in oil content, while yellow seeded varieties are rich in oil and poor in protein.

TABLE 17.5: Composition of soybean grain (Values/100 g edible portion)

Moisture	Protein	Fat	Minerals	Fibre	Carbohydrates	Energy
8.1	43.2	19.5	4.6	3.7	20.9	432

Ca	P	Iron	Carotene	Thiamine	Riboflavin	Niacin	Vitamin C
240	690	11.5	426	0.73	0.39	0.32	0.3

17. POST HARVEST TECHNOLOGY

The grain should be fully dried in sun until moisture percentage 10-12.

18. REFERENCES

Chandal. A. S and Saxena, S C. (1986) New improved cultural practices for soybean. Ind. Fmg. 35 (10) 6.

FAO Prod. Year Book Vol. 51: 1997.

Gopalan, C, Ramasastri, B.V. and Balasubramanian, S. C. (1977). Nutritive value of Indian foods.

Joshi, O. P. and Billore, S.D. (1998) Herbicidal weed control in soybean (*Glycine max*) Indian J. Agron 43(1): 126-128.

Mishra, J. S. (1997). Herbicidal weed control in soybean. Ind. Fmg. 47 (3): 17-20.

Rajput, R. L., Kaushik, J.P. and Verma, O.P. (1991) Yield and nutrient uptake in soybean (*Glycine max*) as affected by irrigation, phosphorus and row spacing.

Reddy, P.S. and Patil, D. P. (1998) Indian Oil seeds: Present Status and Future Prospects. Indian J. Agric. Sci. 68 (8, special issue): 453-9.

Venugopal, K. (1986) Soybean cultivation in Sikkim hills. Ind. Fmg, 35 (10).

18. Pigeonpea

D. S. Yadav

Pigeonpea (*Cajanus cajan* (L) Millsp.) is a backbone of nutritional security of our country. India, has virtual monopoly in its production by bagging 90 percent of world's total production. Pigeonpea like other pulses, is considered a subsidiary crop. It is often grown on marginal lands and is usually inter-cropped with crops such as sorghum and cotton etc. As a crop of secondary importance in many of these systems, it receives little or no purchased inputs. However, farmers in some pigeonpea growing areas are growing more sole crops of pigeonpea and the crop is increasingly gaining status as a cash crop (Ryan, 1994). Besides, it supplements the important component of protein particularly to Indian population where their diet is based on cereal it self.

Pigeonpea is mainly used as *dal*. It is rich in iron, iodine and the essential amino acids like lycine, cystine and arginine. It is also used as ration for milch cattle. Its straw is also palatable and green leaves may be used as fodder. Sticks of pigeonpea are used for various purposes such as thatch and basket making, Pigeonpea being a leguminous plant is capable of fixing atmospheric nitrogen and thereby restore lot of nitrogen in the soil. Its deep root system helps in extracting the nutrients and moisture from deeper soil layers. Deeper root system also help in breaking the plough pans and help in improving soil structure. Extensive ground cover of pigeonpea prevents soil from wind and water erosion, encourages infiltration, minimises sedimentation and smothers weeds. It is also grown on mountain slopes to utilise the poor soils and protect it from erosion.

1. ORIGIN

Pigeonpea is the most widely grown crop in the country and has been under cultivation for over three thousand years. It has been reported to occur in wild state in the upper region of Nile river and the coastal districts of Angola in Africa. Therefore, Africa may be its original place and from where it might have been introduced to India. However, according to Vavilov (1951), India is the origin place of Pigeonpea. De (1974) and Vernon Rayes (1976) also prepared reviews on origin of pigeonpeas. The latter considered the dispute settled in favour of Indian origin.

2. DISTRIBUTION

India is the major pigeonpea growing country in the world where almost 90% of the world pigeonpea is cultivated. Other regions where pigeonpea is grown are south east Asia, Africa and the America. There are substantial areas of pigeonpea in Kenya, Uganda and Malawa in Eastern Africa, and in the Dominican Republic and Puerto Rico in central America (Nene *et. al.*, 1990). In India, pigeonpea is grown on an area of about 3.36 million hectares with the production of 2.18 million tonnes. The national productivity of pigeonpea is only 651 kg per hectare. It is second most important crop after chickpea among different pulse crops in the country.

The data on the trend of area, production and productivity of pigeonpea are given in Table 18.1. It is extensively in the states of Maharashtra, Uttar Pradesh, Madhya Pradesh, Gujarat, Karnataka and Andhra Pradesh. It is evident that Maharashtra and Uttar Pradesh occupy large area under pigeonpea. These states occupy about 31 and 15 per cent, respectively of the total area under pigeonpea in the country, Maharashtra produces maximum of 0.53 million tonnes of pigeonpea which is about 24 per cent of total production in the country.

TABLE 18.1: Area, production and productivity of pigeonpea in different states (1994-95)

States	Area (,000, ha)	Production (000 t)	Productivity (kg/ha)
Madhya Pradesh	389.0	357.5	919
Gujarat	361.0	231.0	640
Maharashtra	1057.6	526.5	498
Uttar Pradesh	505.5	519.1	1027
Andhra Pradesh	279.0	99.4	335
Karnataka	304.5	110.7	364

3. CLASSIFICATION

Cultivated pigeonpea plants may be classified into two groups. All the plants contain 2n = 22 chromosomes.

3.1. *Cajanus cajan* variety *bicolor*

Mostly late maturing varieties are included in this group. Plants are tall, bushy with yellow flower, purple streaped, pods dark coloured containing four or five seeds. Cultivated extensively in northern India.

3.2. *Cajanus cajan* variety *flavus*

Early maturing varieties of comparatively smaller plants, yellow flower and pods containing only two or three seed included in this group. In is mainly grown in Deccan and southern India.

4. CLIMATE

Pigeonpea is mainly grown in tropical and sub-tropical climate. Cool and bright sunny days during flowering and ripening are required for setting of pods. Crop is raised in areas having 20-30° C temperature during *kharif* season and 17-22° C during winters. It is highly susceptible even to light frost. Though, it can tolerate heavy rains provided water logging does not takes place. It has the capacity to tolerate moisture stress to a great extent because of its deep root system.

5. SOIL AND SOIL PREPARATION

Pigeonpea requires light textured, well drained soil, Though, it is grown on a wide range of soils ranging from sandy to clay. Soil should be neutral in reaction and well drained. Production of this crop varies greatly depending upon the depths and moisture holding capacity of soil. Pigeonpea crop generally thrives well on Entisols but suffers from excess water on vertisols and moisture deficit on Alfisols and Inseptisols (Reddy and Virmani, 1981). Pigeonpea is susceptible to salinity as compared to other crops. Similarly fields where pigeonpea has been grown for few seasons should be avoided as the crop may suffer due to a build up of nematodes and increased incidence of soil borne diseases.

Pigeonpea being a deep rooted crop respond well to proper tilth. Deep ploughing to a depth of 15 cm is sufficient. However, where a hard pan exists in the rooting zone, sub-soiling has been found beneficial in improving root development and infiltration rate (Reddy *et. al.* 1978). Harrowing followed by cross operation by cultivator/*deshi* plough and cross planking is sufficient to get the good tilth. Well tilled field may promote better root and nodule development in seedlings. Pigeonpea is very susceptible to water-logging, Particularly during seedling stage. Excess of moisture may cause yellowing of seedlings, perhaps due to inhibition of nitrogen fixation. Standing water even for new days may result in partial or complete loss of stands depending upon the growth stage of the crop. Under such situations, a crop sown on ridges gave 26-31 per cent higher yield as compared with flat beds (Chaudhury and Bhatia, 1971). Proper drainage is, therefore, essential for successful cultivation of pigeonpea. In the areas where ridging is not possible, a gentle slope is desirable.

6. CROPPING SYSTEMS

Long duration varieties of pigeonpea engage the field for whole year. These varieties are sown with the onset of monsoon in June-July and are harvested in the month of March and April. Development of short duration varieties of pigeonpea maturing in 130-160 days have made it possible to grow a second crop in sequence during *rabi* season. These varieties if planted in early part of June are harvested by first week of December. Therefore, several crops like wheat, gram, lentil, sugarcane, etc. may be grown successfully after these varieties of pigeonpea. Rotations adopted with pigeonpea are: pigeonpea-wheat, pigeonpea + urdbean-wheat, pigeonpea-wheat-mungbean, pigeonpea-sugarcane, mungbean (summer) + pigeonpea-wheat, and maize-pigeonpea (post monsoon).

6.1. Mixed/InterCropping

Traditional mixing of pigeonpea with maize, sorghum without consideration of planting geometry and space, crops suffer to a great extent and their yields are reduced. Intercropping and parallel multiple cropping is the improvement over mixed cropping where the crops of dissimilar types are grown together in such a way that they do not affect the performance of main crop and there exists zero competition in between main crop and parallel crop. The example of such cropping may be pigeonpea + urdbean or mungbean. The urdbean/ mungbean being early maturing, is harvested by the time, the space occupied by it is required for pigeonpea. The pigeonpea is planted at wider row spacing and its growth during early period is very slow.

At Pantnagar (U.P.), experiments were conducted to find out the most suitable parallel crop for short duration pigeonpea and to find out the possibility of increasing the profits by changing the planting pattern. Pigeonpea variety T-21 was planted on June 25 in rows 75 cm apart with a plant spacing of 20 cm and in between two rows of pigeonpea one row of inter crop was planted in one case whereas in another case two rows of inter crop were planted in between the paired rows of pigeonpea. The data on grain yield of pigeonpea and mixed crop as well as net profit under each treatment are presented in Table 18.2.

The grain yield of pigeonpea was not affected significantly by any of the mixed crops except maize.

Two planting methods did not differ significantly. Extra yield of 4 to 7 quintals of grain legumes was obtained without affecting the grain yield of pigeonpea. Yield of maize was 29 to 33 quintals per hectare but it adversely affected the grain yield of pigeonpea, however, net profit with inter cropping of maize was highest followed by normal planting of urd bean (Table 18.2).

TABLE 18.2: Yield of pigeonpea as influenced by different inter cropping treatments

Treatment	Grain yield (q/ha)	
	Pigeonpea	Intercrops
Pure pigeonpea (N) (T-21)	23.99	-
Pure Pigeonpea (P)	21.86	-
Pigeonpea + Mungbean (N) (T-1)	18.66	3.83
Pigeonpea + Mungbean (P)	19.46	4.63
Pigeonpea + Urdbean (N) (UPU-1)	24.37	4.58
Pigeonpea + Urdbean (P)	19.99	7.30
Pigeonpea + Cowpea (N) (Pusa Dophasli)	20.63	3.78
Pigeonpea + Cowpea (P)	15.03	5.11
Pigeonpea + Soybean (N) (Lee)	18.50	4.79
Pigeonpea + Soybean (P)	21.70	4.39
Pigeonpea + Maize (N) (Dwarf)	15.83	33.19
Pigeonpea + Maize (P)	15.73	29.22
	Inter crops	Methods
C.D. 5%	4.37	NS
N = Normal rows	P = Paired rows	

7. MANURE AND FERTILISERS

Pigeonpea requires heavy doses of the nutrients, hence care should be taken that the crop does not suffer due to lack of nutrients. A crop of pigeonpea yielding 2 t/ ha grains and 5.4 t/ha sticks removed 115 kg N and 16 kg P on sandy loam soils of IARI (Singh, 1973). Various agronomic factors (Cultural, manurial and cropping system) influence the uptake of nutrients (Ahlawat, 1980).

7.1. Nitrogen

Being a legume, it does not require much nitrogenous fertilisers. However, there is general agreement that a starter dose of 10-30 kg N/ha gives positive response in most of the soils depending upon the initial N status of the soil. On sandy loam soils which are poor in N and organic matter, a starter dose of 20-25 kg N/ha has increased the yield. On the red and laterite soils of Medak, red soils of Sundergarh, mixed red and black soils of Phulbani districts, the responses of pigeonpea to 20 kg N/ha ranged from 1.40-5.20 q/ha over control yields ranging from 3 to 6 q/ha. However, Kulkarni and Panwar (1980) reported a response of 6.70 q/ha over control yield of 4.90 q/ha in Medak with variety Hy- 3c. Roysharma *et. al.* (1980) reported positive response of *rabi* pigeonpea 'Bahar' to 40-60 kg N/ha in Bihar.

7.2. Culture

Soils with high fertility do not response to culture. However, light textured soil with low in organic matter content, newly reclaimed, eroded and flooded lands show positive response to culture. The necessity of seed inoculation with culture for legume which is grown for the first time in a field has been known for a number of year. Erdman (1953) reported that a more effective strain of Rhizobia can increase yield and protein content of legume to an extent of 20 per cent.

7.3. Phosphorus

The importance of adequate supply of phosphorus for higher yields of pigeonpea has long been recognised. In one of the earliest experiments, Krauss (1932) obtained good response of pigeonpea to phosphorus fertilization in Hawaii specially in upland and also recommended a dose of 350 to 1000 kg/ha of acidulated rock phosphate.

In India, responses to phosphorus have been generally positive and in some cases highly significant (Chowdhary, 1968). Ahlawat (1976) realised that phosphorus improved all growth and yield attributes, root and nodule development. Yadav (1984) recorded 15.0 and 27.5 q/ha of grain yield without and with 60 kg P_2O_5/ha, respectively. However, further increase in the level of phosphorus did not increase the yield.

7.4. Potassium

Pigeonpea normally do not respond to potassium application unless grown on the soils low in available potassium. However, on cultivators fields, the response to 20 kg K_2O/ha over 20 kg N + 40 kg P_2O_5/ha ranged from 0.90-2.0 q/ha except in Jalaun district where there was no response to potassium (AICARP, 1985).

7.5. Micro Nutrients

The work on the response of pigeonpea to micro nutrients has been very limited. Most of pigeonpea cultivars show susceptibility to zinc deficiency. Soil application of 2-4 ppm zinc or foliar spray of 0.5 per cent zinc-sulphate with 0.05 per cent lime have proved effective in controlling zinc deficiency (Saxena and Singh, 1970). Agrawal (1973) observed varietal differences in pigeonpea to Zn deficiency. He found that varieties Pant A-3 and Prabhat were more tolerant to Zn deficiency than other varieties.

8. VARIETIES

Several varieties of pigeonpea have been developed. At present emphasis is being laid on the development of short duration high yielding varieties permitting second crop of wheat or other *rabi* crop in the winter. These varieties mature in about 130-180 days. They are harvested by December and escape the risk of frost in the end of December and January which commonly occur in northern India. Earlier longer duration varieties taking 240-300 days to maturity were gown, occupying field for whole of the year. These varieties were usually damaged by frost in the month of December and January under northern Indian conditions.

Following varieties of Pigeonpea have been recommended for cultivation in different states.

1. **Andhra Pradesh:** PDM-1, T-21. Vishakha-1 (TT 6), HY-1, HY-2, HY-3 A, HY 3 C, HY-4, HY-5, PCPL 87, BDN-1, C-11,
2. **Bihar:** Laxmi (Kanke 3), Bahar, Basant, BR 183, BR 65, AS 71-37, Mukta.
3. **Delhi:** NPWR-15, Mukta, Pusa Ageti, Pant A3, Sharda, Pusa 74, Manak, Sagar, Pusa 85, Pusa 33, Pusa 84.
4. **Gujarat:** T 15-15, No. 148, Pusa Ageti, BDN 2, Pusa 74, Pusa 33.
5. **Haryana:** UPAS-120, Prabhat, Pusa 84, Manak, Sagar, T-21, Pusa 74, Pusa 85, Pusa 33.
6. **Himachal Pradesh:** HPA-1, UPAS 120.
7. **Karnataka:** F-52, T-21, Pusa Ageti, Sharda, Hy 3 C, GS-1, PT-221, ICPL-87, BDN-2, TT 6.
8. **Kerela:** SA-1, Prabhat, Pusa Ageti, Sharda.
9. **Madhya Pradesh:** Gawlior-3, Khargone-2, JA-3, T-21, NPWR-15, Pusa Ageti, No. 148, AS 71-37, Sharda, C 11, Pusa 74, JA 9-19, Prabhat, Pusa 33.
10. **Maharashtra:** C-11, No. 148, PT 301, BDN-1, BDN-2, NPWR-15, Vishakha-1, UPAS-120, T-21, No. 290-21, ST-1, No. 84, Hyderabad 185, Pusa 74, Pusa Ageti, TAT 10, Pusa 33.
11. **Orrissa:** Pusa Ageti, UPAS-120, Sharda, Mukta.
12. **Punjab:** Type 21, AL 15, Pusa 84, Manak, Sagar, Pusa 33.
13. **Rajasthan:** UPAS-120, Prabhat, T-21, Gwalior 3, Pusa 74, Manak, Pusa 33.
14. **Tamul Nadu:** SA-1, Co 1, Co 2, Co 3, Co 4, Co 5, Prabhat, T 21, ICPL 87, TT 6.
15. **Uttar Pradesh:** Type 7, Type 21, UPAS-120, Prabhat, Pusa 84, Bahar, Pusa 74, Manak, Pusa 33, Pusa ageti, Pusa 85, Sharda, Mukta.
16. **West Bengal:** Sweta, Chuni, C 11, WB 20, Prabhat, Mukta, S-20.
17. The characters of some of the important varieties are given below:
18. **Bahar:** It is a late maturing variety of about 250-270 days. It was developed in Bihar through selection from a local material. It is suitable for post monsoon planting in the month of September in Bihar and eastern parts of Uttar Pradesh. It is resistant to sterility mosaic. It can yield up to 25-30 q/ha.
19. **BDN-1:** It was developed through a selection from local material of Maharashtra. It is a medium maturing variety of 150-160 days. Seeds are brown with the weight of 90 g/1000 seed. It is tolerant to wilt.

20. **Basant:** It is similar to Bahar. It is also suitable for post monsoon planting in Bihar and eastern U.P.
21. **BR-183:** It was developed through a selection from local material in Bihar. Plants are spreading and early in maturity taking about 160-180 days. Weight of 1000 seed is about 90 g.
22. **BS-1:** It was developed at IARI, New Delhi. It is an early maturing variety of about 150 days.
23. **CO-3:** It was released in 1977. Seeds are brown with the test weight of 70 g/ per 1000 seed. It matures in about 130 days in Tamil Nadu conditions.
24. **Gawlior-3:** It is a late maturing variety of about 240 days, seed colour is light brown and weight of 1000 seed is about 70 g.
25. **JA-3:** It is a medium duration variety developed from a cross of No. 148 and Gwalior 3. Duration is about 180 days and weight of 1000 seed is about 90 g.
26. **Laxmi:** It wa released in 1974 for Chota Nagpur plateau. It is a medium duration variety of about 200 days. It is tolerant to wilt and is suitable for mixed cropping.
27. **Manak:** This variety has been developed from the cross of T-21 and UPAS-120 at Hisar. It is indeterminate and Semi-spreading with medium in height of about 120-130 cm. It is an extra early maturing variety (120-125 days). Seeds are small and the weight of 1000 seed is about 65 g. It is most suitable variety for double cropping and is more popular in northern western plain zone.
28. **Mukta:** It is a medium maturing variety of 180-200 days. Weight of 1000-seed is about 100 g. Its yield potential is 20-25 q/ha. Seeds are medium bold and brown.
29. **No. 148:** It is a medium duration variety of about 160-170 days. Seeds are brown with a test weight of 100 q per 1000 seed.
30. **Prabhat:** This variety was developed at Kanpur. Plants are determinate, dwarf and compact. Fruiting is confined to the top of the plants in large clusters. Seeds are light brown in colour and similar to UPAS-120 in size. This variety is most suitable for mixed cropping with sorghum pearl-millet and other *kharif* crops. It is an extra early maturing variety of about 120-125 days. It can yield up to 15 q/ha. It has been recommended for western, central and Bundelkhand region of U.P.
31. **Pusa Ageti:** It was developed at IARI, New Delhi. Plants are erect and dwarf. Fruiting is confined to the top of branches in the clusters. Seeds and bold with a test weight of about 90 g per 1000 seed.
32. **S-20:** Developed in 1976 from the cross of Kanke-1 and B-7. It is a bold seeded variety with the test weight of 110 g/1000 seed. It matures in 180-190 days. Yield potential is 20-25 q/ha.
33. **Sharda:** It was also developed at IARI, Plants are bushy and semi-spreading types. Seeds are medium bold and brown in colour. It can yield up to 20-25 q/ha.
34. **Type 7:** It is a late maturing variety of 250-270 days. Plants are erect, tall with dark green foliage. It is suitable for mixed cropping with other crops in the plants of U.P. Standard petal is yellow with red streaks. Seeds are bold, brown with a test weight of about 120-130 g per 1000 seed. It has a potential to yield about 20-25 q/ha.

35. Type 17: Tall, semi-spreading, late maturing variety of about 270 days. Seeds are light brown in colour and medium in size. Test weight 1000-seed is about 80 g. It has been recommended for plains of U.P. It has a yield potential of 20-22 q/ha.

36. T 21: It is the first short duration variety developed in 1961 at Kanpur through hybridization from a cross of T-1 x T 190. It matures in 150-170 days. Plants are semi-spreading, medium tall, indeterminate. Seeds are light brown, medium with about 70-80 g weight of 1000 seed. Yield potential is 25-30 q/ha. It is suitable for double cropping with wheat particularly in the northern-western parts of the country.

37. UPAS-120: It was developed at Pantnagar through a selection from P 4758. It matures in 130-140 days and is most suitable for double cropping. Plants are medium tall, semi-spreading. Seeds are smaller than type 21. Its yield potential is about 20 q/ha.

One of the reasons often cited for the low yields of pigeonpea is the lack of high yielding improved genotypes available to the farmers. Farmers generally grow land races of medium and late-maturing types that are suitable for a single cropping sequence only. These genotypes have greater risk of being exposed to abiotic stresses such as frost and drought. Moreover, they do not fit into intensive cultivation systems, e.g., in rotation with wheat in northern India. In keeping with the changing requirements of pigeonpea cultivation, a number of short duration varieties have been developed in India (AICPIP, 1986). Table 18.3 gives a list of short-duration varieties that are currently recommended for different parts of India.

TABLE 18.3: Short-duration pigeonpea varieties recommended for different zones in India, and their time to maturity in those zones

Varieties	Maturity in different zones (days)			
	North-west	North-east	Central	Peninsular
Prabhat	120	-	-	-
Pusa Ageti	150	-	140	-
Pusa 74	160	-	150	-
Pusa 33	160	-	114	-
Pusa 84	140	-	-	-
UPAS 120	140	-	130	-
Pant A3	120	-	-	-
Sagar (H77-208)	145	-	-	-
Manak (H77-216)	140	-	-	-
ICPL 87	-	-	-	120
TT5	160	-	-	-
TT6	-	-	-	120
T21	160	140	140	135
ICPL 151	140	-	120	-

Source: AICPIP (1986).

The production of experimental hybrids using genetic male sterility and their evulation led to release of four pigeonpea hybrids for cultivation. ICPH 8 was the first pigonpea hybrid released during 1991 for central zone of the country. The release of PPH4 in Punjab (1994) and COH-1 (1994) and COH-2 (1997) in Tamil Nadu played the way for popularisation of the pigeonpea hybrids in northern and southern belts of the country. A brief comparative account of hybirds released so for is given in Table 18.4.

TABLE 18.4: **TABLE 18.4:** Hybrids released in Pigeonpea

Character	ICPH 8	PPH 4	COH-1	COH-2
Adaptability zone	Central state	Punjab	Tamil Nadu	Tamil Nadu
Year of release	1991	1994	1994	1997
Parentage	ms Prabhat	ms prabhat	ms T21 x	ms CO5 x
ICPL 161	AL 688	ICPL -	ICPL -	
87107	83027			
Plant type	Indeterminate	Indeterminate	Inteterminate	Indeterminate
Days to maturity	125 (C2)	137 (NWP 2)	117 (S2)	120-130 (S2)
Seed yield (q/ha)	17.8	19.3	12.1	10.5
Superiority	30% over	14% over	22.3% over	13% over
over check	TAT 10,	UPAS 120,	Vambam 1,	COH 1 and
	41% over	14% over	19.4% over	35% over
	UPAS 120	H82-1	ICPL 87	CO5

9. SOWING

9.1. Time of Sowing

Pigeonpea is a traditionally *kharif* crop sown in June-July with the onset of monsoon in various agro-climatic zones of India.

Short duration varieties of pigeonpea are now becoming popular in the irrigated areas. These varieties are harvested much earlier than the occurrence of frost and can also be fitted well in to multiple cropping systems with wheat (Yadav, 1975). Planting of early pigeonpea before the onset of Monsoon in the month of June recorded the maximum yield in multi-location trails conducted in U.P. (Panwar and Mishra, 1973). Rathi and Tripathi (1978) recorded the highest yield of pegeonpea when sown on June 10. They also recorded significant reduction in yield after June 20 and recommended June 10-20 as the optimum time for sowing of pigeonpea 'T 21' in central U.P. (Table 18.5). Similarly in Punjab, Kaul and Sekhon (1975) and Faroda and Singh (1980) in Haryana reported the superiority of June planting over late planting in July.

TABLE 18.5: Grain yield of pigeonpea under different dates of sowing at Kanpur

Sowing date	Grain yield (q/ha)
June 10	18.6
June 20	17.7
June 30	15.5
July 10	13.8

Saxena and Yadav (1975) reported that early variety T-21 and medium duration varieties like BR-183, Khargaon 2, R. G. 72, Sharda and Mukta varieties must be planted up to first fortnight of June. In case of extremely early maturing varieties (120-130 days) like UPAS-120, Planting with onset of monsoon was the optimum because the earlier plantings emulated the reproductive phase during the period of heavy rainfall which ultimately caused drop of flowers and pods. Delay in planting reduced not only the yield of pigeonpea but drastically reduced the yield of wheat in Punjab (Dhingera *et. al.*, 1983).

Singh (1988) recorded higher yield of different short duration cultivars of pigeonpea with the sowing done on June 25 as compared to July, 10 under Faizabad conditions (Table 18.6). He also observed that extra early maturing varietly Manak yielded higher under July, 10 sowing as compared to June, 25 sowing.

TABLE 18.6: Grain yield (q/ha) pigeonpea varieties as influenced by sowing date

Sowing date	Varieties		
	T-21	UPAS-120	Manak
June 25	22.97	19.48	14.42
July 10	20.57	17.87	15.52

9.2. Methods of Sowing

Pigeonpea is generally broadcasted. Experiments conducted at various places have revealed that line sowing is superior over broadcasting. Broadcasting results in uneven plant population which ultimately results in low yield. Optimum depth for seeding pigeonpea seed is 4-5 cm (Khan and Ashley, 1975). Tays (1983) found that emergence of pigeonpea declined from 84 to 42 per cent if seeding depth increased from 4 to 8 cm. Seeding depth is, therefore, most important to obtain optimum plant stand, seedling vigour and yield of pigeonpea. Seeding in rows may be done behind the *deshi* plough or in the furrows opened by tractor. In such areas where temporary water-logging take place, planting on ridges have been found superior. Ridge planting gave 6.7 q/ha higher yield over planting on flat beds at IARI. Similarly at Bandnapur, ridge-furrow system of planting gave 16.07 q/ha grain yield against 12.17 q/ha obtained from planting on flat beds (Ali, 1989).

9.3. Spacing, Seed Rate and Crop Geometry

Long duration varieties of pigeonpea are tall, spreading and occupy the field for about 250-270 days. These varieties are planted at wider row spacing of 90-120 cm and about 30 cm in between plants particularly under rainfed conditions. Under irrigated conditions, early maturing varieties are more popular as they fit well in double cropping systems with wheat and other crops. These varieties are planted at a row spacing of 50-75 cm and plant to plant spacing of 15-20 cm. Singh *et. al.* (1971) also found 75 cm as the optimum row spacing for short duration pigeonpea under Jabalpur (M.P.) conditions.

In case of April planted pigeonpea, a row spacing of 90-120 cm is recommended, as the vegetative growth is much higher than June planted pigeonpea.

Depending upon the size of seed and spacing, 15-25 kg seed/ha of pigeonpea is sufficient. Saxena *et. al.* (1971) did not find much variation in yield of pigeonpea due to the plant population ranging from 50 to 100 thousand plants per hectare.

Seeds produced in the immediate previous season should be preferred as no dormancy has been reported in pigeonpea. Based on the limited work on pigeonpea, bold seeds give better germination and seedling vigour than smaller seeds from the same lot (Khan and Ashley, 1975). Off coloured seeds should be discarded that could be due to mixture of geno-types, or due to infection by *Rhizoctonia solani* and *Alternaria* sp. (Karivaratharaju *et. al.*, 1982).

In pigeonpea, short duration genotypes are very much responsive to plant population as they have small plants that have a shortened growth duration. Plant population requirements for short-duration genotypes vary with location, time of sowing, and genotype, depending mainly on the extent of vegetative growth. The interactions between plant population and sowing date, and plant population and genotype are often significant (Ahlawat *et. al.*, 1985; Chauhan *et. al.*, 1987).

10. IRRIGATION

The intensively managed pigeonpea systems that involve short-duration pigeonpea have a higher water requirement because they are grown at high densities (Mehrotra *et. al.*, 1977). They estimated water use by T 21 to be in the range of 55-60 cm. Bhan and khan (1979) recorded significant responses to one or two supplemental irrigations on sandy loam soils at Kanpur, India. In this study a single irrigation applied at the pod-filling stage gave a better response than application on the basis of a cumulative pan evaporation demand of 80 or 120 mm. In Rajasthan, water applied at the early vegetative stage reduced yield by 14%, whereas when it was applied at the branching stage, it increased yield up to 34% (Makhan Lal and Gupta, 1984).

11. WEED CONTROL

Among the pulses, pigeonpea is the only crop that has a characteristically slow initial growth rate. This makes pigeonpea less competitive with weeds. Weeds grow profusely in the rainy season due to the favourable moisture conditions and, if not timely controlled, can cause up to 90% reduction in seed yield (Saxena and Yadav, 1975 a). Some of the common weeds associated with pigeonpea are *Cyperus rotundus*, *Echinochola* sp, *Digitaria* sp, *Dectyloctenium aegyptium, Setaria glauca, Amaranthus* sp, *Celosia argentea, Commelina benghalensis, Phyllanthus niruri, Digera arvensis* and *Euphorbia* sp. (Shetty, 1981). Their relative importance, however, varies with location.

Short-season crops can also take advantage of high plant density even when grown alone, It has been found that the ability of pigeonpea to suppress weed markedly improved at higher plant population.

Timely weed control is also important. For short-season crops, the first 30 days appear critical, although this period may vary with genotype and time of sowing.

One hand weeding given 45 days after sowing (DAS) resulted in lower yields than weeding twice at 25 and 45 DAS (Singh *et. al.* 1980). However, incessant rains may not permit hand weeding in time since the crop is normally sown in the rainy season. Pre-emergence herbicides can control weeds for the first 30-40 days. A number of herbicides have been found useful for pigeonpea. Singh *et. al.* (1980) found the alachlor at 1 kg al ha^{-1} gave better control, and controlled a wider range of weeds than nitrofen applied at the same rate. Metolachlor and pendimethalin also gave 85-95% weed control at 1.0 kg ai ha^{-1} when used alone or in comination. A hand weeding at 10-15 days after pre-emergence herbicide application kept fields generally weed free. Singh and Singh (1985) found oxadiazone and pendimethalin at 0.75 kg ai ha^{-1} controlled weeds satisfactorily, and resulted in Similar yields to those in the weed free treatment. At Faizabad, integrated method of weed control i.e. pre-plant incorporation of fluchloral in 1.0 kg/ha followed by one hand weeding at 30 DAS gave seed yield comparable to weed free check (Vaishya and Khan, 1989).

12. PLANT PROTECTION

Pigeonpea is attacked by a number of pests. Depending upon the degree of infestation, the loss to the crop may be even up to 75.80 per cent as in case of wilt and sterility mosaic diseases. The important diseases and insects of pigeonpea have been described here as under:

12.1. Diseases

12.1.1. Wilt

Wilt is caused by the fungus *Fusarium oxysporum* (Butter) snyder and *Hansen* sp. Udum. It is the most important disease of pigeonpea mainly prevalent in the parts of Maharashtra, Madhya Pradesh, Uttar Pradesh and Bihar. The leaves of the affected plants become yellowish in colour followed by dropping and finally whole plants dries up. Being a soil-borne disease, its chemical control is not possible. However, some precautionary measures such as collecting and burning the plant trashes left after harvesting, following a 3-4 years crop rotation and mixed cropping with sorghum and selection of disease resistant varieties of pigeonpea may be helpful in reducing the incidence of this disease.

12.1.2. Stem Rot

This disease is caused by fungus *Phytophthora dreschsleri* Tucker, var. *cajani*. The major infestation takes place in Delhi and central Uttar Pradesh. The affected plants show formation of brown to dark brown lessons on the stem near the soil surface. Selection of disease resistant varieties and cultivation of pigeonpea on well drained fields may be taken as precautionary measures.

12.1.3. Cankers

Cankers are caused by fungi like *Dilopdis cajani*, *Collectotrichum cajani* and *Macrophoma cajanicola*. The disease is not of much economic importance. The affected plants show formation of cankers on stem and twigs. Spraying of Dithan M-45 at the rate of 3 kg per hectare may be helpful in controlling the disease.

12.1.4. Leaf Spots

It may be caused by several fungi of which *Cerespora indica* is most common. There are several types of spots on the leaves and petioles of affected plants. These leaves start drying and in severe cases defoliation may take place. Spraying of Dihane M-45 of 0.3 per cent concentration may be helpful in controlling the disease.

12.1.5. Sterility Mosaic

This disease is caused by sterility mosaic virus. It is transmitted from one plant to another by eripohyid mites. The affected plants become pale green in colour and the size of leaves is reduced. Diseased plants fail to flower and no yield is obtained from such plants. Seth (1965) reported that removal of old and volunteer plants well in advance of the sowing is quite helpful in reducing the incidence of disease. Suitable control measure has not so far been worked out. However, repeated sprayings of systemic insecticide such as metasystox etc. is recommended to control the mites.

12.2. Insect-Pests

Pigeonpea is damaged by various insects from seedling stage till maturity. Most of the early maturing varieties particularly which are compact and determinate types are more severally damaged by pod boring insects. The important insects damaging the crop are as under:

12.2.1. Pod Borer (Heliothis armigera)

The young larva of this insect enter the pod and feed the developing grains. It is the most serious pest of pigeonpea and causes on an average 10-80 per cent damage every year. It may be controlled by spraying of 0.07 per cent endosulfan or 0.04 per cent monocrotophos.

12.2.2. Pod Fly (Metanagromyza obtusa mall)

It is a most serious pest of pigeonpea in north India. Damage varies from 20-60 per cent every year. Spraying the crop with 0.04 per cent monocrotophos or 0.03 per cent dimethoate emulsion have proved effective in controlling the insects.

12.2.3. Pod Bug (Clavigralla sp.)

The adults and nymph of this insects suck the sap from the leaves, buds, flowers and pods. Pods become shrival and deformed. It can be controlled by spraying of 0.04 per cent monocrotophos or 0.03 per cent dimethoate. It can also be controlled by dusting 4% endosulfan dust.

12.2.4. Leaf Binders and Webbers

Cydia critica is commonly found in pigeonpea. The thin caterpillars bind the leaves on the plants during early growth period. The webbers mainly *Maruca* sp. and *Adisura* sp. appear at the time of flowering and use to the inflorescence and damage the young pods. The caterpillars of these insects remain in the webbed mass and feed upon the flowers and developing pods. Damage is more sever in Delhi, Uttar Pradesh, Bihar and Madhya Pradesh (Saxena 1974). It has observed that these insects are automatically controlled by insecticides used against pod borers and pod fly etc.

13. HARVESTING AND THRESHING

Pigeonpea is an indeterminate growth type and thereby the growth is continued with reproductive phase. The best time for harvesting of the crop is when over 75 per cent of the pods turn brown. Harvesting may be done with the help of *gandasa* which is a domestic multipurpose heavy cutting blade. After harvesting, plants are left in the sun for drying. Threshing may be done by beating the pods with sticks. Threshing may also be done with the help of small threshers.

14. STORAGE

After threshing, grain should be dried in the sun so that moisture content is reduced to around 10 per cent. Dried grains should be store in dry clean stores.

15. YIELD

Generally, pigeonpea yields range from 15-20 quintals per hectare. Experimental evidences have shown that a yield of 25-30 quintals per hectare can be obtained with the adoption of improved varieties and production technology. In intercropping system, pigeonpea yield is either unaffected or reduced but overall production is higher.

15.1. Constraints of Production

Improved varieties of pigeonpea have high yield potential, but farmers are not able to realise the potential due to various constraints. The results of frontline demonstrations and on farm conducted at various places revealed that the yield potential of pigeonpea up to 25-30 q/ha. But the average productivity of this crop is very low in comparison to its potentiality. Panwar and Yadav (1980) recorded yield as high as 26.0 q/ha from April planted pigeonpea at kanpur. Singh (1988) recorded the yield of pigeonpea to the extent of 25.62 q/ha at Faizabad. Sharma and Rajput (1995) realized an yield of 26.2 q/ha pigeonpea at Bichpuri (Agra). It is, therefore, evident from the above results that pigeonpea has high yield potentiality and is also more profitable because of its comparatively less cost of cultivation and high prices.

The following have been realized as the major constraints in productivity of pigeonpea

1. **Unfavourable weather conditions:** Pigeonpea crop is subjected to a great loss due to adverse weather conditions. This crop is very much susceptible to water-logging conditions. In India, about 85 per cent of rains are received with in 2-3 months in a year. Continuous and heavy rains cause great damage in rainy season. Rains just after sowing adversely affect the germination. Continuous rains do not allow the fields to come in physical condition to perform interculture operations.
2. **Abnormal soil condition:** Abnormal soils such as saline-alkali soils, acidic soils, waterlogged soils which occupy a large area in the country are not at all suitable for the cultivation of this crop.

3. **Varietal constraints:** Lack of short duration varieties, flower drop, indeterminate growth habit, poor response to inputs, instability in performance and high susceptibility to diseases and insect pests are the major constraints in pigeonpea.
4. **Seed constraints:** Poor quality, inadequate availability, high price and untimely availability of seed are the major problems.
5. **Agronomic constraints:** Improper sowing time, inadequate plant population, defective method of sowing, inadequate inter-culture, inadequate use of phosphatic fertilisers and improper method of they application are the major agronomic constraints.
6. **Neglected use of Biofertilisers:** Non-availability of efficient strains, inadequate production, ignorance about the advantages of cultures, improper storage facilities and non-awareness of techniques of application are mainly responsible for minimal use of bio-fertilisers.
7. Lack of plant protection measures
8. Training constraints
9. Socio-economic constraints
10. Inadequate research infrastructure
11. Lack of machinery
12. Losses during storage
13. Losses during processing

16. REFERENCES

Agrawal S C. 1973. Progress report. All India Coordinated scheme on micro-nutrients on soil and plants. Lucknow University (U.P.).

Ahlawat IPS. Saraf CS and Singh A. 1975. Studies on the performance of pigeonpea varieties to dates of planting and row spacing. Indian J. Agron. 20 (3): 245-7.

Ahlawat IPS. 1976. Ph.D. Thesis, Agra University, Agra

Ahlawat IPS. 1980. Proc. Intl. Pigeonpea workshop. ICRISAT, Patancheru (A.P.), India, 15-19 Dec. 1980.

Ahlawat IPS. Singh A. and Mohta NK. 1982. Studies on weed control in solid and mixed stands of pigeonpea and soybean, Indian J. Agron. 27: 191-193.

Ahlawat IPS. Safar CS. and Singh A. 1985. Production potential of summer and *kharif* season pigeonpea intercropped with cowpea and greengram. Indian J. Agricultural sciences. 55: 565-9.

AICARP. 1985. Ann. Rep. AICARP (1984-85).

AICPIP. 1986. Improved varieties of pulse crops ion India. All India Coordinated pulse Improvement Project. DPR Kanpur, 68 pp.

Ali Masood. 1989, Consolidated Rep. of *kharif* pulses Agronomy DPR, Kanpur (UP.)

Baldev B. 1988. Cropping pattern. In: Baldev, B Ramanujam S and Jain HK. (EDs). Pulse crop (Grain legumes). Ox Ford and IBH Publishing, New Delhi pp 513-57.

Bhan S and Khan SA. 1979 Response of *Kharif* crops to irrigation in light textured alluvium of Uttar-Pradesh. Indian J. Agron. 24: 410-16.

Bhargawa JK. Saxena MC. and Tilak KVBR (1972) Proc. of legume inoculants Sci. and Tech Symp. Indian Nat. Acad., New Delhi.

Chandra S. and Ali M. 1986. Recent achievements in pulse production. Tech. Bulletin No. 1 DPR. Kanpur, India, pp. 17-8.

Chaudhary SL. 1968. Proc. II annual workshop on pulse crops. ICAR, New Delhi.

Chaudhary SL. and Bhatia PC. 1971. Ridge planting *kharif* pulses yield high despite waterlogging. Indian Farming 21 (3): 8-9.

Chauhan YS. Venkataratnam N. and Sheldrake AR. 1987. Factors affecting growth and yield of short duration pigeonpea and its potential for multi harvest. Journal of Agricultural Sciences. Cambridge. 109: 519-29.

Dahiya Dr. 1983. IPN 4, Mar. 1985: 23-24.

Dahiya Dr. Singh RC. Rao P. and Kairon MS. 1984. Programme and Abstracts. National Seminar on New Dimensions in Pulse Res. & Dev. May 21-24, 1984.

De DN. 1974. Pigeonpea. In: Hutehinson, J. (ed), Evolutionary studies in world crops. Diversity and changes in the Indian Sub-continent. Cambridge University press, London, pp. 79-87.

DPR. 1988. Technology of Increasing Pulses Production in India. DPR; Kanpur, Indian pp. 12.

Dhingera KK. Sekhon HS. and Sri S.S. 1983. Indian Fmg. 32 (11): 9-10.

Dowson RC. 1970. Pl. Soil. 32: 655-673.

Erdman LW. 1953. Farmers Bull. U.S. Deptt. Agric. Washington, DCUSA.

FAI. 1988. Fertiliser statistics, FAI, New Delhi.

FAI. 1996. Fertiliser statistics 1995-96. The Fertiliser Association of India, New Delhi.

Faroda As. and Singh RC. 1980. Indian Fmg. 32 (2): 15-16.

Jagadale VR. and Daftardar SY. 1985. J. Maharashtra Agric. Univ. 10 (1): 85-88.

Jeswani LM. 1979. Pick a pulse to fit in your cropping pattern. Intensive Ariculture. 27 (3): 15-7.

Kanvaratharaju TV. Ramakrishnan V. Vadivelu KK. and Ramasswamy Kr. 1982. Effect of seed size and seed coat colour on seed quality and productivity of redgram. Madrass Agriculture Journal. 69: 421-30.

Khan TN. and Ashley JM. 1975. Factors affecting plants stand in pigeonpea. Experimental Agriculture. 11: 315-22.

Kaul NJ. and Sekhon HS. 1975. Indian J. Agron 20: 276-277.

Krauss FG. 1932. Hawaii Agric. Exp. Stn. Bull. 64: 1-46.

Kulkarni KR. and Panwar KS. 1980. Response of pigeonpea to fertilisers in India. A critical review. In: proceeding of the International workshop on pigeonpeas. Volume 1, 15-19 December, ICRISAT Centre India. Patancheru, AP, India: ICRISAT pp 212-20.

Lal S. 1976. Improved varieties of arhar. Indian Farming 27 (5): 1-3.

Lal S. 1989. Pulses Production in India: Perspective and Possibilities for 2000 A.D. Paper presented in Naional Symposium on New Frontiers in Pulses Research and Development. Nov. 10-12-1989 DPR, Kanpur (U.P.)

Lenka D. and Satpathy RK. 1976. Indian J. Agron. 21 (3): 2172-20.

Makhan Lal and Gupta OP. 1984. Effect of irrigation schedules on pigeonpea in Rajasthan, India, International pigeonpea news letter. 3: 25-6.

Mehrotra On. Pal M. and Singh GS. 1977. A note on comparative yield and water use efficiency by *kharif* crops under rainfed conditions. Indian J. Agril. Research 1: 247-248.

Manjhi S. Chowdhury SL. and Kavitdhar AG. 1973. Indian J. Agric Sci. 43 (1): 998-1001.

Nane YL. Hall Susan D. and Sheila VK. 1990. The pigeonpea. CAB International ICRISAT.

Panwar KS. and Mishra As. 1973. Proc. 9th All India Pulses Research workers. Workshop held at IARI, New Delhi.

Pathak GN. and Singh KP. 1961. 'T-21' out does 'T-1'. Indian Fmg 11(8): 15.

Ram Sewa and Giri G. 1973. A note on response of redgram varieties to varying fertility levels. Indian J. Agron. 18 (1): 103.

Rathi SS. Singh L. and Malik RS. 1974. Fertiliser News 19 (2): 27-31.

Rathi SS. and Tripathi HN. 1978. Indian J. Agron. 21 (3): 305-306.

Reddy DS. Reddy DR and Acharya CVN. 1978. Studies on water intake and hydraulic conductivity in shallow and deep ploughing operating under rainfed Agriculture at Anantpur. Annals of Arid zone. 17: 291-7.

Reddy SJ. and Virmani SM. 1981. Pigeonpea and its climatic environment. In: proceeding of the International workshop, volume, 15-19 December 1980, ICRISAT centre India. Patancheru, AP, India: ICRISAT pp 259-270.

Reddy Shankara 1974. Cited by Hukkeri SB. and Pandey SL. 1977. Water Tech Centre IARI, New delhi.

Reddy PR. Subbarao IV. and Madhusudan VN. and Rao L. 1978. Sci. Culture 44: 36-37.

Roy Sharma RP. Thakur Hc. and Sharma HV. 1980. Proc. Intl. Workshop Pigeonpea, 15-19 Dec. 1980 ICRISAT, Hyderabad (A.P.)

Ryan JG. 1994. A Glabal perspective on pigeonpea and chickpea sustainable production systems: Present status and future potential, Abstracts. International Symposium on Pulses research. Indian Soc. Pulses Res. & Dev. Kanpur April 2-6, 1994 pp. 1-2.

Saxena HP. 1974. Entomologists Newsletter 4(3): 21.

Saxena MC. and Singh Y. 1970. Proceedings, VII annual workshop of Coordinated scheme on micro-nutrients held at Ludhiana (Punjab).

Saxena MC. and Yadav DS. 1975. Some agronomic considerations of pigeonpea and chickpeas. In: Internaional workshop on grain legumes, 13-16 January 1975, ICRISAT centre, India. Patancheru, AP. India ICRISAT pp. 31-61.

Saxena MC. and Yadav DS. 1979. Parallel cropping with short duration pigeonpea under the humid subtropical conditions at Pantnagar. Indian Journal of Agricultural Sciences. 49: 95-9.

Saxena MC. and Yadav DS. 1976. Pulses are Profitable as short duration summer crop. Indian Farmers Digest. IX (2): 39-41.

Saxena MC. Singh HP. Yadav DS and Sharma RP. 1971. Report of Agronomic expt. on *kharif* pulses at Pantnagar, Nainital (UP.)

Saxena MC. Singh HP. Yadav DS. and Sharma RP. 1974. Report of the Agronomic experiment on *Kharif* pulses carried out at Pantnagar during *rabi* 1974-75.

Sharma RK. and Rajput OP. 1997. Production potential of pigeonpea intercropped with groundnut followed by wheat in sequence. Indian J. Agron. 42 (1): 53-5.

Shetty SVR. 1981. Some aspects of weed management in pigeonpea. In: proceeding of the International workshop on pigeonpea, volume 15-19 December 1980, ICRISAT centre, India. Patancheru AP, India ICRISAT, pp. 137-146.

Shukla UC. 1976. Ann. Progress report of Coordinated Scheme on micro-nutrients in soils and plants for HAU, Hisar.

Singh A. 1988. Agronomic management of pigeonpea in cropping. System with wheat. Ph. D. Thesis. NDUAT, Kumarganj, Faizabad (UP).

Singh A. Prasad R. and Saraf CS. 1981. Effect of plant type, plant population density and application of phosphates fertilisers on growth and yield of pigeonpea. Journal of Agricultural Science. Cambridge. 97, 103-106.

Singh HP. Saxena MC. Yadav DS. and Sharma RP. 1980. Chemical and mechanical weed control in pigeonpea under humid sub-tropical conditions of Pantnagar. Legume Research 3: 22-6.

Singh K. 1973. Ph. D. Thesis. IARI, New Delhi.

Singh L. Maheshwari SK and Sharma D. 1971. Indian J. Agric. Sci. 41 (6): 535-538.

Singh ON. and Singh RS. 1985. Chemical weed control in pigeopea. International pigeonpea News letter 4: 26.

Srivastava DP. Lal S. and Asthana AN. 1997. Research and Development Efforts in hybrid pigeonpea. IIPR (ICAR), Kanpur (UP).

Tay TO. 1983. Growth, development and yield of pigeonpea in the lowland tropics. 4. affect of sowing depth Journal of Agricultural Sciences, Cambridge 101: 435-40.

Tilak KVBR. and Saxena MC. 1972. Proc. legume inoculants. Sci. and Tech. Symp. Indian Natl Sci. Acad. New Delhi.

Vaishya RD. and khan AH. 1989. weed management with herbicides in pigeonpea. International pigeonpea Newsletter 9: 14-16.

Vavilov, NI. 1951. The origin, variation, immunity and breeding of cultivated plants. Chronica Botanica 13(1/6) 1-366.

Vernon Royes, W. 1976. Pigeonpea. In: Simmonds, N.W. (ed), Evolution of crop plants. Longmans, London and New York, pp. 154-156.

Yadav DS. 1975. Ph. D. Thesis, GBPUAT, Pantnagar, Nainital U.P.

Yadav HL. 1984. Pigeonpea Newsletter 3: 31-32.

19. Greengram

S. K. Sharma

Greengram contains about 25 per cent protein of high digestibility and quality. It is used for various purposes either as a whole or in a variety of ways. Sprouted seeds of greengram synthesize ascorbic acid (Vitamin C) in them. Greengram is also a good source of riboflavin and thiamine. Greengram provides very good quality green fodder to animals. Seed coat, cotyledons and broken parts of embryo, after soaking in water, can be used as high quality cattle feed. Being a leguminous crop, greengram fixes 30-40 kg nitrogen per hectare. After picking of pods, greengram plants may be used as green manure. It is a spreading type crop and helps in preventing soil erosion.

Greengram is suitable for growing in areas where turn around period is about 60-70 days, for example, rice-wheat rotation area of north-west India. Field remain vacant for 60-70 days and 90-110 days after harvest of wheat in April and potato, mustard and sugarcane in the last week of March, respectively. This period may be beneficially utilized through summer greengram cultivation. Development of short duration and photoinsensitive varieties have further offered excellent opportunity of greengram cultivation both as a catch crop as well as for summer cultivation.

1. ORIGIN

According to Vavilov (1926), greengram is native of India and central Asia. The basis is that closely related species, *Vigna radiata* var. *Sublobata*, considered as mung-bean is found in wild state in India and central Asian region.

2. DISTRIBUTION

Mungbean is an important crop of India and south east Asia. It is widely grown in India, Pakistan, Bangladesh, Sri Lanka, Thailand, Combodia, Vietnam, Indonesia, Malaysia and China. It is also grown in parts of Africa, America and Australia.

India is the major producer of greengram in the world. India has about 3 m.ha. area and produces about 1 m.t. greengram. Major states which grow greengram in India are Maharashtra, Orissa, Andhra Pradesh, U.P., Rajasthan,

Madhya Pradesh, Bihar and Karnataka. Area, production and productivity of greengram have been variable in different years.

3. CLASSIFICATION

Greengram is also known as moong, mungbean and golden gram. Earlier, the scientific name of greengram was *Phaseolus aureus*. Recently, the genus of greengram has been changed from *Phaseolus* to *vigna*. *Vigna* is distinguished from *Phaseolus* by distinct projections from the large stipules below their point of attachment to the stem.

Purseglove (1991) classified cultivars of mungbean into two types. Distinguishing characters of two are given as under:

Characters	Golden gram	Greengram
Seed colour	Yellowish, shining	Bright green
Nature of seed production	Shy producer	Prolific producer
Tendency of shattering	Yes	Very less
Use	For hay, silage, pasture and cover crop	As pulse

4. GROWTH STAGES

Greengram belongs to family leguminoseae. Its somatic chromosomes (2n) number is 24. The crop is fully self-fertile & self-pollinated. Test weight of seed ranges between 30 to 40 g. Germination of seed is epigeal type. Seeds after germination develop into small herbaceous annual plant with height of 30 to 100 centimeter with slight tendency of twinning in the upper branches.

Life cycle of greengram varies from 60 to 110 days depending upon the type of varieties and growing season. Most of the cultivated varieties available, complete their life cycle in 60 to 90 days. Greengram is an indeterminate type grain legume and period of each growth stage vary according to variations in length of day and day and night temperatures. Further, except vegetative stage (including juvenility), other growth and developmental phases, for example, reproductive, maturity and senescence phases occur simultaneously in different parts of individual plants. Therefore, it is very difficult to fix period for each growth phase.

In greengram, growth rate is very slow up to four weeks after sowing. After this period, greengram plant rapidly covers the ground and dry matter accumulation increases fastly. Finally, pods form and grain development takes place. Thus growth phases may be classified into three parts which are:

- Initial slow growth phase (up to 4 weeks after sowing)
- Fast growth phase (4 weeks after sowing to flowering)
- Phases of dry matter accumulation (pod formation and maturity)

Pandey (1981) has proposed following growth stages for *kharif* greengram with growing period of July to September.

Growth stage	Description
Seedling (1-10 DAS)	– One pair of unifoliate leaves.
Early growth (10-20 DAS)	– One trifoliate leaf formed, second one opening; plant height 12-16 cm.
Rapid leaf formation (25-30 DAS)	– 3 leaves formed, fourth one unfolding; plant height 20-22 cm.
Flowering (35-45 DAS)	– Flowering begins (1-2 flowers/plant); 5 leaves developed, sixth developing; plant height 25-30 cm.
Early pod formation (45-50 DAS)	– 7-8 leaves formed; plant with 1-2 pods, plant height 35-50 cm; branches just appearing.
Mid-pod formation (50-65 DAS)	– Plant height 45-52 cm., 9-10 leaves formed; 1-2 leaves fallen, few more branches appeared, 4-5 pods/raceme, 5-6 racemes/plant.
Late pod formation (65-75 DAS)	– Plant height 60-70 cm.; rapid seed-filling; lower leaves yellow and senescence; older pod brown
Maturity (75-90 DAS)	– 90% pods brown; seed development almost complete; 40-50% leaves yellow or fallen.

5. CLIMATE

Greengram prefers tropical and subtropical climate. In India, it is grown in three seasons, namely; *kharif*, *rabi* and summer. However, areas where temperature is too high as well as too low are not suitable for its cultivation. Temperature is the most important deciding factor for greengram cultivation. The optimum mean temperature is 28 to 30°C. The critical temperature for germination is 15°C. Temperature range for growth and development varies from 20 to 40°C. Greengram is a short day plant and required 12 to 13 hours of photoperiod for flowering. Increase in photoperiod above this range delays the flowering. Warmth environment and bright sun shine are suitable for flowering.

Cultivation of greengram is suitable for regions recessive receiving rainfall between 600 and 1000 mm. Water logging and cloudy weathers are harmful for the crop

6. SOIL AND SOIL PREPARATION

Well drained loamy and sandy loam soils are best suited for greengram cultivation. The optimum soil pHis 6.5 to 7.5. It can tolerate slight soil salinity and threshold EC values are 6 dS/m. Soil moisture level should not go below 50% of available water. Otherwise, number of pods, seeds per pod and seed yield are adversely affected.

Seedbed is prepared by one or two ploughings followed by two or three harrowings. Field should be well leveled and free from weeds. Optimum bulk density is 1.2 to 1.3 kg/m^3. For summer greengram. Pre-sowing irrigation should be given to ensure proper field preparation. It will also ensure proper contact of seed with soil.

7. CROPPING SYSTEM

Greengram is mainly grown as rainfed crop during *kharif*. In Maharashtra, Andhra Pradesh and Tamil Nadu, it is also cultivated as *rabi* crop after paddy. However, recently, summer greengram cultivation is becoming very popular in northern

India to utilise period available in between the harvest of *rabi* crop and sowing of *kharif* crop.

Greengram varieties maturing into 60-70 days have been evolved and which can be fitted between *rabi* and *kharif* cereal crops. Some promising cropping-systems identified for higher land use efficiency and monetary returns are being mentioned below:

Region	Crop rotation			Period
	Kharif	Rabi	Summer	
North and North-western region	Rice	Wheat	Greengram	1 year
Punjab, Haryana, Western U.P., Rajasthan,	Maize	Wheat	Greengram	1 year
Himachal Pradesh, J & K	Maize	Potato	Greengram	1 year
	Pearl millet	Wheat	greengram	1 year
Eastern region	Rice	Potato	Greengram	1 year
Eastern U.P., Bihar, West Bengal, Orissa, Assam	Maize	Wheat	Greengram	1 year
Central zone	Sorghum	Wheat	Greengram	1 year
Madhya Pradesh, Gujarat Maharashtra	Sorghum	Potato	Greengram	1 year
	Pearl millet	Wheat	Greengram	1 year
Southern region	Rice	Rice	Greengram	1 year
Andhra Pradesh, Tamil Nadu, Karnataka, Kerala)				
Central plain zone	Maize/Rice	Mustard	Greengram	1 year
U.P.				

7.1. Mixed/Intercropping

Greengram is a promising inter or mixed crop for *kharif* as well as long duration crops. Short duration greengram suits best as a companion crop of sugarcane (1:1). It vacates the field within 55-60 days and inter-culture operations can be carried out conveniently. In Guntur region of Andhra Pradesh, greengram is an important inter-crop in cotton plantations. For dryland regions, Chetty and Rao (1979) has recommended following intercropping systems:

Region	Intercropping system	Row ratio	LER
Jodhpur	Pearl millet+greengram	1:2	1.24
Akola	Sorghum+greengram	2:1	1.12
Jammu	Maize+greengram	1:2	1.13

Summer greengram may be intercropped with sugarcane, cotton and pigeon pea. Pigeon pea intercropped with greengram in 1:1 ratio resulted in higher yields. In Rajasthan, Pearl-millet + greengram is a promising rotation. Several experiments recommended that planting pattern of 30/70 cm. or 30/90 cm paired row system should be adopted for taking inter-crops.

8. MANURE AND FERTILISERS

One ton of greengram removes about 40 kg N, 4-5 kg P, 10 to 12 kg K, 1 to 1.5 kg Ca, 1.5 to 2 kg each of S and Mg per ha. So for maintaining higher yield level, these nutrients must be replenished through organic and inorganic sources. A dose of 20-25 kg N applied at the time of sowing is sufficient to meet nitrogen requirement of the crop. For the growth and development of root nodules, phosphorus is essential and 40 kg P_2O_5/ha has been found optimum. Phosphorus is applied through diammonium phosphate (DAP) in furrows below the seed before sowing. Spraying DAP (2%) at flowering and pod filling stage has been found advantageous. Responses of mungbean to potassium vary from location to location and are negligible at most of places. However, where soil are low and medium in K_2O kg k has been recommended. It should be applied at the time of sowing. If available, 5-10 tonnes of FYM or compost should also be applied 15 days before sowing. Response of phosphorus varies from soil to soil. The magnitude of response is generally in the order of red soil>alluvial soil>black clay soil.

In greengram, 20 to 40 kg S/ha enhanced the nodulation by increasing number and dry weight of nodules, grain yield and uptake of NPK and S; Gypsum, elemental sulphur and SSP are effective S sources. In paddy-follows or soils deficient in Zn, applicaion of 15-25 kg $ZnSO_4$ per ha improve the yield. Foliar application of $ZnSO_4$ (0.5%) with 0.25% lime has commonly alleviated the Zn deficiency. VAM inoculation increased phosphorus, Zn and Mo uptake in greengram and improved grain yield. To produce one tonne of dry matter greengram requires about 170 g Fe, 38 g Mn, 32 g B, 13 g Zn, 11 g Cu and 1.05 g Mo.

9. VARIETIES

In 1948, Type 1 was released as the first mungbean variety from a local selection from Muzaffarpur (Bihar). Since then different selections from well adopted local germplasm were utilized for release of varieties. A list of greengram varieties commercialized in different years in India is mentioned in below:

Period of release	Varieties
-	BR2, ML1, Mung 305, ADT 1, Type 10
1948-70	Type 1, Co 1, Kopargaon, Jalgaon, Khargone, Krishna 1.
1971-78	Type 44, Jawahar 45, Pusa Baishaki, PS-16, Sunaina, Varsha, T5 1, ML series
1979-87	Mohini (S8), Dhauli (TT 9E), Kondaveedu (LGG 127), Sujata, Jyoti Pant Mung 3, PDM 54, Pusa 105, PS 10, PS 7
1988-96	Basanti (PDM 84-143), ADT 3, RMG 62, MUM 2, Samrat (PDM 84-139)

Year 1980's brought a revolution in the summer greengram acreage and production. In this period Pusa baisakhi, PS 16, PS 7, PS 10 and recently Pusa 105 were released and commercialized. Most of the commercial varietes of greengram have been developed through complex crosses. A brief mention of commonly utilized greengram varieties in hybridization in India is given below:

Parent strain	Derivative varieties
Type 1	K 851, Pusa Baisakhi, Sunaina Type 44, Pant Mung 1, Pant Mung 3, Mohini
Type 44	Pusa Baisakhi, Sabarmati
ML 1	M 131. ML 267, ML 337
Madira and Local	Jawahar 45, ML 1, ML 5, ML 131, ML 267, ML 337
Lam M 294-1	Pant Mung 1, Pant Mung 3

9.1. Mutant Varieties

Important Commercialized mutant varieties of greengram are given in following table:	
Mutant variety	**Parent**
Dhauli (TT 9E)* (Released in 1980)	Type 51 x Local selection
MUM 2	K 851
CO 4	CO 1
Pant Mung 2	ML 26

* First mutant variety of greengram (Orissa)

9.2. Varieties Suitable for Summer Cultivation

Two varieties, viz.; Pusa baisakhi and PS 16, evolved at the IARI, New Delhi are suitabe for growing during summer and spring throughout the country. Some other important varieties in this category are PS 7, PS 10, PDM 11, PDM 54, Panna, Sunaina and SML 32. They are early maturing and can tolerate high temperature of summer. Recently released summer greengram varieties are Pusa 9072 and TARM 1. Availability of such a large number of varieties resulted in increasing spring -summer greengram, area to more than one million ha. in India.

Varieties resistant to viral disease:

1. **Yellow Mosaic Virus (YMV):** In most of YMV resistant varieties, the source of disease resistance is wild relative of greengram, viz.; *Vigna radiata* var. *sublobata*. The varieties are MUM 2, Samrat, Pant Mung 2, Pant Mung 3, Pusa 9072, ML 1, ML 5 etc.
2. **Mosaic Mottle:** Type 2, Jawahar 45
3. **Cucumber mosaic virus:** PI 27, 1401- a germplasm donor.

Varieties suitable to different zones are as follows:

Important varieties of Greengram	
Zone	**Varieties**
North-Western zone (Punjab, Haryana, Rajasthan, Western U.P., Himachal Pradesh, J & K)	
Early	Pusa Baisakhi, Type 44, PS 16, Pusa 105, K 851, Pant Moong 3
Medium to mid late	Jawahar 45, RS 4, Varsha, ML 1, ML 5, ML 9
North – Eastern zone (Eastern U.P., Bihar, West Bengal, Orissa, Assam)	
Early	Pusa Baisakhi, Type 44, PS 16, K 851, Pant Moong 1, Pant Moong 2, Pant Moong 3.
Medium to mid late	Sujata, Panna, Dhauli, Sabarmati, Mohini
Central zone (Madhya Pradesh, Gujarat, Maharashtra)	
Early	Khargone 1, Gujarat 1, PS 16, Pusa Baisakhi
Medium to mid late	Jawahar 45, kopergaon, Mohini, Krishna 11, Gujarat 2.
Peninsular zone (Andhra pradesh, karnataka, Tamil Nadu and Kerala)	
Early	Pusa Baisakhi, PS 16, Pusa I, Pusa 10 Kondaveedu.
Medium to mid late	Jawahar, PDM 1, PDM 2, Kopergaon, Mohini, ADT 2

Chief characteristics of important commercialized varieties are as follows:

1. **Type 44:** It was developed by a cross between T 1 x T 49 in 1962, semi spreading and compact plants, maturing in 60-70 days, hence can be accommodated in high densities, protein 26%, test weight 36 g, average yield 8-10 q/ha., suitable as a catch crop in all seasons.
2. **Pusa baisakhi:** It is a selection from T 44, matures in 60 to 70 days, specially suited for cultivation in summer, seeds bright green, average yield 10-15 q/ha.
3. **Jawahar 45 (Hybrid 45):** A selection from a cross Madira] U.P. local at JNAU, Jabalphur (M.P.) Also called as hybrid 45, matures in 70-85 days, suited to *kharif* seasons and for intercropping with sugarcane in peninsular India, average yield 10-12 kg/ha., test weight 30 g.
4. **K 851:** It was developed at C.S.A. U.A & T, Kanpur by a cross 4453 -3 Type - 1, matures in 60-65 days, suitable for spring and summer sowing, semi spreading and synchronous maturity, average yield 10-12 q/ha., dark foliage with purple pigmentation.
5. **PS 16:** Developed at IARI, New Delhi by a selection from an Iranian germplasm in 1978, erect and profusely branched plants, suitable for both *kharif* & summer season all over the country, matures in 60-65 days, average yield 10-12 q/ha
6. **Pant Moong 1:** Yellow mosaic resistant variety released in 1982 for growing in U.P. in *kharif* and zaid seasons, matures in 60 days (summer), and 75 days (*kharif*), test weight 35 g, average yield 10-12 q/ha.
7. **Mohini (S-8):** It is a yellow mosaic virus and Cercorpora leaf spot tolerant variety developed at IARI, New Delhi from a cross BR-2 x Type 2, best suited for NW and NE plains, known for erect and well branching habit, matures in 70-75 days, suitable for *kharif* cultivation, average yield 10-12 q/ha.

10. SOWING

10.1. Time of Sowing

Season	Time of sowing	Region
Kharif	Second fortnight of June to first fortnight of July	North-west India
Rabi	Oct.-Nov.	Central & southern India, Madhya Pradesh, Andhra Pradesh, Tamil Nadu and Orissa
Summer	Middle of march to middle of April	Uttar Pradesh, Punjab, Haryana, Delhi, Bihar and West Bengal

10.2. Method of Sowing

Greengram is sown either by broadcast or behind the plough method. In *kharif* season, ridge sowing is also beneficial in high rainfall area.

10.3. Spacing and Seed Rate

Season	R x R spacing	P x P spacing	seed rate
Kharif	45 cm.	5-10 cm.	15 kg/ha
Rabi	30 cm.	5-10 cm.	20 kg/ha
Summer	30 cm.	5-10 cm.	25-30 kg/ha

10.4. Seed Treatment

Greengram seeds are inoculated with *Rhizobium* spp. (Cowpea type) @ 4-5 g per kg seed regardless of seed size. For virgin soils at least 105 to 106 viable rhizobia per seed are required.

11. IRRIGATION

Water requirement of greengram is 15-30 cm. Number of irrigations for greengram vary according to seasons and duration crop. *Kharif* season crop require irrigation only when monsoon rain fails or long dry spell is observed at flowering or grain development stage. *Rabi* and summer greengram should be grown in areas where irrigation facilities are available. Minimum daily water use is 3 mm/day but can be as high as 4.5 mm/day depending upon the season. The critical stages of water supply are flowering and pod development stage. One or two light irrigation (5-7 cm. water per irrigation) are recommended at critical stages to boost the yield. However, for summer greengram 3-5 irrigations are optimum and water should be applied at 50% soil water depletion from its maximum root zone depth. The recommended IW/CPE ratio is 0.6 to 0.8. At least three irrigations, the first at pre-flowering stage (20-25 days), second at flowering (25-40 days) and third at grain filling stage are necessary. Pre-sowing irrigation for summer greengram is a pre-requisite for good seed germination and seedling establishment.

Water stress affects all aspects of greengram growth and development. In terms of seed yield, this crop is most sensitive during late flowering and early pod filling stages. Greengram is an intermediate crop between water saver & water spender categories. However, under high temperature conditions, greengram shows mechanism of paraheliotropy i.e., the plane leaf surface is oriented along the direction of sun's rays, thus minimizing radiation interception. This crop under severe stress condition almost acts like an ephemeral plant. Therefore, it is an important pulse crop for arid and semi-arid regions. Water logging should be avoided at flowering and pod filling stages. It may cause reduction in yield to the extent of 50 to 60%.

12. WEED CONTROL

Greengram crop should be free of weeds upto initial 35 days. Crop is infested by *kharif*, *rabi* and summer weeds. *Trianthenra monogyna*, *Portulaca oleracea*, *Boerheavia diffusa*, *Amaranthus viridis* (all broad leaf weeds) and *Cyperus rotundus*, *Cyanodon dactylon*, *Dactyloctenium aegyptium* and *Digitaria songuinallis* (all narrow leaf weeds) are important *kharif* weeds of greengram. *Cyperus rotundus* is a major problem in summer greengram. In *rabi* season,

Chenopodium album, Melilotus indica, Asphodelus tenuifolius, Anagalli arvensis, Circium arvense and *Convolvulus arvensis* are major weeds. Problem of weeds is more severe in *kharif* as compared to *rabi* and summer seasons.

For weed control in greengram, integrated weed management approach should be followed. For good yields, two hand weedings, first at 25 days after sowing and the second after 45 days after sowing are done. Pre-plant application of fluchloralin or pendimethalin at 1.0 kg/ha followed by one hand weeding give effective control of both annual grasses and broad leaf weeds. Barnyard grass in greengram can be effectively controlled by pre-emergence application of Flauzifopbutyl @ 0.75 kg/ha. Intercropping of greengram with pigeonpea also suppresses weed growth. Maintenance of higher plant population *kharif* greengram reduces the weed growth.

13. PLANT PROTECTION

Important diseases and insect pests of Greengram

Name of disease	Causal agent	Important symptoms	Control	
			Non-Chemical	Chemical
Yellow mosaic virus	Virus transmitted by whitefly	On young leaves yellow diffused round spots & later affected leaves completely turn yellow	Grow resistant varieties e.g. Pant Mung 1, 2 & 3, ML 1, ML 131, T1, T 44	Whitefly can be controlled by spraying dimethoate (0.03%) malathion (0.06%) or endosulfan (0.07%)
Cercospora leaf spot	*Cercospora* fungus	Small violet red colour rounded spot with grey coloured centre on leaves & pods	Use healthy seeds	Spray diathane M-45 @ 2 kg in 100 litre water
Anthracnose	Colletotrichum capsici	Dark brown circular spots with concentric ridges on leaves	Use of healthy seeds	Spray of diathance Z-78 @ 2 kg in 1000 litre water.

Other important diseases of greengram are rust, seed and seedling rot and charcoal rot.

13.1. Insects Pests

Symptoms and control measures of important insect pests of Greengram

Name of insect	Damaging form	Stage of attack	Symptoms	Control
Pod-borer	Cater pillar	Young leaves & fruiting stage	Larval feed on leaves & defoliate the plants. Also bores the green pods & feeds pods & on developing grains	Spraying with endosulfan (0.03%) or carbaryl (0.05%) when attack of pest occurs. Deep ploughing of field & expose the pupae to hot sun. Use of *Bacillus thurin giensis* (Bt) strain aizawi, when pod borers appears on the crop.

Name of insect	Damaging form	Stage of attack	Symptoms	Control
Hairy caterpiller (Bihar hairy cater pillar & red hairy cater pillar)	Larval (cater pillar)	Young leaves (seedling & young plant)	Larval feed on green matter of leaves	Collect & destroy the eggs & young larvae
Jassids	Nymph and adult	Severe damage in young stage of plant	Insects suck sap from leaves. Leaves turn brown, curl from edges and dry	Spray the crop with 0.4% monocrotophos 40 EC
Whiteflies	Adult	Adult colonise on leaves & suck the sap	Yellowish discoloration of leaves	Spray the crop with diamethate (0.03%), malathion (0.06%) or endosulfan (0.06%)
Beetles	Adult	Feed on plant parts especially pollen and petals of flowers	Scraping of leaf portion & small holes on leaves	Soil application of phorate 10 G @ 1 kg/ha or monocrotophos 0.75 kg/ha monocrotophos or metasystox spray at 30 & 50 DAS
Aphids	Adult	Tender parts of plants including inflorescence & pods	They suck sap from plant leaves. Leaves crinkle, inflorescence wither & deshaped the pods Honey dew deposition on leaves	Spray the crop with monocrotophos 40 EC

14. HARVESTING

Greengram should be harvested at its physiological maturity stage. This stage is recognized by uniform shedding of leaves and brown colour of pods with hard seed inside them. Practically, crop is harvested in subsequent stages. First picking is done when pods are moist during morning hours to avoid pod shattering. Subsequent pickings are done at weekly intervals. Varieties with synchronous maturity require only one or two pickings and may even be harvested with sickles. Green crop after picking may be fed to cattle or may be incorporated into the soil. Pods after complete drying under sun are threshed manually. Finally seeds are winnowed to separate the chaff.

15. YIELD

Under improved package and practices *kharif* greengram may yield 12 to 15 q grains /ha, while summer greengram may yield 15-20 q grains/ha

16. POST-HARVEST TECHNOLOGY

There is considerable losses during post-harvest operations of pulses viz.; threshing (0.5%), transport (0.5%), processing (1.0%) and storage (7.5%) with a total loss to the extent of 9.5%. Therefore, due attention should be given to the knowledge of post-harvest technology. PHT includes drying, storage and milling operations. Moisture content of grain at the time of harvesting of greengram crop is about 18 to 20 per cent. For safe storage, the seed moisture per cent should not exceed 12 to 13 per cent. Seeds can be safely stored if the storage is scientifically designed, equipped and managed to provide required dry, cool and protective conditions. Milling of greengram means removal of outer husk and splitting the grain into two halves. Milling may be done either by traditional method or by alternate drying and wetting method. After proper de-husking and splitting, polishing of split pulses is done by treating them with a small quantity of oil and or water. Yield of split pulses in traditional mills is only 65 to 70% in comparison to 82 to 85% potential yield. Almost 80% of greengram is consumed in the form of *dal*. The seed coat accounts for about 15% of seed weight and the remaining 85% is cotyledons. The average yield of *dal* from greengram is 72% of raw material.

17. QUALITY

Composition of Greengram grain

Particulars	Content
Total carbohydrates (%)	62.6
Crude protein (%)	23.86
Fat (%)	1.15
Crude fibre (%)	5.27
Ash (%)	3.32
Lysine (mg/g)	436
Methionine (mg/g)	75
Cystine (mg/g)	55
Food energy (calories)	347
Digestibility (%)	79
Biological coefficient (%)	72

Greengram contains about 2.4% protein. Its protein is rich in lysine and poor in S-containing amino-acids like methionine and cystine. It is rich in Ca, Fe, K and is a good source of vitamins such as thiamin, niacin and vitamin A. Qualitatively, mungbean has higher diagestibility coefficient and biological value. Toxic constituents like tannins (polyphenols) and trypsin inhibitors are very low in greengram. Haemagglutonins are not present in greengram. It is generally lower in flavus production as they contain relatively less oligosaccharides.

18. CONSTRAINTS OF PRODUCTION

Greengram has lowest mean productivity of around 3.8 q/ha on the national basis. The important causes of low greengram, production in India are ecological factors, lack of better plant type and appropriate production and post-harvest technology, basic research and socio-economic constraints along with non-availability of quality seeds in required amounts. Specific constraints causing low production of greengram may be discussed in following points.

1. Varieties of greengram under cultivation are late maturing and low yielding.

2. Yellow Mosaic Virus is a deadly disease of this crop and causes 30 to 70% losses in the grain yield. There is a need of development of suitable disease resistant varieties and control measures.
3. Stem-fly and pod bores cause serious damage to the greengram at early and late stages of the crop respectively. Available insecticides to control these pests are of limited use for farmers.
4. In rainy season, water logging in field for a temporary period causes heavy mortality of plants. Farmers so far has not started giving due attention on this problem.
5. Non-synchronous maturity of traditional varieties of greengram require large number of pickings. This amounts to heavy loss in shattering and heavy labour cost at harvesting.
6. About 90% of greengram in India is under rainfed conditions. So crop faces severe moisture stress at grain filling stage. It results in considerable yield reduction.

19. SUMMER GREENGRAM

Summer greengram is grown mainly for efficient utilization of land and other resources to achieve high levels of production. Improvement in irrigation facility and crop management technology for efficient land utilization have boosted the farmers to grow summer greengram. Availability of short-duration and photo insensitive varieties has enhanced crop intensity. Crop intensification in the predominant rice-wheat system in northern India. Greengram will give dual benefit of pulse grains and utilizing its fresh straw as green manure for *kharif* crop. Research study showed that in addition to 11.0 q/ha of pulse grains, the incorporation of greengram straw with 50% NPK through chemical fertilisers saved 60 kg N, 30 kg P_2O_5 and 15 kg K_2O/ha. Incorporation of greengram straw also mobilized the availability of N and micronutrients. Thus inclusion of summer greengram in and exhaustive cropping system can aid to the sustainability of the system. Its nitrogen contribution may range from 25 to 30 kg N/ha. Green-manuring during month of May-June add 50 to 60 t/ha of fresh biomass to the soil. It helps in maintaining physical, chemical and biological properties of soil. In summer season, there is very less or no infestation of insect-pest and diseases on greengram because of high temperature and desiccating winds.

20. REFERENCE

Chetty, Cok. R. and rao, U.M.B. (1979). Proc. Intl. Workshop on intercropping. ICRISAT: 277-281

Gautam, R.C. (1995). Intercropping in Pearl millet: a remunerative practice of rainfed farming Indian Farming. 45 (4): 21-22 July (1995)

Nutman, P.S. (1976). In: Symbiotic nitrogen fixation by plants (Ed. P. S. Nutman), Cambridge Univ. Press. 211-237.

Pandey, R.K. (1981). Pulse crops Newsletter (IARI Regional Institute, Kanpur). 1 (1): 72: 73.

Poehlman, J.M. (1991). The Mungbean. Oxford & IBH Pub. Pvt. Ltd., New Delhi: 99.

Purseglove, J.W. (1991). Tropical crops: Dicotyledons. Longman Group UK Ltd., England: 290-304.

20. Black Gram

P. S. Rathore

Black gram, commonly known as *urd* or mash, is one of the important *kharif* pulse crop grown in India. The whole or splitted grain is mainly used as *dal*. The *dal* of *urd* is rich in protein (24%) and phosphoric acid (385 mg/100 g). The grain after soaking in water is used for making *mugodi* or *bari*. Many South Indian dishes as *dosa* and *idli* have *urd* as main ingredient. The *urd* flour is also used in making *papad* and *dahi-bara*. The sprouted seeds are also etable after frying. The plant parts and pod husk is a good nutritive fodder to milch animals. The crop also builds soil fertility. It fixes about 70-90 kg N/ha.

1. ORIGIN

Black gram appears to have originated in India from wild progenitor *Phaseolus sublobatus* Roxb. The description of *urd* is also available in ancient literature specially in *Arthosasthra* and *Charak Samhita*. Samples of grains have been found at Chalcolithic site and Navdatoli – Maheshwar in India around 1660-1440 B.C.

2. DISTRIBUTION

Black gram is grown in India, Bangladesh, Burma, Sri Lanka, Indonesia. In India it is grown in Madhya Pradesh, Uttar Pradesh, Punjab Maharashtra, West Bengal and Andhra Pradesh. In India it is grown in 2.70 mha area with production of 0.94 mt. About 1/3 of India's black gram is produced in Madhya Pradesh, Maharashtra and West Bengal.

3. CLASSIFICATION

Black gram, *Vigna mungo* belongs to the family leguminoseae and sub family Papilionaceae. It posses 2n = 22 and 24 chromosomes. Bose (1932) classified it into two sub species.

***Vigna mungo* var. *virdis*:** Plant types of this sub species mature late and seeds are small with greenish look.

***Vigna mungo* var *niger*:** Plant types of this group mature early with bigger and bold seeds of black colour.

4. SOILS AND SOIL PREPARATION

Black gram performs well on well drained black cotton soils, but it can also be grown on sandy loam soils. The soils should be free of salts and pH 4.7-7.5. After harvest of *rabi* crop field is ploughed with mould board plough. One and two harrowings are required to bring soil under good tilth. All the weeds and crop residues must be thrown outside the field to check termite problem. If problem of termites, mix 20-25 kg of endosulfan 4% dust just before last harrowings.

5. CLIMATE

It is a hot weather crop, grown in *kharif* season in north and central parts of India. In eastern states it can be grown in winter where as in southern parts of India it can be grown in both *rabi* and *kharif*. Crop is grown from sea level to 6000 ft elevation. Crop requires well distributed rain fall. Heavy rain fall at flowering is detrimental. Crop is also sensitive to photoperiod. It flourishes well under short day conditions.

6. CROPPING SYSTEM

Black gram is grown either mixed or inter-cropped other *kharif* crops specially maize, sorghum, cotton, pearlmillet and Arhar.

Black gram is commonly inter-cropped with Arhar. Black gram (T9) may be sown in between two rows of arhar (50 cm). Both the crops are sown on same day with 50% seed rate of Black gram. This system gave 3-4 q additional yield of *urd* without any adverse effect on Arhar. Similarly it can also be inter-cropped with maize. After every 5th row of *urd*, at 30 cm apart maize may be sown.

Black gram is generally grown in rotation with following crops.

Black gram - mustard	North
Black gram - wheat	North and north west.
Black gram - Maize - Green gram	North east and east.
Black gram – wheat	Central parts.
Black gram – safflower	Southern parts.

7. MANURE AND FERTILISERS

As it is a leguminous plant, its nitrogen requirement is low. On very poor soils 10-20 kg N/ha as as starter dose is required to boost nodulation. Soils on which it is cultivated have strains of bacteria responsible for nitrogen fixation.

TABLE 20.1: Effect of phosphorus on yield of black gram.

Levels of phosphorus	Grain yield (q/ha)	
	1992	1993
0	1.57	2.24
20	3.60	3.56
40	4.97	5.72
60	4.94	4.90
CD (P = 0.05)	0.49	0.57

Rhizobium inoculation is essential if it is grown in new field. Phosphorus and potassium requirement should be based on soil test values. In absence of test, a general dose of 40 kg P_2O_5/ha and 30-40 kg K_2O/ha is recommended for the crop. Ramamoorthy (1997) noted appreciable increase in yield with application of 40 kg P_2O_5/ha at Pudukkottai (Table 20.1). The full dose of fertiliser may be drilled 2-3 cm below the seed by country plough.

8. VARIETIES

Selection of variety for region is important for getting higher yield. Improved varieties of black gram are given in table with their area of adaptability.

TABLE 20.2: Characteristics important black ground cultivars

Variety	Maturity (Days)	Yield (q/ ha)	Characteristics
T9	75-80	8-10	Adaptable in whole country, photoinsensitive, suitable for *kharif*, spring and summer. Test weight 40 g. medium sized, dark green.
Pusa 1	80-85	12-15	Resistant to yellow mosaic, suitable for Haryana, Rajasthan, Uttar Pradesh, Bihar, Gujarat, Madhya Pradesh, Andhra Pradesh, and Maharashtra., good for all seasons, black coloured.
Krishna	90-100	8-10	Recommended for Rajasthan, Punjab, and Haryana, bold and black seeds.
Khargone-3	85-90	12-15	Bold, black seeds, suitable for Madhya Pradesh.
Pant U-30	68-75	12-15	Resistant to yellow mosaic, recommended for central and peninsular India, medium and brown seeds.
Pant U-19	80-85	10-15	Medium sized, brown coloured seeds, resistant to yellow mosaic.
Mash-48	85-90	10-15	Bright, black and bold seeds, drought tolerant.
HPU-6	100-120	-	Suitable for Kangra and Kulla valleys of Himachal Pradesh and western Uttar Pradesh.
Pant-430	80-90	10-12	Resistant to yellow mosaic, sutable for all seasons.
Type-56	-	12-15	Good for *kharif*, medium bold seeds.
Naveen	90-95	10-12	Medium, light yellow-green seeds, sutable north Bihar, west Bengal and Orissa.
Mash-1-1	85-90	8-10	Medium bold, black seeds, suitable for Punjab.
Ujjain-4	80	10-15	Black seeds, suitable for Madhya Pradesh.
ADT-2	70-75	8-9	Bold, black seeds suitable for Tamil Nadu.
ADT-1	85-90	12-15	Suitable for peninsular India.
PDU-1	-	-	Recommended for North western and eastern plain zone, central zone, good for spring and tole rant to yellow mosaic virus.
DPU 88-31	-	-	Suitable for North eastern plain zone, good for *kharif* season, to tolerant to yellow mosaic virus.
UG218-PS1	-	-	Suitable for Punjab, Haryana and Rajasthan.
Kalindi	-	-	Suitable for Bihar and West Bengal.
Sarla and BR 68	-	-	Suitable for Assam, eastern Uttar Pradesh and Bihar.
TAU-1, D-75, D 6-7	-	-	Suitable for Madhya Pradesh, Gujarat and Maharashtra.

Variety	Maturity (Days)	Yield (q/ ha)	Characteristics
KM-2, CO-3 and PDM-2	-	-	Suitable for Andhra Pradesh, Tamil Nadu, Kerala and Karnataka

9. SOWING

Spring crop of black gram should be sown in second fort night of February summer crop can be sown in mid March to beginning of April. The rainy season crop is sown in last week of June-first week of July with the onset of monsoon. If good quality irrigation water is available crop is sown with pre sowing irrigation. Rainfed crop should be sown with the onset of monsoon, delayed sowing or delayed onset of monsoon beyond 15 July heavily reduces yield.

9.1. Spacing

Summer crop of black gram is sown at 20-25 cm row space while for rainy season crop, row to row space of 30 cm is adequate. Plant to plant space of 7-10 cm is adjusted at the time of thinning.

9.2. Seed Rate

Seed rate of 12-15 kg/ha is required for rainy season crop. For summer crop 15-20 kg seeds/ha are needed. Bold seeds should be preferred for sowing.

9.3. Method of Sowing

The crop is sown by *pora* or *kera* method. Seeds are best sown by seed drill.

9.4. Seed Treatment

The seeds should be treated with thiram 3 g/kg seed for protecting crop from seed borne diseases. Seeds should be also treated with *Rhizobium leguminosarum* for enhanced N fixation, particularly if crop is taken first time in the field.

10. IRRIGATION

Summer crop of black gram need irrigation at an interval of 10-12 days depending on soil and climatic conditions. In all 4-6 irrigations are needed for summer crop. Generally *kharif* crop does not need any irrigation. One or two irrigations may be required if dry spell exists for more than a fortnight. Crop can withstand available soil moisture depletion upto 75%. In case of excess rainfall, adequate drainage should be provided. Irrigation should be stopped at least two weeks before harvesting for uniform maturity.

The total water requirement of crop is 150-250 mm depending on climatic conditions. The crop may be irrigated at IW/CPE ratio of 0.6-0.9 with best results at 0.6. Flowering is the most critical stage of irrigation.

11. WEED CONTROL

Kharif crop of black gram is severely infested with weeds at initial stages of crop growth. Losses due to weeds are of the magnitude of 40-50%. First 30-40 days from sowing are considered critical for weed crop competition. The main weed flora of black gram consists of doob, (*Cynodon dactylun*), motha (*Cyperus rotundus*), satha (*Trienthema partulatcastrum*), hazardana (*Phyllanthus niruri*) and cholai (*Amaranthus viridis*).

One hand weeding at 30-35 days after sowing takes care most of weeds as later growth of weeds is suppressed by smoothering.

Weeds may also be controlled by herbicides. The herbicides and their mode of application is given below:

Herbicidal weed control in Blackgram

Herbicide	Dose kg/ha	Mode of application
Fluchloralin	0.5-1.5	Pre-plant ncorporation
Isoproturon	0.5	Pre-plant
Pendimethalin	0.5-1.0	Pre-plant
Trifluralin	1.0	Pre-emergence
Nitrofen	1.5	Pre-emergence

12. HARVESTING AND THRESHING

Kharif crop is ready for harvest after 110-115 days and summer crop after 70-75 days. When 90-95% pods become blackish in rainy season crop and 75-80% in spring and summer season. The crop should be harvested. Harvesting is done by cutting the plants with sickle. The pods should be dried in sun for 4-5 days and there after it is threshed by beating with sticks or by bullocks or by thresher. The grain are separated by winnowing. The grains are stored when moisture contents in grain are 8-10%.

13. YIELD

With improved cultivation practices, one may harvest 10-12 q/ha grain for *kharif* crop and 8-10 q/ha for summer crop. The effect of adoption of cultivation practices is given in Table 20.3 with full adoption 188% increase in yield was recorded over farmer's practice in Srivastava and Srivastava (1995).

TABLE 20.3: Contribution of production factors in black gram

Production factor	Grain yield q/ha	Increase in yield over farmer practice (%)	Increase is yield over full package (%)
Farmer's practice	4.8	-	65
Fertiliser only	10.2	113	26
Weed control only	8.1	69	41
Plant protection only	6.9	44	50
Fertiliser + Weed control	12.3	156	11
Fertiliser + Plant protection	10.3	115	25
Weed control + Plant population	8.6	79	38
Fertiliser + Weed control + Plant protection	13.8	188	-
CD (P = 0.05)	0.61	-	-

14. PLANT PROTECTION

Plant protection is important step to realize good crop. The insect pest and diseases of black gram are as follows:

14.1. Insect-Pests

Red hairy caterpillar: (*Amsaita nsorei*) Adult of this insect feeds on leaves and tender twigs. Insect may be collected in kerosine water. Use of light traps also attract insect. Young caterpillar may be killed by dusting 4% endosulfan dust @25 kg/ha. or by spraying 0.07% endosulfan.

White fly (*Bemisia tabaci*): It is soft tiny insect, causing severe damage by sucking cell sap. It also acts as a vector for virus causing yellow mosaic. The insect is controlled by applying 10% granules of thimet @ 10 kg and 0.1% malathion after 4-6 week of applying thimet.

14.2. Diseases

Yellow mosaic: It is a viral disease, young leaves show mild scattered yellow specks at initial stages. As the severity of disease increases yellowness on leaves increases, even the apical leaves turn completely yellow. The diseased plant look stunted with lesser number of pods. The disease is transmitted by white fly. To have a less incidence use resistant varieties like Mash 1-1, UPU-1, UG 117 and Pant U 26. To prevent white fly, spray 0.1% metasystox + 0.1% malathion at 10-15 days interval as and when even small yellowness is observed.

Cercospora leaf spot (*Cercospora canescens*): Leaves of infected plant develops brown lesions. At later stages pods also develop brown spots. On appearance of first symptom spray dithane Z-78 at 2 kg/ha. or bavistin at 0.5 kg/ha in 1000 L of water. Repeat the spray 8-10 days after first spray.

Seedling rot: It is a fungal problem caused by number of fungi i.e. *Fusarium* sp., *Pythium* sp., *Rhizoctonia* sp., and *Macrophomina* sp. The germinating seedling start dying resulting into poor stand. The disease is causing severe damage. To manage, do not grow black gram on same piece of land year after year, sow the seeds treated with seed-dressing fungicide captan 2 g/kg of seed.

15. POST HARVEST TECHNOLOGY

In India, pulses are mainly consumed as *dal*. *Dal* milling is an important industry. Breakage of grain is the major drawback of wet and dry milling. The composition of grain is given below:

Composition of Blackgram grain

Constituents	Value
Moisture %	10.9
Protein %	24
Fat %	1.4
Carbohydrate %	59.6
Minerals %	3.2
Fibre	0.9
Calorific Value 1 kg.	3470

16. CONSTRAINTS OF PRODUCTION

Poor soil fertility, inadequate plant population, unavailability of quality seed of high yielding varieties are the main factors contributing to its low productivity. Lack of adequate and timely plant protection measures and failure to conserve optimum moisture further intensifies the problem. To have a sustained production organic farming with use of improved crop production technology and timely and adequate plant protection should be given emphasis.

17. REFERENCES

Asthana, A.N. (1998). Pulse crop research in India. Indian J. Agric. Sci. 68 (8): 448.

Ramamoorthy, K, Balasubramaniam and A. Kianj (1997). Response of rainfed black gram. (*Phaselous mungo*) to phosphorus and sulphur nutrition in red/latertic soils. Indian J. Agric 42 (17): 191-193.

Saini Gangasaran, (1987) *Urd ka adhikatum utpadan lane ki vidhi*.- Khad Patrika 28 (7): 73.

Yadav, R.L., Singh P; Prasad R. and Ahlawat, I.P.S.(1998). Fifty years of Agronomic Research in India. ISA, New Delhi.

21. Kidney Bean

P. S. Rathore and R. S. Rathore

Kidney bean belongs to genus *Vigna* and sub family papilionaceae and family leguminaceae. It is popularly known as moth bean or dew bean. The plant is normally bushy to semi erect with large number of leaves and short stem. The plant is very hardy and can tolerate high temperature and drought.

It is multi-purpose crop. The grains are rich in protein and mostly used as a pulse in arid and semi arid regions of Rajasthan, and to lesser extent in the adjoining parts of Gujarat, Haryana and Punjab. *Namkin* and *papad* are also made from the grains. The seeds contain about 9.3 per cent moisture, 23-26 per cent protein, 59 per cent carbohydrate, 4 per cent each fibre and ash and 0.7 per cent fat. India has exported 5225 mt papad of RS. 1667 lakh in 1992-93. The whole grain is also eaten after soaking and frying. The hay is fed to animals. Green pods are used in preparing vegetable. Growing of moth also enriches soil by fixing atmospheric nitrogen and protect soil against wind erosion.

1. ORIGIN

Plateau of central India is considered as the primary centre of origin of moth bean. Vavilov (1926) and Jain and Mehta (1980) described that both wild and cultivated species are found in India.

2. DISTRIBUTION

Moth bean is cultivated in India, Sri Lanka, Malaysia, Myanmar and South China. India is the major producer of moth. It occupies 15-18 lakh hectares with production of 2.45-4.0 lakh tonnes. Crop is distributed from north western Himalaya to Karnataka in south and from hills of N.E. Himalayas to Saurastra in west. Rajasthan is the largest state in the country growing moth.

Approximately 85% area and 56 % production of India's moth is derived from Rajasthan. Besides Rajasthan, other important moth producing states are Maharashtra, Gujarat, Jammu and Kashmir, Punjab and Haryana. State wise area and production of moth is given in Table 21.1.

TABLE 21.1: Area and production of moth bean in major states

States	Area (ha)	Production (t)
Rajasthan	1085.95	72.25
Maharashtra	129.12	35.20
Gujarat	66.07	9.60
Jammu & Kashmir	8.77	9.40
Uttar Pradesh	1.75	0.35

3. CLASSIFICATION

Two species are found

1. *Vigna acontifolia* (Jacq). Marechal - cultivated
2. *Vigna trilobata* (L.) Verdc. It is wild and endemic to India.

4. CLIMATE

Moth bean is deep rooted drought tolerant plant. It is grown from sea level to 2480 m altitude in arid and semi-arid parts of India specially Rajasthan. It can be grown in areas receiving less than 500 mm rainfall. The average temperature requirement for growth and development of crop is 25-30° C. High humidity areas are not at all suitable for raising crop. Bright sun shine and normal humidity are considered good for growth and high yield of crop. Crop does not grow well in high rainfall areas.

5. SOILS AND SOIL PREPARATION

Crop can be grown successfully in sandy loam soils of low to moderate fertility. It can be grown on very sandy to sand dunes of poor fertility, low in organic matter and poor in water retention. The crop could be grown in arid western Rajasthan, where other legumes are grown with difficulty.

One or two ploughing with plough or harrow followed by planking are sufficient to loose the soil. The field should be prepared with onset of monsoon without much delay to avoid loss of soil moisture. On sand dunes crop can be sown without any tillage operations directly.

6. CROPPING SYSTEM

In extremely low rainfall areas, moth is grown alone as pure crop while in areas receiving adequate rains it may be grown as an inter-crop with pearlmillet, cotton and green gram. Yadav (1992) recommended intercropping of pearlmillet + moth in ratio of 4:8. Whereas Mishra and Das (1963) recommended intercropping of grass + moth in 6:3 ratio for western parts of Rajasthan.

7. MANURE AND FERTILISERS

Soils of moth growing areas are generally poor in organic matter and essential nutrients. To harvest high yield, well decomposed 8-10 t. FYM should be applied 15-20 days prior to sowing. Addition of organic matter not only supplies essential nutrients, but also improves physical condition and water retention in soils. Crop also responds to fertilisers. Being a leguminous crop its nitrogen requirement is low and is full filled through atmospheric nitrogen fixation. About 60-90 kg N/ha may be fixed with 21-30 nodules/plant and 70-110 mg dry wt./plant. However, for initial requirements 20 kg N/ha is recommended for the crop. Application of 40 kg P_2O_5/ha further enhanced the yield. Bhati et. al. (1980) also reported a dose of 40 kg P_2O_5/ha for crop. (Table 21.2). Sulphur is getting importance, crop respond well to application of 20 kg S/ha through gypsum.

TABLE 21.2: Influence of phosphorus on grain yield of moth bean

Levels of phosphorus (kg P_2O_5/ha)	Seed yield (q/ha)
0	4.18
20	5.04
40	5.73
60	5.33
CD (P = 0.05)	0.49

8. VARIETIES

Improved variety occupies an important place in harvesting good crop. A number of varieties of mothbean with high yield potential and resistance to yellow mosaic has been identified. The important varieties are given below:

Variety	Adaptability	Days to maturity	Average grain yield (q/ha)	Other
Jadia	Rajasthan & M.P.	80-90	5-8	-
Jwala	North western Rajasthan	80-90	6	Resistant to yellow mosaic virus
IPCMO 880	Rajasthan	90-100	4-5	Moderately resistant to yellow mosaic virus.
RMO 40	Rajasthan	62-65	6-7	Synchronous maturity & drought resist. Resistant to yellow mosaic virus.
IPCMO 912	Semi arid regions of Rajasthan	75-80	5	Tolerance to yellow mosaic & bacterial blight.
RMO 257	Rajasthan	62-65	7-9	Drought escaping & resistant to yellow mosaic
FMM 96	Rajasthan	58-60	5-7	Escape mid season drought, resistant to yellow mosaic
Baleshwar 12	Gujarat	110-115	4-5	Susceptible to yellow mosaic
Moth Gujarat-1	Gujarat	-	4	Highly susceptible to yellow mosaic virus.
Type 3	-	-	4-5	Fodder variety 20-25 q/ha dry matter.

9. SOWING

Crop is mostly raised under rainfed conditions. Therefore, it is generally sown with onset of monsoon. Result of experiment conducted at Agricultural Research Station (ARS), Mandore (Jodhpur) indicate that crop sown in first rain produced

3.13 q more over crop sown on second rain and about 4.25 q more over crop sown in third rain. Gill and Singh (1980) reported that moth can be sown in spring season for seed as well as fodder purpose in March or April.

9.1. Seed Rate

Pure crop of moth may be seeded with 8-10 kg/ha (Bhati, 1986). For fodder, the optimum seed rate is 20 kg/ha. In intercropping system seed rate of 2-5 kg/ha is sufficient. However, seed rate must be higher under delayed onset of monsoon.

9.2. Spacing

Crop should be sown in lines 45 cm apart, but early maturing, erect type RMO 40 may be sown at 30 cm apart. Plant to plant distance should be adjusted at 10 cm during thinning.

9.3. Seed Treatment

For control of several seed borne diseases, seeds should be treated with seed-dressing fungicide. Moth seeds be treated with 2 g Bavistin or 3 g captan/kg seeds. Result of experiment conducted at ARS, Mandore (Jodhpur) revealed that seed treatment not only increased the yield but also reduced plant mortality.

10. IRRIGATION

Crop is mostly grown as rainfed. Therefore, the yield depends upon onset and withdrawl of monsoon as well as spell of dry days in between rains. Flowering is considered as the most critical stage. Moisture conservation practices increase the yield. Shekhawat and Singh (1992) reported increase in yield by applying 10 kg *jalsakhti*/ha. Spring crop of moth is taken as irrigated crop. Crop requires irrigation at 10-15 days interval. Irrigation at 75% available soil moisture gave more yield over 25 and 50% available soil moisture (Anon. 1986).

11. WEED CONTROL

First 20-30 days are very critical from point of view of crop weed competition. During this period weed infestation causes considerable decline in crop yield because of competition for soil moisture and space specially. After 30 days, crop have smoothering action on weeds because of dense foliage. Weeds which grow luxuriantly in the crop are - *Cynodon dactylon*, *Cyperus rotundus*, *Trienthema partulactrum*, *Phyllanthus niruri*, *Amaranthus viridis* and *Tribulus terrestris*. Losses as high as 30-40 % have been reported by various workers due to weeds. One hoeing and weeding at 20-30 days after sowing takes care most of the weeds. Weeds can also be controlled by using herbicides. Pre plant incorporation of fluchloralin at 1.0 kg/ha checks the weed infestation in pure crop of moth as well as in intercropping system with pearlmillet. The influence of weeds and herbicidal control in moth is given in Table 21.3.

TABLE 21.3: Herbicidal weed control in moth

Treatments	Seed yield (q/ha)	
	Mandore	Fatehpur
Weedy check	4.39	1.44
One hand weeding 30 DAS	6.77	2.29
Fluchloralin Pre plant (1 kg/ha)	6.44	1.44
CD (P = 0.05)	1.18	0.57

12. PLANT PROTECTION

Important insect pests and diseases of moth bean are mentioned below:

12.1. Insect Pests

Galerucid beetle and jassids: This feed on leaves and suck their cell sap. For control, spray disulfoton at 1.0 kg/ha in soil at sowing and spray of lindane 0.01% is beneficial.

White fly: This insect is vector of virus causing yellow mosaic. Spraying endosulfan 0.07% keeps this insect away.

12.2. Diseases

Yellow mosaic virus: Disease is causing severe damage in moth bean. In initial stages, small yellow patches develop on leaves, these extend throughout the leaf and whole plant may look yellow. Infected plant bear less number of pods and small immature seeds. Two sprays of dimethoate (0.2%), one when yellow spots are observed in field and second 15 days later (Rathore and Agnihotri, 1983; 1985).

Dry root rot: Rotting of seed inside soil or seedling death may occur due to presence of *Rhizoctonia bataticola*, or *Macrophomina phaseolina*. These are both seed borne and soil borne. Seed treatment with bavistin 2 g/kg is only solution.

Root knot: Nematode *Meloidogyne incognita* forms knots on roots of moth there by affect metabolic processes including water absorption and nitrogen fixation. Verma et al (1981) suggested aldicarb at 1 kg/ha or carbofuron at 2 kg/ha to control nematodes of moth.

13. HARVESTING

Yellowing of leaves and drying up of pods indicate the maturity stage of moth. Crop should be harvested with no delay to avoid dropping of pods and shattering losses. Harvesting is done by uprooting of plant. Up rooted plants are kept on threshing floor for 4-5 days for drying. After sun drying the grains are separated from pods by beating the pods with sticks and grains are cleaned by winnowing. Threshing may also be done by multi-crop thresher.

14. POST HARVEST TECHNOLOGY

Threshed grains are dried in sun till 8-10% moisture and stored.

15. YIELD

Well managed crop of moth bean may yield from 4-8 q with fodder yield of 12-15 q/ha.

16. CONSTRAINTS OF PRODUCTION

Similar to other grain legumes it is also grown on marginal and sub marginal lands under scarce conditions. Lack of improved and desirable genotypes. Harvest index is also low. This is due to high shedding of flowers. Improved varieties have 2-2.5 times more harvest index. None availability of quality seed and poor attempts for plant protection, weed control and poor plant stand are the major constraints in high production of moth bean.

17. REFERENCES

Anon. (1986). Annual report, State moth improvement ARS, Mandore, RAU.

Bhati D.S., J.R. Mathur & R.C. Sharma 1987. Effect of row spacing and seed rates in the yield of dew bean. Annals of arid zone. 26(2):115-116.

Bhati, D.S. (1986). Rajasthan dalhan - moth. Unnat krishi, Dir. ext. p.p. 20-22.

Gill, P.S. and K.Singh (1980). Performance of cowpea and moth for forage yield under different irrigation levels & dates of planting. Haryana Agric. Uni. 5 Res. 10:532-35.

Jain, H.K. and Mehta, K.L. (1980). Evolution, adaptation and use of the species of vigna cultivated in India. Advances in legume science - summer field, R.J. and Bunting, A.H. Eds. Royal Botanical Garden, Kew, U.K. 459-468.

Mishra, D.K. Das (1963). Rate of legumes in crop husbandry of western Rajasthan. J. soil water conserv. India. 9:124-34.

Pundir, P & Dahama, A.K. (1997) Moth bean in Rajasthan. Publication of directorate of research, RAU.

Rathore G.S. and J.P. Agnihotri (1985). Effect of insecticides on yellow mosaic of moth bean (VI) ha

Vavilov, N.I. (1926). Studies on the origin of cultivated plants. Chro.Bot. 13:1-16.

Yadav, N.D. (1992). Studies in strip cropping of pearlmillet + arid legumes in rainfed arid western Rajasthan. Curr. Agric. 16:89-92.

22. Cowpea

S. K. Sharma

Cowpea or lobia is a most versatile *kharif* pulse because of its smothering nature, drought tolerating characters, soil restoring properties and multi-purpose uses. Cowpea is called as vegetable meat due to high amount of protein in grain with better biological value on dry weight basis; cowpea grains contain 60 per cent carbohydrates, 22-28 per cent protein, 1.8 per cent fat, 3-4 per cent fibre, 3-3.8 per cent ash and 9-11 per cent moisture. It is a rich source of calcium and iron. The nutritive value of cowpea varies according to climate, soil fertility status and genotype.

The crop forms excellent forage. The feeding value of cowpea forage is high and quite comparable to the Lucerne. It forms a palatable fodder when mixed with cereal fodder.

Cowpea gives a heavy vegetative growth and covers the ground so well that it checks the soil erosion in problem areas. Apart from this, being a leguminous crop, cowpea can fix about 70-240 kg/ha. of nitrogen per year. On an average, a single season crop under ideal condition can fix about 80-90 kg/ha. nitrogen.

Cowpea is often grown as a green manure for soil improvement. Westcott and Mikkelsen (1987) reported that cowpea a typically contains 54 kg N/ha. in the tops and 4.9 kg N/ha. in the roots.

Short duration, fast growth and low water requirement characteristics of cowpea are well-suited for growing it in dry land areas.

1. ORIGIN

According to Vavilov (1939), the cultivated cowpea originated in the Indian sub continent. However, evidences support, central Africa to be the centre of origin of cowpea as most of the wild species have been found here. It is considered that cowpea spread to rest of world from Ethiopea and west Africa. In India, cowpea is raised as a mixed crop mostly grown in Gujarat, U.P., Rajasthan, West Bengal, Tamil Nadu, Andhra Pradesh, Kerala and Orissa.

2. CLASSIFICATION

Cowpea belongs to family leguminosae and genus *Vigna*. Purseglove, J. W.(1968)

has classified all cultivated types of cowpea into following three groups.

- *Vigna unguiculata* subspecies *unquiculata*
- *Vigna unguiculata* subspecies *cylindrica* (catjang)
- *Vigna unguiculata* subspecies *sequipedalis* (yardlong, asparagus bean)

Subspecies *unguiculata* has early maturing, almost upright and day neutral plants. It originated or evolved somewhere in transition zone between rainforest of west Africa and Savanna. It is chiefly grown for forage purpose. Other two subspecies *cylindrica* and *sequipedalis* evolved through selection from subspecies *unguiculata* after its introduction in India. *Cylindrica* has determinate and upright plants and is chiefly grown for forage. *Sequipedalis* has spreading habit and is mainly grown for long (30-100 cm) pods and is used as vegetables. Pod and seed size are the chief distinguishing features of three subspecies which are as follows:

TABLE 22.1: Distinguishing features of three cowpea subspecies

Particulars	*Vigna sinensis* (Cowpea)	*Vigna sinensis* sub. sp. *cylindrica* dr. *V, unguiculata* (Indian cowpea, catjang)	*Vigna sinensis* sub. sp. sesquipedalis *(Yardlong green)*
Plant characters	Annual erect and trailing type, 10-30 cm. long leaves	Annual erect or ascending when green	Annual, trailing type 2-4 m. height
Pod length	20-30 cm.	7.5 – 12.5 cm.	Larger than 30 cm.
Seed length	6-9 mm	5-6 mm	8-12 mm
Pods on drying	Neigher flabby nor inflated	Deflaxed	Inflated flobby and shrinking

3. CLIMATE

Cowpea can be grown in all tropical and subtropical climates. The best temperature for germination of cowpea is 12-15° C. It thrives well between 21 and 35° C. The optimum temperature for primary and secondary root nodulation is 24° C and 33° C, respectively. At higher temperatures hemoglobin content gets reduced. Cowpea is a short day plant and requires a minimum critical day-length of 12.5 hours. For higher production of lobia, 27° C day temperature and 22° C night temperature is required. 60 cm. rainfall. Areas receiving 60 cm rainfall are suitable for short duration varieties and 60-100 cm, for medium and long duration varieties. It can tolerate drought and, therefore, is an important crop of dry areas. It is very susceptible to water-logging and is not suitable for heavy rainfall areas. The time of occurrence of first is often cited as the climate criterion limiting the distribution of this crop. Uniform rainfall distribution is very important for growth and development of cowpea.

4. SOIL AND SOIL PREPARATION

Well drained loam to sandy loam soils are best suited for cowpea. Soil should be neutral in pH. Saline and sodic soils are unfit for cultivation of cowpea. Cowpea prefers deep ploughing for better primary roots development which grows upto 100-150 cm, straight way into the well-drained soil. In normal soils primary roots penetrate upto 50-60 cm, depth. Field should be prepared by giving two or three cross harrowings followed by planking. For summer crop give a pre-irrigation immediately after harvesting of *rabi* crop. When the field comes in condition, prepare it by giving two or three harrowings.

5. CROPPING SYSTEM

Cowpea can be grown in rotation with many crops. In general, cowpea is grown as a mixed crop with Arhar, Maize, Sorghum and Bajra. In spring and summer season, cowpea is grown as a pure crop. As a forage, it is grown mixed with maize for higher production. The most important rotations are as follows.

For grain or vegetable purpose	
Maize - Wheat - Cowpea (summer)	1 year
Cowpea – Wheat	1 year
Cowpea – Oat	1 year
Cowpea – Sugarcane	1 year
Cowpea – Wheat	1 year
Maize - Potato – Cowpea	1 year
Sorghum + Cowpea – Oat	1 year
Cowpea - Berseem – Maize + Cowpea	1 year

For fodder purpose	
Jowar + Cowpea - Berseem - Maize + Cowpea	1 year
Maize - Oat - Maize + Cowpea	1 year
Maize - Berseem - Maize + Cowpea	1 year
Jowar - Berseem - Maize + Cowpea	1 year
Rice – Cowpea	1 year
Rice – Mustard	1 year

6. MANURE AND FERTILISERS

Similar to other pulse crops, it needs a small quantity of nitrogen for early growth period on those soils which are poor in organic matter. The recommended NPK dose for cowpea is 20-50-30 kg/ha, respectively. All fertilisers should be applied as basal by drilling 5-7 cm below the seed. Micro-nutrients like zinc, molybdenum, boron and iron should be applied accordingly to soil test values. Through research work, it has been found that highest yields were obtained. With 50 kg P_2O_5/ha. applied to the soil at sowing time and 25 kg P_2O_5/ha foliar application at the first flush of flowering. Phosphorous should be applied through DAP. The application of Ca aids nodulation as an important nutrient of legumes and also frees some of Mo which may be unavailable as a result of the low pH of the soil. In cowpea plant zinc concentration below 20 ppm produces deficiency symptoms and concentration above 400 ppm causes toxicity. In cowpea application of Mo gives response. 60-80 kg/ha molybdenised super-phosphate every five years on the same soil is suggested in place of acid/basic phosphate application.

7. VARIETIES

According to use and suitability of agro-climate conditions, different varieties have been recommended. However, same variety may be grown for more than one purpose, grain, fodder, and vegetable. The varieties recommended for use as pulse grain are, C-152, FS-68, Cowpea 74, Co-3 etc. for use as fodder are type-2, Pusa Rituraj, Pusa Barsati etc. and for use as vegetable are Pusa phalguni. Pusa dofasli, Pusa barsati and FS 68. Most of grain varieties showed shy growth habit, late maturing and susceptibility to viral and bacterial diseases. Now most of the breeding programme throughout country is developing short duration, drought resistant, high yielding and disease resistant varieties.

The recommended varieties for different zones are as follow:

Zone	States	Name of varieties
North-western zone	Punjab, Haryana, Western U.P. Rajasthan, Himachal Pradesh, J & K	Pusa 152, Pusa sawani (T5269) FS 68, RS 9, V '16' (Amba)
North-eastern zone	Eastern U.P., Bihar, West Bengal, Orissa. Assam, Maharashtra	Pusa 152, Pusa sawani, V '16' (Amba), RC 19, FGS-1, Gomati
Central zone	Madhya Pradesh, Gujarat, Maharashtra.	Pusa 152, V '16', 'V 240' K11, K 14, 'Gujarat cowpea 1', 'Gujarat cowpea 2'
Peninsular zone	Andhra Pradesh, Karnataka. Tamil Nadu	Pusa 152, V '16' (Amba), 'CO 1', 'CO 2', 'CO 3', 'CO 4', Krishnamani

The important characters of promising varieties of cowpea are given below:

1. **C-152:** Developed by selection from Iranian variety P-42 at IARI; matures in 105-110 days, suitable for *kharif* season planting in Punjab, Harayana, Delhi, U.P., Bihar, M.P., and Maharashtra, yields about 15-20 quintals grain/ha, medium bold brown seed. Good for fodder, for pure and mixed cropping.
2. **Pusa phalguni:** A selection from 'Doloque on Tonkin' from Canada. It is a busy dwarf variety (50-60 cm plant height), best suited for sowing in Feb. - March in northern plains, matures in 60-65 days and yields 10-12 q/ha grains and 50-60 q/ha green pods.
3. **Pusa Dofasli:** A cross between Pusa phalguni Philippines selection. It is an improvement over Pusa phalguni for Pod quality. Suitable for growing in *kharif* and summer season in northen plains. The Pod is about 18 cm long and gives the first picking about a week later than Pusa phalguni. It is photoinsensitive variety and matures in 60-65 days and produces 30-40 q/ha Pods/ha and 10-12 q grain/ha.
4. **Pusa Rituraj:** It is a recent selection, developed by NBPGR, New Delhi. A highly photoinsensitive variety grown both in summer and *kharif* season in most parts of north India. The plant is bushy in nature and matures in 60-65 days. It yields pods so profusely, as to cover foilage appearing like a fan. Average green pod yield is about 80-85 q/ha.
5. **FS - 68:** It is sown both during rainy and spring season. It takes about a week later than pusa-dofasli in maturity, suitable for both grain and vegetable, pods small and green; seeds small and white, yield 10-12 q/ha.
6. **CO.3:** Developed by selection from C-152. It is a semi-erect variety, matures in 80-85 days and yields 10-15 q grain/ha.
7. **Cowpea-74:** Suitable for growing in Punjab and Haryana both for grain and fodder purpose. It is resistant to viral diseases. It yields 300 q/ha green fodder and 12-15 q grain/ha.

8. **Gomti (5286-3):** Developed by selection in 1974 at Kanke (Ranchi). It matures in 80-85 days and can be grown throughout the year. Best suited for growing in Chhotanagpur region. It yields 15-18 q/ha grains and 30-40 q/ha pods.
9. **RC - 19:** A cross between Virinia New Era. Best suited for growing in NE plain zone and Rajasthan. It matures in 60-65 days and yields 10-12 q grains/ ha.
10. **Type - 2:** Developed by selection in 1955 from a material collected from Jhansi. Matures in 125-130 days, spreading type, good for grain, fodder and green manure. Average yield is 15-20 q/ha.
11. **Pusa barsati:** It is selection seed material imported from Philippines. Earliness is the special characteristic of this variety. It is suitable for growing in monsoon season in Punjab, Haryana, U.P., Delhi and M.P. It matures in 75-80 days and yields 85-90 q green pods/ ha.
12. **CO 1:** Developed by selection in 1970 from local germplasm. Good for grains and fodder purpose in Tamil Nadu. It is a semi-spreading bold seeded variety which matures in 100-105 days. Average grain yield is 5-6 q/ha.

8. SOWING

8.1. Time of Sowing

In north India, cowpea is grown in *kharif* and Spring season. In Spring only photo-thermo insensitive varieties are grown. In south India, this crop is also grown in *rabi* season. *kharif* cowpea should be sown with the monsoon rains. Late sowing affects plant growth adversely which results in poor number of flowers and less grain and pod yield. Summer crop is sown after harvesting of *rabi* crops in the first week of April. Delayed sowing than this date reduces grain yield and sometimes early monsoon rains may destroy the crop. Spring crop in north India is sown in last fortnight of February or first fortnight of March. Early sowing in February results in poor germination due to low temperature and sowing late in March affects grain yield adversely due to forced ripening.

8.2. Method of Sowing

Cowpea is sown by broadcast or line sowing method. The row spacing of 30-45 cm. (30 cm for summer and 45 cm for *kharif* cowpea) and 10-15 cm of plant spacing are maintained.

8.3. Seed-Rate

Optimum plant population of cowpea for one hectare is 4-6 lakhs. However, seed-rate of cowpea varies according to the purpose, it is grown. For grain or vegetable 20-25 kg, forfodder 35-45 kg, and green maturing 35-40 kg seeds are required.

8.4. Seed Treatment

Cowpea seeds are inoculated with *Rhizobium* spp (cowpea type) at the rate of 5 g culture/kg seed, regardless of seed-size. For the protection of seed against soil born diseases, treat the seeds with thiram @ 2 g/kg seed.

9. IRRIGATION

Cowpea is a deep rooted pulse crop with tap root system. In *kharif* season, it is grown with monsoon rains and utilise the rain moisture effectively. Both summer and spring cowpea respond to irrigation water. Early sown rainy season cowpea needs one or two pre - sowing irrigations (70 mm water for irrigation). In summer, number and frequency of irrigations depend upon the soil-type and prevailing weather. Two irrigations are optimal after a pre-sowing irrigation. One irrigation can be given at flowering stage (30-35 DAS) and the second at grain development stage (42-45 DAS). Water requirement of cowpea is 300-400 mm. The best criteria for irrigation in cowpea is the range of soil-moisture between medium dry to wet; crop should be irrigated at 25-50% depletion of available water from its maximum root depth.

10. WEED CONTROL

Crop yield losses due to weeds competition in cowpea have been estimated to range from 30 to 50%. The weed flora infesting this crop is give below:

Nature of weeds	Name of weeds
Annual	
Narrow leaf weeds	*Paspalum sanguanalis* *Digitaria sangnanalis* *Cyperus iria* *Dactyloctenium aegypticum* *Setaria glauca* *Echinochloa colonum*
Broad – leaf weeds	*Amaranthus viridis* *Celasia argentia* *Amaranthus spinosus* *Trianthema monogyna* *Cleome viscosa*
Perennials	*Cyperus rotundus* *Sorghum halepense*

Cowpea has a smothering effect on weeds. Because of fast and early growth, it attains full ground cover stage at 30 DAS. Thus, the crop needs protection from weeds for initial 15 to 20 DAS. Two weedings at 15 and 30 days of crop growth ensures an effective weed control. During rainy season, weeds can be controlled by use of chemicals. Use trifluralin @ 1 kg/ha as pre -emergence application for effective weed control. In cereal-legume mixture, alachlor 1 kg/ha as a pre-emergence application gives reasonable control of weeds. Pre-emergence application of prometryne at 0.75 kg/ha is found to be promising. In sorghum inter-cropped with cowpea use basalin 1 kg a.i./ha in 800-1000 liters of water as pre-planting spray. It should be well incorporated in the soil before sowing.

11. PLANT PROTECTION

Cowpea being rich in protein is attacked by a number of pests (viz., stem boring maggots, aphids, jassids and white flies) fungal, bacterial and viral diseases, which are given below:

11.1. Diseases

Common name	Causal organism	Symptoms	Source of infection	Control
Cercospora leaf spot	Cercopora- 1. canescens 2. c. cruenta	Angular brown or red spots with grey or brown centre and reddish purplish border on leaves, stalk and pods.	Plant debris	Spray with benomyl (0.2%) or diathane m-45 (0.2%) at 15 days interval. Follow up crop rotation. Collect crop residues and destroy.
Bacterial blight and canker	Xanthomonas-vignicola	First cotyledon become red, then seedlings become shrivelled. Brown spots on trifoliate leaves; Cankers develop on stem; Pods infection leads to shrivelling of pods.	Seed borne	Seed treatment with streptomycine sulphate (500 ppm) for 30 minutes. Spraying the crop with 0.2% carboxin when disease appears.
Rust	Urontyces appendiculatus	White pustules on lower surface of leaves. Affects new shoot and pods also.	Soil and seed borne	Spray the crop with dithane m-45 @ 2 kg/ ha in 1000 liters of water.
Cowpea-mosaic	CMV (Cowpea Mosaic Virus) transmitted by Aphid (*Aphis gossypii*)	Leaves pale yellow and exhibit mosaic, vein bending symptoms. Pod size reduces and become twisted.	Sap and seed transmissable	Use healthy seeds. use metasystox 0.1% spray for aphid control.

11.2. Insects and Pests

Common name	Symptoms	Control
Pod borer *Heliothis armigera* *Adisura atkinsoni*	Larvae first feed on pod's surface and then bore into pods and then feed on seeds.	Collect and destroy damaged pods larvae. Spary with sevin at 0.2% if damage is high, or spray endosulfan at 0.05%.
2. Aphids *Aphis* spp	Suck the cell sap from all tender parts of plants. Leaves turn brown and crumpled, plants look sick.	Spary the crop with 0.1%, metasystox or 0.04%, monocrotophos 40 EC. Apply thimet 10% G @ 10 kg/hec as a basal dose.
3. Hairy caterpillar	Most important Pest and eat away all green matter of the leaves at seedling stage.	Spray endosulfan 35 EC in 1000 liters of water per hectare. Collect and destroy the eggs and young larvae.
4. Jassids Amrasca Kerri	Both nymphs and adults pierce the plant tissues and suck the cell sap. Leaf margins become yellow and curled.	Same as for aphid control

12. HARVESTING

Green pods for vegetable purpose can be harvested 45-60 DAS depending upon nature of variety. It is desirable to harvest pods when they are tender and half-grown otherwise at later stage they become 'puffy' and develop fibers. The grain crop may mature in 60-110 DAS and harvesting should be done when pods are fully matured. In summer and spring crop, pods should be picked up first and then plants are harvested separately for green forage. Fodder crop should be cut when it attains the age of 40-45 days.

13. YIELD

Cowpea produces about 12 to 15 quintals of grains and 50-60 quintals of straw per hectare. Vegetable cowpea produces 50-60 quintals of green pods per hectare, while fodder crop yields 250-300 quintals of green fodder per hectare.

14. QUALITY

Nutritive analysis of cowpea grain (per 100 g of edible portion) is given below:

Composition	Amount
Water	11 %
Calories per 100 g	342
Protein	23.4%
Fat	1.8 %
Carbohydrates	60.3%
Calcium	76 mg/100 g
Iron	5.7 mg/100 g
Thiamine	0.92 mg/100 g
Riboflavin	0.18 mg/100 g
Nicotinic acid (niacin)	1.9 mg/100 g

15. POST HARVEST TECHNOLOGY

After harvesting, pods should be dried and threshed. The threshed grains should be dried in sun to bring moisture content in seed upto 8-9% for safe storage. The following table shows the post-harvest standard of good quality seed of cowpea.

Criteria	Value
Minimum germinable seed	70 %
Minimum pure seed	98.8 %
Weed seed maximum	0.2 %
Inert matter maximum	1.2 %

After ensuring full standards of cowpea seeds, treat them with captan 75% WDP or thiram 75% dust @ 0.3 g/kg seed. Then bag and tag the seed lot.

16. CONSTRAINTS OF PRODUCTION

Following are important constraints for poor productivity of cowpea in India:

1. *Lack of better plant type:* African and Asian scientists have developed early matureing (61 days) cowpea varieties which had shown a potential of 1.8 t/ha yield. Unfortunately, such transformation has not been made yet in India.

Still farmers are growing long-duration lodging and shattering susceptible and photothermo-sensitive varieites of poor harvest index.
2. Poor or no fertilization by farmers results into low yields.
3. Area under irrigated cowpea is almost negligible and crop is grown as rainfed as either mixed or as cash crop.
4. Insect, pest and disease problem is serious in cowpea and once pest attack is there, it becomes uncontrollable.
5. Crop is mostly grown as a secondary crop, with mentality of 'They Take Their Own Care'.

17. REFERENCES

Pandey, R. K. 1961. Pulse Crop News Letter, IARI, Regional Institute, Kanpur. 1(1): 72-73.
Purseglove, J. W. 1968. Tropical crops: Dicotyledons, Longman Group, U.K. Ltd. England: 290-304.

23. Bengal Gram

I. P. S. Ahlawat

Bengal gram commonly known as gram or chickpea is the most important winter season pulse crop of India. The scientific name *Cicer arietinum* has been derived from Roman work. Romans used to call it *Cicer* because of the resemblance of the seed to the head of a 'ram'. The word *arietinum* originated from the word *aries* which also means 'ram'. In ancient Indian literature of the 1st to 4th century AD, the *Puranas*, the gram was known as *chennuka* in Sanskrit. In most Sanskrit-derived languages of the Indian subcontinent, gram is known as *chana*. The Anglo-Indian name of gram (Bengal gram) was derived from the Portuguese grao (grain). Bengal points to the location of the earliest British conquests in India, without further implication in regard to origin of chickpea, just as the Royal Bengal Tiger does not really imply Bengali origin.

Gram is mainly used for human consumption as well as for animal feeds. It is utilized as whole seed, *dal* (decorticated dry split cotyledons) and flour (powdered *dal*). Gram based foods are prepared using a wide range of recipes. Soaked, sprouted, fermented, boiled, roasted for cooked, salted or unsalted, parched, fried, steamed as well as green foliage and grain as vegetables are the important forms in which it is consumed by the people. Soaked grain and husk are fed to horses and cattle as concentrate and roughage, respectively. The straw of gram is an excellent fodder for cattle. Gram is considered to have medicinal effects such as for blood purification, germinated seeds are recommended to cure scurvy, malic and oxalic acids collected from green leaves are prescribed for intestinal troubles. Gram being an important source of protein and calories is also rich in minerals and vitamins.

1. ORIGIN

The gram is said to be one of the oldest pulses known and cultivated from ancient times both in Asia and Europe. Gram most probably originated in an area of present-day southeastern Turkey and adjoining Syria. Three wild annual species of *Cicer*, closely related the gram are found there. These are *Cicer bijugum, Cicer echinospermum* and *Cicer reticulatum*. The last can be considered as a progenitor, or perhaps had a common ancestor with gram. De

Candolle (1982) traced the origin of gram to an area south of Caucasus and northern Persia. Vavilov (1926, 1949-50) designated two primary centers of origin (now centers of diversity), southwest Asia and Mediterranean, a secondary one, Ethiopia. He noticed that like other grain legumes, large seeded cultivars abounded around the Mediterranean basin, whereas small-seeded cultivars predominated eastwards. There are lignuistic indications that the large-seeded, cream coloured gram reached India two centuries ago, apparently through Afghanistan, as its Hindi name *Kabuli chana*, an illusion to the Afghanistan capital Kabul. The small-seeded, dark coloured gram is called *desi chana* (local) and these denominations are now quite used to distinguish the two main groups of cultivars.

2. DISTRIBUTION

The distribution of gram area and production in the world is very uneven. Out of 207 countries, only 33 report growing and only 18 out of 33 gram growing countries plant more than 20,000 hectares. Around 92% of the area and 89% of the production of gram are concentrated in semi-arid tropical (SAT) countries. Referring to the international situation, Asia accounts for the bulk of area (92%) and production (89%) of gram. The important countries in the world with respect to area are India, Pakistan, Mexico, Turkey, Burma and Ethiopia. In terms of production, India ranks first followed by Pakistan, Turkey, Mexico, Ethiopia and Burma. Gram growing centers in India are located between 15 and 34 N latitude. In India, Madhya Pradesh, Rajasthan, Uttar Pradesh, Maharashtra, Haryana, Gujarat and Bihar are main gram growing states. The area and production of gram in different states are given Table 23.1.

TABLE 23.1: Area and production of gram in India

State	Area (000 ha)	Production (000 t)
M.P.	2401.2	1727.5
Rajasthan	1449.0	793.8
U.P.	1028.8	919.7
Maharashtra	591.4	301.5
Haryana	387.0	259.0
Gujarat	123.4	77.7
Bihar	122.6	100.7
All India	7201.0	6010.0

3. CLASSIFICATION

Cicer was originally classified in the tribe Vicieae; but its position is suffficiently distinct to consider the genus a tribe of its own, the cicereae (Kupicha, 1977 and Nozzolillo, 1985). There are 43 species of *Cicer* (cultivated and wild) throughout the world. The Indian gram have been classified in two broad categories:

1. ***Desi* or brown gram** (*Cicer arietinum* L): It is most widely grown group. The colour of seed ranges from brown to yellow. The flower colour is white, pink or blue. Seed size ranges from 5-12 mm. Stems are erect to semi-spreading.

Leaflets are 3-7 pairs. Plants have good branching ability. Chromosome number is 2n = 14,16,24,32

2. ***Kabuli* or white gram** (*Cicer kabulium*): Plants of this group have poor yield potential than *desi* type. Grains are bold and attractive and usually white in colour. Plants are tall and erect with moderate branching ability. The chromosome number is 2n=16.

4. GROWTH STAGES

The growth stages of gram have been described on the basis of morphological and development changes occurring in the plant during the growing season.

1. **Seedling:** At this stage first leaf opens, second leaf just attaining full size, vegetative nodes 1-2.
2. **Primary and secondary branches:** At this stage, 4-6 leaves are opened, primary and secondary branches are started, vegetative nodes 5-6 and height of the plant is about 6-7 cm.
3. **Tertiary branching**: 9th leaf opening, vegetative nodes 8-9, height of plant 10-12 cm.
4. **Flowering:** At this stage, plant bears 10-15 leaves, lower two leaves turn yellow, plants bear 1-2 flowers, vegetative nodes 11-19, reproductive nodes 1-2.
5. **Early pod formation**: Lower 1-2 leaves are fallen, flowering on 5 nodes and 1-2 nodes with pods.
6. **Mid pod formation and seed filling:** Plants have flowers on 8-10 nodes, fast seed formation, active seed filling, primary branches with 4-6 pods and 3-4 flowers, senescence of lower 3-4 leaves, plant height 25-30 cm.
7. **Late pod formation:** Flowering on 10-12 nodes, 8-9 nodes with pods, 4-5 pods and flowers dried, terminal pod formed, fast leaf senescence (10-30%), yellowing up to 30-40 cm, 95% pods turn brown.

5. CLIMATE

In India, gram is cultivated in winter season. In temperate countries, it is mainly cultivated in spring and summer season because of the availability of favourable temperature in this season. It requires cool climate for its growth and development and high temperature for maturity. The optimum temperature for its growth ranges from 15 to 25°C. Severe cold and frosts are injurious for its growth and development. Frost at the time of flowering causes flower drop. Low temperature at the time of germination reduces the germination percentage. Heavy rains at the time of germination, flowering and hail storm at and after flowering causes severe damage to the crop. Gram required cloud free days for its normal growth.

Being leguminous crop, gram needs less rainfall. The deep-rooted system makes it a preferred crop in dry tracts receiving an annual rainfall of 60-100 cm. Water logging at any stage of growth may destroy the crop. However it responds to light irrigations at flowering and grain filling stages.

6. SOIL AND SOIL PREPARATION

Gram is able to grow on a wide range of soils. It is commonly believed to be a hardy crop, This can be grown on marginal soils which are not suitable for crops such as wheat. The soils is major production areas range from sandy (dunes in Thar Desert) to sandy loam (Entisols of north India) to deep black cotton soils (vertisols of central India). Gram thrives best on deep loams or silty-clay loams free from excessive salts and may retain upto 200 mm of moisture in a profile to a depth of 1 m. However, too light and too heavy soils are not good for the gram because on light soils with low moisture retention capacity, growth of gram is retarded and on heavy soils having high moisture retention capacity, vegetative growth is abundant which restrict the light penetration in crop canopy and makes the crop prone to diseases. The soil should be near neutral in reaction, through a pH range 6-8 is favourable for growth of gram. The soil should be well drained.

Gram needs clody and rough land for good aeration in the root zone. This could be achieved by deep ploughing followed harrowing. In rainfed areas where gram is grown with conserved moisture, deep tillage at the start of the rainy season is found beneficial as it opens up the ground, ensures greater conservation of moisture and reduces physiological wilt, particularly on soils that tend to develop hard pans in the root zone. Bunding of the soil, the so-called *Haveli-system* in central India on heavy soils, and stirring of the soil during the rainy season in Indo-Gangetic plains to promote infiltration have also been found beneficial.

7. CROPPING SYSTEM

Improved cropping systems and alternate cropping practices can increase and stabilize the yield. The adoption of a cropping system in a particular region results from ecological factorsm, such as soil and climate, as well as non-ecological factors, such as consumer demand, market, socio-economic policies of the region, and technological capabilities. Generally successful intercropping of two or more species should result in efficient use of resources leading to enhance total crop production and returns compared to sole cropping. The patterns of crop rotation and mixed cropping are highly variable in different production areas in India. The most efficient crop rotations (Table 23.2), and mixed/intercropping systems of gram in different parts of the country have been give here under.

7.1. Crop Rotation

TABLE 23.2: Most efficient crop rotations in different regions

Region	Duration of rotation (years)	Crop rotation
Indo-Gangetic plains	One	Pearlmillet-gram
	One	Sorghum-gram
	One	Maize-gram
	One	Short duration cotton-gram
	One	Early rice-gram
	One	Guar-gram
	One	Sesame-gram

Region	Duration of rotation (years)	Crop rotation
North-East India	One	Rice-gram
Punjab, Haryana, Rajasthan	One	Fallow-gram
	One	Pearl millet-gram
Central & Peninsular India	One	Fallow-gram
	Two	Winter sorghum-gram
	Two	Durum wheat-fallow-gram
	Two	Sugarcane-gram

7.2. Mixed Cropping

Mixed cropping of gram with other cool season crop is quite common. The common crop mixtures include:

Gram + wheat, gram + barley, gram + linseed, gram + rapeseed, gram + taramira, gram + safflower, gram + *rabi* sorghum

7.3. Intercropping

In dry lands of central India, gram proved an economic intercrop with *rabi* sorghum. Intercropping of gram between two rows of sugarcane to improve the land use efficiency during the early phase of sugarcane growth is common both in Indo-Gangetic plains as well as in peninsular India. Relay cropping of gram in standing crop of rice is practiced by the farmers in northeastern rice fallow areas to make use of soil moisture, which is critical for good stand establishment of gram in these areas.

8. MANURE AND FERTILISERS

In India, gram is grown on a variety of soils, ranging from coarse-textured sandy soils to heavy clay soils. The crop is seldom fertilized and hence often suffers from deficiency of one or more nutrients. Total uptake of nutrients by a chickpea crop has been estimated to range from 60 to 200 kg N ha^{-1}, 5 to 15 kg P ha^{-1}, and 60 to 170 kg K ha^{-1}. The response of crop to mineral nutrients largely depends on various soil and environmental factors. The specific deficiency symptoms of important nutrients are given in Table 23.3.

TABLE 23.3: Foliar symptoms of mineral nutrient disorders in gram

Mineral nutrients	Symptoms
Nitrogen	Chlorosis first on older leaves, pink pigmentation on stems and in a defuse pattern on the upper surfaces of older leaves.
Phosphorus	Dark green foliage, red-purple pigmentation on stem and upper surface of leaflets of the lower leaves, leaflets later become yellow-green or buff-green.
Potassium	Chlorosis of margins and tips of older leaves, reddish pigmentation on leaflets. Necrosis developing first on the tips of leaflets and covering the whole leaflet turning it lights brown in colour.
Calcium	Development of rachis and leaflets of young expanding leaves arrested, growing tip of rachis, and tip of unexpanded leaflets turn brown. Necrosis of leaflets. Death of growing points enhances auxiliary bud development older plant parts remain dark green.

Mineral nutrients	Symptoms
Magnesium	Young and middle-aged leaves initially show light green chlorosis, which becomes more severe on the distal half of the leaflets of fully expanded leaves. Necrosis of the leaflets tips or margins in severe deficiency.
Sulphur	Chlorosis first appearing on younger leaves, extending over the entire plant in severely deficient plants. Red anthocyanin pigmentations on stems.
Zinc	Chlorosis of younger leaves, red-brown pigmentation on the margins of upper surfaces of leaflets, and lower portions of stems. Stem plant growth. Stipules of younger leaves become chlorotic.
Iron	Chlorosis first on the terminal 3-4 leaves of each stem. White straw-coloured necrotic patches develop on the distal half of leaflets and stipules of young leaves.

8.1. Manure

Farm Yard Manure (FYM) is used at 12-15 tons ha^{-1}. This helps in better root development. FYM is spread on the ground and ploughed down deep with a soil turning plough 10-20 days before sowing, so that proper decomposition of the applied manure takes place in the soil.

8.2. Fertiliser

Nitrogen: Gram being a leguminous crop meets its major nitrogen requirement (about 75%) meets through the process of symbiotic nitrogen fixation. However, soils with low organic matter and poor nitrogen supply may require 15-20 kg N ha^{-1} as starter dose, which can meet plant requirement before the formation of nodules.

Phosphorus: The responses of gram to phosphorus are conspicuous (Table 23.4 and 23.5). In general, the responses to applied P ranging from 40 to 60 kg P_2O_5 ha^{-1} have been positive. The effects of P are usually more pronounced when applied in conjunction with starter N, *Rhizobium* or irrigation. Placement of P 3-5 cm below the soil surface showed better response than broadcast application.

TABLE 23.4: Effect of phosphorus on gram yields (t ha^{-1}) at various locations

Added P_2O_5 (kg ha^{-1})	New Delhi 1985/86	Rahuri 1984/85	Indore 1983/84
0	1.3	1.0	1.2
20	--	1.1	--
40	1.7	1.2	--
46	--	--	1.4

TABLE 23.5: Mean yield of gram as influenced by different fertiliser treatments under rainfed condition

Fertiliser dose (kg ha^{-1})	Grain Yield (q ha^{-1})	
	1978-79	1979-80
Control	5.3	12.6
20 P_2O_5	6.3	14.8
40 P_2O_5	7.3	16.8
10N	6.4	14.1

Fertiliser dose (kg ha^{-1})	Grain Yield (q ha^{-1})	
	1978-79	1979-80
10 N + 20 P_2O_5	7.4	16.7
10 N + 40 P_2O_5	9.7	18.5
20 N	6.3	14.8
20 N + 20 P_2O_5	7.9	18.2
20 N + 40 P_2O_5	9.1	20.5
20 N + 40 P_2O_5 + 20 K_2O	9.5	21.3
CD (P=0.05)	1.1	0.6

Potassium: The responses to K are generally small and seldom significant because of the higher K status of the Indian soils. In soils deficient in K, application of 20 kg K_2O ha^{-1} is recommended. In trials conducted on cultivators fields under rainfed conditions in Jaunpur (UP), the highest yield was obtained with the application of all the three nutrients at their highest dose (Table 23.5). Applications of nitrogen and phosphorus separately as well as in combinations resulted in significantly increase in yield (Yadav and Singh, 1981).

Micro nutrient: Among the micronutrients, zinc and iron play an important role in gram production. Application of 25 kg $ZnSO_4$ ha^{-1} at the time of sowing along with the macro nutrients is recommended. The effect of Fe is more pronounced in the presence of Zn. One spray of 2% $FeSO_4$ solution 30 days after sowing is recommended for improving the productivity of gram.

9. VARIETIES

The productivity of pulse crops in our country is low due to lack of high yielding photo-insensitive varieties. However, several improved varieties of gram have been developed in different gram growing states. A list of recommended varieties is given in Table 23.7.

The description of some of the important varieties is given below:

TABLE 23.6: Recommended varieties of gram for various states

State	Irrigated and unirrigated	Irrigated late sown	Unirrigated	Kabuli gram
Uttar Pradesh (All)	KL 468, K 850 Pant G 114, H 82-2 Ajay (Pusa 408) Haryana gram 1	Amar (Pusa 230) Gaurav, Pusa 209, Pusa 261, T-3	T-3, H 208, C 214, Pusa 212, Radhey	K4 (White) K5 (Green), C 104, Lcc-32
Haryana and Punjab	Ajay, Pusa 372, G 140, G 543, Haryana gram 1	Amar, Pusa 209, Gaurav (H 75-35) C 235, G 130	Pusa 212, H 208, G 24, C 214	L 144, L 550, C 104
Bihar	Pusa 240, Pusa 372, Pant G 104, T-3, BR-77, BR-78	Amar, Atul, C 235, Pusa 209, BR 65, BR 78	H 208, G 114, Radhey	C 104, NP 2
Rajasthan	Ajay, Udai (KPG 59), T-3, RSG 44, Pusa 372, Haryana gram 1	Amar, Gaurav, Girnar, Pusa 209, C 235, RS 11	Pusa 212, RS 10, RSG 2	L 550, C 10, Lcc 32

State	Irrigated and unirrigated	Irrigated late sown	Unirrigated	Kabuli gram
Madhya Pradesh	Udai, Pusa 482, T-3, JG 62, JG 74, Chaffa, Adhartal 5	Amar, Pusa 209, Girnar, Gwalior-2, Gwalior-3, Ujjain 24	Ujjain 21, Radhey, JG 221, JG 74, Pusa 212	C 104, Lcc 32
Maharashtra and Gujarat	Chaffa, BDN 9-3	BDN 9-3	--	C 104
Karnataka	-	Annegiri	-	-
Tamil Nadu	CO 2	-	-	-
West Bengal	B 108, B 110, B 115, B 124			

9.2.1. Desi or Brown Gram Varieties

1. **Radhey:** Developed from a cross T 197 x 76. Plants are tall, semi-spreading with light green foliage. It matures in 155-160 days. Seeds are bold and light brown with the test weight of 200 g. The average yield is 25-30 q/ha.
2. **Type 3:** Developed in 1959 from a cross T 197 x 195. It is a tall (75-90 cm), semi-spreading and late maturing variety (160-165 days). Seeds are light brown, round and bold (240 g/1000 seed). It is suitable for Uttar Pradesh and parts of Madhya Pradesh.
3. **Pant G 114:** It was developed in 1981 from a cross G130 x 1540. Plants are medium tall, seed yellowish brown, small (130 g/1000 seed). It is a late maturing variety and adapted for northern and eastern zones of country. The variety is moderately resistant to blight with a yield potential of 30-35 q/ha.
4. **H 208:** It was released in 1977 and developed from (S 26 x G 24) x C 235. This is a medium tall, semi-spreading and medium maturing (155 days) variety. It is suitable for rainfed areas with good moisture retentive soils. Average yield is 20 q/ha.
5. **C 235:** Its pedigree is IP 58 x C 1234. It was released in 1960. Seeds are yellowish brown and medium in size. It is tolerant to *Ascochyta* blight and is suitable for rainfed areas. The average yield is 20-25 q/ha.
6. **Hare Chole 1:** This variety was developed from S-26 x GG Bijapur and released for general cultivation in 1973. This is the first seeded variety having medium sized seeds. It is a late maturing variety (170 days). Average yield is 12 q/ha.
7. **RS 11:** It is a white flowered mutant from RS 10. Seeds are medium bold and brown in colour. It is drought resistant and medium in maturity. It is suitable for Rajasthan and has yield potential of 20-25 q/ha.
8. **Pusa 209:** It is a cross of P 827 x C 235 and was released in 1980 from IARI, New Delhi. Seeds are attractive, medium bold and light in colour. Plant height is about 65 cm and it matures in 145-165 days. Its yield potential is 25-30 q/ha.
9. **Pusa 408:** It is a mutant of G 130 and was released in 1985. Seeds are yellowish brown in colour. It matures in 145-165 days. It is moderately resistant to wilt. Its yield potential is 20-25 q/ha.

10. **ICCC 4:** It was developed from H 208 x T3. It was released in 1983 from ICRISAT, Hyderabad. Seeds are light brown, medium bold in size. Plants are semi erect with basal branching. Average yield is 18 q/ha.
11. **B 124:** It was developed from a cross N 31 x B 75. It is small seeded variety. Its average yield is about 24 q/ha. It is suitable for growing in West Bengal.

9.2.2. Kabuli Gram Varieties

1. **ICCC 32:** It was developed by the cross of L550 x L 2. It was released in 1984. Grains are medium bold and salmon white in colour. It is a semi-spreading variety and matures in 145-165 days. It is resistant to wilt and suitable for cultivation in *Kabuli* growing areas.
2. **L 144:** It was developed from across of S 26 x rabat. It was released in 1975. It is a tall and quick growing variety with a light green elongated stem. The seeds are bold and salmon white in colour with a test weight of 350 g. The average yield is 12-15 q/ha.
3. **L 550:** It is tall, semi-spreading, medium maturing variety with light green foliage. Seeds are bold (250 g/ 1000 seeds), creamy white and round. The average yield is about 20 q/ha. It is adapted to irrigated areas.
4. **K 4:** It is a selection from local germplasm. It was released in 1949. It is a late maturing (165 days), tall, spreading variety. The flowers are white. Seeds are round, bold (370 g/1000 seeds), and white in colour. It is suitable for culinary purposes. Its yield potential is 8-10 q/ha.
5. **K 5:** It was developed by the cross of 312 x K 4 and was released in 1959 for general cultivation. It is a late maturing (170-175 days), tall, semi-spreading and having white flowers with greenish wing. Seeds are round, smooth and bold (370 g/1000 seeds). The yield potential is 18-20 q/ha.

10. SOWING

One of the major factors responsible for low yields under field conditions is inadequate plant stand. Poor seed quality, seedbed condition and improper seeding method are largely responsible for poor plant stand. Use of optimum good quality seed, proper method and depth of seeding and plant spacing are important agronomic considerations in the production of gram.

10.1. Sowing Time

The time of sowing of gram is more critical in determining the yield of unirrigated gram than the crop, which is either irrigated or raised in milder environments, permitting a long growing season. The date of sowing has a profound influence on the crop performance because it determines the kind of environmental conditions to which the various phenological stages of the crop will be exposed. Sowing of chickpea under rainfed conditions is mainly determined by availability of conserved moisture in the seeding zone. In such areas, sowing should be done earlier in the first fortnight of October (Panwar *et. al.* 1980). In case, adequate moisture is available in the seeding zone, sowing in the second fortnight may ensure better crop.

A significant yield variation in gram varieties due to variation in sowing date was recorded by Kumar *et. al.* (1983) at Faizabad (UP). Crop sown on 30

October gave the highest yield (Table 23.7). Delay in sowing beyond 30 October significantly reduced the yield. The yield reduction was also noticed in early sowing of gram before 15 October.

TABLE 23.7: Mean grain yield (q/ha) of gram varieties as influenced by dates of sowing

Dates	Varities						
	Radhey	C235	K 468	Type 1	Type 3	Pant G 110	Mean
1 October	21.3	22.1	18.6	18.9	21.9	17.4	19.8
15 October	27.5	25.1	21.9	24.6	27.7	22.7	24.9
30 October	30.1	23.8	25.0	23.1	31.0	22.9	26.0
15 November	22.3	22.2	20.8	21.9	21.7	20.5	21.6
30 November	16.5	17.8	13.8	15.1	15.3	14.4	15.5
Mean	23.6	22.2	20.0	20.7	23.5	19.3	-

10.2. Method and Depth of Sowing

In dry lands, seeds should be placed by traditional plough at the depth where soil moisture is adequate for germination. *Pora* method using country plough, or using seed drills is common for sowing of gram at the depth of 10-12 cm from the soil surface in sandy loam and loam soils in the Indo-Gangetic alluvium plains.

Depth of seeding is an important factor affecting the germination and final plant stand in areas where the crop is raised on conserved moisture from the previous season. Studies on loamy sand soils in northern India showed that seeding at 10 cm depth was better than shallower seeding (5 cm depth) if the gram crop was raised without irrigation. Mortality of seedlings in the unirrigated crop was 42.7% with 5 cm depth of seeding as against 33.8% with 10 cm depth (Kaul *et. al.*, 1974). In situations where the soil moisture for germination is not limited, a depth of 5-7 cm seems to be ideal for the emergence of gram.

10.3. Seed Rate and Spacing

The seed rate is determined by growth habit of plant and seed size. Large number of varieties is available in gram with wide variations in seed size. A seed rate of 50-75 kg ha^{-1} has been found sufficient for *desi* varieties and 75-100 kg ha^{-1} for *kabuli* types (Moolani and Chandra, 1970). At Faizabad, Singh and Yadav (1985) found that K 468, a small seeded variety, yielded highest at 75 kg ha^{-1}. Varieties Type 3 and Radhey yielded the highest at 100 kg ha^{-1} (Table 23.8).

TABLE 23.8: Effect of seed rate on yield (q/ha) of gram varieties

Variety	Seed rate (kg/ha)			
	50	75	100	Mean
Type 3	18.1	19.8	20.0	19.3
K 468	22.5	24.2	21.6	22.8
Radhey	20.3	21.8	23.0	21.4
Mean	20.3	21.9	21.5	

The optimum level of plant population appears to differ depending upon environment conditions and the plant type. In a favourable environment that permits an adequate period for vegetative and reproductive growth, chickpea show a high degree of plasticity. From the results of several studies the optimum plant population appears to be about 33 plants/m^2 (Singh, 1983). Under the late sown conditions significantly higher yield was obtained at a plant population of 500×10^3 plants/ha over 333×10^3 plants/ha. A row spacing of 30 cm with a plant spacing of 10-12 cm is optimum for sowing of gram under normal conditions.

10.4. Seed Treatment

The seeds should be treated with fungicide brassicol + thiram (1:1 ratio) @ 2.5 g/kg of seed or bavistin @ 0.5 g/kg of seed. Seeds should also be inoculated with *Rhizobium* and phosphobactrin cultures.

11. IRRIGATION

The amount of water needed by a crop throughout its growth is the sum of evaporation from the soil surface and transpiration needed to achieve potential growth and yield. This sum depends on the climate and the cultivar grown.

TABLE 23.9: Effect of irrigation of yield of gram

Treatment	Grain yield (q/ha)				
	1970-71	1971-72	1972-73	1973-74	1974-75
No. irrigation	19.3	7.58	15.31	39.29	12.09
Irrigation at 45 days	20.48	14.95	18.28	38.87	16.09
Irrigation at 75 days	--	16.75	18.44	35.90	13.32
Irrigation at 45+75 days	18.88	13.18	23.43	39.55	17.42
Irrigation at 105 days	--	--	20.46	--	--
Irrigation at 45+75+105 days	21.45	--	--	--	--
Irrigation at early podding	--	--	--	38.16	16.20
Irrigation at 45 days+early podding	--	--	--	37.01	17.51
CD at 5%	NS	3.63	3.74	NS	2.49

Gram is mainly grown under rainfed conditions. It has a deep tap root system and can make an efficient use of the profile-stored water in the well drained alluvial soils of northern states and retentive clay/clay-loam soils of mid-west and southern states. Although water requirements of the gram are relatively low, it often suffers from periodic water dificits owing to low and uncertain rains in the *rabi* season. Supplemental irrigation to release the crop from soil moisture stress at critical stages of crop growth has been found to result in substantial yield increases in gram. Supplementary irrigation early during vegetative growth on light soil with a small water holding capacity or later during vegetative growth and early pod filling stage on heavier and deep soils gave increased yields (Saxena, 1980). The late flowering and pod setting stages appear to be the most sensitive to soil moisture stress (Palled *et. al.*, 1985). In the absence of winter precipitation, the crop responds well to supplemental irrigation. One irrigation at 45 or 75 days after sowing or at early pod filling stage generally increased the yield (Table 23.9). The absence of responses in 1970-71 and 1974-75 were due to well-distributed rainfall during crop season (Saxena, 1977).

In a three year study on sandy-loam soils in Punjab, Sandhu *et. al.* (1978) reported that one post-sowing irrigation to gram on the basis of IW/CPE ratio of 0.4 appeared a better practice for obtaining high yields.

The care in irrigation is needed because of the high sensitivity of gram to lack of soil aeration. Proper land leveling and the use of ridges or narrow flat beds for the purpose of uniform distribution of irrigation water and avoiding water logging is required for optimum plant growth and development and for getting higher yields of gram.

12. WEED CONTROL

The gram crop is sensitive to weed competition than many other crops. The first 4 to 6 weeks are most critical for weed competition and the reduction is grain yield may vary 40 to 90% depending on the intensity and type of weed flora. About 60 weed species are reported to infest gram fields. *Chenopodium album* is the most important weed of gram in India.

12.1. Weed Flora in Gram in India

- **Annuals:** Chenopodium album, Melilotus indica, Lathyrus aphaca, Medicago denticultla, Trigonella polycerata, Polygonum plebijum, Asphodelus tenuifolius, Anagallis arvensis, Trichodesma indicum.
- **Perennials:** *Convolvolus arvensis*

In order to control the weeds in gram fields, following methods may be adopted.

12.2. Mechanical Method

Weeding by mechanical methods include the use of hand chisel (khurpi), hand hoe and wheel hoe etc. Since the competition is more in the initial growth stage, two hands weeding (3 to 4 weeks and 6 to 8 weeks after sowing) should be done. Under the conditions where weeds reappear following late rains or delayed irrigation, a third weeding may also be needed around 9 to 12 weeks after sowing.

12.3. Chemical Method

In view of the limitations of mechanical methods of weed control, chemical control merits appropriate consideration. The herbicides that are effective in controlling the weeds in gram are given in Table 23.10.

TABLE 23.10: Herbicidal weed control in gram

Time of application	Herbicide	Dose (kg/ha)
Pre-plant incorporation	Methabenzthiazuron	1.0
	Terbutryn	1.2
	Trifluralin	1.0
	Fluchloralin	1.0
Pre-emergence	Prometryn	0.5
	Terbutryn	0.75
	Methabenzthiazuron	1.5
	Alachlor	1.5
Post-emergence	Fomesafen	0.25
	Methabenzthiazuron	1.5

13. PLANT PROTECTION

13.1. Diseases

More than 50 pathogens have been reported so far on gram from different parts of the world. Some of the internationally serious diseases, in order of importance are: Ascochyta blight, Fusarium wilt, dry root rot, stunt (pea leaf roll virus), Botrytis gray mould, black root rot, Phytophora root rot, and Pythium root and seed rot.

Ascoshyta blight: It is the most serious disease of gram between the latitudes 30° and 45° where relatively low temperature (15°-25°C) prevail during the crop season. It is caused by fungus *Ascochyta rabici*. The disease develops whenever the winter-sown gram in north-west India receive rains during the crop season. Diseased debris and infected seeds are the main source of inoculum.

The pathogen attacks all above ground parts of the plant. When source of inoculum is seed borne, the seedlings, under favourable conditions, develop dark brown lesions at the basal part of the stem. In case, the inoculum source is airborne spores, the first symptom of blight usually appear as small necrotic specks in the newly formed leaves. The necrosis in a susceptible cultivar is gradually down wards killing the plant.

13.1.1. Control Measure

1. Removal and destruction of dead debris
2. Crop rotation
3. Deep sowing
4. Intercropping gram with wheat, barley, and mustard
5. Advancing the sowing date in autumn results in 50-100% yield increase
6. Seed treatment with thiram or captan (0.25%) or with bavistin-thiram combination @ 250 g/q of seed
7. Foliar spray - Four protective sprays with captan or zineb @ 2.5 kg/ha at 10-15 days interval. or Bordeaux mixtures spray
8. Use resistant varieties namely, BG 261, BG 280 and H 75-35 (Gaurav)

Fusarium wilt: It is prevalent in the entire gram growing areas of India and is caused by fungus *Fusarium oxysporum*. The young growing tips of diseased plants become limp and drooping. The leaves start withering due to inadequate water supply and excessive transpiration. On splitting open the taproot lengthwise, dark-brown fungus streaks can be seen in the root near the pith region.

In February-March, soil temperature and reduced soil moisture conditions favour the growth of the fungus. The pathogen enters the root system and xylem vessels, thereby reducing water supply to the plant.

13.1.2. Control Measures

1. As far as possible sowing of gram should not be done before third week of October.
2. Deep planting of about 8-10 cm deep in the light soils reduces the wilt incidence.
3. Resistant varieties such as ICC 202, ICC 391, Pant G 114, Pusa 209 and C 235 should be grown.
4. Seed treatment with bavistin 25% +TMTD 50% @ 250 g/q of seed

Dry root rot: The disease is of economic importance in non-irrigated areas of Haryana, western UP, MP, AP and Rajasthan. The taproot on pulling out is found to be devoid of bark and lateral roots. The dead root is dry and brittle towards the tip. Black sclerotic bodies of the fungus can be seen on the main root below the bark.

13.1.3. Control Measures

1. Early sowing of early maturing cultivars with timely irrigation minimizes the disease.
2. Seed should be treated with captan or thiram or PCNB @ 2.5 g/kg seed.

Bortrytis gray mould: Botrytis gray mould (BGM) causes considerable damage in gram, and its importance has only recently been realized. The regions between 25 to 30° are most prone to BGM.

The symptoms appear on stems, leaves inflorescence and pods as gray or dark brown lesions covered with mouldy sporophores. Stem lesions are 10 to 30 mm long and girdle the stem completely. Affected leaves and flowers turn in to a rotting mass.

It is seed borne and has a wide host range. The inoculum is almost always present in the environment waiting for the right weather to become active.

13.1.4. Control Measure

1. Seed treatment with Ronilan @ 100 g/q of seed or with bavistin-thiram combination @ 200 g/q seed
2. Foliar spray of bavistin+thiram or Ronilan (0.1% suspension) at seven days interval protects the crop.
3. Disease tolerant varieties like G 558, G 866 and BG 292 should be grown. *Kabuli* types are generally less susceptible than *desi* types.

Black root rot: Incidence of black root rot under field conditions was reported in 1974 from India. The fungus *Fusarium solani* causes it. Plants may display the disease at any stage and affected plants turn yellow and show wilting. The root system is rotten, most finely roots disappear and the remainder turns black. Affected plants dry prematurely. The disease occurs commonly when soil moisture content is high in light soils.

13.1.5. Control Measures

1. Grow resistant lines of gram (18 lines of gram resistant to black root rot have been identified by ICRISAT)
2. Modifying irrigation schedules should help to reduce the disease.

Rust: The fungus *Uromyces ciceris-arietini* causes it. The rust appears first mainly on the leaves as small, round or oval, cinnamon brown powdery pustules. Occasionally these can be seen on the stems and pods. The rust develops with cool and moist weather; rain is not necessary.

13.1.6. Control Measures

1. Grow only resistant varieties
2. With the appearance of first symptom, spray the crop with dithane M-45 (0.2%) followed by two more sprays at 10 days interval.

Stunt (Pea leaf roll virus): It is a viral disease and infection during the early stages of plant growth leads to a total loss. *Acyrthosiphon pisum* and *Aphis craccivora* vectors the virus. The affected plants can be easily spotted in the field by their yellow, orange, or brown discoloration of foliage and stunted growth. The stunting is due to shortened internodes. The leaflets are smaller.

13.1.7. Control Measures

1. Delayed sowing reduces the disease in northern India
2. Grow resistant varieties.

13.2.B. Insect Pests

Gram has relatively a few insect-pests and generally suffers little pest-caused loss when compared with most other semi-tropical leguminous crops because:

1. The plants are covered with glandular hair which exude acid-droplets containing high concentration of malic acid, so insects appeared to be deterred by this very acidic (pH 1.3) exudate and,
2. As this crop is normally sown before or just after the winter in almost all areas of the world in which it is of importance, it grows during the period when insect activity and population are at a minimum.

The important insect-pests of gram are:

Gram pod borer: The gram pod borer (*Heliothis armigera*) is the most serious pest of gram and causes the yield reduction up to the extent of 75 per cent. *Heliothis* lay eggs on plant under the surface of the terminal leaflets. The eggs hatch after 2-10 days, according to temperature.

The first instar larva, translucent with a black head, is found feeding on the upper surface of leaflets. The fully-grown larva is about 40 mm long, the head is mottled brown, and the colour pattern of the body consists of a narrow, dark, medium dorsal band. The underside of the body is uniformly pale. The moths feed on nectar and are active only during night. The total life cycle may be as short as 30 days under ideal conditions.

13.2.1. Control Measures

1. Use of resistant varieties such as ICC 506, ICC 6863, PDE 2, AKG 33, Pant CE 1, BG 256.
2. Early sowing in north and central India (15-30 October)
3. Intercropping gram with mustard, wheat, safflower and barley
4. Use of insecticides, spray of endosulfan (0.07%) or monocrotophos (0.04%) or chloropyriphos (0.05%) solution. Use of bio-pesticides such as Neemark, Achook, Nimbecidin, Neem guard @5% solution.
5. For biological control, use of NPV @250 a.i./ha and *Bacilluis thurigiensis* available by different names as dipel, delfin, biolep @ 1 to 1.5 kg ha^{-1} can be effective against *Heliothis*.

Gram cutworm: The gram cutworm (*Agrotis* sp.) is a serious pest in low-lying areas where fields are clody. The larvae of this insect remain hidden under these clods during the day time and cause damage during the night. The caterpillars cut the plants at ground level. The pest is sporadic in nature.

13.2.2. Control Measures

1. Dusting 4 per cent endosulfan @ 20 kg/ha is effective in controlling the pest.
2. Mixing of 4 per cent endosulfan in the soil @ 20-25 kg/ha

Termites: There are remarkably a few reports of termite (*Odontotermus obesus*) damage in gram. However, in northern India there can be considerable loss by termites, particularly in sandy soils. Termites often act as saprophytic feeding on the roots of plants that have been killed, or are dying, from diseases such as *Fusarium* wilt.

13.2.3. Control Measures

1. Soil and seed treatment with chloropyriphos is effective against termites.
2. Maintaining the optimum soil moisture because attack of termites is severe in dry soils.

14. HARVESTING AND THRESHING

Harvesting at correct time is essential to get good quality grains and higher yield. The crop should be harvested at physiological maturity or at full maturity (7 days after physiological maturity). The moisture content in grain at harvest should be around 20%. Harvesting is done manually either by plucking out by hand or by cutting down the plants with sickle.

The harvested plants should be allowed to dry in the sun for about five to six days on threshing floor. The threshing is done either by beating the plants with sticks or by trampling under the feet of bullocks or tractor tyres. Winnowing follows threshing. The threshed grains should be cleaned by winnower and again for a week to reduce the moisture.

15. YIELD

The average grain yield under rainfed conditions is about 700-800 kg/ha with 1,000-1,300 kg straw/ha. Under ideal conditions, a good crop of *desi* varieties gives about 2,000-2,500 kg grain yield/ha were as *kabuli* varieties gives about

2,500-3,500 kg grain yield/ha. The harvest index is about 38 to 40%. The yield in case of mixed or intercropping depends upon the plant population of gram in the mixture.

16. QUALITY

Seed size, shape, colour, texture (biting), fiber and protein content are the important quality parameters of gram. The genotype, growing location and season, sowing time and irrigation practices are the factors, which affect the quality of the grain. The important factors affecting the quality parameters are listed in table given below.

TABLE 23.11: Nutritive value of gram grain

Constituent	g/100 g	Constituent	mg/100 g
Protein	23.0	Phosphorus	342.9
carbohydrate	63.5	Calcium	185.6
Fat	5.3	Magnesium	141.0
Calorific value (cal/100 g)	347.6	Iron	6.6
Vitamin A	0.19		
Thiamin	0.29		
Riboflavin	0.20		
Vitamin C	3.87		
Niacin	2.25		

TABLE 23.12: Factors affecting quality parameters in chickpea

Factor	Quality parameter affected
1. Genotype	Seed size, shape, colour, protein
2. Growing conditions	
i. Rainfall	Protein (%), seed size
ii. Temperature	Protein (%), seed size, seed texture
3. Cultivation practice	
i. Sowing time	Protein (%), seed size, colour
ii. Irrigation	Protein (%), seed size
4. Processing	
i. Dehulling	Fiber content
ii. Roasting	Texture (biting)
iii. Soaking and drying	Texture (biting), protein

17. POST HARVEST TECHNOLOGY

Efforts to increase the production of gram need to be supplemented by efforts to minimize losses at various stages. There is a sizeable loss of grain quantity and quality during post-harvest operations, like threshing, transport, processing and storage.

17.1. Drying

Drying is the most efficient and economical of all measures taken to save grain from spoilage. By this operation, the moisture content of produce can be reduced to a level which is safe for storage. At the farmer's level, the drying process of most grain is done in the traditional way by spreading it on the ground as a thin layer under sun. Such a method of drying though cheap, but is not considered good. Therefore mechanical dryers operated either with liquid fuels or with solar energy developed at Indian Grain Storage Institute, Hapur should be used for drying purposes.

17.2. Processing

1. **Dehulling:** The endosperm of gram is covered by a single fibrous seed coat (hull) on a wrinkled surface. Before cooking or other processing operations it is often necessary to remove this seed coat. In addition to reducing the fiber content, this improves the appearance, texture, cooking qualities and palatability. In India, dehulling is generally done in two major steps. The first step of loosening the hull is either by (i) prolonged sun drying until the hulls are loosened, or (ii) application of water followed by several hours of sun drying and tempering, or (iii) heating the grains in a pan with or without sand and with vigorous stirring, or (iv) a combination of these techniques. Dehulling is done either by (i) pounding using a mortar and pestle, or (ii) grinding in hand operated wooden or stone shellers ('chakki'), or (iii) grinding on a rough flat stone with a stone rolling pin usually operated by the legs.
2. **Grinding:** The whole gram or the dehulled seeds are ground dry to a flour (popularly known as *Besan*) or ground wet to a batter for a number of sweet, savory or breakfast food preparations, either alone or in combination with cereals or millets.

17.3. Storage

To prevent losses during storage, farmers adopt various methods but they succeed only partially owing to the storage of grains in traditional storage structures. These structures are not rat, moisture and insect proof. An ideal storage structure is one, which is constructed in such a manner as to prevent the entry of elements detrimental to the grain and to prevent the exit of elements beneficial to the storage grain.

The Indian Grain Storage Institute, Hapur has developed a number of designs of farm level modern storage structures for capacities ranging from 0.09 to 14.5 metric tons to suit the requirements of the farmers in different agro climatic conditions. The farmers should use these structures for safe storage of the grain. The moisture content in seed at the time of storage should not be more than 10 per cent.

18. CONSTRAINTS OF PRODUCTION

The several constraints limiting the gram productivity are as follows:

18.1. Agro-Climatic Constraints

The main agro-climatic constraints to production and productivity in south and south-east Asia are summarised in Table 23.13.

TABLE 23.13: Main agro-climatic constraints

Soil group	Crop growth stages		
	Seedling establishment	Vegetative period	Flowering and seed formation
Vertisols	Hot and/or dry seed beds	Supra-optimal temperatures (>30 °C) and/or drought	Supra-optimal temperatures; drought (exacerbated b soil cracking)
Entisols	Hot and/or dry and/or saline seed beds	Drought (if winter rains fail); soil salinity	Drought; soil salinity; frost damage.

18.2. Genetic Constraints

1. Low genetic yield potential of local cultivars-the present dry varieties of gram are adapted to marginal conditions and poor responsive to applied inputs i.e., irrigation, fertilisers etc.
2. Poor plant type - the present day cultivars lack both form and frame to support a more productive plant type.
3. Lack of stability - the yields of the gram cultivars fluctuate from year to year due to susceptibility to insect-pest and diseases.

18.3. Management Constraints

1. Adoption of poor agronomic practices-poor seedbed preparation, unavailability of moisture at critical stage, inadequate seed rate, limited or no use of inputs like fertilisers, *Rhizobium* culture adversely affect the production of gram.
2. Improper method of sowing-the age-old practice of broadcast of seed is still followed by the farmers to a large extent. This results in uneven competition between plants under field conditions.
3. Lack of plant protection measures-It is because of low yields in traditional agriculture that farmers lack both will and resources to undertake plant protection measures.
4. Adverse effects of cultivation under rainfed conditions-Most of the sowing of crop in India is done on conserved moisture. If farmers are not able to conserve moisture in time, inadequate plant stand is inevitable in the fields.

18.4. Socio-Economic Constraints

The low support price, lack of organised marketing system for pulses unlike cereals, lack of improved implements and technology with the farmers and their poverty are important socio-economic constraints.

19. REFERENCES

Ahlawat, I.P.S. 1986. Annual Report Division of Agronomy, Indian Agricultural Research Institute, New Delhi 110012, India. p.27-28.

De Candolle, A. 1983. Origine des plantes cultivees, Paris p. 258-260

Joshi, A.C., Aparadh, V.H., Rao, K.V. and Umrani, N.K. 1988. J. Maharastra Agric. Univ. 13(2): 168-170

Kaul, J.N., Sekhon, H.S. and Dhingra, K.K. 1974. In: proceedings of the All India Coordinated *Rabi* Pulses Workshop. UPIAS, Kanpur, India.

Kumar, R., Yadav, H.L. and Yadav, D.S. 1983. Indian J.Agron. 28(1): 87-88

Kupicha, F.K. 1977. Botanical J. Linnean Soc. 74: 131-162.

Moolani, M.K. and Chandra, S. 1970. Gram Cultivation in Haryana, HAU, Hisar, p.15.

Nene, Y.L.1980. In: Proceedings of International workshop on Chickpea Improvement. ICRISAT, Hyderabad, A.P. India. p.171-178.

Nozzolillo, C. 1 985. Canadian J. Botany 63:1-6

Palled, Y.B., Chandrashekharaiah, A.M. and Radder, G.D. 1985. Indian J. Agron. 30:104-106

Panwar, K.S., Faroda, A.S., Malik, D.S., Ahlawat, I.P.S., Dhingra, K.K., Rao, M.R., Singh, R.C. Dahiya, D.R. Singh, Atar and Roysharma R.P. 1980. A Review on Quarter Century Research Work on Grain Legumes Agronomy in India. In: Quarter Century of Agronomic Research in India (1955-80) (eds. Rajendra Prasad, K.S. Prashar and R.P. Singh), Indian Society of Agronomy, New Delhi 110 012, India. p.176-202

Sandhu, B.S., Parihar, S.S., Khera, K.L. and Sandhu, K.S. 1978. Indian J. agric.Sci. 48: 486-92

Saxena, M.C. 1977. Research on winter pulses, GBPUAT, Pantnagar, Nainital, U.P.

Saxena, M.C. 1980. In: proceedings of the International Workshop on Chickpea Improvement, ICRISAT, Hyderabad, India. p.96-98.

Singh, B.P. 1983. Indian J. Agric. Sci. 53: 543-549

Singh, S.C. and Yadav, D.S. 1985. Indian J. Agron. 30(4): 414-416.

Thakur, N.S., Raghuvanshi, R.K.S. and Sharma, R.A. 1989. International Chickpea Newsletter 20: 19-20.

Vavilov, N.I., 1926. Studies on the origin of cultivated plants. Leningrad. p.129, 238.

Vavilov, N.I. 1949-50. Chronica Botanica, New York. 13(1/6):26-38, 75-81,151.

Yadav, D.S. and Singh, A. 1981. Pulse Crops Newsletter 1(2): 44.

24. Lentil

P. S. Rathore

Lentil is one of the oldest nutritious *rabi* pulse crop. The whole or splited seeds are used in preparing *dal* and delicious *namkin* commonly called *dalmoth*. The splited seeds are also used in making soup and porridge. Green pods are used as vegetables. The leaves and stalks are used as cattle feed.

1. ORIGIN

Lentil is believed to be originated from south western Asia and Mediterranean region. It spread to Europe, India, China and Ethiopia.

2. DISTRIBUTION

The important lentil growing countries of the world are India, Tajakistan, Canada, Bangladesh and Nepal. It is cultivated over 3287 thousand hectare area and production is around 2.89 mt (1997). In India it is cultivated through out north, and Maharashtra. The leading states include Madhya Pradesh Uttar Pradesh, Bihar, Maharashtra and West Bengal. Lentil in India is grown over 1.0 mha area with a production of 5 lakh tonnes. During the year 1992-93 India exported 7.89 thousand tonnes of lentil. Area and production of lentil in world is given in Table 24.1.

TABLE 24.1: Area and Production of lentil in world

Countries	Area 000 ha	Production 000 mt
India	1200	920
Turkey	553	570
Canada	329	379
Bangladesh	206	171
Nepal	160	124
Syria	120	88
Total	3287	2896

3. CLASSIFICTION

Lentil is a member of family leguminoseae with 2n chromosome number 14. Two species of genus *Lens* have been identified as cultivated species, i.e., *Lens esculenta* Moench Medit, *Lens culinaris* Medit. Specis *esculenta* has been further divided into two as follows:

1. *Lens esculenta macrosperma* Moench, commonly known as *masur*. Seeds are bold and flattened about, 6-9 mm diameter. Pods are large and flat. Flowers are mostly white and large.
2. *Lens esculenta microsperma,* commonly known as *musri*. Seeds are small about, 3-6 mm in diameter, and convex. Pods are small and convex. Flowers are small with violet blue to white colouration.

4. GROWTH STAGES

The following growth stages have been identified in lentil.

1. **Seedling stage:** Plant is small, tender and with one leaf
2. **Branching:** Plant with fully developed three leaves, 4th leaf in developing stage and primary branches begins to develop.
3. **Flowering:** Plant with 10-11 leaves, 1-2 flowe nodes and 11-15 vegetative nodes.
4. **Pod formation:** Pod begins to develop.
5. **Maturity:** 90-95% leaves starts withering.

5. CLIMATE

Lentil is cultivated in *rabi* season. Crop needs cool temperature during vegetative phase and warm temperature at maturity. The optimum temperature for its growth is 18-30° C. The plant can tolerate frost, but sensitive to high temperature.

6. SOIL AND SOIL PREPARATION

Crop can be grown on any type of soil ranging from light loam to black cotton soils. The crop is generally grown after harvesting paddy in low lying areas. For good yield crop should be grown in well drained, neutral loam soils. Field should be prepared soon after withdrawal of monsoon. One to two ploughing followed by 2-3 cross ploughing for heavy textured soils and one ploughing followed by 1-2 cross ploughings are sufficient for light soils. The soil should be loose and friable. To protect the crop from soil borne insects specially termite apply. 25 kg endosulfan dust 4% just before last ploughing.

7. CROPPING SYSTEM

Most common crop sequences with lentil practiced in different parts of India are: Paddy – lentil, cotton – lentil, sorghum – lentil, maize – lentil, fallow - lentil (conserved soil moisture). The crop can also be grown as mixed crop with mustard, toria and barley. As an inter-crop it grown with mustard taramira and sugarcane.

8. MANURES AND FERTILISERS

Addition of manures and fertilisers increased the growth and yield of lentil. Most of the workers suggested a starter dose of 20 kg N/ha along with 50 kg P_2O_5/ha as basal at the time of sowing by drilling. The nitrogen requirement of later stages are full filled by N fixation by inoculating seeds with *Rhizobium leguminosarum*. On zinc deficient soils, application of 25 kg $ZnsO_4$ gave good response.

9. VARIETIES

Use of improved variety seed according to suitability enhances crop yield. The lentil varieties with states for which it suits along with characteristics are given in Table 24.2.

TABLE 24.2:

Variety	Recommendation	Characteristics
Type 36	U.P., NE Hills	Brown, Small seeded matures in 130-140 days yield 15-20 q/ha
Pant L-406	UP, Bihar, Punjab NE hills and plains of M.P.	-
BR - 25	Bihar, M.P.	Reddish brown, Small seeded matures in 125-130 days Av. yield 15-20 q/ha.
JLS - 1	M.P.	Brownish gray mature in 120-130 days
L-9-12	Punjab	Brownish with black Spot, Av yield 15-20 q/ha
Asha	West Bengal	Brownish grey small Seeded, 27% protein matures in 125-130 days, Av. yield 14-16 q/ha
B-235	West Bengal.	Small seeded, 28% protein matures in 116-120 days Av. Yield 16-18 q/ha.
B-177	West Bengal.	Small seeded 29% Protein, Matures in 125-130 days, Av. Yield 15-20 q/ha
Ranjan.	West Bengal.	Semi spreading 28% protein, Matures in 125-130 days, Av. yield 15-17 q/ha
DPL 15	North western plain zone.	Bold seeded
DPL 62	North western plain zone	Bold seeded

10. SOWING

Time of sowing is a key factor in realizing high yield of any crop. Research conducted on the crop indicated second fortnight of October as an appropriate time of sowing. The optimum time of sowing being 15-20th October. Delay in sowing result in severe reduction in yield as well as quality of grain.

10.1. Spacing

Crop is sown at row to row space of 30 cm by seed cum fertiliser drill or can be sown behind the desi plough. Plant to plant space is kept 10-12 cm and adjusted at the time of thinning.

10.2. Seed Rate

The optimum seed rate for timely sown crop is 30-35 kg/ha. Under late sown conditions closer spacing of 20-25 cm with seed rate of @50 kg/ha be used.

10.3. Depth of Sowing

Seed should not be sown more than 3-4 cm deep. Deep sowing result in poor establishment of crop.

11. IRRIGATION

Generally, the crop is cultivated under unirrigated areas on conserved soil moisture. As it is a deep rooted crop. It can make use of stored soil moisture, hence crop can tolerate moisture scarcity to some extent. In case of poor moisture storage or ill distributed winter precipitation crop performance could be improved by supplementing the crop at least for two times, once at pre-flowering stage and second at pod filling stage. The water requirement of crop is being 200 mm. Depth of each irrigation is 6-8 cm. Scientifically irrigations in lentil may be scheduled at IW/CPE ratio of 0.4-1.0 with best being at 0.6. Bhan (1975) at Kanpur obtained best performance when crop was irrigated at pre-flowering and pod formation while Panwar and Paliwal (1975) obtained high yield of late sown lentil when irrigation were applied at pre-flowering and early pod formation. The effect of irrigation on lentil is given in Table 24.3.

TABLE 24.3: **TABLE 24.3:** Effect of irrigation in yield of lentil

Irrigation	Grain yield q/ha	
	Hissar	Ludhiana
No irrigation	10.26	7.98
Irrigation at pre-flowering	22.38	12.82
Irrigation after flowering	20.35	11.31
Irrigation at pre-flowering after flowering	23.87	15.85
CD (0.05)	4.41	1.62

12. WEED CONTROL

Lentil is a slow growing crop particularly at early stages, hence infested badly by the weeds. It takes about 45-60 days to spread on ground. The first 30-60 days are considered critical for crop weed competition. The losses of 25-30% are recorded in the crop, when infested with weeds. The major weed flora associated with the crop are *Chenopodium album, Fumaria parviflora, Melilotus alba, Chenopodium murale, Anagallis arrvensis, Anagallis arvensis, Asphodelus teunifolius* and *Vicia sativa* as broad leaf weeds and *Cynodon dactylon* and *Cyperus rotundus* as grassy weeds. Weeds must be controlled at right stage. Two manual weedings, first at 25-30 days after sowing and second at 45-50 days after sowing keep the crop free of weeds. Herbicidal weed control consists of fluchloralin 0.75 kg/ha in 800-1000 L of water as pre-plant and should be incorporated into soil. Ahlawat et al (1979) obtained impressive control of weeds by using prometryne at 0.5 kg /ha.

13. PLANT PROTECTION

13.1. Insect Pests

Hairy caterpillar and pod borer- The hairy caterpillar feed on the leaves and flowers and pod borer make hole in to pod and grain. The incidence may be reduced by spraying monocroptophos 36 WSC 500 ml or endosulfan 35 EC in 500 L of water.

13.2. Diseases

- **Wilt:** The disease is caused by fungus *Fusarium oxysporum* sp *lentis*. The diseased plant appear yellowish and in severe cases plant dries up and ultimately die. The disease also affect plant roots. Roots of infected plant are brown and spread of root is also small. The disease is difficult to control. Some of the preventive measures check its spread. Follow crop rotations. Do not grow lentil in field infested with wilt at least for 3-4 years. Sow resistant varieties Pant L-416, Sihore 74-7 and always treat the seeds with seed dressing fungicides.
- **Rust:** The affected plant show pink to brownish pustules on stem and leaves. In advance stage, black pustules also develops. If disease is not controlled, the affected plant may wither and die. This is a fungal disease caused by *Uromyces fabae*. To check the damage, cultivate resistant varieties L-9-12, Pant L-406. Spray dithane M-45 (0.2%) at 15 days interval. Depending upon severity of disease 2-3 sprays may be required.

14. HARVESTING AND THRESHING

The crop should be harvested, when pods turns brownish. Over ripening of pods leads to shattering, hence plant should not be over dried. After harvesting, the crop is dried in sun on the threshing floor and threshed by beating the plants with sticks or by bullocks. The grains are dried till 12% moisture and stored.

15. YIELD

Crop cultivation by using improved technologies may yield 20-25 q/ha grain and 30-40 q of straw.

16. QUALITY

Seeds are brown, grey or brown in colour. It is most nutritious pulse with high digestibility. The chemical composition of grain is given in Table 24.4.

TABLE 24.4: Chemical composition of lentil grain

Consituents	Moisture	Protein	Fat	Minerals	Fibre	Carbohydrate	K cal
Value (%)	12.4	25.1	0.7	2.1	0.7	5	343

17. REFERENCES

Asthana, A.N. (1998) Pulse crop research in India. Indian J. Agric sci. 68 (8): 448

Bhan S (1975) Water requirement and irrigation management of crops in India (Ed. Hukken' *et. al.*).

FAO Production year book Vol. 51, 1997

Panwar, K.S. and Paliwal H.S. (1975) Indian J. Agron. 20: 194-195.

Prasad, R, Parashar K.S. and Singh R. P. (1981) Quarter century of Agronomic Research in India (1955-1980). Indian Soc. Agron.

Saxena, M.C and Singh, H.P. (1977) Research in *rabi* pulses. Tech Bull GBPUAT Pantnagar.

Yadav R.L., Singh, P. and Prasad, R. (1998) Fifty years of Agronomic Research in India. Indian Soc. Agron., New Delhi.

FIBRE CROPS

25. Cotton

M. V. Venugopalan

Cotton is the most important cash crop of India. It is the backbone of our sprawling textile industry contributing 7.0% to our GDP, fetching an export earning besides providing employment to 60 million people engaged in the production, promotion, processing and trade of cotton. Cotton accounts for 45% of the world fibre and supplies 10% of world edible oil.

Cotton is primarily cultivated for its lint, which is spun into yarn, which is used for textile and several other industrial uses. Raw cotton is also used for medical and surgical purposes. The linters are used for stuffing, cushions, pillows, mattresses etc. It is also used in the manufacture of high grade paper, rayon, films, explosives etc. The stalk is used as fuel, for making paper pulp, for cultivating mushrooms etc. Low grade fibres, waste materials, wadding and upholstry. The seed is crushed for obtaining seed oil-an edible oil and the residues-hulls, press cakes and press meal are excellent cattle feed. Cotton seed cake is also a valuable concentrated organic manure.

1. ORIGIN

The origin of the genus *Gossypium* is a subject of considerable controversy. Based on the evidences it can be postulated that the 'A' genome carrying old world cotton was evolved in the southern Ethiopia. However, Hutchinson *et. al.* (1947) assumed that the brushwood Savana, between Ngambiland and southern Mozambique is the original home of *G herbaceum* var. *africanum* and the race *acerifolium* arose by the domestication of race *africanum* in southern Arabia. *G. herbaceum* var. *africanum* was introduced into India via the Arabian and Persian coasts along the trade route.

The perennials related to race *acerifolium* are the most primitive cultivated *herbaceum* and is found in the belt across West Africa to Makran in Pakistan. Their differentiation gave rise to race *persicum* in Persia (Iran) and Kuljianum in Turkey. These annual races were taken northwards and had to flower before the onset of winter. The annual races were introduced into India around 780 were it differentiated into race *wigtianum*.

The *G. arboreum* race *indicum* is closely related to *G herbaceum* race *acerifolium* and is believed to have developed from it. The oldest known specimens of Asiatic cotton dating back to 3000 BC found in the excavations of Mohen-jo-daro in Pakistan belonged to this group (Santhnam and Hutchinson, 1971). *G. arboreum* later spread along the Indo-gangetic plains giving rise to the northern perennial forms. These spread eastwards to Burma, Indonesia and Philippines where race *burmanium* originated. The western spread of the perennial races into Sudan gave rise to race *sudanese*. From the northern spread arose the race *sinense* in China. The cernuum cotton is distributed in Assam.

FIGURE 25.1: Open cotton ball.

The new world cottons *G. barbadense* and *G. hirsutum* originated through allopolyploidy between two diploids, one of which is related to the 'A' genome carrying old world cotton and the other related to new world species carrying 'D' genome (Phillips, 1974).

The wild species, *G. raimondii*, is the closest relative of 'D' genome that has entered into the ancestory of the new world cotton. *G barbadense* probably originated in northern Peru and the *G hirsutum* probably originated from *G barbadense* itself in the highlands of Guatemala and southern Mexico.

2. DISTRIBUTION

Cotton is grown in over 80 countries from 47□ N latitude in Ukraine to 32□ S in Australia and South America. However, the twelve major countries enlisted in Table 25.1 account for 80-85 per cent of the acreage and contribute 85-90 per cent of the world production. India ranks first in acreage under cotton. Besides, it is the second largest consumer and the third largest producer of raw cotton.

Based on the agro-climatic conditions and spices distribution, three distinct regions are recognized-

- The northern *hisutum-arboreum* zone comprising of Punjab, Haryana, North West Rajasthan and Western Uttar Pradesh.
- Central *hirsutum-arboreum-herbaceum* region comprising of southern Rajasthan, Gujarat, Madhya Pradesh and Maharashtra.
- Southern *hirsutum-arboreum-barbadense* zone comprising of Karnataka, Andhra Pradesh and Tamil Nadu.

India cultivates all the four spinnable species of cotton as well as their hybrids. The species distribution in terms of acreage for hybrids, *G. hirsutum, G. arboreum, G. herbaceum* and *G. barbadense* is 40%, 36%, 16%, 8% and 02% respectively. the cotton production increased from around 23 lakh bales in 1947-48 to 171.2 lakh bales in 1996-97. Today, cotton is cultivated in about 9.0 m ha in 9 major states (Table 25.2).

TABLE 25.1: Area, production and productivity in the major cotton producing countries 1996-97)

Country	Area (000 ha)	Production (000 tonnes)	Productivity (kg lint/ha)
USA	5208	4125	792
China	4722	4203	890
Uzbekisthan	1500	1034	689
India	9166	2977	327
Pakistan	3200	1590	497
Turkey	744	784	1054
World	33825	19416	574

TABLE 25.2: Area, production and productivity of cotton in different states in India (1996-97)

State	Area (lakh ha)	Production (lakh bales of 170 kg)	Productivity (kg/ha)
Punjab	7.42	16.00	367
Haryana	6.47	13.50	354
Rajasthan	6.54	14.00	364
Madhya Pradesh	5.27	18.80	605
Maharashtra	30.95	30.00	182
Gujarat	15.24	34.30	382
Andhra Pradesh	10.07	26.50	477
Karnatake	6.68	9.00	29
Tamil Nadu	2.60	5.50	360
Others	0.42	1.00	405
Total	91.66	171.50	318

3. CLASSIFICATION

Hutchinson *et. al.* (1947) recognized twenty species under the genus *Gossypium* based on cytological, genetical, geographical and archeological evidences. However, there are only 4 spinnable species. Two of them viz. *G. arboreum* and *G. herbaceum* and their sub species have haploid chromosome number n = 13 are classified under section VIII-*Herbacea*. The new world cottons viz. *G. hirsutum* and *G barbadense* have haploid chromosome number n = 26 are included in the section IX-*Hirsuta* (Phillips, 1966). A consolidated classification, embodying the views of various authorities in this subject is furnished in Table 25.3.

TABLE 25.3: Classification of the Genus *Gossypium*

Section	Species	Races
A. Capsules with hairs on sutures, seed without lint (n=13)		
I Strutiana	G. Strutianum, G. australe, G. bickii, G. Robinsoni, G. populifolium, G. custulatum	
II Erioxyla	G. aridum, G. armourianum. G. harknessi, G. lobatum	
III Klotzschiana	G. klotzschianum, G. davidsoni, G. raimondii	
IV Thurberana	G. thurker, i G. trilobum, G gossipoides	
V Anomala	G. triphyllum, G. anomalum	
VI Stocksiana	G. stocksii, G. incanum, G. somslense, G. areysianum	
VII Longicalyciana	G. longicalyx	
B. Capsule with hairless sutures, seed with lint		
OLD WORLD (n=13)		
VII Herbacea	G. arboreum	indicum, burnanicum, cernuum, sinense, bengalense, soudanense
	G. herbaceum	africanum, acerifolium, persicum, kuljianum, wightianum
NEW WORLD (n=26)		
VIII Hirsuta	*G. tomentosum* (wild species)	
	G. hirsutum	mariegalante, punctatum, latifoluim, morilli, palmeri, yucatanense, richmondii
	20 barbadense	brasilense

The characteristics of the four cultivated species of cotton are as follows:

- ***Gossypium hirsutum***: These are small shrubby plants, 60-150 cm tall with few vegetative branches. The stem is green or brownish in colour, bearing large chordate, 3-5 lobed, hairy or glabrous leaves. Flowers are large with pale yellow to cream coloured petals. The fruits, called bolls are large, rounded with small inconspicuous oil glands. These capsules (bolls) have 3-5 locules with each locule containing up to 11 seeds. Seeds bear lint hair and have a thick seed coat covered with fuzz.
- ***Gossypium barbadense***: These are fine and long linted cottons. Plants grow tall upto 22-70 cm and put forth few to many ascending vegetative branches (monopodia). Leaves are usually glabrous with 3-5 lobes. The lobes are deeply incised. Flowers are large with characteristically deep yellow petals with a prominent purple or red spot at base. Bractioles are long and divided at the apex to from 10-15 acuminate teeth. The capsules (bolls) are large and

have 3 locules each bearing 5-8 seeds. The whole testa is covered with long and shining lint.

- ***Gossypium herbaceum*:** These are hardy shrubs growing 60-130 cm tall. The main stem is thick, rigid and has few monopodia. The leaves are lobed, often hairy and flat. Possess large bracts that flare widely from flowers or capsules. Bracts have a serrated margin with 6-8 teeth. Bolls are generally round and beaked. They contain 3-4 locules each containing about 11 seeds. Seeds bear long lint hair and short fuzz.
- ***G. arboreum*:** They are basically perennial natured. These are branchy shrubs growing 150-200 cm tall. They may also be annual sub-shrubs which grow 50-140 cm tall. Leaves are often hairy, contain 5 to 7 lobes with linear stipules. Bractioles are triangular with several teeth at the apex. Flower may be creamy, white, yellow or purple coloured with a long staminal column. Bolls are trilocular with pointed tips and fitted with oil glands. Each locule may contain up to 17 seeds.

4. GROWTH STAGES

Cotton is an indeterminate plant exhibiting overlapping vegetative and reproductive and reproductive growth phases and these phases cannot be clearly demarcated. The duration of the different phases depends upon the variety, latitude, climate and management practices. These phases can broadly be differentiated into:

- **Germination phase:** Under optimum ambient conditions, the germination and the emergence is completed in 4-7 days.
- **Early vegetative phase:** This phase is characterized by a rapid root growth, slow growth of he main shoot and emergence of the first true leaves. This is followed by the growth of main stem and the differentiation of early monopodial branches.
- **Squaring:** The first square (flower bud) appears between 35 and 70 days after germination depending on the variety, location and crop husbandry. The squaring continues till maturity.
- **Flowering:** The first flower appears about 20-35 days after the appearance of the first square and flowering continues for about 60-80 days or more depending upon he availability of soil moisture. However, the peak flowering period is from 70-100 days after germination.
- **Boll development:** This period commences from flowering and continues till maturity. During the first 15-18 days after the fertilization of the flower, bolls attain 90% of their final size. The elongation of lint fibres in the boll is completed in 21-24 days. During the subsequent phase of approximately the same duration, the cell wall thickens due to the deposition of cellulose. The boll cracks on maturity along the locules and the fibres ultimately dry out and become ready for picking.

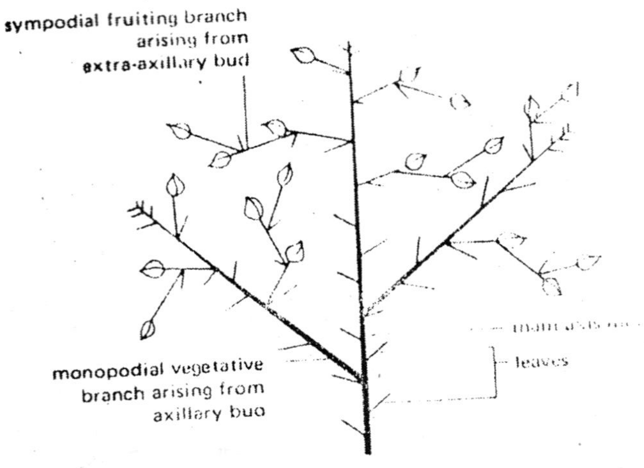

FIGURE 25.2: Branching in cotton.

5. CLIMATE

Climatic features, notably temperature, light and rainfall have a profound influence on the distribution, growth, productivity and quality of cotton.

5.1. Temperature

Cotton is basically a tropical plant needing warm temperature for normal growth. However, cotton as an annual crop has acclimatized itself to the tropics, sub-tropics and warm temperate zones where ever, at least 180-200 frost free days are available. Soil temperature ranging from 20 to 30° C is considered optimum for germination. Germination is seriously delayed at temperature below 18° C. During the growth phase, mean daily temperature of 21-27° C is optimum. Day temperatures ranging from 27 to 32° C with cool nights favour fruiting and boll development. The growth of cotton plant and fibre development is seriously limited when the minimum temperature falls below 15° C. Under adequate soil moisture, cotton withstand high temperatures of 43-45° C for short periods.

5.2. Rainfall

Cotton being a tropical plant of arid origin prefers moderate rainfall but highly susceptible to waterlogging. A minimum rainfall of 500 mm is adequate. The quantum and more importantly the distribution of rainfall determines the production and productivity of cotton in rainfed areas. During the vegetative phase, moderate rainfall is good. Heavy rains during flowering and fruiting accentuates bud and boll shedding. Dry period during ripening and boll bursting phases ensures good quality lint.

5.3. Light and Photoperiod

Abundant sunshine is ideal for cotton production and cloudiness adversely affects boll setting. Cloudy weather lasting 2 to 8 days cause heavy bud and boll shedding. Cotton, because of its tropical origin is a short day plant. However, its migration and domestication at higher latitudes, its variability and continuous genetic improvement have made them adaptable to long day and day neutral conditions. Berkeley (1931) concluded that cotton is a day neutral plant and the influence of exposure to sunshine is apparently related to temperature. Photo insensitive varieties are now widely available.

The climatic conditions in the cotton growing regions of India show considerable variations. High temperature of about 45° C during sowing and seeding emergence and low temperature accompanied by occasional frost coinciding with the picking period and moderate rainfall ranging from 300-700 mm are the features of the north zone. In the southern and central zones, the climate is equal. The maximum temperature ranges from 32 to 40° C and the minimum temperature from 10 to 20° C Rainfall ranges from 500-1250 mm, predominantly through the south west monsoon from June to Sept. In Tamil Nadu and Andhra Pradesh a part of the rainfall from the north-east monsoon is received during Sept. to December.

6. SOIL AND SOIL PREPARATION

Cotton is grown on a variety of soils across the world. Deep, fertile soil with adequate humus and high water holding capacity and good internal drainage is best suited for growing cotton. Cotton is sensitive to excessive moisture and water logging. Under rainfed conditions, it is generally grown on soils with high water holding capacity provide better internal drainage and greater productivity (Hake *et. al.* 1996). Where irrigation is available it is cultivated on a variety of soils ranging from sandy to sandy loams and clay loams. Cotton being a deep rooted crop, soil depth of less than 60 cm is considered unfavourable. Cotton grows best on neutral to slightly alkaline soils and a pH of 7.0 to 8.0 is considered optimum (Muller, 1968). The minimum pH for successful cotton cultivation is 5.3. Cotton is sensitive to salinity only during germination and early crop establishment. Established crop is extremely tolerant to salinity with a threshold E.C. of 7.7 ds/m and 50% yield reduction at 17.7 ds/m E.C. (Ayers, 1977).

The predominant cotton growing soils of India are sandy to sandy loams (Entisols and Inceptisols) in the north zone black soils (Vertisols) in Central India and in a variety of red (Alfisols), alluvial (Inceptisols) and mixed red and black soil in the southern zone.

Proper and timely soil preparation conserves soil moisture, provides optimum tilth for germination, controls weeds, breaks hard pans if any, in the subsoil and exposes pupae of pests to scorching heat and predatory birds. However, in most cotton growing regions, the time available for satisfactory tillage is limiting.

In the north zone, the delay in the harvest of the preceding wheat crop and subsequent availability of canal water for pre-sowing irrigation delays the soil preparation operation. In the central zone and rainfed areas of the south zone, cotton crop season is extended into the early summer months. The soil, particularly vertisols, become difficult to till after the harvest of cotton. In these regions, tillage operations begin with the onset of monsoon. Under both the situations, tillage is limited just to ensure a satisfactory seedbed for timely sowing.

Singh *et. al.* (1974) recommended deep ploughing up to 22 cm either every year or alternate year for the soils of Punjab. In the north zone, the field is irrigated after the harvest of the *rabi* crop and then ploughed. A light irrigation is then given and the field is harrowed a couple of times and planked. In black soils of central and southern zone, deep ploughing is recommended once in 3-4 years. Field is prepared by repeated ploughing with a blade harrow (*bakhar*) after the pre monsoon or monsoon showers.

7. CROPPING SYSTEMS

Cotton a long duration crop (170-210 days) with slow initial growth rate up to 50-60 days after germination, is traditionally sown by adopting a wide inter-row spacing. The recent technological advances resulted in a number of semi-compact varieties with greater rejuvenation capacity. These factors permit several intercropping systems involving cereals, pulses/legumes, oilseeds and vegetables. Among pulses, cowpea, green gram, moth bean, black gram, cluster bean and pigeon pea (as strip crop) are commonly inter-cropped. Oilseed like soybean and groundnut can also be profitably inter-cropped. Cereals (*ragi* and setaria) and vegetables (onion, garlic, chilies) are also being successfully inter-cropped in the south zone.

7.1. Sequential Cropping

1. **North Zone:** Traditional cotton was a long duration crop which was usually monocropped or depending upon the availability of water, methi or senji was either sown after cotton or even sown in the standing crop of cotton after the initial pickings as a relay crop. With the advent of short duration varieties like Bekaneri nerma, Ganganagar ageti, H 777, F 414 etc., cotton-wheat emerged as a dominant system. This system is practised in 80-85% of the cotton area in this zone. Cotton-mustard is the next dominant system.
2. **Central Zone:** Cotton was traditionally monocropped. Cotton-jowar, cotton-soybean/groundnut are some dominant rotations of the rainfed tract. Wherever assured irrigation is available, sequential cropping with wheat, gram, groundnut or safflower is practiced.
3. **South Zone:** Groundnut, rice, ragi, wheat, maize, pulses etc. are grown in sequence depending upon the availability of water. In rice fallow tracts of Andhra Pradesh and Tamil Nadu, summer cotton is grown sequentially to single as well as double crop of rice.

TABLE 25.4: Dominant cotton based cropping systems

State	System
Punjab, Haryana and Rajasthan	– cotton-wheat, cotton-mustard
M.P., Maharashtra and Gujarat	– cotton (monocrop, cotton-jowar (two year rotation), cotton-wheat intercropping with blackgram, greengram, soybean, groundnut and red gram.
Tamil Nadu	– cotton (monocrop), rice-cotton, rice-rice-cotton, cotton-jowar, cotton-pulse-jowar, intercropping with onion, groundnut, blackgram.]

State	System
Andhra Pradesh	– cotton (monocrop), cotton-rice (1 year), cotton-chilli or cotton-tabacco (two year rotation)
Karnataka	– cotton (monocrop), cotton-wheat, intercropping with chilli, groundnut, blackgram, greengram

8. MANURE AND FERTILISERS

Based on the relative occurrence of nutrient deficiencies world wide, Hearn (1981) classified the nutrients into the following groups:

Very common	N
Common	P, K
Occasional	Mg, S, Zn, B and Mn
Rare	Ca, Fe, Cu
Unknown	Mo, Cl, Na

Under Indian conditions, deficiencies of N, P and K among major, Mg and S among secondary and Zn and B among micro nutrients have been recorded under field conditions in cotton.

8.1. Nutrient Deficiency Symptoms

1. **Nitrogen:** Plants expressing N deficiency show stunted growth with small, pale, chlorotic leaves and poor root system. Wilted leaves dry out early and shed. Rate of squaring and flowering is slowed down, bolls do not mature to full size and shed before reaching maturity. N deficiency symptoms appear only after the petiole N falls below 2000 μ g/g.
2. **Phosphorus:** P is a key component of nucleic acids and plays an active role in energy transfer. P deficient cotton plants appear stunted with dark leaves. Accumulation of anthocyanin pigments may give a purplish colours to leaves. Petiole P values below 1000 ug/g indicates P deficiency.
3. **Potassium:** Specific deficiency symptoms in cotton plants begin with a yellowish white mottling of leaves, accompanied by yellow spots between the veins. Veins appear darker. As the deficiency becomes more severe, yellow spots become whiter. Later the tips margins as well as the spots become brown and the leaves curl downwards while lamina becomes bronze and is shed. Fibres become shorter and less firm K deficiency aggravates wilt (*Fusarium* sp.) and Alternaria leaf spot diseases.
4. **Sulphur:** Cotton deficient in S produce yellow leaves at the apex while older leaves remain green in contrast to N. Also, unlike N, S deficiency, the veins remain green.
5. **Magnesium:** Mg deficient cotton plants develop purple red leaves with striking green or white veins. Older leaves are affected first and they senesce prematurely.
6. **Zinc:** Zinc deficient cotton plants have cup shaped small leaves with interveinal chlorosis. Chlorotic tissues become brown and broned as tissues die.

7. **Boron:** B deficiency results in the shedding of squares, flowers and young bolls. The petioles of young leaves are irregularly thick red and on splitting open longitudinally, exhibit dark bonds with necrotic pith. The terminal bud often dries.

8.2. Manures

Cotton crop grown on black clay soils experience poor aeration due to the swelling of these soils on wetting. Similarly, cotton grown on alluvial soils encounter crust problem adversely affecting germination. Repeated tillage only provides temporary solution to these problems but in reality aggravates the problem by deteriorating the soil structure. To restore the health of such soils, adequate organic matter supplementation is very essential. Application of farm yard manure or compost partially reduces the inorganic fertiliser dose, improves soil physical properties, cation exchange capacity and nutrient storage and release. Wherever FYM or compost is available, they should be applied @ 3-5 tonnes/ha at the time of land preparation, at least 2 to 3 weeks prior to sowing.

8.3. Fertilisers

Nitrogen: Both rainfed and irrigated cotton responds well to N application. The response is in the range of 8-10 kg seed cotton/kg N. For irrigated cotton, 100-200 kg N is adequate although specific responses up to 300 kg N/ha has been obtained in Gujarat for hybrid cotton. For rainfed cotton varieties, 50-60 kg N/ha is adequate Hybrids respond to a slightly higher dose of N and 90 kg N/ha is recommended for rainfed conditions.

Losses of N due to leaching can be reduced by split application of N in 2-3 splits, 1/2 at sowing and 1/4 at each squaring and flowering. Delay in the application of the last split after peak flowering would cause fake vegetative growth and delays maturity with little yield advantage. However, if specific deficiency symptoms appear during the post flowering period, foliar supplementation using 2% urea or 2% DAP is recommended. Volatilization losses of N due to surface application by broadcasting can be reduced by applying urea into the moist zone either by ring or column method.

Use of boifertilisers can reduce the fertiliser requirement as well as cost of cultivation. Inoculation of cotton seed with *Azotobacter chrococcum* (strains C2 or M4) or *Azospirillum brazilense* @ 200 g/kg seed results in saving of fertiliser N to the extent of 20 kg N/ha.

Phosphorus: Response of cotton to P application is generally observed when the soil available P is below 14 kg/ha. P application improves root growth, boll weight and seed oil content. Present day high yielding cultivars respond significantly to P application and the optimum dose ranges from 30-80 kg P_2O_5/ha under rainfed or irrigated conditions depending upon the species/varieties grown (Mannikar, 1993). In most cases 30-40 kg P_2O_5/ha has been found adequate. In the north zone, under cotton-wheat rotation if the preceding wheat crop receives the recommended P dose, based on soil test, the subsequent cotton crop does not require P application (Singh *et. al.*, 1988). The entire does of P must be applied as a basal dose by placement 5-10 cm below the soil i.e. in the moist zone since P has limited mobility in the soil.

Potassium: K regulates water in the plants and the translocation of cellulose during fibre development. The response to K application has been low to negative at many locations due to high status of soil available K. However, the deletion of K results in mining of non-exchangeable K reserves in the soil (Chand and Kapur, 1994). Annual application of K is essential to achieve potential yield levels and for long term sustainability of the system (Venugopalan and Pundarikakshudu, 1999). In most cases 30-40 kg K_2O/ha is adequate. Zone wise fertiliser recommendations are presented in Table 25.5.

TABLE 25.5: Fertiliser recommendations for cotton in different zones

Fertiliser	North zone	Central zone	South zone
Dose	N-60-80 kg/ha. P & K as per soil test Zn-5.6 kg/ha once in two years in cotton wheat rotation.	NPK-40-20-20, 50-25-25, 80-40-40 for varieties. 100-50-50, 160-80-80, 240-120-120 for irrigated hybrids and 90-45-45 for rainfed hybrids	NPK-40-20-20, 60-30-30, 90-45-45 for varieties 100-50-50, 120-60-60 and 150-60-60 for hybrids.
Time/method	Half N at thinning + half at first flowering. Foliar application of 2% urea, if needed. Entire P and K to be drilled at sowing.	N-3 equal splits at thinning, squaring and flowering, P & K at sowing/thinning. Foliar application of 2% urea and/ or DAP at early boll development, if needed. Fertiliser application by ring or column method.	Half N at first square and half at peak flowering Entire P & K at sowing in Karnataka rainfed cotton-entire NPK at planting. Irrigated hybrid cotton-N in 4 splits. Spot application of fertilisers.
Bio-fertilisers	-	Seed treatment with *Azotobacter* and *Azospirillum*	Seed and soil treatment with *Azotobacter* and *Azospirillum* in Tamil Nadu

8.4. Secondary and Micronutrients

S deficiency is increasing in the intensive cotton growing regions. In soils having low S status, response of up too 20 kg S/ha has been observed (Mannikar, 1993).

In India, since cotton is grown on soils with 6.0 to 8.5 pH, the availability of Ca and Mg is not limiting. However, sometimes excess exchangeable Ca may lead to Mg deficiency. When the Mg content in the leaves at flowering stage, falls below 0.3%, reddening of leaves is observed. Mg efficiency can be corrected by soil or foliar application of Mg SO_4.

The response of cotton to Zn application ranges from 0 to 8 q/ha. About 30 to 35 ppm Zn in mature cotton leaves is considered adequate in sandy loams of Punjab (Brar and Sekhan, 1990). Zinc deficiency can be corrected by soil application of 20-25 kg $ZnSO_4$/ha or foliar application of 0.5% $ZnSO_4$. Zinc sulphate has been included in the fertiliser recommendations for cotton in Andhra Pradesh, Punjab and Gujarat.

Boron deficiency has been reported from the Nagarjuna Sagar Canal Project areas in Andhra Pradesh. B deficiency is associated with boll rot in cotton and can be corrected by foliar application of 0.1% boric acid. Soil application of 10 kg borax/ha is recommended for cotton on B deficient soils.

9. VARIETIES

Research carried out under the All India Coordinated Cotton Improvement Project has yielded more than 110 varieties/hybrids which have been released for different parts of the country. A list of the currently recommended varieties is furnished in Table 25.6. A few of them which have made significant impact on the national production or in widening quality spectra need special mention. These are enlisted in Table 25.7.

TABLE 25.6: **TABLE 25.6:** Improved cotton varieties and hybrids of India under cultivation

States	G. hirsutum	G. arboreum	Hybrids
North Zone (NZ)			
Punjab	F 286, F 414, F 505, F 846, F 1378, LH 886, LH 900, LH 1134, F 1084, LH 1556	LD 230, LD 491, LD 327	Fatch (LHH 31), LDH 11
Haryana	H 777, HS 6, H 974, HS 45, H 1098	DS 1, DS 5, HD 107	Dhanlaxmi (HHH 81), Omshankar (CSHH 29) (entire N.Z.)
Rajasthan	G. agethi HS 6, RST 9, B. Nerma, RS 875	RG 8	Raj HH16 (Maru Vikas)
Western U.P.	Vikas	Lohit Shamli	-
Central Zone			
M.P.	Khandwa 3, Vikram, KC 94-2	Maljari	JKHY 1, JKHY 2 (no wilt)
Maharashtra	PKV 081, LRA 5166, CNH 36, LRK 516, LRK 516, Rajat	PA 183, AKH 4, AKA 5, AKA 8401, Y1, Jyoti, Eknath, Rohini	PKHy2, NHH 44, CICR HH1, DCH 32, PKVHy3, NHB 12, PKVHy4
Gujarat	G.Cot 12, G.Cot 14, G.Cot 16, LRK 516, CNH 36	G.Cot 15, G.Cot 11, G.Cot 13	H 8, H 10, DH 7, DH 9
South Zone			
A.P.	LRA 5166, NA 920, LK 861, L 389, Kanchana	Srisailam Mahandi NA 1315	JKHy-1, DCH 32, HB 224, DDH2
Karnataka	Sharda JK 119, Abhadita	G 22, AK 235, Raichur 51, DB 3-12	DCH 32, DHB 105, DDH2, DHH 11, HB 224
Tamil Nadu	ECU 5, MCU 7, MCU 10, LRA 5166, MCJ 5VT, MCU 9, MCU 11, LRK 516, Suvin, Surabhi (Anjali)	K 10, K 11	Savita, Surya, DCH 32, HB 224, TCHB 312, RCH 2

TABLE 25.7: Significant features of some important cotton varieties

Name	Year of release	Yield potential (q/ha)	Ginning percentage (%)	Staple length (mm)	Potential for spinning (HSC)	Recommended state/region
G. hirsutum						
MCU 5	1968	40	34.0	29.5	60s	South zone, Orissa
DHY 286	1975	15(R)	36.0	28.0	50s	Maharashtra
F 414	1976	30	34.2	22.5	30s	Punjab
H 777	1978	30	34.6	22.5	30s	Haryana
G.ageti	1979	25	36.0	23.5	30s	Rajasthan
Sharada	1981	30	38.5	28.0	40s	South zone
LRA 5166	1982	30	35.0	24.5	34s	South & central zone
F 286	1983	25	33.0	22.5	30s	Punjab
RST 9	1992	26	36.0	23.8	30s	Rajasthan
Anjali	1992	30	36.0	25.0	36s	Central zone
G. barbadense						
Suvin	1974	20	34.0	35.5	120s	Tamil Nadu, AP
G. arboreum						
AKH 4	1975	8(R)	39.0	23.0	20s	Maharashtra
AKA 8401	1987	10(R)	38.0	27.0	40s	Maharashtra
RG 8	1988	25	40.0	16.3	-	Rajasthan
LD 327	1987	35	42.0	16.0	-	Punjab
K 10	1985	7(R)	36.0	27.0	30s	Tamil Nadu
G. herbaceum						
G.Cot 11	1979	10(R)	39.0	23.5	24s	Gujarat, Rajasthan
G.Cot 13	1981	10(R)	39.0	23.0	30s	Gujarat
DB 3-12	1983	8(R)	34.0	29	24s	Karnataka
Hybrids						
H 4	1970	40	34.0	29.0	50s	Gujarat, Maharashtra
JKHy1	1976	30	33.0	27.0	40s	MP, AP, Maharashtra
H6	1980	28	35.0	29.0	60s	MP, AP, Maharashtra, Gujarat
H8	1988	30	34.0	26.0	50/60s	Gujarat, Maharashtra
PKV Hy2	1981	25	35.0	25.0	40s	Maharashtra
NHH 44	1983	30	35.0	25.0	30/40s	AP, Maharashtra
Savitha	1987	30	34.0	30.0	60s	Tamil Nadu, AP, Karnataka, Orissa

Name	Year of release	Yield potential (q/ha)	Ginning percentage (%)	Staple length (mm)	Potential for spinning (HSC)	Recommended state/region
DCH 32	1981	30	34.0	33.0	80s	Tamil Nadu, AP, Karnataka
C.Cot DH 7	1985	15(R)	37.0	22.0	24s	Gujarat
C.Cot	1988	15(R)	34.0	28.0	50s	Gujarat

10. COLOUR COTTON

The renewed interest in the research on colour cotton is an off shoot of the growing consumer driven movement. There is a world concern about the growing health hazard associated with the use of textiles dyed with azo-group and other synthetic dyes. In India, brown, linted varieties of Asiatic cotton, Coconada-1 Coconada-2 and Red Northerns were commercially grown during the first half of this country. Besides the germ plasm accessions, wild species are a rich repository of colour linted cotton. A brown linted variety KC 92-2 has been released for commercial cultivation in Madhya Pradesh and another variety DDCC-1 is being released for cultivation in Karnataka.

Profile of colour linted Germplasm

Species	No. of accessions	Lint colour range
G. hirsutum	40	Brown (different shades) and green
G. arboreum	6	Brown, red brown, kakhi and almond
G. barbadense	2	Brown
Wild species	17	Brown, brownish tan, gray, cream

11. TRANSGENIC COTTON

The infestation of cotton pests led to the total vanishing of the crop in many parts of the world. Transgenic cottons with Bt insecticide genes and other biotic stress has been developed in the west and is in cultivation. Bt transgenic cotton minimizes damage to cotton caused by lepidopteran pests. Genetically engineered varieties are also available with tolerance to Glyphosate and Bromoxynil herbicides. These varieties would enable over the top applications of a wider range of herbicides previously unavailable to cotton growers to control noxious weeds. The major advantage is that the insecticide is produced within the plant all the day and accordingly the procuring of pesticides, transporting, mixing, handling and the associated hazards, drift and wash out due to rains can be mimimised under western conditions. The apprehension is that whether non-target pests will flare up, wild crossing with native crops will increase the weediness potential, gene flow, development of immunity to Bt genes in target pests, multinationals hold on seed production, pricing and distribution especially in case of transgenics with terminator genes and its implications and impact on cotton farming dominated by small farms.

12. SOWING

12.1. Time of Sowing

Cotton is sown practically around the year in one region or the other.

North Zone: The optimum time of sowing is the first week of May. Newly developed early and compact plant types can be sown up to the last week of May.

12.2. Central Zone

Irrigated: The optimum sowing time is mid March to mid April in the Deccan canal tract and 15th to 25th May in Vidarbha and Khandesh region of Maharashtra and in Madhya Pradesh.

Rainfed: Rainfed cotton is grown with the onset of monsoon i.e. from the first fortnight of June to first week of July in Khandesh and Vidarbha, second fortnight of July in Marathwada (in Maharashtra) and the first fortnight of July in South Gujarat and in the rainfed herbaceum track of Rajasthan.

12.3. South Zone

Irrigated: The optimum sowing time ranges from 15th April to 15th May in Ghataprabha project area and 15th July to 15th August in the Tungabadra project area of Karnataka. In rice fallow of Andhra Pradesh and Tamil Nadu, December and January-February respectively is the ideal sowing time. In the winter Cambodia tract of Tamil Nadu, mid August to early September is the ideal sowing time, while in summer Cambodia tract of Tamil Nadu mid February to early March is the optimum time.

Rainfed: The optimum sowing period ranges from the third week of June in assured rainfall areas of Karnataka (Dharwar, Belgaum, Chitradurga and Mysore), June-July in Krishna-Godavari deltaic zone of Andhra Pradesh. In black soils of Telangana region of Andhra Pradesh July is the optimum sowing period. Mid September is the optimum sowing time in the black soils of Tamil Nadu, where the north-east monsoon is the main source of water.

12.4. Seed Rate and Spacing

Plant population and geometry varies with the plant type, canopy architecture, soil moisture, soil depth and moisture availability. Nevertheless, optimum plant population is a prerequisite for realising high yields. The recent plant breeding efforts has resulted in the development of short duration, compact, sympodial genotypes. These genotypes respond to a higher population density than the traditional genotypes requiring wider spacing. The present recommendations, based on the work conducted under AICCIP, are summarized in Table 25.8.

Under late sown conditions, yields can be improved by reducing the spacing and increasing the plant population by 20-25%.

The concept of Ultra Narrow Row (UNR) cotton is becoming very popular in the US (Hake et al, 1996). This system facilitates early fruiting and growth control. The UNR cotton is spaced at 19 x 19 cm spacing resulting in 277000 plants/ha. This system saves water, pest control and harvesting costs.

TABLE 25.8: Recommended seed rate and spacing in different zones:

Species	Growing condition	Zone	Seed rate (kg/ha)	Spacing (cm)	Plant population/ha
G. hirsutum	Irrigated	Northern	20-22	75x15	88900
			20-22	67.5 x 15	99000
			16-18	100 x 30	33300
		Southern	10-15	75x30/45	44444/ 29630
	Rainfed	Central	18-20	45x45	49383
			18-20	60x30	55555
		Southern	18-20	60x30	55555
G. abroreum & *G. herbaceum*	Irrigated Rainfed	Northern Central	10-12	60 x 30	55555
			12-15	45 x 30	74074
			12-15	60 x 20	83333
			10-15	60 x 30	55600
G. barbadense	Irrigated	Southern	8-10	90x30	37037
			12-15	75 x 30	44444
Hybrids (Intra and Inter specific)	Irrigated	Northern	2.5-3.0	67.5 x 60	29690
		Southern	2-3	90 x 30/45/60	37037/ 24691/ 18578
		Central	2-3	120 x 40/60	20833/ 13889
				90 x 60/75	18518/ 14815
	Rainfed	Northern	3.0	67.5 x 67.5	21948
		Central	2.5-3.0	150 x 60	11111
			3.0-3.5	90 x 60	18518
		Southern	2.5-3.0	120 x 60	13889

12.5. Method of Planting

In the north cotton zone, cotton is sown using a tractor or a bullock drawn seed drill. In rainfed areas, hybrid cotton is usually hand dibbled using 2-3 seeds/ hill. This system ensures proper depth of seed, uniform geometry, improves, germination and reduces seed requirement. Paired row planting preferred in such zone to facilitate furrow irrigation and for more convenient interculture.

Depth of sowing is crucial to ensure timely and uniform germination. Seeds should be placed 3-4 cm deep in a loose moist soil. However, when dry sowing is done in anticipation of monsoon, seeds must be placed 5-6 cm depth. Thinning is done when the seedlings are 10-15 days old, by removing weak, diseased and off type plants. Gap filling, if necessary should be done 2-3 days after germination.

13. IRRIGATION

About 95% of the cotton grown in the north zone and 40% of the cotton grown in south zone is irrigated, with canals being the main source of water.

A wide variation in the water requirement of cotton ranging from 600-1000 mm has been reported (Table 25.9). The water requirement during seedling to first flower is low ranging from 2.5-3.8 mm/day. The requirement is maximum during peak bloom (8-10 mm/day) and declines steadily till about 4-5.0 mm/day during the boll maturation phase. The most critical period for moisture stress is between the first flower opening and peak boll development and stress during this phase causes shedding of buds and bolls, reduces boll number, boll weight, lint index and fibre strength.

Several methods of scheduling irrigation based on CPE (Cumulative Pan Evaporation), IW/CPE and available soil moisture depletion have been suggested by differed workers (Table 25.10). However, scheduling based on these techniques are location specific and farmers have no access to such information during the current season to decide on the time of application. Moreover, canal water system being a supply driven are, its availability during the critical stages is not guarantied.

In the north zone, cotton is sown after a pre-plant irrigation. The next irrigation is delayed upto 50-60 days to permit root development and also to reduce weed growth. Depending on the soil types 4-5 irrigations are recommended in light sandy soils and 2-4 irrigations on heavier soils. On red sandy loams (Bhavanisagar) 9-12 irrigations, 6-8 irrigations on clay loams of Coimbatore and clayey soils of Siriguppa have been recommended (Bonde 1992).

Surface irrigation by flooding is the common practice in north India. Heavy and frequent irrigations causes salinity and must be discouraged. In Maharashtra, Tamil Nadu, Andhra Pradesh and Karnataka, furrow or alternate furrow irrigation is becoming popular. This system saves 30-50% water besides facilitating early crop establishment which improves yield by 15-70% (Patil and Angadi, 1997).

TABLE 25.9: Water requirement of cotton in different zones

Region/State	Consumptive water use (mm)
North India	710-736
Delhi	1066
Gujarat	965-1143
Madhya Pradesh	660-685
Karnataka	838-889
Andhra Pradesh	770
Tamil Nadu	635-762
Rahuri (Maharashtra)	780-950

TABLE 25.10: Criteria for scheduling irrigation in cotton

Place	Schedule
Sri Ganganagar (Raj)	100 mm CPE
Sri Ganganagar (Raj)	0.8 IW/CPE
Hissar (Haryana)	25% ASMD
Parbhani (Maharashtra)	125 mm CPE
Rahuri (Maharashtra)	75 mm CPE
Coimbatore (Tamil Nadu)	75% ASMD

Place	Schedule
Bhavanisagar (Tamil Nadu)	50% ASMD
Siriguppa (Karnataka)	75% ASMD

Rain water management is an important component of rainfed cotton cultivation. Rainfed cotton often experiences excessive rainfall during vegetative growth phase and terminal drought situation during boll development phase. Soil and water conservation measures like sowing on ridges with 0.2-0.4% gradient, adoption of broad bed and furrow system etc. help to provide surface drainage during the initial period as well as to conserve moisture during the receding phase of monsoon. Application of bio-mulches like *Sesbania* sp. between cotton rows or preparation of soil or dust mulch, light harrowing prevents development of cracks and reduces evaporative losses. Collection of excess rainwater into farm ponds located down stream and recycling it during critical early boll development phase invariably improves seed cotton yield.

14. WEED CONTROL

Trianthema sp., *Echinochola crusgalli*, *Digera arvensis* and *Cyperus* sp, are some important weed species of the north zone (Balyan *et. al.*, 1983) *Celosia argentina*, *Cyanotis axillaris*, *Digitaria sanguinalis*, *Dinebra retrafelxa*, *Euphorbia* sp., *Ipomoea* sp. and *Cyperus* sp. are the dominant weed flora. In the southern zone, *Trianthema portulacustrum* and *Cyanadon* sp. are the major weed species, while on the rice fallow of this zone, *Portulaca oleracea*, *T. portulacustrum* and *E. crusgalli* are dominant. (Solaippan, 1992).

The critical period for weed competition is from sowing to 60 days after sowing, after which the crop canopy covers the inter-spaces. The weed menace is aggravated in rainfed vertisols where incessant rains during the early season of the crop, leaves the soil sticky unworkable, making mechanical weeding impossible. In northern zone, high soil and ambient temperature during sowing season, reduces the efficacy of pre-emergence herbicides.

In the northern zone, interculture through mechanical methods is a usual method to contro weeds. However, a combination involving pre-emergence application of Pendemethelin @ 1.5 L/ha and one hand weeding at 35 DAS is very effective (Prasad, 1994).

In the central and south zones, pre-emergence/pre-plant incorporation of pendemethalin or fluchloralin and trifluralin @1.0-1.5 kg/ha depending upon the soil type is effective. On heavy clays and in soils with higher organic carbon, the herbicide dose would be higher. The above groups of herbicides are very effective against monocots. Against broad-leaved weeds, diuron is very effective. However, due to its poor selectivity, for post-emergence application, only protected spray of duiron is recommended. In the central zone, weed control is accomplished by repeated hoeings using locally fabricated blade harrows, either unidirectional (in case of rectangular planting geometry) or bi-directional (in case of square planting). junior hoe is widely used for interculture in Tamil Nadu.

15. PLANT PROTECTION

15.1. Insect Pests

Although more than 130 species of insect pests are known to attack cotton the list of important insect pests along with the nature of damage are presented in Table 25.11 given below:.

TABLE 25.11: Insect Pests and their damage on cotton crop

Name	Nature of damages
Aphids, *Aphis gossypii* Glover	Nymph and adults pierce the tissue and feed on plant sap from the underside of plant parts. Honeydew secreted favours the development of sooty mould, which reduces photosynthetic efficiency.
Jassids, *Amrasca davastams* (Distant)	Nymphs and adults feed from the under surface of leaves. When damage becomes severe, the leaf edges curl downwards, leaves become red and brown before drying.
Whitefly, *Bemisia tabaci* Genius	Both nymphs and adults feed on cell sap. Chlorotic spots develop on the affected areas on the leaves. Such leaves shed prematurely.
Thrips, *Thrips tabaci*, Lindeman, Frankliniella schultezi (Trybom)	Nymphs and adults rasp the green tissue of leaves from dorasal and ventral sides and feed on the lacerated tissue.
Spotted boll worm, *Earias vittella* Fabricius Spiny boll worm, *E. insulana* Boisdual	Larvae bore into the stem from the terminal bud leading to the death of terminal portion of main shoot. Lateral branches grow out and plants have bushy growth. During Flowering phase, larvae go into the flower buds and feed on their contents, causing shedding of squares. They also damage the young fuits (bolls), causing shedding.
American boll worm, *Helicoverpa armigera* Hubner	Larval damage to squares, flowers and bolls cause their extensive shedding.
Pink boll worm, *Pectinophora gossipiella* Saunders	Larvae enter into flowers. Such flowers do not open properly due to internal webbing of petals, causing characteristics rosette formation of flowers. Larvae enter very young fruits, feed on the seeds, destroying their viability. In northern zone, they also join two seeds and hibernae as larvae.
Spodoptera litura Hubner	Larvae cause damage by feeding on leaves as also on fruiting bodies.

TABLE 25.12: Recommended pesticides in cotton

Pests/Insecticides	Dosage Rate (G.A.I./Ha)	Stage Of The Crop
Jassids/Aphids/Thrips		
Neem products (300 ppm)	2.5 l/ha	Early phase of crop growth
Methyl-o-dominion 25 EC	300 g.a.i/ha	Early phase of crop growth
Phosphamidon 85 WSC	200	-do-
Acephate 75 SP	290	Above 60 days
Whitefly		
Neem products (300 ppm)	2.5 lit	As and when ETL is crossed
Triazophos 40 EC	600-800	-do-
Acephate 75 SP 290	-do-	
Boll worms		

Pests/Insecticides	Dosage Rate (G.A.I./Ha)	Stage Of The Crop
Neem products (300 ppm)	2.5 lit.	40-60 days
Endosulphan 35 EC	875-1050	-do-
Phosalone 35 EC	700-900	-do-
Quinalphos 20 AE	50-700	During fruiting stages
Chloropyriphos 20 EC	500-700	-do-
Carbaryl 50 WP	1000-1250	-do-
Profenofos 50 EC	500	As ovicide
Deltamethrin 1.8 EC	10-12.5 g ai/ha	Above 75 days only once
Alphamethrin 10 EC	15-25 g ai/ha	or twice in the cropping period
Cypermethrin 10 EC	40-60 g ai/ha	
Fenvalerate 20 EC	75-100 g ai/ha	
Spodoptera and others		
Chloropyriphos 20 EC	500	
Quinalphos 20 AF	500	
Diflubenzuron 50 WP	75	For early instars only
Poison bait using monocrotophos 36 SL Seed dresser	250-500	Early and grown up larvae
Imidacloprid 70 WS	5-10 g/kg of seed	Early sucking pests
Carbsulfan 20 SP20 g/kg of seedEarly sucking pests		
Soil insecticides		
Carbofuron 3 G750-do-		
Phorate 10 G1000-do-		

The insecticides should be sprayed as per the recommended doses. Improper selection of insecticides and their dosage, indiscriminate use of insecticide and improper application techniques have contributed to the development of resistance in insects particularly boll worms and whitefly against insecticides. Bio agents like *Trichodramme chilonis* (an egg parasite) and H-NPV (*Helicoerpa* Nuclear Polyhedrosis virus) should be used to manage *Helicoverpa armigera* during the early stage. Synthetic pyrethroids, against whom the resistance level is high in *H. armigera* should be used with caution and not more than twice in the season. They should be alternated with organo-phosphate insecticides.

15.2. Diseases

The cotton crop suffers from a number of diseases which can conveniently be classified into foliar and soil borne diseases. The important foliar diseases are bacterial blight caused by *Xanthomonas axonopodis* cv. *malvacearnum*, leaf spots, caused by *Alternaria macropora*, *Myrothecium roridum* and *Helminthosporium speciforum*. Grey mildew, a foliar disease caused by *Ramularia areola* which was initially restricted to *G. arboreum*, is also affecting *G. hirsutum*. Leaf curl is another devastating foliar disease. The soil borne diseases include root rots caused by *Rhizoctonia solani* and *R. batalicola* and wilts caused by *Fusarium oxysporum* and *Verticillium dahliae*. Root rot and wide spread in the north zone, fusarium wilt in the black soils of Gujarat, Madhya

Pradesh, Maharashtra, Karnataka and Gujrat and Verticillium wilt in Tamil Nadu. Detailed descriptions on various diseases affecting cotton can be obtained from Hillocks 1992. Disease management strategies are presented in Table 25.13.

TABLE 25.13: Disease Management Strategies

Bacterial blight and fungal leaf spots	–	Spray the crop using streptocycline (0.01%) and copper oxychloride (0.25%)
Grey mildew	–	Spray wettable S (0.2%) or carbendazim (0.1%)
Boll rot	–	Spray copper oxychloride (0.25%) or carbendazim (0.1%) along with the recommended insecticide
Root rot	–	Treat seed with carbendazim 2-3 g/kg seed before sowing. Drench the plants in the affected patch with carbendazim (0.1%). Intercropping with moth bean and application of $ZnSO_4$ kg/ha should be recommended where the disease is prevalent. Seed treatment with antagonists like *Trichoderma viride* and *T. harzianum.*

Leaf Curl: This plant disease is now threatening cotton cultivation in north zone. Leaf curl was first noticed in Nigeria in 1912. It is causing considerable damage in cotton fields in north India since 1993.

The disease is caused by 'Gemini virus' which is transmitted by whitefly (*Bemecia tabaci*). The initial symptoms begin on the top newly emerging leaves with the darkening and thickening of leaves. As rains get thickened the leaves either cup upwards or curl downwards. Later leaf shaped enations appear on the lower side of the leaves from the main vein. Plants remain stunted. Excessive shedding of buds and bolls occur.

The *G. arboreum* cottons are not affected by this disease. Growing *G. arboreum* cotton or resistant varieties like RS 875, LRA 5166, LH 144 are some reliable methods to curtail this disease. Removal of alternate hosts like *Hibiscus esculentus*, *Abutilon* sp, *Sida alba*, *Althea rosea* and *Zinnia* sp, should be recommended. The vector *Bemesia tabaci* should be contained by timely spraying to check further spread of the disease.

16. HARVESTING

In India cotton is harvested by picking fully opened bolls by hand in 3-4 pickings. Picking is done when bolls burst fully and kapas hand down. In develop countries cotton is almost exclusively mechanically harvested using spindle type or brush type strippers. Chemical defoliants like Droop 50 wp (Thiadiazarm) @ 100 g/ha or Ethrel @ 5000 ppm can be sprayed at 50% boll opening stage to ensure more uniform boll opening and to hasten maturity. Care should be taken to pick only fully opened bolls without dried bur, leaf bits etc. as clean cotton with little or no trash content fetches higher price.

17. QUALITY

The quality of cotton fibre is basically a genetic character. However factors like climate, soil, moisture and crop production practices strongly influence the quality of fibre. The following quality parameters influence the market value of the produce.

1. **Ginning percentage (also called Ginning Out Turn, GOT):** It is ratio of the weight of lint to that of seed cotton, expressed as percentage. The GOT is low for *G barbadense* 28-30%, intermediate for *G. hirsutum* 33-38% and high for *desi* types (36-42%) Ginned lint is pressed into bales of (170 kg each).
2. **Grade:** The grading of cotton is a composite rating based on the colour (tinge), trash content (non lint foreign materials like leaf bits, broken seeds etc.) and texture of lint. The commercial grades are extra superfine, superfine, fine, good and fair.
3. Fibre Quality parameters include length, fineness, strength and maturity
4. **Fibre length (span length):** The spinnability of a fibre is largely dependent on its length. It is the arithmetic mean of all fibres present in a sample of cotton. Based on the staple length lint is classified into short (20 mm and below), medium (20.5-24.5 mm), medium long (25.0-27.0 mm), long (27.5-32.0 mm) and extra long (32.5 mm and above). Staple length of 25-26 mm is suitable for blending with viscose for medium counts. For finer counts, with viscose, span length of around 28.0 mm is required.
5. **Fibre Fineness;** The fineness is another parameter deciding the spinnable count and quality of yarn. If the same count of yarn is spun from two varieties, the yarn spun from the finer cotton will have a larger number of fibres in its cross section, will be stronger and more uniform than the one span from a coarse variety. Based on the micronaire values fibres are classified into very fine (<3.0 mm), fine (3.0-3.9 mm) average (4.0-4.9 mm), coarse (5.0-5.9 mm) and very coarse (>6.0 mm) Coarse fibre with micronaire values above 5.0 are used for producing rough fabrics, stuffing and for medical purposes.
6. **Fibre strength:** The durability of the textile depends upon the fibre strength. It is expressed in terms of tenacity or the load or force required to break a fibre of unit linear density (g per tex or centi Newtron per tex)
7. **Fibre maturity:** It is an index of the extent of fibre development. A high percentage of immature fibres causes wastage during processing.

18. POST HARVEST TECHNOLOGY

To maintain the quality of fibre for the textile industry and good quality seed for oil extraction, care should be taken to handle the picked cotton. Kapas picked from open bolls should not be contaminated with dried leaves and other foreign matter. Kapas from the last picking should be kept separately for marketing as it is generally of lower grade. It should never be mixed with good kapas. Kapas should be heaped in a clean place either on raised platforms or cemented floor where there is no direct contact with sunlight. Covering the heaps with polythene or tarpaulin with reduce contamination.

Seed cotton sold in the market is taken to ginneries where pre cleaning is done to remove extraneous impurities. Cleaned seed cotton is then ginned either in roller gins or sawgins. Extra long staple cotton can be ginned only in roller gins. However, for short medium and superior medium seed cotton either saw or roller gins are equally efficient. The lint obtained from ginning is pressed into bales and sent for further processing. Cotton seed oil is extracted from the seeds.

19. REFERENCES

Ayers, W. E. 1997. Arkansas Agric Exp. Stn. Bull No. 198.

Balyan, R. S. Bhan. V. M. and Singh S. P. 1983. Chemical and cultural weed control studies in cotton. Tropical Pest Management 29: 567-69

Basset, D. M., Stockton, J. R. and Dickens, W. L. 1970. Root growth of cotton as measured by ^{32}P Uptake. Agron. J. 60: 200-203.

Berkeley, E. E. 1931. Studies on the effect of different length of day with variations of temperature and vegetative growth and reproduction in cotton. Ann. No. Bot Garden. 18, 4

Bonde, W. C. 1992. Achievements in cotton production technology. In. Basu, A. K. and Narayanan, S. S. eds. Achievements of AICCIP (1967-92), CICR, Nagpur, pp 73-97.

Bonde, W. C. and Shanmugam, K. 1990. Cotton production technology for maximizing yields. In Basu, A. K. Mannikar, N. D. and Narayanan, S. S. eds. Cotton Scenario in India ICAR, New Delhi, pp: 74-80.

Chand, M and Kapur, M. L. 1998. Response of cotton and sugarcane to applied K. In. Dev, G. and Sandhu, P. S. eds. Proc. Workshop on the Use of K in Punjab Agriculture pp 73-88.

Fryxell, P. A. 1979. Natural history of cotton tribe (Malvaceae, Tribe Gossypieae) Taxas A & M Univ. Press, College Station, Texas.

Hake, S. J., Kerby, T. A. and Hake, K. D. 1996. Cotton production Manual. Univ. California. Pub. No. 3352. p 330

Hearn, A. B. 1981. Cotton nutrition. Field Crop Abstracts 34(1) 1-33.

Hillocks, R. J. 1992. Cotton Diseases. CAB International, Oxon, UK.

Hutchinson, J. B., Silow, R. A. and Stephens, S. G. 1947. The evolution of *Gossypium* and the differentiation of cultivated species. Oxford Univ. Press, New York.

Mannikar, N. D. 1993. Fertiliser management in cotton. In Tandon, H. L. S. ed. Fertiliser Management in Commercial Crops, FDCO, New Delhi, pp 26-46.

Müller, G. 1968. Cotton- Cultivation and Fertilization. Monograph on Tropical and Sub-tropical Crops, Ruhr-Stickoff Aktiengesellschaft, Bochum, West Germany, pp 25-43.

Nicholson, G. E. 1960. The production, history, uses and relationships of cotton in Ethiopia. Econ. Bot. 14:3-36.

Patil, S. S. and Anagadi, S. S. 1997. Drip Irrigation. Univ of Dharwad, Dharwad, pp 15-9.

Phillips, L. L. 1974. Cotton. In Simmonds ed. Evolution o Crop Plants, Longman Group Ltd. Essex, England, pp 196-200.

Prasad, H. 1994. Effect of herbicides on seed cotton yield, dry matter of weed and yield attributing parameters of American Cotton. J. Cot. Res and Dev. 8(1) 50-4

Raymond, E. F., El-zik, K. M. and Wilson, L. T. 1989. Intergreated Pest Management Systems and Cotton Production. Wiley Interscience, New York

Santhanam, V. and Hutchinson, J. B. 1974. Cotton In: Hutchinson, J. B. ed. Evolutionary Studies in World Crops,. Cambridge Univ. Press. pp 89-100

Sethi, B. L., Sikka, S. M., Dastur, R. H. Gadkari, P. D., Balasubramanyan, R Maheshwari, P., Rangaswamy, N. S. and Joshi, A. B. 1960. Cotton in India-a Monograph, Vol-I, ICCC, 474p.

Singh et. al. 1974. J. Res. PAU. 10:329-83.

Singh, R, Virk, J. S., Brar, A. S. and Tiwana U. S. 1988. Response of cotton to N and P in cotton wheat rotation. J Ind. Soc. Cot. Improvement 13:154-8.

Solaiappan, U., Mani, L. S. and Sharif, N. M. 1992. Weed management in cotton. Ind. J. Agron. 37:878-88.

Suryavanshi, G. B., Khade, K. K. and Tendulkar, A. V. 1992. Progress of agronomic research on agronomic research on irrigated cotton in Western Mahashtra. In Proc. First Vasantrao Naik Memorial Nat. Sem. Agric Sci., Nagput pp 77-90.

Venugopalan, M. V. and Pundarikakshudu, R. 1999. Long term fertiliser experiment in cotton based cropping systems in rainfed vertisol. In Swarup, A., Reddy, D. D. and Prasad, R. N. (Ed) Proc. Nat. Workshop Long Term Fert. Management Through IPNS, PEIL 2-4 1998, IISS, Bhopal, pp 283-291.

26. Jute

V. N. Saraswat

Raw jute in the trade and industry refers to both jute and mesta. The fibre has commercial importance next to cotton and used for manufacture of packaging materials. The major jute products are jute twine, yarn, hessian, sacking, carpet backing etc. Despite tremendous competition from synthetic materials as well as bulk handling system in global market, jute continues to be dominating as vital packaging material for wide range of commodities like food grains, sugar, cement, fertilisers, cotton packaging, salt, vegetables etc. India's share of export of jute goods, particularly in sacking has come down over the years on account of changes in history and geography of the region but India continues to have a vital stake in export of jute goods. Though, demand for sacking has come down in the external market, but there is demand on an increasing scale for sacking material on domestic market on account of increasing production of agriculture products and their export. On export front of jute goods, demand is expected to rise in respect of items like special quality hessian, jute blended yarn, jute decorative fabrics, jute felts, geo-jute and such other diversified jute products. Jute is biodegradable and is renewable source of ligno-cellulose. In the recent years rising concern over ecology and impact on the environment of the use and disposal of synthetic materials has led to new interest in the natural fibre like jute which has many advantages over synthetics and jute has been recognized as environment friendly.

The fibre is obtained from the stem (bark) of two species belonging to Tiliaceae family *viz. Corchorus capsularis* L., known as white jute and *Corchorus olitorius* L. or tossa jute. While mesta fibre is obtained from two species belonging to Malvaceae family viz. *Hibiscus cannabinus and Hibicus sabdarffa* var. *altissima* known as mesta or bimli in India and as kenaf in other producing countries. The fibre obtained from the bark of jute is also known as bast fibre or phloem fibre.

Jute cultivation started in India in the early 19th century. Towards the end of 18 th century, the East India company engaged Mr. Roxburg (a botanist) to identify some fibre crops which can be substitute of hemp fibre for manufacture of packaging materials. Roxburg, the great Botanist, then Superintendent of the Botanic Garden of the East India company at Sibpur near Calcutta, first used the word *jute*. It is probably the anglicized form of the word 'Jhout' or 'Jhot' or 'Jute' used by Oriya's gardeners who were employed in the Botanical Garden. The first commercial mention of the word 'Jute' is in the customs returns of the exports for 1828, when 364 Cwt. of raw jute were sent to Europe. However, Roxburg in search of a suitable substitute for European hemp in India, for manufacture of cordage for company's ships and a cheap material for sacking exported 100 tons of raw jute in the name of 'Pat' in 1793 (Wallace, 1909). By the beginning of the 19th century, the word 'Pat' was however completely superseded by jute in all commercial correspondences. At present, it ranks in importance only next to cotton. Though, jute has faced severe competition from synthetics during last two decades, but it has now reviving its importance and has got a wider prospects in national as well as international perspectives.

FIGURE 26.1: A jute plant (*Corchorus capsularis*).

1. ORIGIN

Royale (1855) is of the opinion that the ancient Greeks used to call a poherb 'Korkboros' from which the generic name of *Corchorus* is derived. The *Corchorus* (kordhoros), still cultivated in neighbourhood of Aleppo and described by traveller in the East as a pot-herb in Egypt and Arabia as well as in Palestine, is supposed to be *Corchorus olitorus*. It is a small and herbaceous plant in dry soils of Syria, a little taller in north India and attains it maximum size only in warm moist climate of Bengal. The occurrence of *C. olitorius* under mid and cultivated conditions in tropical Africa has been reported by Olive (1863) in his 'Flora of Tropical Africa'. Masters (18740 and Prain (1908) recorded it as indigenous to India. Walt (1889, 1908), states that it is indigenous to many parts of India from where it has spread to all tropical countries. From an analysis of the number of species of *Corchours* found all over the world, it is evident that the number is the maximum in Africa, i.e. 36 (according to index kewensis), in tropical Africa alone there are 20 species; the rest are distributed in sub-tropical regions of that country, where as in India only eight species are found. Although wild *olitorius* is found in India, its races so far recorded are less numerous than *Capsularis* in India. Kundu (1951) has discussed in detail all the aspects of the origin of jute and has stated that the primary centre of origin of *C. olitorius* is Africa and secondary centre may be India on the Indo-Burma region.

C. capsularis is not found in Africa or in Australia. It has been regarded by many early botanists as Chinese origin. Walt (1908) was of the opinion that *C. capsularis* was not wild in India, Malaya and China. There are however, some specimens of wild *C. capsularis* collected from such places as Akola, Bombay, Madras, Travancore, Gwalior, Surat, Banda, N.W.F.P. etc. in Calcutta Herbarium. After careful study Kundu (1951) stated that *C. capsularis* is not an immigrant but its centre of origin is Indo-Burma, which according to Vavilov's map includes South China.

2. DISTRIBUTION

The genus *Corchorus* (family: Tiliaceae) includes about 40 species distributed throughout the tropics. The largest number of species under the genius is however, found in Africa. In India, only eight species are found, these being *Corchorus capsularis* linn., *C. olitorius* Linn., *C. trilocularis* Linn., *C. urticaefolius* Linn., *C. fascicularis* Linn., *C. depressus* Christen (*C. abtucgirys* Roeusch), *C. tridens* Linn. and *C. aestuans* Linn. (*C. acutangulus* Linn.), of all these species *C. capsularis* and *C. olitorius* were selected by man in the wake of civilization as economic plants and are commercially important. The other species are found wild in nature (Kundu, 1951).

Jute is principally grown in India and Bangladesh. The other country grow jute are Nepal, China, Taiwan, Thailand, Burma, Peru, Brazil, Kampuchea, South Vietnam and Indonesia. But information on area under jute, as separate from kenaf, rossele and *Abutilon* (grown in China) is often unclear. The Table 26.1 below presents the available information.

TABLE 26.1: Area and production of jute in the world

Countries	Area 1000 ha	Productivity q/ha	Production 1000 MT.
Bangladesh	598	13.89	830
India	1270	11.77	1495
China	270	40.66	1098
Brazil	60	10.54	70
Nepal	46	10.00	4
Burma	68	15.35	88
Thailand	197	12.69	250
Indonesia	4	27.95	12
Peru	4	11.55	4
Vietnam	15	20.00	30

In the post partition years of the country, the jute was mainly grown in Bengal and Assam and scattered cultivation for home consumption was being made in other states. Since most of the efficient jute growing areas went to the then East Pakistan (now Bangladesh), the Indian jute industries had to depend on imported raw jute from Bangladesh. Due to sustained efforts, India has become one of the major producer of raw jute but jute cultivators are always reluctant to invest more on monetary input due to uncertainties associated with the price of jute/mesta in different years. The area and production of jute has been given in Table 26.2.

TABLE 26.2: State-wise area, production and yield of jute

State	Jute					Raw Jute Production	
	Area	%share	Production	%share	Yield	Production	%share
Andhra Pradesh	-	-	-	-	-	530.0	6.0
Assam	103.0	12.8	895.9	11.8	1566.94	10.6	
Bihar	148.0	18.5	1132.5	15.0	1369	1320.8	14.9
Meghalaya	4.8	0.6	37.5	0.5	1406	60.0	0.7
Orissa	35.9	4.5	326.1	4.3	1635	545.2	6.2
Tripura	2.9	0.4	22.0	0.3	1366	79.0	0.9
Uttar Pradesh	4.3	0.5	45.7	0.6	1911	45.7	0.5
West Bengal	502.7	62.7	5102.4	67.5	1827	58.7	
Others	0.2	-	0.6	-	540	129.7	1.5
All India	802.8	100.0	7562.6	100.0	1696	8838.9	100.0

(Area in '000 ha, production '000 bales, yield in kg/ha).

3. CLASSIFICATION

Though the two species from which the jute fibre of commerce is obtained are similar in general appearance, the leaves of *C. olitorius* has a shinning upper surface and a rougher under surface and is are almost tasteless, when chewed. The leaves of *C. capsularis* contains a bitter glucosides corchorin, and taste bitter on chewing, hence it is known as tita (bitter) part where as *C. olitorius* is mitha (sweet) part. There are, however some non-bitter types are available of *C. capsularis*. A detailed description of the two species is presented below for comparison in Table 26.3.

TABLE 26.3: *C. capsularis* vs *C. olitorius*

	C. capsularis	C. olitorius
General	– Plant 5-12 feet tall, can withstand water-logging in later stages; herbaceous, annual, duration 3-5 months depending on the time of sowing	– Plant 5-15 feet tall, cannot normally withstand water-logging; herbaceous, annual, duration 4-5 months depending on the time of sowing; flowers prematurely if sown very early

	C. capsularis	C. olitorius
Stem	– Cylindrical; the colour of the stem varies from full green to dark-red through various intermediate shades; periderm on the basal portion, the extent of which varies at different stages of maturity	– Cylindrical; may be green or coloured light-red or deep red. Shades of colour fewer than in *capsularis*. Periderm either absent or developed poorly; lenticels present in later stages
Branching habit	– Brached or unbranched; auxiliary buds may or may not develop into branches	– Branched, but branches normally develop less vigorously
Leaves	– Glabrous, usually 5-13 cm x 2.6 & 8.2 cm. ovate oblong, accuminate, coarsely toothed, lower most pair of serrations enlarged and end in filiform appendages, petiole 4-8 cm. varies in colour from green to dark-red in different varieties. Stipule usually 0.5-2 cm. or more; foliaceous in some varieties; tip coloured except in the full green types; base coloured or green	– glabrous, ususally 7-18 cm x 4-8 cm. oblong, acuminate pair of serrations more enlarged than in capsularis and filiform appendages longer and very prominent petiole 4-9 cm may be green, light-red or dark-red, Stipule usually 0.5-1.5 cm or slightly more; tip coloured or green; base coloured or green; base coloured except in the full green types.
Flowers	– Small, 0.3-0.5 cm. in length and 0.5-0.6 cm. in bredth, in leaf opposed cymes in groups of 2-5 or more; sepals 5, yellow or pale yellow; stamens 20-30, another coloured yellow or pale yellow. Ovary rounded, 5 carpelled, syncarpous; ovules axile, usually 10 in each loculus in two rows giving about 50 ovules in each ovary; style 2-4 mm; stigma 2-3 fid, pubescent	– In leaf opposed cymes in groups of 2-5, abolut 1 cm. in length, about 2-2.5 times the size of those of capsularis; sepals 5-6, coloured or green; tips prolonged in flower buds; petals 5-6, yellow, entire or split, stamens 30-60, anthers yellow. Ovary elongated, 5, rarely 6 or more carpelled, syncarpous; ovules axile, usually 40 in each loculus in 1 row giving about 200 ovules in each ovary; style 3-5 mm stigma globular, entire pubescent
Anthesis	– 1-2 hours after sunrise	– An hour or less before sunrise
Capsule	– Rounded, 1-1.5 cm in diameter, wrinkled, rarely smooth, muricate and 5 locular. Seeds 7-10 in two rows in each locule without transverse partitions, 35-50 in each fruit	– Elongated, 6-10 cm. long, 0.3-0.8 cm. in diameter, ridged length-wise and 5-6 locules, seeds 25-40 in single row in each loculus, with transverse partitions between each seed, 127-200 in each fruit
Seed	– Small, chocolate brown, 4 to 5 faced, weight about 300 seeds per gram	– Smaller than those of *capsularis*, bluish green to steel-gray or even black in colour. Weight about 500 seeds per gram

The two species different in quality of fibre. The olitorius fibre is finer, stronger, more lustrous than *C. capsularis*. It is either yellowish, reddish or greyish colour depending upon the variety and nature of retting water. The *C. capsularis* fibre is ordinarily whitish and therefore called 'white jute' in trade.

4. CLIMATE

The most important weather factors which determine the efficiency of plant growth, development and fibre yield are temperature, relative humidity and rainfall. Both heat and water must be available to the plant in a regulated fashion, excess or deficit of either or both affect the crop growth and development and ultimate yield. A hot and humid climate is most suitable for growth of jute crop. The crop shows optimum growth when the maximum temperature ranges between 27°C and 40°C. Air humidity requirement for jute range between 65-95%. Annual well distributed rainfall of about 150 cm of which 25 cm precipitates during first 45 days of growth is most suited. Premature flowering mainly observed in *C. olitorius* is found to be due to short day length and low temperature (Rao, 1995).

5. JUTE GROWING REGIONS

Depending upon soil conditions and rainfall pattern jute growing tracts of India may be classified into nine agro-climatic zones as noted below:

1. **Zone-1:** Districts of West Bengal like Murshidabad, Malda Nadia, 24-Parganas, Hooghly, Howrah, Burdwan and Birbhum which are situated on both sides of the Ganges and lower basin on Damodar comprise this zone. These are predominantly *olitorius* belt with moderate to late rainfall. Soil nitrogen is medium to low. Phosphorus and potassium are medium to high.
2. **Zone-2:** Jalpaiguri, Coochbehar, Siliguri and Maldah district of West Bengal cover this zone. The area are situated at both sides of the rivers Teesta and Mahananda, receiving early and very heady rainfall. Soil is acidic, Lime is applied to bring the soil pH between 6.5 to 7.0. Nitrogen is medium to high and deficient in P, K and Mg. *Capsularis* jute is predominant here. Organic matter 4-6 t/ha along with N, P & K are to be applied on soil test and crop response values basis for profitable cultivation of jute.
3. **Zone-3:** The areas are Golpara, Kamrup, Darang, the expending areas of North bank of Brahmaputra and the lower Assam valley receiving heavy to moderate rainfall. Soils are sandy loam in texture and acidic in reaction. Liming is essentially needed. These are generally rich in organic matter and nitrogen but are devoid of bases and organic phosphate. Predominantly *capsularis* jute is grown here. Medium and high lands can be selected for olitorius jute.
4. **Zone-4:** Garo Hills, South Kamrup, Nagaon, expending areas of Sibsagar South of the Brahamputra and areas under Surma Valley come under this zone. Soils are mostly acidic with medium rainfall. These are marked by red laterites and new alluvial clay soils. Application of organic matter and lime are needed. Soils are rich in nitrogen and medium to poor in P, K and Mg. This is predoninantly a *capsularis* belt.
5. **Zone-5:** The areas are Catcher and Tripura. The nature of soils are same as in zone-4 capsularis jute is grown here.
6. **Zone-6:** West Dinajpur, Purnea and Saharsa receiving medium to heavy rainfall comprise this zone. These soils are acidic with nutrient level is low to medium. Liming is essential. Soils are located on the north of the Ganges and on both sides of the Koshi and the Ganges rivers. This is predominantly a *capsularis* belt.

7. **Zone-7:** The area consists of Balasore and Cutack with late and medium rainfall. These are transported yellow and red soils having low to medium in nutrient content. The soils are deltaic alluvium of Mahanadi and its tributaries. This is mainly *Capsularis* area with *olitorius* grown in some portions.
8. **Zone-8:** The area under this zone is the low lying basin of Midnapore district of West Bengal. Annual inundation is general phenomenon of this area. Soil is buried laterite with transported alluvium from Chota Nagpur. This is acidic soils with low N, P and K contents. *Capsularis* Jute is grown here.
9. **Zone-9:** Terai region at Uttar Pradesh along the river Ganga with very low and late rainfall. These are acidic soils with rich organic matter. This is exclusively a *Capsularis* belt.

6. SOIL AND SOIL PREPARATION

The two species of jute i.e. *Corchorus olitorius* and *C. capsularis* have preferences for their land situations and they differ in their requirements for climate and edaphic conditions and also cultural practices to be followed. The olitorius jute prefers well drained, fertile, light textured soils while capsularis jute is grown in heavy textured, moist to wet lands which one annually renovated by deposits of self. A few varieties of olitorius jute recently develop have better adaptability under wide range of situations. Even they could be grown in midland conditions and can tolerate water stagnation at later stages of growth to some extent

However, both the species of jute sown with the pre-monsoon showers and the intercultural practices are completed before the onset of monsoon. However, with the introduction of improved photo-insensitive *olitorius* jute varieties, if has now become possible in grow *olitorius* jute early by supplementing one or two irrigation for facilitating second crop of paddy to plant by first week of August in the multiple cropping system.

A well uniform land with good drainage is best suited for cultivation of jute. Since *capsularis* jute can with stand water logging at later phases of its growth, efforts therefore are made to sow *capsularis* jute early in low lying areas which are subjected to inundated conditions during monsoon, so that it could attain good height by mid of June and may be harvested before heavy rains or flood water accumulates for prolonged period.

A soil pH between 6.0 to 7.5 is best suited for its cultivation. The land is to be well prepared to get a fine tilth. Usually 4-5 ploughing followed by laddering are given. Good land preparation ensures proper and uniform germination and makes the seed drill sowing possible. On less fertile soil, FYM or compost.4-5 tonnes/ha is to be thoroughly mixed one month before sowing this is to be followed by inorganic fertilisers viz., P and K based on STCR values (Soil Test Crop Response Values) are to be applied as basal before sowings.

7. MANURES AND FERTILISERS

Responses to nitrogen has been universally observed, while of phosphate and potassium have been observed in a few isolated cases. The official recommendation till 1965 was 22.4 kg and 45.0 kg N/ha for *Olitorius* and *Capsularis* respectively jute (Pandey and Goswami, 1965).

The uptake of nutrients by jute crop depends on inherent fertility status of soil, fertiliser application, variety of crop grown and agro-climatic conditions. On an average the uptake of major nutrients by jute crop is 40-80 kg N/ha, 20-40 kg P_2O_5/ha and 100-150 kg K_2O/ha. The above figure includes the leaves remain attached at the time of harvest (Dargan, 1971).

The responses of fertilisers varies in different soil types and the cropping system adopted. The low lying areas inundated with floods and soil deposits every year need hardly any fertilisers. Contrary to it in the uplands where soils are light textured, need high fertilization for optimum yields. In fields, where jute is taken after potato which is judiciously fertilized, requires a little or no fertilization for higher yields and most farmers grow jute without any fertilisers. The general requirement for nitrogen varies from 20-60 kg N/ha for *Olitorius* jute and 40 to 80 kg N/ha for *Capsularis* jute in the inorganic form. This may be applied as Ammonium Sulphate, Urea or Calcium Ammonium Nitrate. The responses of fertilisers obtained in multi-location trial in different states of the Indian Union are given in Table 26.4.

The table indicated that application of nitrogen 40 kg/ha or 80 kg/ha in *Capsularis* jute increase the fibre yield significantly at all cases except in low land situations of northern parts of West Bengal, where soils contains medium to high level of nitrogen. The responses of P and K were only seen in acidic soils of Assam. A dose equivalent to 80 kg N, 40 kg P_2O_5 and 80 kg K_2O could be used for higher profits in Assam conditions; while at there place nitrogen at 40 kg/ha or 80 kg/ha is advisable.

Similar trends of increasing yields with increasing levels of nitrogen were seen in *olitorius* jute variety (Table 26.6). JRO 632 at places except in Assam where application of P and K enhanced the fibre yields (Saraswat, 1996). The studies made at JARI during 1953 to 1958 to see the comparative performance of different forms of nitrogen indicated that Ammonium Sulphate was found to be the best. The nitrates i.e, sodium nitrate, potassium nitrate, ammonium sulphate nitrate, chilean nitrate gave lower yield than ammonium sulphate. Potassium Sulphate was better that other nitrates (Dargan, 1971). In separate study different methods of application of nitrogen fertiliser was studied result indicated that top dressing in two halves i.e. at 3 and 5-6 weeks crop age gave significantly higher fibre yield over single application as top dressing or half as basal application and half as top dressing or full dose as basal application.

TABLE 26.4: Fibre yield of Jute (JRC 212) q/ha under different fertiliser combinations (average of 3 years)

Treatments (kg/ha)	Locations						
	Nagaon Assam	Sorbhog Assam	Bud-Bud West Bengal	Cooch-behar West Bangal	Katihar Bihar	Bahraich U.P.	Coimbatore T.N.
Control	22.94	2.51	15.62	15.70	13.11	12.74	4.04
N40	27.23	2.61	26.13	14.72	16.75	19.06	4.59
N80	31.90	3.44	31.10	15.28	19.57	23.30	5.02
N40 P40 K0	29.36	4.97	25.99	17.84	17.80	17.30	5.45

Treatments (kg/ha)	Locations						
	Nagaon Assam	Sorbhog Assam	Bud-Bud West Bengal	Cooch-behar West Bangal	Katihar Bihar	Bahraich U.P.	Coimbatore T.N.
N40 P40 K40	30.99	7.63	26.59	12.96	17.97	19.52	5.74
N40 P40 K80	31.30	7.48	27.47	14.45	16.14	21.19	4.57
N40 P40 K160	27.15	5.96	30.98	13.26	16.20	20.06	5.04
N80 P40 K0	29.62	3.58	34.52	16.72	18.01	23.25	4.37
N80 P40 K40	37.54	5.16	33.02	19.14	17.84	22.85	4.49
N80 P40 K80	37.65	6.58	31.65	18.48	18.48	22.02	4.31
N80 P40 K60	35.62	7.49	31.29	19.38	18.76	22.63	5.08
N80 P80 K80	36.19	7.09	34.82	16.44	20.55	25.58	4.28
S.E.(m)	2.63	1.14	1.65	1.76	1.33	1.13	0.66
C.D. at 5%	7.58	3.30	4.74	-	-	3.24	-

TABLE 26.5: Fibre yield of Jute (JRO-632) q/ha at different centres (average of 3 years)

	Treatment (kg/ha)	Bahraich mean 1972-74	Nagaon mean 1972, 73 & 75	Sorbhog mean 1972, 74 & 75	Katihar mean 1973, 74 & 76	Coimbatore mean 1975-77
T1	Control	17.83	24.44	17.48	17.13	19.14
T2	N30	25.59	26.96	19.62	20.40	21.30
T3	N60	27.67	30.48	20.91	23.63	21.37
T4	N30 P30 K0	26.33	25.75	18.27	21.23	21.72
T5	N30 P30 K30	24.19	26.18	24.16	19.99	20.98
T6	N30 P30 K60	24.63	28.57	23.24	22.38	21.29
T7	N30 P30 K120	25.11	33.90	22.39	20.54	27.60
T8	N60 P30 K0	28.90	28.03	20.92	23.81	21.17
T9	N60 P30 K30	28.77	32.53	21.07	23.62	21.37
T10	N60 P30 K60	29.79	30.84	23.17	21.79	21.39
T11	N60 P30 K120	28.26	33.38	23.59	21.38	21.86
T12	N60 P60 K60	30.24	29.01	23.25	19.85	20.47
C.D. 5%	3.38	1.63	4.19	3.32	0.57	

The studies on application of fertilisers indicated that 10-15 kg of nitrogen per hectare could be supplemented as foliar application. However, the responses vary from place to place and under different ecological conditions. In general *capsularis* jute is more responsive to foliar application of nitrogen than *olitorius* jute. Urea was found to be the best form of nitrogen fertiliser for foliar application. The further study indicated that best time of application is the grand growth period i.e. 6-7 to 10-12 weeks stage. A dose of 10-15 kg N/ha as foliar application could be applied in 2 to 3 installments at 5-6 weeks stage followed by 7-8 weeks and 9-10 weeks stage. Urea could be sprayed in combination with insecticides like endosulfan to protect the crop against insect pests. Urea could be sprayed using hand sprayer (high volume spray) at a concentration of 1-2% or upto 20% concentration with low volume power sprayers (Mitra and Saraswat, 1983).

8. VARIETAL IMPROVEMENT

In early attempts R.S. Finlow, fibre Expert, Dept. of Agriculture, Bengal and his Co-workers initiated work on plant breeding and genetic aspects in 1900. They worked for 25 years to develop better varieties mainly by selection from local types. Their work gave two good varieties viz. Dacca-154 or D-154 (*C. capsularis* L.) and Chinsura Green or C.G, a olitorius type. After the partition of the country the development work was carried out at Jute Agricultural Research Institute, Barrackpore and gave rise to varieties like JRC 212 and JRC 321 of *capsularis* jute and JRO 753 and JRO 620 *olitorius* jute through selections.

The next phase of varietal improvement brought in new recombinants by extensive inter varietal hybridization and mutation breeding to combine desirable characters of the parents like high yields, non branching habits, moderate disease resistance, response to high fertility land etc. were considered. By irraddiating some of the varieties with X- rays, a physical mutagen, a number of new promising varieties have been evolved of which JRC 7447 was released. Gamma rays were also used in getting better varieties and Anobike Tossa or IR-1 was evolved from JRO 623, which did well in certain areas (Saraswat, 1995).

Since genetic base was two narrow, attempts were made to broden it by installing African genes through 'Sudan Green'. This has brought phenomenal change in the cultivation of mainly olitorius jute. The varieties so evolved were resistant to premature flowering and could be sown in midland conditions which was earlier not possible. Some of these premature flowering resistant varieties are JRO 878, JRO 783 and JRO 524. Tetraploides of both *C. capsularis* and *C. olitorius* using the standard and wild strains, have been produced by treatment of seeds and seedings with colchichine. The tetraploides have some advantages. However, the plants were short and the number of fibre strands in the fibre wedge is usually low; they also had greater pollen sterility, low seed set and coarser fibre. Hybridization of the two cultivated species at the tetraploid level or in diploid-tetraploid combinations has not been very successful.

The different varieties which have been released for cultivation have been tabulated below. However, among the *capsularis* varieties JRC 321 is recommended for low land conditions, while JRC 7447 is vary high yielding specially at higher ferrtility conditions. This variety is suitable for upland and mid land conditions. Amongst the *olitorius* varieties, JRO 632 is most suitable for upland condition and around 15th April sowing, if it is sown early it tend to flower prematurely and results in poor fibre yield. The varieties JRO 7435 could be sown early by mid March or late upto May both in midlands and upland conditions. The variety JRO 524 has advantages over JRO 7835 as its peeling of bark is easier than JRO 7435 and yields better under certain conditions.

TABLE 26.6: Promising varieties of white jute (*C. capsularis*)

Name of variety	Pedigree	Year of release	Developing agency	Average yield q/ha	Special Characters
JRC-321 (Sonali)	Selection from 'HEWTI'	1954	C.R.I.J.A.F., Barrackpore	25	Coppery red stem suitable for early sowing and low-lying double cropped areas; good quality fibre.
JRC-212 (Sabuj Sona)	Straight selection from local type	1954	C.R.I.J.A F., Barrackpore	28	Full green plant, with higher levels of N (above 60 kg N/ha), the yield is likely to be about 30-35 q/ha.
JRC-7447 (Shyamali)	X-ray derivative of JRC-212	1971	C.R.I.J.A.F., Barrackpore	29	Full green pigmentation, gives better performance at higher fertility level specially nitrogen.
JRC-4444 (Baldev)	JRC-212 x D 154	1980	C.R.I.J.A.F., Barrackpore	33	Full green, broad and long leaves, higher yield of fibres.
UPC-94 (Reshma)	JRC-321 x JRC-212	1983	Jute Research Station, Bahraich	27	Coppery red stem, green leaves, higher yield, especially suitable for low lying areas.
Hybrid-C (Padma)	JRC-6165 x JRC-412	1983	C.R.I.J.A.F., Barrackpore	28	Stem coppery red, non-branching. Seed bold and chocolate in colour, suitable, suitable for low lying areas of Assam and North Bengal.
KC-1 (Joydev)	Gamma-ray derivative of JRC-4444	1992	Jute Research Station, Kendrapara	36	Higher yield, suitable for Orissa.

TABLE 26.7: Promising varieties of tossa jute (*C. olitorius*)

Name of variety	Pedigree	Year of Release	Developing agency	Average yield (q/ha)	Special characters
JRO-632 Baisakhi Tossa	Selection from local type	1954	CRIJAF, Barrackpore	30	Full green type, suitable for high and midland, sowing time after 15th April
JRO-878 Chaitalo Tossa	JRO-620 x Sudan Green	1967	CRIJAF, Barrackpore	32	Red stem, premature flowering resistant, non-lodging and non-shattering pods, suitable for mid-March sowing and harvest by mid to end July to accommodate paddy.
JRO-7835 Basudev	JRO-632 x Sudan Green	1971	CRIJAF, Barrackpore	34	Full green stem, other characters are same as JRO-878.
JRO-524 Navin	Sudan Green x JRO-632	1977	CRIJAF, Barrackpore	36	Better fibre quality, easier fibre extraction and least susceptible to yellow mite.

Name of variety	Pedigree	Year of Release	Developing agency	Average yield (q/ ha)	Special characters
TJ-40 Mahadev	Seclection from an intermutant cross	1982	B.A.R.C., Trombay	35	Full green, high yield, suitable in jute belt of Orissa.
JRO-3690 Savitri	Tobacco leaf x Long internode	1985	CRIJAF, Barrackpore	30	Fullgreen with long and broad leaves, better quality fibres.
KOM-62 Rebati	Gamma-ray derivative of JRO-878	1992	Jute Research Station, Kendrapara	35	High yield and good quality fibre.

9. SOWING

In jute, higher seed rate is always given to facilitate germination and to maintain proper population. Since many of the jute plants die during early stage of growth due to lack of moisture and adverse soil and climatic conditions. Further, the seed size of the two species of jute differs hence, the requirement of seed also varies to maintain uniform plant populations at the final stage. The seed requirement in case of *capsularis* jute is 10.0 and 7.0 kg/ha for broadcast and line sown crop respectively while for *olitorius* jute, it is 7.0 and 5.0 kg/ha respectively. The line sowing is done with the help of single row or multi-row seed drills. The inter-row spacing is maintained in *olitorius* jute at 25 to 30 cm and in *capsularis* jute it is 30 cm apart. The broadcasts crop is sown by scattering the half seeds along the field and the remaining half across the field to ascertain uniform distribution of the seed all over the field. This is followed by laddering to cover up the seed with soil

10. INTERCULTURAL OPERATIONS

The interculture practices adopted in jute crop are mainly raking in broadcast sown crop while wheel-hoeings in row cropped jute and 2 to 3 manual weeding and thinning to maintain uniform plant population and weed free conditions and to promote tilth and aeration in soil. The further detail are given below:

10.1. Raking

Two raking/harrowing are done in broadcast sown crop by bullock driven implement locally called *bida* which is given along and across the field. The first raking is done when plants are 5-6 cm tall followed by weekly or biweekly intervals. The raking improves the tilth of the soil and at the same time removes some of the weeds and excess plant populations. On an average 20 lakhs to 30 lakhs seeds germinate, while the total requirement for plant population between 3 to 4 lakhs/ha. Therefore, the excess plants are removed or thinned out during weeding operation. The high seed rate is considered necessary in case of jute because of following reasons

1. The germinating seedling is very delicate and often can not singly break through the soil crust that may form when there is rain soon after sowing and therefore requires the help of other seedling for emergence.

2. Until regular monsoon sets in some time in June, the young crop often subjected to long period of drought as the northwester shower on which it depends are irregular; therefore, in the ev ent of mortality of seedling during drought an adequate population is ensured.

The thick stand in the initial stages helps to check the growth of weeds in the field. Later population after racking adjusted along with the weeding operations.

10.2. Wheelhoeing

In row cropped jute, 3-4 wheelhoeings are given. The first hoeing is done when plants are 10-15 days old. It is followed by at weekly or bi-weekly intervals but before 6-7 weeks crop age. Initially wheelhoeings are done with 1-2 times and later with 2-3 times fitted with the wheelhoe. This also much depend upon the inter space between two rows. The wheelhoe uproot lot of germinating annual weeds at the early stages of growth and improve tilth. The wheelhoeing is done when soil is not very hard or wet. The wheelhoes are also given just after top dressing of fertiliser to mix with in the soil.

10.3. Weeding

The first weeding and thinning is done at 3-4 weeks after sowing depending upon the intensity and growth of weeds. If the weed infestation is high, the first weeding is done 2 weeks after sowing and repeated at weekly or biweekly intervals. Usually 2 manual weeding are recommended. The thinning operation i.e. reduction in plant population is done simultaneously so as to adjust the space to 8-10 cm apart in case of broadcast sown crop and 5-7 cm between plant to plant in case of row sown crop at the final thinning. The dense population produces thin and week fiber while under spacing produces course fiber. For optimization of yield, spacing adjustment is very essential right from early stages of growth. The successful chemical weed management mainly depends upon the knowledge of floristic composition of the weeds and edaphic conditions of the field. In case of jute, there is always dominance of grassy weeds followed by sedges, mainly nusedge and a few broad-leaved weeds. For control of annual grasses, application of fluchloralin 1.0 kg/ha as post emergence directed application in row cropped jute. For control of *Cyperus rotundus*, post-emergence application of MSMA 4.5 kg/ha as directed application at 3 weeks cropage is to be applied in row cropped jute. Field infested with both grasses including Cynodan dactylon and Cyperus rotundus are to be applied with tank mixed MSMA and dalapon at 2.25 and 3.0 kg/ha respectively as directed post-emergence application. Field badly infested with weeds some times requires one additional manual weeding between 5-6 weeks stage (Saraswat, 1998).

11. IRRIGATION

For early sown and in the absence of pre-monsoon showers, one presowing irigation after sowing at an interval of 15-20 days are given. Usually 2-3 irrigations for early sown crop is required. Jute germinate at 18-22 % soil moisture and irrigations are required when available moisture of the field has reached to 20%. From farmers point of view, irrigations are given when plants tips droops in the day and recovers in the night. Any set back to crop due to water stress during early stage of growth does affect the final yield and growth recovers with the onset of monsoon.

12. PLANT PROTECTION

Root rot and stem rot are common diseases of jute which are caused by fungus, *Macrophomina phaseolina, Rhizoctonia bataticola*. For controlling these diseases, the seeds may be treated with agrosan G.N. or ceresan 5 g/kg of seed or bavistin 2 g/kg of seed. dichloran + captan 5 g/kg seed or PCNB (quitozene) 5 g/kg is also suggested. Foliar application of Copper oxichlodide (50% Cu) was also found quite effective. However, bavistin (0.1%) spray proved superior to dithane M-45, macupraux and benlate in controlling jute diseases. Maximum control of *M. phaseolina* by seed treatment with benomyl followed by 2 foliar spray of the same compound is also achieved.

Jute is attacked by a number of insect pests mainly, jute stem weevil, jute semilooper, bihar hairy caterpillar, yellow and red mites. Most reliable control method at present is chemical control using pesticides but it has many detrimental effects such as environmental pollution from pesticide residue, health hazards, resurgence of pest, evolution of resistant pest strains, toxic to non-target organisms like beneficiary parasites, honey bee, predators, microbes and earth worms. Three to four sprayings of endosulfan (0.07%) during the cropping season can control major pests. These may be applied at 15-20 days interval starting from 20-30 days crop age. Spraying with kelthane 0.04% is effective for controlling red mite of *capsularis* jute where as endosulfan at 0.04% is effective against yellow mite.

13. HARVESTING

Jute may be harvested any time between 100 and 150 days. Earlier harvest gives lower yield but better quality of fiber. The maximum yield is obtained when harvested at 150 days but fiber is coarser. The ideal harvest is when crop is in small pod stage. Since, now a second crop of paddy is to be adjusted it becomes necessary to harvest by 1st week of August. For optimum fiber yield and of good quality, the crop could be harvested between 120 and 135 days stage.

The plants are cut close to the ground with the sickle. In flooded areas, particularly in North Bengal and Assam and some parts of Tarai areas of Uttar Pradesh, sometimes the workers have to dive under water to cut the plants. In many low lying areas where the crop remains 2-4 feet in water at the time of harvest, plants are pulled out and the root portion are usually cut out before steeping.

In highlands, the harvested plants are left in group at different places in the field for 2-4 days, when most of the leaves dried up. After this period the plants are tied up in bundles of not exceeding 25 cm in diameter and plants are shaken so that all leaves fall on the ground. The bundles are removed for retting to a nearby pool or ditch. The bundles are laid flat in water at least 2-3 feet in depth and arranged side by side so as to form a regular platform, which is usually called 'JAK' or float. The layers of bundles or float sinks partially by its own weight. However, it is covered with a layer of leaves and wooden/concrete blocks are kept to dip the float 2-3 inch below the water surface.

14. POST HARVEST TECHNOLOGY

14.1. Retting

Process of separating fiber from the bark. This is one of the most important Post harvest operation which governs the yield and quality of jute fiber.

During retting, the disintegration of tissues starts from cambium and extends to the ray cells, phloem and cortex and finally liberating the fiber bundles from the wood. In retting, by the time the top portions of the stems are retted the thick lower portions still remain unretted. If lower portion are allowed to ret properly, the top portion become over retted. If lower portion are allowed to ret properly, the top portion become over retted. To ensure a uniform retting of the whole plant, bundles of stem should be placed upright in about 2 feet of water initially for 2-3 days before they are completely steeped in water. The retting process is completed in 10-18 days depending upon several factors such as maturity of the plant, temperature and other water conditions. In the process of retting, the plant tissues absorb water, swell up and release soluble salts of organic substances beside other compounds. The soluble compounds thus released from plants, creates favourable environment for the retting medium for the growth of microbes. There micro-organisms develop and multiply by utilizing free sugars, pectins, hemicellulose and proteins released from the plants. The process generally accelerate at higher temperature. The bacteria becomes active on the surrounding soft tissues and help in exposing fibre bundles. The neighbouring soft tissues are generally phloem, parenchyma, sieve tubes, companion cells, medullary rays and cambium. Retting process is associated with bio-chemical changes due to the activity of the aerobic and anaerobic bacteria, apart from these organisms some fungi are also associated. It is reported that aerobic microbes dominate in first half while anaerobic in later half of the retting.

14.2. Extraction

The end point of retting is judged by feel of fingers and on exhibition of mashiness. A few randomly selected reeds from top and bottom of bundles are taken and visually observed. Fibers after visibly loose and easily detachable on single wash are extracted manually after removing the bundles from tank. They are either extracted single plant basis or a few plants together and break the wood from middle called 'beat-break-jerk' process. The fiber then washed and depth for drying bomboo poles.

15. YIELD

With good management practices and suitable varieties, on an average 30-35 q/ha of dry fiber in case of *olitorius* jute and 25-30 q/ha may be obtained.

16. JUTE SEED PRODUCTION

The bulk of the jute seed is produced in Vidharva region of Maharastra, Karnataka, Andhra Pradesh and Gujarat are transported to different jute growing state for fiber crop before the onset of monsoon in February to March. The productivity of seed in drier states is 5-9 q/ha for *olitorius* jute and 4-6 q/ha in *capsularis* jute, while seed productivity in jute growing areas is only 2-4 q/ha. The crop is sown with the onset of monsoon in these state by late June to last week of July. The seed crop is raised on any type of soil but well drained loamy soils are ideal for higher seed production. Acidic soils are avoided as the crop is subjected to attack of stem rot. Other practice followed are as for fibre crop only while wide spacing of 40-45 cm is used for seed crop. To avoid more unfavourable growth the crop is sown late as to get a short but bushy crop. Sometimes, topping of the apical bud become necessary to make the plant branchy. In years of drought 1/2 irrigation are given to save the crop from drought. Depending upon the fertility of the soil a dose equivalent of 20-40 kg N/ha is top-dressed after the first weeding.

The crop is harvested at full maturity but before the pods starts shattering. The colour of the pods change to dark brown or sometimes to blackish is appearance at maturity. The crop after harvest is kept on clean floor and dried stem are beaten with wooden sticks. The seed then winnowed and cleared. The seed is often treated with organomercuric compounds and stored in gunny bags lined with alkathene cloths. The seed should be kept in dry places as prolong contact of damp air looses the viability.

16.1. Mesta

The term *Kenaf* is applied to *Hibiscus cannabinus* L (HC) and H. sabdariffa L. (HS) is known as roselle, roselle hemp, java jute or Thailand mesta. In India both are in general called mesta. Both the species of *Hibicus* have already been recognized as important source of long vegetable (soft) fibre, closely allied to jute which could supplement the raw fibre supply to mills. HC mesta yields fibre of better quality much similar to jute. The HC mesta is sown is May after jute sowing are over on marginal soils in areas where jute is cultivated. However, HS mesta is grown as rainfed in relatively drier regions of the country i.e. Andhra Pradesh, Maharashtra, Karnataka, Tamil Nadu etc. Since both the mesta are grown on marginal soils with low fertility levels, the weed intensity and growth is also less.

The land is prepared by ploughing the field twice or thrice followed by laddering. The seed is broadcast 18-20 kg/ha on moist soil and laddering is given. One or two combined weeding and thinning given to remove the weeds and adjust plant population at 10 to 12.5 cm apart. A nitrogen dose equivalent to 20-30 kg/ha is applied as top dressing. The crop is harvested at 50% flowering stage in September in case of HC mesta and by middle November to middle of December in case of HS mesta. In both the cases, the fiber yield varies between 15 and 22 q/ha. However, for better yields the crop is to be sown on well pulverized and properly levelled land so that there may not be any stagnation of water. Application of 5 tones of FMI or compost to be applied at least a month before sowing. The crop (HC) may be sown by May while HS in late June to first fortnight of July with the onset of monsoon. To facilitate intercultural operation, the crop should be sown in rows 30 cm either behind the plough or using drill. The thinning should be done when crop is 10-12 cm tall and again after 15-20 days. The weeding operation can be combined with thinning operations. The crop is tolerant to drought but water logging should be avoided as it favours incidence foot rot and stem rot diseases. The crop can be top dressed with 30-40 kg N/ha in two half after and second weeding and thinning. The other practices are same as for jute. A good crop of mesta could yield 25-30 q/ha of raw fibre.

17. REFERENCE

Biswas, S.K. 1998. Raw Jute Production in India. Problems and Prospects. National Level Workshop-cum-Training on jute, mesta, sunnhemp and ramie. CRIJAF, Barrackpore, July 20-24 1998.

Dasrgan, K.S. 1971. Use of fertilisers in Jute Production. ICAR, Tech. Bull. No. 30.

Kundu, B.C., K.C.Basak and P.B. Sarkar 1959, Jute in India. ICJC, Calcutta.

Masters, M.P. 1974. In Hookers Flora of British India, Vol. I: 376-87.

Mitra, P.C. and Saraswat, V.N. 1983. Cultivation of Jute for higher fibre production. Nepal Jute (June-August): 18-21.

Pandey, S.N. and Goswami N.N. 1969. Multiple Cropping can maximize food and fibre production in West Bengal. Ind. Fmg. 17(72): 18-19.

Prain, D. 1908. Vegetation of the districts, Hooghly, Howrah and 24-Parganas. Rec. Bot. Survey India 73.

Royle, J.F. 1885. Fibrous Plants of India.

Saraswat, V.N. 1995. Varietal Improvement Programme on Jute & Allied Fibres (1970-1994). CRIJAF, Barrackpore.

Saraswat, V.N. 1975. Chemical Weed Control in Jute. Ind. Fmg. 24(7): 17-20.

Saraswat, V.N. 1995. Production Technology Development Programme on Jute Allied Fibres (1997-1994). CRIJAF, Barrackpore.

Wallace, D.R. 1909. Romance of Jute.

Walt. G. 1989. Dicitionary of economic products of India. 2:534-62.

Walt. G. 1908. Commercial Products of India, London: 405-427.

SPICES AND MEDICINAL CROPS

27. Fennel

P. S. Rathore and D. S. Bhati

Fennel, *Foeniculum vulgare* Miller is an annual herbaceous crop cultivated in *rabi* season for seeds. Plant belongs to family Umbelliferae. The crop is mainly cultivated for its seeds. The dried green fruits are used for flavouring soups, meat dishes, sauces, pickles and other confectionery. The aroma in seeds is due to presence of volatile oil (0.7-1.2%) which is relatively higher in smaller seeds. The fruits are carminative and considered useful in chest, spleen and kidney diseases. Fennel oil is used in fautile colic and flatulence. The seeds are also used in preparation of allopathic and aryurvedic medicines to cure constipation, dysentery and diarrhea.

1. ORIGIN

Fennel is believed to be originated in Southern Europe and Mediterranean region.

2. DISTRIBUTION

Fennel is cultivated throughout the temperate and sub tropical region. It is mainly cultivated in Rumania, Russia, Germany and France. In India it is cultivated in Gujarat, Rajasthan, Karnataka, Uttar Pradesh and Punjab. In India fennel is cultivated on 16,687 ha. with production of 17,438 tonnes. During 1995-96 India exported 2573 mt amounting Rs. 746 lakh. In India, crop is mainly cultivated in Kaira, Baroda, Ahmedabad and Banas Kantha of Gujarat and Tonk, Sirohi, Jodhpur, Ajmer and Pali districts of Rajasthan.

3. CLASSIFICATION

There are two cultivated species of genus *Foeniculum*.

Foeniculum vulgare Mill: Cultivated in India, Russia, Rumania, Hungry, Germany, Italy, and Argentina. The oil of this species is bitter and 2n chromosome number are 22.

Foeniculum dulce: Mainly grown in France, Italy and South Europe. The oil is sweet and 2n chromosome number is 22.

4. CLIMATE

Raising of fennel, requires cool and dry climate. Crop is very sensitive to frost particularly at flowering. Moderate cool and dry weather from January-March is good for its high yield and quality produce. Heavy cloudy weather as well as high humidity at the flowering and seed formation stages, increase incidence of aphid and Ramularia blight. Rainfall at maturity causes deterioration in quality of fruit, the colour become dull. High velocity wind specially after irrigation at maturity result in heavy lodging.

5. SOIL AND SOIL PREPARATION

Crop can be cultivated on any type of soil except very sandy soils. Well drained sandy loam to loamy soils and black cotton soils are best suited for its cultivation. Crop needs 3-4 ploughings in heavy and 2-3 ploughings in light soils upto a depth of 15-20 cm. Field should be planked after ploughing to have a fine tilth. In order to control termites and other soil borne insects, apply 25 kg endosulfan dust (4%) before last ploughing.

6. CROPPING SYSTEM

Commonly followed cropping systems are mentioned below:

Maize - Fennel	Green gram - Fennel
Paddy - Fennel	Fodder (sorghum) - Fennel
Ground nut - Fennel	As a mixed crop it is grown with chilies.

7. MANURE AND FERTILISERS

Crop should be supplied with 10-15 t of FYM at 15 days prior to sowing. Fennel also respond well to fertilisers. Application of 90 kg N/ha, 1/2 as basal, 1/4 when crop attends 30 days of growth and 1/4 at flowering just after irrigation (Table 27.1) in sandy loam soils (Bhati, 1987). While a dose of 45 kg/ha is found sufficient for heavy soils. Besides N crop should also be supplied with 40 kg P_2O_5/ha through drilling at the time of sowing.

TABLE 27.1: Effect of N on Fennel yield.

Nitrogen levels (kg/ha)	Seed yield (q/ha)
0	9.30
30	10.19
60	11.66
90	13.77
CD (P = 0.05)	0.75

8. VARIETIES

Mostly, local variety is under cultivation in all the states. The varieties of fennel available are S 7-9, PF 35, Guj. Fennel 1, Co 1, Rajendra Sourabha. UF 101 was identified for release in 1995 for Rajasthan. This variety matures in 150-160 days and gives 15-16 q yield. Other entries like, UF (M)-3, UF 90, UF 112 and UF 131 are under advanced testing stage.

9. SOWING

Time of sowing is important to get high yield and quality seeds. The optimum time of sowing of fennel is from mid September to mid October. Delay in sowing reduces yield as well as quality. However crop could be sown upto 15th November.

Seed rate of 10-12 kg/ha and row to row space of 50-60 cm in heavy soils and 45 cm in light soil and plant to plant space of 20 cm is recommended for the crop. The depth of sowing is kept 1.5-2.0 cm. Crop at many places is sown by broadcast, specially when it is sown mixed with chilies. After seedlings of chilies are established, seeds of fennel are broadcast. However, line sowing is better.

9.1. Transplanted Crop

Nursery is raised in June - July. In 100 sqm area 3-4 kg seeds are sown. Seedlings are ready for transplanting in 45-60 days. Seedlings of 50-60 days are transplanted in well prepared field in the evening and irrigation is applied.

9.2. Seed Treatment

Germination is a problem in fennel. Seed soaking for 5 hours improves germination (Patel and Jai Saini, 1964). Seed should also be treated with seed dressing fungicide to protect crop from seed borne diseases. Fungicide bavistin at 2 g/kg seed may be used for treating the seeds of fennel.

10. IRRIGATION

Crop requires frequent irrigations. Crop is raised by giving pre sowing irrigation. A light irrigation is given soon after sowing. Second irrigation is given at 10-15 days after first depending upon soil type and temperature. Subsequent irrigation may be scheduled at an interval of 15-25 days based on weather conditions and soil. Crop is sensitive to moisture stress at flowering seed formation and milk stage. Hence, there should be no moisture stress at these stages. Heavy irrigation should be avoided, as it may cause lodging, especially when high velocity wind prevails in March-April.

11. WEED CONTROL

Weeds grow luxuriantly in fennel because of its initial slow growth and wider row spacing. The most common weeds found in fennel are *Asphodelus teunifolius*, *Anagallis arvensis*, *Chenopodium murale*, *Meillotus alba*, *Sonchus arvensis*, *Spergula arvensis* and parasitic weed *Cuscuta reflexa*. Bhati (1994) noted 70% infestation with dodder in Tonk (Rajasthan). First weeding and hoeing is done at 30 days after sowing. One to two more weedings are required depending upon severity of weeds. Weeds in fennel can also be controlled by using herbicides. Pendimethalin at 1.0 kg/ha as pre-emergence gives satisfactory control of weeds in fennel.

12. EARTHING

Earthing of plants, at the time of weeding and hoeing is found beneficial in fennel. Earthing saves the plants from lodging due to high velocity winds. It also improve quality of fennel fruits.

13. PLANT PROTECTION

13.1. Insect-Pests

Aphids: Aphid (*Hyadaphis coriander*) feed on plant sap and flowers. Affected plants turn yellowish and produce shrivelled seeds of poor quality. It is controlled by spraying 0.03% phosphomidon or monocrotophos or 0.05% endosulfan. Repeat the spray at 10-15 days interval depending upon severity. Spraying should be done in the evening to protect honey bees.

Mites: Mite (*Petrobia lateus*) infest young inflorescence and leaves. Leaves become whitish yellow. In case of prolong attack yield is severely reduced. Quality of seed is hampered. It is controlled by spraying acaricide ethion (50 EC) 0.02%.

13.2. Diseases

Ramularia blight: This disease is caused by *Ramularia foeniculie*. The disease affect lower leaves, form dark brown spots. Flower buds turn yellowish brown and dries up. When infestation occur early, no seed formation takes place, but if infestation occur at later stages under developed seed is produced.

Alternaria blight: Disease is caused by fungi *Alternaria tenius*. The disease mainly infest inflorescence. Flower buds become yellow or brownish and if persist whole inflorescence dries up. Spray 0.1% bavistin or 0.2-0.3% indofil M-45 immediately after appearance of first symptom. Repeat the spray at 10-15 days.

Powdery mildew: This disease result with the infestation of fungi *Erysiphe polygoni*. Disease first appears as white powdery mass on the leaves and later spread on whole plant. Dusting sulphur at 20-25 kg/ha or spraying 0.1% karathane in 800-1000 L. of water controls the disease. Repeat spray after 15 days, if necessary.

Gummosis: Disease is marked by oozing of gummy secretion from inflorescence, resulting into shrinking and drying of inflorescence without production of seed. The cause of disease is still not clear. To prevent the occurrence of disease, use disease free seeds and a follow crop rotation.

14. HARVESTING

Crop reaches maturity in 170-180 days. Crop does not mature at one time. The crop is harvested by plucking umbels 2-3 times when umbels are fully developed, but are still green. Over ripening gives more yield, but quality of fruits is poor. Bhati (1990) obtained higher seed yield when umbels were picked at full length seed and full seed turning yellow. However he obtained more net returns when umbels were harvested at half seed length and green. (Table 27.3). After harvesting, umbels are dried in sun for 1-2 days in shade for 10-12 days. Drying seeds in sun for long time may deteriorates seed quality by making seed colour dull. The grain harvested at half seed length should not be used for sowing as it has poor germination. After drying the seed is are threshed by beating with sticks and cleaned by winnowers.

TABLE 27.2: Effect of time of harvest on fennel yield and net returns

Stage of harvesting	Seed yield q/ha	Net returns Rs./ha
Half length fruit	9.30	15377
Full length yet green	12.26	15035
Fully matureed yellow fruits	12.13	7102
CD (P = 0.05)	0.65	830

15. YIELD

Average yield of fennel is 7-8 q/ha but using improved cultivation technique, one may get 12-15 q/ha.

16. STORAGE

The harvested fruits after cleaning are stored at 8-9% moisture in cool dry storage at 25-28° C. in gunny bags. No insecticide should be mixed directly with seeds, material may be fumigated.

17. POST HARVEST TECHNOLOGY

Quality of spice seed depend on processing. Sun drying result in contamination with dust and others. Mechanical dryers may be used for quality product. Dried seeds are cleaned by cleaning equipment and graded to *Agmark* specification. The composition of Fennel seed is given below:

Composition of Fennel seed (%).

Constituents	Value
Moisture	6.3
Protein	9.5
Fat	10.0
Crude fibre	18.5
Carbohydrate	42.3
Minerals	13.4

18. REFERENCES

Anonymous (1996) Seed Spices in Rajasthan, Rajasthan State Agric Marketing Board. Jaipur.

Bhati, D.S. (1990) Effect of stage of umbel picking and nitrogen fertilization in fennel. (*Foeniculum vulgare*). Indian J. Agron 35 (4): 375-379.

Bhati, D.S. 1994. Dodder (*Cuscuta reflexa* Roxb) a severe parasitic weed of fennel (*Foeniculum vulgare* Mill).

Bhati, M.S. (1987) Effect of nitrogen levels and stages of umbel picking in yield and quality of fennel M.Sc (Ag) Thesis. Sukhadia University, Udaipur.

Peter, K.V. (1998) Spices Research in India. Indian J. Agric. Sci. 68 (8): 527-532.

Pundir, P and Dahama, A.K. (1996) Seed Spices Research in Rajasthan. Publication of Directorate of Research, RAU Bikaner.

28. Fenugreek

D. S. Bhati and P. S. Rathore

Fenugreek or *methi*, *Trigonella foenum-graecum* Linn. occupies a prime position among the seed spices grown in India. It is cultivated mainly for seeds. It is also cultivated as a leafy vegetable or some times as a fodder crop. The seeds are used as a spices and condiments to improve the flavour and the nutritive value of foods. Grains have a strong odour and slight bitterness in taste. Grains are used as a seasoning agent for pickles. Fried seed with a small quantity of oil is used for sesoning vegetables. Beside being an important spice the fenugreek posses an important medicinal properties. There are mucilaginous demulcent diuretic carminative, astringent, emollient and aphrodisiac, therefore seeds are used in preparation of several ayurvedic medicines. Seeds are consumed to increase the appetite, correct digestive system and also to relief pain of joints particularly, in old age. The importance of fenugreek seed has further increased for the important steroids *diosgenin*.

Green tender leaves of young plant form a popular vegetable through out the country. In Rajasthan dried leaves are also used to flavour vegetables.

The specific odour with bitter taste is mainly attributed to olceresin content. Analysis of different fenugreek samples at CFTRI, Mysore indicates that seeds contain 7-11% moisture, 27.7-38.6% crude protein, 3.4-6.8% mineral matter (total ash), 5.2-8.2% petro-ether extract, 16.6-28.8% alcohol extract and 0.2-2.3% acid insoluble ash. It may be noted that fenugreek seeds are rich in protein, oleornesins extraction of which by alcohol is almost three times more efficient than petroleum ether extraction and certain varieties in diosgenin. Other important constituents are the alkaloids (Trigonelline 0.38%) and nicotinic acid (Chopra et.at. 1956).

Methi is widely used in India as a pot-herb. The analysis of leaves for their proximate composition gave the following values (g/100 g edible mater): moisture 86.1 per cent, Protein 4.4 per cent, fat 0.9 per cent, fiber 1,1 per cent, other carbohydrate 6.0 per cent and ash 1.5 per cent. The mineral compounds (mg 100 g edible matter) are: Co 395, Mg 67, P 51, Fe 16.5, Na 76.1, K 31.0, Cu 0.26 S 167, Cl 165 and traces of strontium and have been reported in some samples. The vitamins present in the leaves includes (per 100 g edible matter) are carotene 2340 μg, thiamine 0.04 mg, riboflavin 0.31 mg, nicotinic acid 0.80 mg, and vitamin C 52 mg.

1. ORIGIN AND DISTRIBUTION

Funugreek is considered to be native of south-east Europe and west Asia. It is now cultivated in India, Argentina, Egypt and Mediterranean countries like southern France, Morocco, Algeria, Ethiopia and Labanan. India occupies a prime position among the fenugreek growing countries of the world. It is grown on 29578 ha with production of 30432 tonnes (1993-94). India exported 7910 q of fenugreek worth RS. 1275 lakhs during 1994-95. The area and production is mainly concentrated in the state of Rajasthan, Gujarat and Tamil Nadu though it is also cultivated in Uttar Pradesh, Himachal Pradesh, Madhya Pradesh and Andhra Pradesh. The main fenugreek producing districts of the Rajasthan are Sikar, Jaipur, Chittorgarh, Jhalawar, Alwar, Nagaur and Pali.

2. CLASSIFICATION

Funugreek, *Trigonella foenum-graecum* Linn. belongs to the sub-family Papilionaceae of the family Leguminosae. There are two types of fenugreek i.e., the ordinary methi or desi methi and scented methi, also known as Kasuri or Kasturi methi, Marwari methi or chompa methi. The former consists of the young plants of *T. foenum-graecum* and the latter probably consist of the tender plants of *T. cornciculartor*. The two species differ in their growth habits. The kasuri methi is initially slow growing and remains more or less in a rosette condition during most of the vegetative period. The flowers of desi methi are white and the pods are straight and 6-7 cm long, while kasuri methi bears bright orange-yellow flowers and smaller sickle-shaped pods.

3. CLIMATE

Funugreek is a cold weather crop. It is cultivated as a *rabi* crop and is fairly tolerant to frost. In south India it is also cultivated as a *kharif* crop in moderate or low rainfall areas. Heavy rainfall area are not suitable for its cultivation. At the time of initial growth it requires moderately humid climate and low temperature. But cool dry climate at the time of maturity favours seed yield. Cloudy weather during flowering and fruiting stages increases pest and disease incidence.

4. SOIL AND SOIL PREPARATION

Funugreek can grow successfully on all types of soils ranging from loamy clay to loamy sand. A satisfactory crop of methi can also be grown on black cotton soils (heavy soils) as well as on sandy soils (Singh and Joshi, 1960). As compared to the other legume crops it is fairly tolerant to salinity (Habib et. al., 1971). Fertile, well drained loam or sandy loam soil is best for fenugreek cultivation. A through soil preparation is necessary for good growth. In heavy soils 3-4 whereas, in light soils 2-3 ploughing followed with planking are sufficient for field preparation. There must be optimum soil moisture at the time of field preparation to ensure proper soil tilth. Soil should be treated with 4 per cent enfosulfan or 1.5 per cent qninalphos or 2 per cent methyl parathion dust at the rate of 20-25 kg/ha to control the white ant and other soil born insect pest.

5. MANURE AND FERTILISER

For good yield apply 10 to 15 tonnes of FYM at the time of field preparation. In addition to this application of 15 kg nitrogen and 25 kg P_2O_5 in the form of fertiliser at the time of sowing (Bhati *et.al.*, 1987). Rathore (1989) reported 20 kg N and 50 kg P_2O_5 for getting high yield of crop under sandy loam soils. Bhati (1993) also recommended 40 kg P_2O_5 with optimum irrigation to achieve higher yield.

6. VARIETIES

The important varieties of fenugreek with their characteristic are given below:

Co1: It is a quick growing dual purpose variety and suitable for inter cropping. It is early maturing and mature in about 90 days. It gives 685 kg seed yield or 4-5 tonnes green leaves. It was released during 1982 and developed at Coimbatore (TNAU). It is suitable for Tamil Nadu State.

Rajendra Kanti: It was developed from RM-16 through mass selection. It was developed at Dholi (Rajendra Agricultural University) Bihar and was released in 1988. It is a high yielding variety and gives 1000-1400 kg/ha seed yield. Plants are bushy with more branched and medium height. It has field resistant to leaf spot disease (Edibon and Kallupthrackal, 1991).

R.Mt. 1: Rajasthan methi-1 was developed at SKN College of Agriculture, Jobner (Rajasthan Agricultural University), 1989. This variety was developed through pure line selection from a local collection of Nagaur district of Rajasthan. This variety was previously named "Parbha", now named as Rmt.1. (Bhati and Sharma 1996). It is suitable for Rajasthan and Gujarat and gives 30-35 per cent higher yield than local cultivators. It gives 1600-1800 kg/ha seed yield under favourable conitions with improved agronomic practices yield as high as 2500-2700 kg/ha can be obtained (Sharma and Bhati 1984, Bhati and Sharma 1995). Seeds are fold with typical yellow colour of methi having 11.70-13.30 g test weight. Its plant are tall and moderately branched. The variety is less susceptible to root rot and powdery mildew which are important diseases of fenugreek. It is also resistant to nemotodes.

Lam Sel-I: It was developed at Guntur (APAU) through selection of germplasm. It is suitable for Andra Pradesh. This is a early maturing type and matures in 68 days. It gives 740 kg/ha seed yield. It plants are bushy.

R.Mt. 143: Rajasthan methi-143 is developed from SKN college of Agriculture, Jobner. It has been identified for release by the 14th National Group Meeting of All India Coordinated Research Project on Spices held at Bangalore, 1997 (AICROS, 1998). It is suitable for Rajasthan and gives higher yield than R.Mt.1 (1900-2100 kg/ha).

Pusa early bunching and E.C. 4911 are developed at IARI, New Delhi. These varieties matures in 125 and 165 days, respectively. The department of Agriculture, Maharashtra has also released two important varieties of Desi Methi viz., No. 47 and No. 14, which respectively yield 21 and 18 per cent more than the local type and its leaves are rich in vitamin C (Bose and Soma, 1986) GF1 from Gujarat and CO 2 from Tamil Nadu has also developed, which gives 10-30% higher yield than local variety (Singh, 1993).

The other important varieties are UM144, UM 302 and UM 303 from Jobner (Rajasthan), HM-103, HM-110, HM-114 and HM-141 from Hissar (Haryana), *T. fenu* 58 and *T. fenu* 102 from Coimbetore.

7. SOWING

The optimum sowing time in north India is last week of October to first week of November. Rathore (1989) obtain highest yield from the 15th October sown crop in Rajasthan. Therefore, crop can be sown from 15th October to first week of November. Bhati (1988) reported that there was significant reduction in yield of fenugreek, if sown after first week of November (Table 28.1). Late sown crop is affected more by insect-pest and diseases particularly aphids and powdery mildew. In late sown crop it is likely that maturity coincides with high temperature which cause forced maturity resulting in poor yield. In south India fenugreek can be sown in both the seasons *rabi* and *kharif.* Optimum sowing time for *rabi* crop is first week of October to mid of October and for *kharif* crop mid June to last week of July.

7.1. Seed and Seed Treatment

To obtain proper plant population 20-25 kg seed per hectare is enough for sowing. Before sowing seed should be treated with bavistin or agrosan G.N. or thiram @ 2.0 g per kg seed. Seed treatment with *Rhizobium* culture improve the growth and seed yield of fenugreek (Singh and Joshi 1980), Behari and Joshi (1991) reported that seed treatment with *Rhizobium melilotii* improve nodulation, fresh weight of nodule and seed yield.

7.2. Sowing Methods

Fenugreek can be sown by broadcasting as well as line sowing. The broadcasted seeds should be covered well by the soil with the help of rack and irrigation is given immediately. Sowing is done by *deshi* plough at 20-30 cm row space (Bhati, 1988). Line sowing at 30 cm is better over 20 cm row space or broadcast sowing. Care should be taken that the seed depth does not exceed 5 cm for proper germination of seeds. The field must have adequate moisture at the time of sowing. Depending upon the soil temperature and moisture, germination starts on third to fourth day and completed within a week. A 10-12 cm plant to plant distance is congenial to high seed yield.

8. IRRIGATION

Fenugreek is cultivated as an irrigated crop. The number of irrigation required for a good crop of course depends upon the type of the soil and evaporation prevailing during the season. Studies conducted at the Jobner centre of AICRPS indicated that the good fenugreek crop needs 8 irrigation at about 15 days interval with IW/CPE ratio of 1.0 (Bhati, 1986). The number of irrigation may be less in heavy soils and 4-5 irrigation are sufficient. Distribution of irrigation during the growing season should be such that the crop does not suffer a water stress during pod formation and seed development stage.

9. WEED CONTROL

Two weeding and hoeing, first at 10-15 days after sowing and second 40-50 days after sowing are normally required to keep the crop free from weeds and soil well aerated (Sharma and Bhati, 1981). At the time of first weeding, plant to plant distance of about 10 cm within the rows should by adjusted. In broadcast sown crop also the excess plant should be removed to minimize plant competition.

Weed control experiment conducted at Jobner indicates that pre plant incorporation of fluchloralin @ 0.75 kg/ha with one hand weeding at 50 days after sowing or pre emergence oxyflurofen @ 0.15 kg/ha with hand weeding is suitable to control weeds and was at par with weed free plot (AICRPS, 1995). Maliwal and Gupta, (1989) obtained better weed control with pre emergence application of pendinethalin @ 0.75 kg/ha and recorded higher net returns over manual weeding.

10. PLANT PROTECTION

The is crop attacked by powdery mildew (*Erysiphe polygoni* D.C.) root rot (*Rhizoctonia* sp.) and downy mildew (*Pernospora trigonellae*). The first two diseases occur regularly while the incidence of the third is sporadic in nature and usually occur in the early stage of the crop growth. Beside the disease, the insect-pest, aphid (*Aphis craecora* Koch) attacks the crop and causes heavy losses. To obtain good yield, it is therefore, necessary to protect the crop from these disease and pest. The control measures along with the symptoms of each disease and the pest are describes below:

10.1. Powdery Mildew

The disease caused by (*Erysiphe polygoni* D.C.) the initial stage of this disease is marked by the appearance of whitish powdery mass on the leaves. If not checked at this stage the disease intensifies and the whole plant becomes whitish with powdery coating, The seed size and the yield is adversely affected.

The disease can be effectively controlled by spraying the crop with 500-600 liters/ha of 0.2 per cent solution of wettable sulphur or 0.1 per cent solution of karathan LC. Dusting the crop with 20-25 kg/ha of sulphur dust can also control the disease. The spray or dusting may be repeated after 10-15 days.

10.2. Root Rot

The disease is caused by *Rhizoctonia* sp. and normally appears at the late stage of the crop. Beginning with the wilting and drying up of the leaves and in the last whole plant dries up. The incidence may however, be reduced through adopting proper crop rotation through and deep ploughing in summer. The incidence is severe if the legume crops such as mothbean or greengram in *kharif* is followed by fenugreek in *rabi*. Therefore, fenugreek should not be sown after mothbean or green gram.

10.3. Downy Mildew

It is caused by *Pernospora trigonellae*. The initial stage of the disease is marked by appearance of whitish downy growth on the lower surface of the leaves. As the disease advances the leaves become yellow and fall. The plant becomes stunted and in the severe form, it dies. The disease spread fast if temperature rises. The disease can be effectively controlled by spraying the crop with 0.2% solution of any copper fungicides namely cupramer, fytolan, blue copper disulfatan. The spray should be repeated if the disease symptoms reappears or the control is not complete.

10.4. Aphids

Aphids (*Aphis carecivola* Koch.) cause heavy losses to the crop. It can however, be easily controlled by spraying of 0.03 per cent solution of endosulfan, dimetheol phosphomidon or monocrotophos. The spray should be repeated if insect reappears or cloudy weather prevails.

11. HARVESTING

The harvesting should be done when the crop has turned yellow and most of the leaves excepting the top once, fallen off. The seed will shatter if the harvesting is delayed. The harvested plant are staked in small bundles. After drying in the sun for 4 to 5 days the seeds are separated by beating the plant at the threshing floor. The seeds are cleaned by winnowing and stored. The seed should not contain more than 7-9 per cent moisture at the time of storage.

12. YIELD

Normally seed yield of 1600-1800 kg/ha is obtained. Yield as high as to 3000 kg/ha can be obtain under favourable conditions

13. REFERENCES

AICRPS (1995a) All India Coordinated Spices Improvement Project, Rajasthan Agricultural University campus, Jobner 303 329 19th annual Report 1994-95.

AICRPS (1995b). All India Coordinated Research Project on Spices, Indian Institute of Spices Research (Indian Council of Agricultural Research). Calicut, India, Annual Report 1994-95

Bhati, D.S. (1986). Response of Fenugreek to irrigation and fertilization. International Seminar on Water Management in arid and semi-arid zone pp. 829-834. At Department of Agriculture Meteorology, Haryana Agriculture University, Hissar, India. November 27-29, 1986.

Bhati, D.S. (1987). The review of Agronomy of minor species. VIII Workshop on All India Coordinated species Improvement Project held at Regional Agricultural Research Station. Lam Guntur, India. From 29th January to 1st February 1987.

Bhati, D.S. (1988). Fenugreek (*Trigonella foenun graecum*) response to sowing dates and spacing. Indian J. Agric. Sci. (6): 437-39.

Bhati, D.S. (1989). Agronomical Research in Seed Species in India. An analysis of two states proceedings of First National Seminar on Seed Spicies. Pp. 71-102.

Rathore, P.S. and Manohar, S.S. (1988). Effect of date of sowing, levels of nitrogen and phosphorus in nitrogen and phosphorous uptake by fenugreek. Madras Agric.J. 75(11-12): 432-33.

Rathore, P.S. and Manohar, S.S. (1989). Effect of date of sowing, levels of nitrogen and phosphorus in growth and yield of fenugreek. Madras Agric. J. 76(11): 647-648.

Rathore, P.S. and Manohar, S.S. (1990). Effect of date of sowing, nitrogen and phosphorus on quality and nodulation of fenugreek. Indian Cocoa Areacanut and Spices J. 131(4): 148.

29. Cumin

G. R. CHAUDHARY

Cumin (*Cuminum cyminum* L.) has got an important place in seed spices. Cumin seeds are used for culinaryl as well as for medicinal purposes as a stimulant and carminative. They are stomachache, astringent and useful in conditioning of diarrhea and dyspepsia. The aromatic flavour in the cumin seed is due 3 to presence of a volatile oil, the per cent content of which varies with the varieties, geographical types and climate. The volatile oil content in Indian varieties varies from 2.52 to 3.0 %. The chief ingredient of volatile oil is cumin aldehyde (cuminol) of which quantity varies from 15 to 40 per cent. In Indian varieties, it is found only up to 16 per cent. Besides it, a sizeable quantity of cumin seed is exported earning foreign exchange of Rs. 1424 lakhs (1995-96).

1. ORIGIN

Cumin is considered to be native of either Egypt or Syria or Eastern Mediterranean region.

2. DISTRIBUTION

It is mainly cultivated in India, Iran, Egypt. Morocco, China, Indonesia, Japan, Pakistan, Syria and Turkey. In India, it is cultivated on 420755 hectare with a production of 166524 tonnes. Cumin is mainly cultivated in the States of Rajasthan (50%) and Gujarat (40%) Jalore, Barmer, Jodhpur, Nagaur, Pali, Ajmer, Bhilwara and Tonk are the main cumin producing districts in Rajasthan.

3. CLASSIFICATION

Cumin (*Cuminum Cyminum* L.) belongs to family Umbelliferae. The genera Cuminum has got only one species *Cyminium* syn. *Odorum*, Now, The plant of cumin (*Cuminum cyminum* L.) has been brought under the family Apiaceae.

4. GROWTH STAGES

1. **Germination:** germination (the formation of the embryonic root and the shoot) takes 10 to 18 days to be completed depending on the prevailing temperature.

2. **Branching:** The crop reaches the branching initiation stage in about 30-35 days after sowing.
3. **Flowering:** Flowering starts in about 10 to 15 days after flower bud initiation e. g. 70 to 75 days after sowing.
4. **Grain formation and development:** Grain formation takes place at 80 to 85 days crop growth stage and completed in 25-38 days.
5. **Maturity:** The crop natures in about 120 to 130 days.

5. CLIMATE

Cumin is a winter season crop. Dry and moderately cool climate is best suited for cumin cultivation. Areas with high humidity and frequent rains particularly after the flowering stage are not suitable for cumin crop. High humidity following the flowering stage increases the incidence of diseases like blight and powdery mildew as well as of the aphid attack which drastically reduce the yield and deteriorate quality of the produce.

6. SOIL AND SOIL PREPARATION

A rich sandy loam soil is best suited for cumin cultivation (Jansen, 1981). It can, however, be successfully cultivated on well drained medium heavy as well as moderately saline soils. Cumin is grown in field on which it has not been cultivated over last three years.

The soil should be brought to fine tilth by ploughing the field twice or thrice followed by planking. If soil moisture level is low, a pre sowing "Plewa" irrigation should be given before field preparation. Insecticides such as 25 kg endouslfan 4% per hectare may be incorporated in the soil at the time of field preparation to control termites and other soil harboured insects.

7. CROPPING SYSTEMS

Cumin is generally growth in pure stand (Mono-cropping). The incidence of wilt infection increases if the cumin is grown year after year on the same piece of land. A three year crop rotation i.e. cluster bean-cumin, cluster bean-Wheat, cluster bean-mustard was found to be a most suitable cropping sequence to reduce the incidence of wilt disease in cumin.

8. MANURE AND FERTILISERS

Cumin crop requires no additional manuring if 10 to 15 tonnes per hectare of FYM or compost has been applied to the preceding *kharif* crop. Application of 20 Kg P_2O_5/ha at the time of sowing and 30 kg N/ha through top dressing in the standing crop. This may be applied in single dose or in two equal splits at 30 DAS and remaining at 60 DAS (Bhati *et. al.*, 1987; Chaudhary, 1989).

9. VARIETIES

RZ-19: This variety is suitable for all areas of Rajasthan and Gujarat. The grains of this variety are bold, attractive and dark brown in colour. The variety matures in about 125 days. It is less susceptible to wilt, powdery mildew and blight diseases as compared to local varieties and yield of about 7 to 8 q/ha can easily be obtained by adopting recommended package of practices.

The other improved cultivars of cumin are Guj-cumin-1 suitable for Gujarat and Rajasthan, S-404, Mc-43 and Guj-cumin-2 recommended for cultivation in Gujarat.

10. SOWING

Cumin crop can be sown from 15th November to 5th December. However, for best results sowing should be done around 15th November. Late sowing results in poor yield (Bhati *et. al.* 1984). Cumin is generally sown by broadcast method. The field, after preparation, is divided into beds of convenient size. The seeds are then uniformly broadcast and lightly covered with soil by running the iron rakes in the beds. For convenience of intercultural operations such as, weeding, hoeing, and spraying, the line sowing is better. However, both the methods of sowing does not differ in respect of seed yield (Chaudhary and Gupta, 1991). For line sowing, shallow furrows are drawn in the beds at a distance of 22.5 to 25 cm with the help of iron or wooden hooks. After putting the seeds in the furrows they are covered by lightly running the rakes in the beds. In either method of sowing care must be taken that all the seeds are uniformly covered with soil such that seed depth never exceeds 1.0 cm to ensure proper germination. If properly sown, an optimum plant population can be achieved with a seed rate of 12 to 14 kg/ha.

11. IRRIGATION

First light irrigation is given immediately after sowing to wet the surface soil. The flow of the water in the beds must be slow so that the seeds are not disturbed. Second irrigation should be applied 8 to 10 days after first irrigation when the seeds have started germinating. If the germination is still incomplete and there is crust formation, a light irrigation may be given to facilitate the germination. Subsequent irrigations are given at an interval of 15 to 25 days depending on the climate and soil type. At the time of grain formation, give the last heavy irrigation. At the time of active grain filling period, irrigation should be avoided, as it increases the incidence of powdery mildew and blight diseases and aphid infestation on the crop. Scientifically irrigation in cumin at ratio of 0.6 was found beneficial (Patel et. al. 1991)

12. WEED CONTROL

The common weed flora of cumin field are Chenopodium murale, C. album, Heliotropium ellipticum, H. ovalifolium. Melilotus indica. M. alba, Brassica kaber, Rumex dentatus, Argemone mexicane, Asphodelus tenuifolius, Cynodon dactylon etc. Because of slow growth and short stature, the cumin crop faces severe weed competition. The critical period of crop-weed competition in cumin is between 15 and 30 days of sowing (Mehta *et. al.*, 1985). Two weeding and hoeing are necessary to control the weeds and for proper aeration in the rhizosphere. If one weeding is to be done, then it should be done between 15 and 30 days after sowing. If two hand weeding are to be done, they should be done on 15 and 30 days after sowing. At the time of first weeding thinning should also be done to remove excess plants.

In cumin, Chemical control of weeds has been found to be very effective. Application of pre emergence terbutryn or oxadiazone at the rate of 0.5 to 1.0 kg/ha has been found to be very effective in controlling the common weeds of cumin (Mustafee, 1990). Fluchloralin has been found to be effective against "Jeeri" (*Plantago pumila* wild.) a crop associated serious weed of cumin (Chaudhary and Gupta, 1989). Care must be exercised that there is sufficient moisture in the soil at the time of herbicide application.

13. PLANT PROTECTION

13.1. Diseases

Wilt: It is caused by *Fusarium oxysporum* sp. *cumini*, which may infect the crop at any stage of its growth. In fields the disease generally occur in patches. It is characterized by wilting of affected plants which dry up. It is difficult to control this disease completely but the incidence of the disease can be reduced by adopting the following measures.

1. Use healthy and disease free seeds obtained from wilt free crop.
2. Seeds should be treated with bavistin @ 1.0 g/kg seeds to check the spread of disease in new areas.
3. A three year crop rotation i.e. cluster bean-cumin-cluster bean-wheat-cluster bean-mustard was found to be a better crop sequence to reduce the wilt disease in cumin.
4. Use of Trichoderma at the rate of 4.0 g/kg or solarisation of field during summer also reduces the disease incidence (Lodha, 1995)
5. Use of 30 kg N+20 kg P_2O_5 + 30 kg k_2O/ha reduces wilt incidence in cumin (Champawat and Pathak, 1988).
6. Use of certified/TFL seed of improved varieties.

Blight: It is caused by *Alternaria burnsii*. Continuous cloudy weather after flowering is almost sure to induce the incidence of this disease. the disease appears in the form of brown spots on leaves and stem and tips bent downward. This disease spreads so fast that it is difficult to control after it has appeared. Therefore, only prophylactic measures are recommended for the effective prevention of the disease.

13.1.1. Control

If at the flowering stage cloudy weather prevails and atmospheric humidity increases, spray 400 to 500 litres/ha of 0.1 per cent solution of bavistin or 0.2 per cent solution of any one of indofil M-45, cupramar, fytolan and blue copper. Repeat the spray after 10 to 15 days or even earlier if there is a rain or heavy dew.

Powdery mildew: It is caused by *Erysiphe polygoni*. Disease appears in the form of white powdery mass on the leaves of plants which spreads on whole plant if control measures are not adopted. If the disease infection occurs at an early stage, seeds are not formed. When disease appears late, the seed formation occurs, but they remain small and discoloured.

13.1.2. Control

1. Dust sulphur powder at the rate of 20-25 kg/ha.

2. Spray 400-500 litres/ha of 0.2 per cent soultion of wettable sulphur or 0.1 percent solution of kerathane L. C. and 0.05 percent of calixin.
3. Dusting or spraying should be repeated after 15-20 days. If the spray is done after the stage of flower initiation, it is always advisable to mix the fungicides recommended for blight so that both the diseases can be controlled.

13.2. Insects and Pests

Aphids: Aphids (*Myzus persicae*) in cumin suck the plant sap from tender parts and flowers. The crop turn yellow with the attack of aphids which result in poor and shriveled grain formation. The yield is reduced and quality of grain deteriorates. Though insecticides which can completely control the aphids are available, their use in controlling the aphids must consider the following points-

1. Insecticidal residual standards are strictly regulated in seed spices particularly in the international markets. No insecticides leaving residues, therefore, should be used.
2. Pollination in cumin is brought about by honey bees, hence care must be taken in choosing the insecticide that it does not damage honey bee or at least causes minimum damage. Endosulfan (35 Ec) has been considered a relatively safer insecticide, which can be sprayed at the rate of 0.07 peer cent concentration in the evening hours as the honey bees activity remains low in the evening. However, careful spray of 500-700 litres emulsion of 0.03 per cent malathion (50 per cent) or.03 per cent methyl demeton per hectare effectively controls this insect. The spray should be repeated if necessary.

Frost damage Incidence of frost damage is frequently recorded on cumin particularly in the frost-prone areas. The crop is most vulnerable to frost damage at the flowering and early seed formation stages. Spray of crop with 0.1 per cent solution of sulphuric acid, irrigation prior to the incidence of frost, wind breaks obstructing against the northern cool waves or creating smoke cover in the early morning before sunrise have been found to reduce the frost damage.

14. HARVESTING

Cumin crop matures in about 120 to 130 days depending upon the climatological conditions and varietal characteristics. The crop is ready for harvest when the plants wither and the seeds loses its dark-green colour. Harvesting can be done by uprooting the plants but cutting by stickles is recommended to prevent the contamination of seed with sand or dung which may spoil the quality of the produce. It also prevents contamination of seed with wilt spores which is a soil borne disease. Harvested plants are stacked on the clean threshing floor or tarpolin for drying in the sun. After drying, the grains are separated by light beating with the sticks or rubbing between pieces of lint free gunny bags or tarpolin pieces followed by winnowing.

15. YIELD

The average yield of cumin is around 5 q/ha. However, when improved varieties and the package of practices are properly followed, yield as high as 7-8 q/ha can be easily obtained.

16. QUALITY

The quality of cumin depends on its cleanliness, colour, luster, grain size and aroma. Presence of mould seed or pesticide residues deteriorate the quality and render the seed unfit for consumption. Proper care should be taken in harvesting and post harvesting operations like threshing, winnowing, cleaning, grading and storage. Adulteration with any extraneous material is effectively prevented and colour, luster and volatile oil content of the grains should be preserved till it is marketed.

India finds its world market share for cumin threatened by the recent entry of some other countries like Turkey, Iran and China. Pakistan, Syria and Egypt are other significant exporters of cumin seed and these countries are slowly penetrating our export markets of France, Saudi Arabia, Singapore, Morocco and Sri Lanka. If India has to sustain its growth in export, it has to ensure excellence in quality as per extraneous matter including dust, stones, lumps of earth, chaff, stem or straw should not exceed 7.0 per cent by weight. The proportion of edible seeds other than cumin should not exceed 5.0 per cent by weight. The amount of insect damage should not exceed five per cent by weight. The seeds should be free from added colouring matter.

17. POST HARVEST TECHNOLOGY

17.1. Drying

The seeds should be dried in sun to exclude moisture, The moisture content in seed should not exceed 11 per cent by weight.

17.2. Processing

The dried seeds should be cleaned to get rid off extraneous matter by various cleaning equipments/machines like sieve, eliminator, scalper, cleaner, and gravity separator and should be graded to Agmark specifications.

17.3. Storage

The cleaned seeds are filled in gunny bags and stored in damp free well aerated store room at room temperature (25-28 °C). Moisture level and relative humidity should not be more than 13-19 per cent and 81 per cent, respectively. No insecticide should, under any circumstances, be used directly on cumin seeds. Stored material should be subjected to periodic fumigation for which only authorised persons should be engaged.

17.4. Marketing

Cumin plays an important role in export earning and internal marketing-Export earning by cumin seed earned appreciation as indicated in the 7th Report on Spices of permanent parliamentary committee (1994-95. Kishangarh, Jaipur and Kekri in Rajasthan are known for cumin trade. The traders including exporters used to visit these places for making bulk purchases. After establishment of Unjha Mandi in Gujarat in 1954, this had developed as a major market of cumin and attracted the farmers of Rajasthan and Gujarat for marketing. Now, cumin from Unjha Mandi is exported via Mumbai to several other countries. The main markets for cumin are Japan, U.S.A., U.K., Yemen Arab Republic, Canada, Singapore, Saudi Arabia and U. A. E. The marketing season of cumin starts from February to June.

18. CONSTRAINTS OF PRODUCTION

The productivity of cumin is low. Some of the important reasons of low productivity are-

1. Cumin is traditionally grown on marginal lands with low fertility.
2. Lack of sufficient number of improved varieties to suit different agro-climatic situations.
3. Lack of proper adoption of improved package of practices and control measures for diseases and pests.
4. Inherent characteristics of the plant e.g., poor and slow germination resulting in poor plant stand.
5. Slow initial growth of the crop resulting into severe crop-weed competition.
6. High incidence of diseases as wilt, blight and powdery mildew.
7. Non-existence of regulated market facilities for cumin and lack of quality awareness in post harvest operations due to which growers receive poor prices.

19. REFERENCES

Bhati, D. S., Jain, M. P., Sharma, R. K. Bhatti, B. N. and Chaudhary, G. R. 1984. Cumin cultivation in Rajasthan. Indian Cocoa, Arecanut and Spices Journal VIII (1): 8-9.

Bhati, D. S., Mali, A. L. and Rathore, P. S. 1987. Response of cumin (*Cuminum cyminum*) to nitrogen and phosphorus. Indian J. Agron. 32 (2): 197-98.

Champawat, R. S. and Pathak, V. N. 1988. Role of nitrogen, Phosphorus and potassium fertilisers and organic amendments in cumin (*Cuminum cyminum*) with incident by *Fusarium oxysporum* f. Sp. Cumin. Indian Journal of Agricultural sciences 58 (9): 728-730.

Chaudhary, G. R. 1989. Effect of sowing methods, nitrogen level and weed control on competition nutrient uptake and quality of cumin. Indian J. Agric. Sci. 59 (6): 397-399.

Chaudhary, G. R. and Gupta, O. P. 1989. Removal of nutrients by weeds and its effect on grain yield of cumin grown with different weed management practices. Indian J. Agron. 34 (4): 396-398.

Chaudhary, G. R. and Gupta, O. P. 1991. Response of cumin (*Cuminum cyminum*) to nitrogen application, weed control and sowing methods. Indian J. Agron. 36: 212-216.

Jansen, P. C. M. 1981. Spices, condiments and Medicinal plants in Ethiopia, Their Taxonomy and Agricultural Significance. Centre for Agricultural publishing and Documentation, Wageningen.

Lodha, S. 1995. Soil solerization, summer irrigation and amendments for the control of *Fusarium oxysporum* f. sp. cumini and *Macrophomina phaseoline* in arid soils. Crop Protection 14 (3): 215-219.

Mehta, H. M., Trivedi, G. C. and Patel, H. R. 1985. Studies on the Crop-weed competition in cumin (*Cuminum cyminum* L.). Ann. Conf. ISWS (Abstr.) GAU, Anand, pp. 88.

Mustafee, T. P. 1990. Effective methods for chemical weeding in close cropping system. Pesticides 24 (1): 11-15.

Pruthi, I. J., Jain, R. K. and Gupta, D. C. 1987. A note on reaction of some spices to root knot nematode (*Meloidogyne javanica*). Haryana Journal of Horticulture science 16: 1-2, 154-155.

30. Coriander

D. S. Bhati and P. S. Rathore

Coriander is one of the important seed spices, mainly cultivated for its fruits. The fruit posses a fragrant odour and pleasant aromatic taste. The aromatic odour and taste of coriander fruits is due to an essential oil, which is made up of hydrocarbons and oxygenated compounds. The major oxygenated compound present is corianderol ($C_{10}H_{17}OH$) a terpene tiartiary alcohol. Besides the essential oil, the seed contains 16.1% fatty oil, 14.1% protein, 21.6% carbohydrate, 32.6% fibre, 11.2% moisture and 4.4% mineral matter.

The essential oil is used chiefly as a flavouring agent for spirituous liquors and in coco and chocolate industries. It is also used in perfumes, soaps, candy, tobacco, meat baked foods, canned foods and to mask offensive odours in pharmaceutical preparations.

The stem, leaves and fruits have a pleasant aromatic odour. The entire plant, when young is used in preparing chutneys and sauces, the leaves are used for flavouring curries and soups. The fruit is extensively used for as condiment, pickling sausages and in seasonings. They are also used for flavouring pastry and cookies. Besides, condiments, coriander also have medicinal value. Fruits are used against stomach ache. The dry fruits are said to have carminative, diuretic tonic. The seeds are also chewed to correct foul breath.

1. ORIGIN AND DISTRIBUTION

Coriander is a native of Mediterranean and near east regions. It is one of the earliest known species to the mankind. It is mainly a crop of tropics and extensively grown in India, Russia, Morocco, Rumania, France, Spain, Italy, Holland, Yugoslavia, Burma, Pakistan, Turkey, Mexico, Argentiana and to some extent in England and USA (Seetharaman and Arjunan, 1979).

In India it is grown throughout the country. The main coriander growing states are Rajasthan, Andhra Pradesh, Madhya Pradesh, Tamil Nadu, Bihar, Karnataka and Uttar Pradesh. Total area under cultivation in India is 4.75 lakh hectares with production of 2.50 lakh mt (1998). During 1997-98 India exported 20 thousand tonnes of coriander seed and earned Rs 5.4 crores of foreign exchange.

2. CLASSIFICATION

The genus *Coriandrum* has about six species. Out of six, only *Coriandrum sativum* commonly known as dhania or coriander is cultivated. Coriander is broadly divided into two types according to size of fruit.

1. Large fruit type-Diameter 3-5 mm. It is mainly grown in tropical and subtropical countries like Morocoo and India, has low volatile oil content (0.10-0.35%).
2. Small fruit type-Diameter 1.5-3.0 mm, grown in temperate regions particularly in Russia. It usually have high volatile oil content (>0.45%)

3. CLIMATE

The time of sowing differ from states to states. In Rajasthan, Gujarat, Bengal and Uttar pradesh the seeds are sown during winter. In Maharashtra it is sown during monsoon and in Tamil Nadu in autumn, while in Karnataka and certain parts of Tamil Nadu it is grown in two seasons viz. May-August and October-January. However, major area of coriander is under *rabi* season.

Coriander is a tropical crop and can be successfully cultivated as a *rabi* crop in area free from severe frost during the month is February-March, when the crop is in flowering and seed setting stage. A dry and cold weather favour high seed production (Bhati, 1989). Cloudy weather during flowering and fruiting stage increases pests and disease incidence. Coriander is generally grown as a rainfed crop either pure or mixed with other crops. It is also grown as an irrigated crop.

4. SOIL AND SOIL PREPARATION

Irrigated crop of coriander can be cultivated on almost all soil types provided sufficient organic manure is applied. Unirrigated crop may be cultivated only on the heavy soils, having good water retention capacity. The preparatory tillage on heavy soils must be started as soon as soil moisture after *Kharif* season is optimum.

One ploughing with soil turning plough followed by 2-3 ploughing with deshi plough or cultivator are normally sufficient to bring the soil to a fine tilth, required for coriander sowing. To avoid loss of soil moisture and to break the clods, the field should be planked immediately after ploughing. In irrigated coriander, if soil moisture is not sufficient preparatory tillage should be carried out after presowing irrigation. On light soils 2-3 ploughings are sufficient for sowing coriander. To control while ant and other soil borne insects mix 4% endosulfan dust or 1.5% methyl parathion @ 20-25 kg/ha.

5. MANURES AND FERTILISERS

To obtain good crop of coriander mix 10-15 tonnes of well decomposed FYM or compost once in 2-3 years. In addition to this 30 kg nitrogen, 30 kg P_2O_5 and 20 kg K_2O/ha are also recommended to coriander under unirrigated conditions. For irrigated crop, amount of nitrogen should be increased to 60 kg/ha (Bhati, 1988 and Bhati, 1991). Full quantity of phosphatic and potassic fertilisers and 1/3 nitrogen should be applied as basal. Remaining nitrogen is applied in two equal doses, the first about one month after sowing and second at flower initiation

(Bhati, 1989 b) with irrigation. Table 30.1.

TABLE 30.1: Effect of nitrogen on coriander yield

Nitrogen kg/ha	Umbels/Plant	Seeds/umbel	Seed yield kg/ha
0	19.8	29.4	516
30	25.1	30.7	622
60	27.6	30.0	690
90	27.0	31.7	706
CD at 5%	1.5	1.8	34

6. VARIETIES

Improved varieties along with their specific characteristics are given in Table 30.2.

TABLE 30.2: Improved coriander varieties

Variety	Special characters	Crop duration (maturity group)	Yield per ha (kg)
Gujarat Coriander-1	High yield, more number of branches: seeds bolder and greenish in colour	112 days Medium	1100 kg
Gujarat Coriander-2	High yield; more secondary branches, dense foliage, fistular stem, umbels large size, grain purpose variety, bold seeds, no lodging	110-115 days Medium	1500 kg
Rajendra Swati	High yield potential, medium height plant; suitable for intercropping; fine seeded rich in essential oil, resistant to stem gall disease, free from attack of fruit fly	100 days Medium	1200-1400 kg
Sadhana (1989)	High yield; suitable for rainfed areas, semi erect type, resistant to aphids and mites; 0.2% oil	95-105 days Medium	1025 kg
Co-3 (1991).	High yield, dual purpose, less susceptible to wilt and grain mould, medium size grain, brownish yellow colour, 0.4% oil	90 days Medium	640 kg
CS-287	Short duration; small, bold grains, tolerant to wilt and grain mould suited for rainfed tracts	78 days Early	510 kg
Lam Sel. CS-2	Medium tall, bushy type more number of branches grain purpose variety, tolerant to pests and diseases, good quality grain 0.4% oil.	110 days Medium	1300 kg
*UD-20	Medium tall plant: bold grains	100-110 days Medium	1175 kg
Co-1	Tall plant, many umbel per plant, suitable for green leaves and grains	110 days Medium	500 kg.
Co-2	High yield, dual purpose, tolerant to drought	90-100 days Medium	1,500 kg.
Rcr-41	High yield, tall erect, round and small seed, 0.25% oil and tolerant to wilt	130-140 days late	1,200 kg.
Swathi	High yield, semi erect, suitable for delayed sowing,.	80-90 days early	885 kg.

7. SOWING

7.1. Sowing Time

Coriander crop is taken once in a year in North India whereas twice in South India. Depending upon the day temperature, optimum sowing time in North India is between 15th October and 15th November, when day temperature has come down below 25°C which does not affect the germination adversely (Bhati, 1989). Best sowing time for higher yield with best quality seed is between last week of October to first week of November (Bhati, 1991; Table 30.3). Early sowing is congenial for better growth; but the high temperature at sowing time may reduce the plant stand by killing the germinating seed crop. The early sown crop may come to flowering stage earlier making the crop more vulnerable to frost damage. Late sowing though affects the crop growth and its yield adversely, it is safer in areas of high frost incidence. Late sown crop is also affected more by insect pest, particularly aphids (Jain and Yadava, 1986) and diseases. In South India first crop is sown in May and second in October (Kumar, 1977). The October sown crop gives better yield along with good quality seeds.

TABLE 30.3: Effect of sowing date on seed yield and essential oil content in coriander seed

Date of sowing	Seed yield kg/ha (Pooled.)	Essential oil content (%) Pooled.
31 October	1711	0.2155
10 November	1561	0.2162
20 November	1361	0.2076
CD AT 5%	52	0.0064

8. SEED RATE AND SOWING

Depending upon seed size and soil moisture availability at sowing, 10-15 kg seeds/ha are sufficient for sowing seed should be splited in two by rubbing (Bhati, 1987 and Bhati, 1994) and treated with bavistin or agrosan GN or thiram @ 2 g/kg seeds.

Seeds should be sown in rows at 30 cm row spacing (Bhati, 1988 and Bhati 1991) at 4-5 cm depth under rainfed conditions and 1.5 -2.0 cm depth under irrigated conditions. In irrigated coriander, dry sowing followed by light irrigation is preferred for better germination. A second light irrigation should be given after 8-10 days to ensures proper germination.

9. IRRIGATION

For irrigated crop, depending upon temperature and soil type 3-4 irrigations are required in addition to those given for germination. The first irrigation is given after 35-40 days from the date of sowing, second after 60-70 days and third 80-110 days when seed formation has started. Care should be taken that adequate moisture is available to crop after flower initiation, particularly at the time of seed formation.

Malpani (1992) obtained 2.29 q grain when crop was irrigated at 30 mm CPE while in crop irrigated at 90 mm CPE only 1.55 q yield was recorded.

10. WEED CONTROL

In unirrigated crop normally there is need of intercultural operations. But if there is an early rain in the standing crop hoeing and weeding should be done to remove the weeds and to provide better soil condition.

In irrigated crop depending upon weed problem and duration of a variety two to three weeding cum hoeing operations are required, the first 30-40 days after sowing and later at intervals of 25 to 40 days. At the time of first weeding and hoeing, thinning should be done to maintain the plant spacings of 10-15 cm with in the rows.

Weeds can be controlled with herbicides where labours are costly and scares. Application of Pendimethalin @ 1.0-1.5 kg/ha as preemergence or fluchloralin @ 1.0-1.5 kg/ha as preplant soil incorporation keep weed free for first 40-50 days (Gupta, 1998). Oxyflurofen @ 150 ml/ha can also be used to control weeds in coriander. Malpani (1992) obtained higher yield by using metalochlor @ 2 kg/ha as pre-emergence (2.39) over weedy check (1.17 q/ha). For effective and economic weed control, herbicide along with one weeding cum hoeing should be done at 40-50 days after sowing (AICRPS, 1995).

11. PLANT PROTECTION

To obtain good yield with better quality, seed crop should be protected from diseases, pests and frost.

11.1. Diseases

Wilt, powdery mildew, stem gall and blight are the important diseases of coriander which adversely affect the yield and seed quality.

Wilt: The disease infest the plant at any stage of crop growth. As the wilt infection occurs in the roots, it is very difficult to control this disease. Though direct control measures are not available, the following preventive steps help in reducing wilt incidence.

1. Deed summer ploughing
2. Do not grow coriander for 2-3 years in wilt infested field.
3. Use disease free seeds obtained from the wilt free crop in the preceding year.
4. Grow wilt tolerant varieties like Rcr 41 and CS 287 in wilt infested fields.
5. Before sowing, treat the seeds with 1.5 g bavistin or thiram/kg seed.

Powdery mildew: The initial stage of this disease is marked by the appearance of whitish powdery mass on the leaves and stem. If unchecked, the disease gets intensified and the whole plant becomes almost white with the powdery coating. Infection in the early stage normally does not permit seed formation. However, if the infection occurs at a late stage, the seed formation may occur but the seeds will be small and shriveled, thus affecting the yield and quality of the produce.

To control powdery mildew spray wettable sulphur @ 2 g/L water or karathane LC 1 mL/L water. Dusting the crop with 20-25 kg of sulphur/ha would also control this disease. Spray or dusting may be repeated after 10-15 days if required. Stem gall: This disease is caused by *Protomyces macrosporus*. Disease almost appear regularly. The spread is more under high moisture conditions. It is very difficult to control this disease. Rcr 41 variety is found to be highly resistant to this disease. Use of disease free seeds limits its spread.

Blight: With the appearance of the disease, dark brown colour spots appear on the stem and leaves. To control blight, spray blitox 1 g or mancozeb 2.0 g/L.

11.2. Insect-Pest

The major insect-pest affecting coriander is aphids. Besides, this mite and cut worm are the other important pests.

Aphids: The attack of aphids usually occur after flower initiation. This insect though can be easily controlled by a number of insecticides, but their use at flowering may kill the pollinating bees. The problem may however partially solved by spraying the crop in evening when bee population is low. endosulfan is safer for bees as compared to other insecticides. Besides this, phosphomidon 85 LC 0.3 mL or malathion 50 EC or monocrotophos 36 LC 1.0 mL/L water can also be used. Spray may be repeated if necessary (Jain, 1984).

Might: This insect suck the cellsap. In area where mite appear regularly, crop should be sown during last week of October to first week of November. It can be controlled by measures described for aphids.

Frost: Frost damages crop frequently in North India, particularly under unirrigated conditions. Burning of waste material is useful in protecting the crop. Spraying 0.1% sulphuric acid at flowering protect the crop from frost.

12. HARVESTING AND THRESHING

Stage of harvesting and handling of harvested crop affect the yield as well as the essential oil content. Over ripening results in deterioration of quality, particularly colour of the fruits. The crop must be harvested when the seeds on the main umbel turn slightly brown (Bhati, 1989 c). Crop is harvested by cutting with sickles. If possible harvested crop should be stocked in shade to avoid loss of volatile oil content. Grains are separated by beating and cleaned by winnowing and dried properly before storage.

13. YIELD

Yield of irrigated crop varies from 10 to 15 q/ha where as unirrigated crop gives only 4 to 5 q/ha. Under favourable conditions with good management yield can be recorded as high as 20 q/ha under unirrigated and 27 q/ha under irrigated conditions.

14. REFERENCES

AICRPS (1995). All India Co-ordinated Research project on spices, Indian institute of species Research (Indian council of Agricultural Research), calicut 673012, Kerala, India, Annual Report 1994-95 pp 59.

Aiyadurai, S.G. (1966). A review of Research on spices and cashewnut in India. ERNAKULAM-6, Indian council of Agric. Research.

Bhati. D.S. (1987). The review of Agronomy of minor spice, VIII workshop on All India co-ordinated spices improvement project, held at Regional Agricultural Research station, Lam (A.P.A.Y.) 522-034 (A.P.), 29th Jan. to 1st Feb. 1987.

Bhati, D.S. (1988). Effect of nitrogen application and row spacing on coriander (*Coriandrum sativum*) production under irrigated condition in semi-arid Rajasthan. Indian Journal of Agri. Sci. 58 (7): 568-569.

Bhati, D.S. (1989 a). Agronomical Research in seed spices in India An analysis of the status. Proceeding of first National seminar on seed spices pp. 71-102.

Bhati D.S. 1989 b). Study on N application schedule in irrigated coriander (*Coriandrum sativum* L.) first National seminar on seed.

Bhati, D.S. (1898 C). Effect of harvesting time on the quality and productivity of coriander (*Coriandrum sativum*). First National seminar on seed spices, October 24-25, 1989; Held at ARS, Durgapora (RAU) Jaipur, pp. 19-20.

Bhati, D.S. Rathore, P.S. and Mali, A.L. (1989) Repons of coriander to soil and foliar application of sulphur faming systems (1&2): 49-51.

Bhati D.S. (1991). Effect of sowing date, row spacing and nitrogen on growth, yield and quality of coriander (*Coriandrum stivum* L.) Ph. D. Thesis, Department of Agronomy, Rajasthan Agricultural University. Bikaner (India).

Bhati, D.S. (1988 b) Effect of leaf plucking on growth yield and economics of coriander varieties under semi arid conditions. Indian J. Agron. 33 (3): 242-244.

Bhati, D.S. (1994). Coriander (*Coriandrum sativum*) response to seed splitting and seed rate. Haryana J. Agronomy 10 (1): 22-24.

Chaudhary, A.R. (1961). Receptivity of Stigma viability of pollen and opening of flowers in *Coriandrum sativum* and *Foeniculum vulgare* Agric. Pak. 12; 29-41.

Gupta, O.P. (1998). Modern weed management. Agro Botanica, Bikaner. pp. 327

Jain, P.C. (1984). Incidence of pests and their control on coriander (*Coriandrum sativum* L.) Ph. D. Thesis Sukhadia University, Udaipur, India.

Jain P.C. and Yadava, C.P.S. (1986). Effect of dates of sowing on incidence of insect pests of coriander. Indian journal of Agricultural sci. 56 (1): 56-59.

Malpani, S (1992) Efficacy of herbicidal weed control and irrigation scheduling in coriander (Coriandrum sativum L) MSc (A) Thesis RAU, Bikaner under guidance of Dr. P.S.Rathore.

nethap Seetharaman, R. and Arjunan, G. (1979). Coriander the wonder spice. Indian farmers digest 12 (9): 33-34.

Ramanujam, S.; Joshi, D.S. and saxena, M.B.L. 1964. Extent and randomness of cross pollination in some umbelliferous spices of India. J. Genet. 24: 62-67.

31. Dill Seed

Avtar Singh and R. K. Mahey

Dill (*Anethum graveolens* L.) commonly known as *sowa* is an aromatic herb belonging to family Umbelliferae. Dill seed is well known for its medicinal properties due to the presence of an essential oil (2.5 to 3.0 percent in seed and 0.55 to 0.60 percent in whole plant). The oil extracted from seed is called dill seed oil and is slightly yellow to brown in color while oil extracted from whole plant (herb) is known as dill weed oil. It is pale yellow to yellow liquid. Its emulsion in water is commonly known as dill water and is considered to be an aromatic carminative especially useful in the flatulence, colic pain, vomiting and hiccup of infants and children. Dill oil is also used in food, flavour and perfume industries.

Seeds constitute as a part of various indigenous medicinal preparations and is also used as a condiment in soups, salads, processed meats, sausasges, spicy table sauces and particularly in dill pickling.

Dill herb is used as a pot herb. Dill stems and blossom heads are used for dill pickles and for flavouring soups. It has a good demand in the country and recently it has also developed a great potential for export in the international market. India earned the maximum amount of Rs. 502.18 lakhs from the export of these seeds during 1995-96. India earned 108 lakhs by export of dill seed/oil, in 1995-96 (Table 31.1).

TABLE 31.1: Export of dill seed and ajawain seed

Year	Quantity (Mt)	Area (ha)	Productivity	Values (Rs. Lakhs)	Export (mt)	Value (lakhs)
1991-92	1282	-	-	251.52	-	-
1993-94	-	9995	4963	-	740.53	167.23
1994-95	-	3988	1985	-	821.31	198.39
1995-96	-	-	-	-	531.27	107.91

1. ORIGIN AND DISTRIBUTION

Dill is indigenous to South Europe. It appears as a wild plants in the Eastern

United States from Connecticut to Virginia. It is also found wild in the West Indies. Dill is cultivated in England, Holland, Germany, Hungry, Rumania, Turkey, U.S.A. and Russia.

2. CLASSIFICATION

Dill commonly known as 'sowa', is an aromatic herb belonging to family Umbelliferae. Anethum genus has three species of herbs, two of which are cultivated for dill oil such as *Anethum graveolens* Linn. (Syn. *Peucedanum graveolens* Linn) is popularly known as European dill and *Anethum sowa* Kurg Syn. *Peucedanum graveolens* Linn is commonly known as Indian sowa. Sometimes it is regarded as a variety of *Anethum graveolens*.

3. CLIMATE AND SOIL

The moderately cool climate has been found to be suitable for luxuriant growth. The hot and dry climate is required at the time of maturity. It can be grown on all type of soil except alkali, saline and water logged soil. The well drained sandy loam to loamy soil is best suited for its cultivation. Singh (1985) recorded the highest seed yield of dill sown on loamy sand soil. A fine seedbed by giving two or three ploughings each followed by planking is preferred for its sowing.

4. CROPPING SYSTEM

Dill seed can fit well in the following cropping system:

1. Paddy dill seed
2. Maize dill seed
3. Mash dill seed
4. Moong dill seed
5. Fodder dill seed

4.1. Manure and Fertiliser

Singh et. al. (1971) found that application of nitrogen @ 45 kg/ha produced the highest number of branches and umbels per plant and also gave the highest seed yield.

TABLE 31.2: Seed yield, Straw yield and different levels of nitrogen

Nitrogen (kg/ha)	Seed yield (q/ha)	Straw yield (q/ha)	Nitrogen uptake Seed (kg/ha)	Nitrogen uptake Straw (kg/ha)
0	5.2	32.6	12.3	18.7
30	6.4	44.4	15.1	25.1
60	7.7	53.7	18.3	31.1
90	8.8	56.9	21.2	33.7
120	7.1	62.0	18.0	35.2
C.D. at 5%	0.5	2.8	1.1	1.9

Singh (1985) and Randhawa et. al. (1987) recorded increase in seed yield with increase in nitrogen level upto 90 kg/ha. One third of nitrogen as calcium ammonium nitrate at the time of sowing and another one third of nitrogen top dressed at 65 days after sowing and remaining one third dose at the initiation of flowering was found appropriate. However, Singh and Randhawa (1990) found significant increase in herb yield with 90 kg N and 20 kg P_2O_5/ha.

5. SOWING

Dill seed has been successfully introduced for cultivation in the plains of North India where the seed is sown during September-October and harvested for seed in April. Singh and Randhawa (1989) at Punjab Agricultural University, Ludhiana and Gupta (1976) in Tarai region of U.P., concluded that dill could be sown in the mid October in order to get the highest seed yield (Table 31.3). Delay in sowing causes reduction in yield.

Singh and Randhawa (1990) reported that in order to get the higher herb and oil yield the crop should be sown on 30th September in Punjab (India). The oil content in seed and herb was not influenced by dates of sowing; Randhawa *et. al.* 1987: Singh and Randhawa, 1990.

TABLE 31.3: Seed yield and Straw yield under different sowing dates

Treatments	Seed yield (q/ha)	Straw yield (q/ha)
30 September	5.1	90.0
15 October	10.2	72.2
30 October	8.4	55.2
15 November	6.3	33.0
30 November	4.5	19.5
15 December	2.3	13.1
C.D. at 5%	0.7	4.0

5.1. Seed Rate

Generally 5 kg seeds are required to sow one ha. Singh (1985) and Randhawa *et. al.* (1987) reported that lower seed rate of 5 and 10 kg/ha produced statistically same seed yields, however the highest seed yield was obtained under 10 kg/ha. The seed yield decreased as the seed rate increased from 10 to 15 kg/ha. Singh and Randhawa (1990) reported that to get the maximum herb and oil yield, the crop should be sown with 10 kg seed/ha. Oil content in seed and herb was not significantly affected due to differences in seed rate. Singh and Randhawa (1990). Similarly different seed rates had no significant effect on quality of the oil. Randhawa et al, (1987).

5.2. Spacing

Crop is sown at 45 cm spacing by Kera method. Singh and Singh (1970) found that yield of seeds was maximum at 45 cm space between rows. Oil content in seed and herb was not influenced by various spacing (Singh and Randhawa *et. al.* 1987; Singh and Randhawa 1990). The different row spacing did not influence the quality of oil (Randhawa *et. al.*, 1987; Singh et al, 1987).

6. IRRIGATION

Apply a pre-sowing irrigation before sowing. Apply first irrigation 10-15 days after sowing to ensure proper seedling emergence. Subsequent irrigation should be applied to ensure proper moisture throughout the growth period, taking care that there is no water stress at flowering stage. In the months of March and April, apply light irrigations on calm days to avoid lodging.

7. WEED CONTROL

Give one or two hoeings depending upon the weeds intensity. The first hoeing may be done 30-40 days after sowing and thinning may be done at this stage, if necessary. Weed flora in dill field comprises *Chenopodium album*, *Solanum sarrachoids*, *Amaranthus lividis*, *Amaranthus spinosus*, *Amaranthus retroflexus*, *Echinochloa crusgalli*, *Aethusa cynapium*, *Solanum nigrum*, *Senecio vulgaris*, *Avena fatua* and *Setaria viridis*.

7.2.1. Chemical Control

In Russia, Burykin *et. al.* (1962) suggested application of pre and post-emergence mineral oil or post emergence spray of CIPC for the satisfactory control of weeds. Yields and vitamin C contents of dill were increased by herbicide treatments.

In Hungary, Foldesi and Svad (1965) reported that prometryn applied @ 2.5 to 3.5 kg/cadastral (0.57 ha), gave satisfactory weed control in dill seed without causing crop injury during both rainy and dry seasons.

Pre-emergence and post-emergence application of 0.56 kg linuron/ha gave lower dill yields than ethalfluralin. bentazon post-emergence application produced burned unmarketable leaves.

Wall and Friesen (1986) in Manitoba Canada determined the effectiveness and selectivity of a range of herbicides for control of annual weeds in dill. They observed that pre-sowing incorporation of 1.1 kg trifluralin and ethalfluralin and 6.5 kg EPTC/ha and post-emergence application of 0.25 kg sethoxydin/ha gave effective control of *Avena fatua* and *Setaria virdis*. Soil incorporated application of 3 kg chloramben, 1.1 kg trifluralin and ethalfluralin and 4.5 kg EPTC/ha effectively controlled *Chenopodium album* and *Amaranthus retroflexus*. Post emergence application of 1 kg linuron/ha controlled several broad leaved weed species but resulted in severe crop injury in some seasons. Fresh weight and oil yields of *Anthum graveolens* were closely related with degree of weed control produced by the herbicide treatments while carvone content decreased and phellandrene content increased at high levels weed competition.

8. PLANT PROTECTION

The Root rot damaged is abundant when spring rains followed by sudden rise in soil temperature. It caused damping off in younger plants which was rather rare, most of the late infected plants began turning yellow and the wilting was partial which continued for a period ranging between 6 to 8 weeks. The affected plants bear normal umbels and the fruits on main axis and some older branches, although did not grow its full size, ripened normally.

The powdery mildew is observed in dry hot seasons of March and April. It has also been reported to infect commonly the east Indian dill crop. The infected plants showed lesions like white bloom on young shoots, leaves inflorescence and fruits. It spreads soon to cover most of the upper exposed parts. The infection spread both by conidia and ascospores, covered large part of fields. Dusting of metallic sulphur @ 25 kg/ha after detection of symptoms has been recommended. It has also noted that dusting during flowering period need very careful handling atlest the inflorescence get charred.

9. HARVESTING

The crop is ready to harvest by the first week of May. Harvesting should be done at the proper stage which vary with the purpose for which the crop is grown i.e. for seed or essential oil. Harvest the crop in the morning and transport to the threshing floor immediately to prevent seed shedding. Zlatev (1978) found that the most suitable stage for harvesting the whole plant of dill seed for oil extraction is milky waxy stage of seeds because essential oil content is maximum at this stage. Kapoor and Abrol (1961) reported that maximum oil content was obtained when the fruits are fully developed but green in colour i.e. 7-9 days after dropping of petals when it changes colour from green to grey and dark grey. However, Singh (1985) reported that for getting seed rich in oil content, the crop should be harvested when the fruits are fully developed but yellowish green in colour.

10. YIELD

An average yield of 8 to 10 quintals of seed per hectare can be obtained from the normal sown crop.

11. CONSTRAINTS OF PRODUCTION

1. **Lack of processing:** Methods of curing and storage facilities which are major problem having great influence on the production of good quality product of seed spices.
2. **Lack of market:** It is main problem which affects its production, because no proper markets are available to procure the produce of seed spices. The government is also not giving the support price which has a direct bearing the production and quality of the produce.
3. **Lack of seed/planting material:** There is no availability of healthy and disease free seed/planting material from different research centres/seed producing agencies. The farmers are forced to use the seed of local varieties resulting in poor quality produce.
4. **Lack of transfer of technology:** It is also the important factors which has the great effect on the quality and production of seed spices because of lack of extension staff. Technology generated on research stations does not reach to the farmers.
5. **Introduction of exotic species:** To produce the good quality product and to increase the production of seed spices and essential oil, introduction of exotic species like European dill (*Anethum graveolens* L.) over Indian dill (*Anethum sowa* L.) should be tried.
6. **Lack of research:** There is lack of organized research on seed spices like the others spices such as pepper and cardamom etc. which is the main cause and farmers are not growing seed spices with the proper use of improved

technology. There is need to standardization of developed technology with respect of various input factors in different agro-climatic zones.

12. REFERENCE

Burykin. O.A., Kran, A.P. and Sajthurhanov, S.R. 1962. Chemical weeding in the extreme north. Horticulture Abstract. 32: 3059.

Foldesi, D. and Svab. J. 1965. Chemical weed control and new cultural practices in coriander and dill (1) chemical weed control and dense row spacing in coriander and dill. Horticulture Abstract. 35 (1-4): 8473.

Gupta, R. 1976. Studies on cultivation and improvement of dill (*Anethum graveolens* L.) in Indian Part. III Agronomic studies. Indian Perfumer. 20: 85-89.

Kapoor, L.D. and Abrol, B.K. 1961. Post fertilization development study of dill fruit in relation to its oil content, Journal of Scientific Industrial Research 20 (2): 54-56.

Precheur, R.J. and Garrabrants, N.L 1984. Weed control in dill. Horticulture Abstract 54: 3887.

Randhawa, G.S., Singh Avtar and Mahey, R.K. 1987. Optimising agronomic requirements for seed yield and quality of dill (*Anethum graveolens* L.) oil. Acta Horticulturae 208: 61-68.

Singh, Avtar and Randhawa, G.S. 1989. Plant population and seed yield of dill (*Anethum graveolens* L.) as influenced by some agronomic factors. Haryana Journal of Agronomy 5(2): 110-114.

Singh, Avtar and Randhawa, G.S. 1990. Investigations on some agronomic inputs influencing oil content, oil and herb yield of dill (*Anethum graveolens* L.) New Botanist: 17 (1-2): 111-115.

Singh, Avtar, Randhawa, G.S. and Mahey, R.K. 1987. Oil content and oil yield of dill (*Anethum graveolens* L.) herb under some agronomic practices. Acta Horticulturae 208: 51-60.

Singh, Avtar. 1985. Yield and quality of dill (*Anethum graveolens* L.) as influenced by some agronomic factors. Ph. D thesis, Punjab Agricultural University, Ludhiana.

Singh, R.S. and Singh, C.P. 1970. Studies of some agronomic factors on the growth, yield and essential oil content of dill (*Anethum graveolens* L.) Indian oil and soap Journal 36 (2).

Singh, R.S. and Singh, L.B. and Singh, C.P. 1971. Response of N and P on yield and essential oil content of dill in non saline alkali soils. Fertiliser News 16 (2): 48.

Timoshenko, M.A. 1973. Prometryn in dill culture. Horticulture Abstract 43 (4): 2315.

Wall, D.A. and Friesen, G.H. 1986. The effect of herbicides and weeds on the yield and composition of dill (*Anethum graveolens* L.) oil. Crop Protection 5 (2): 137-142. Horticulture Abstract 56-8127.

Zlatev, S. 1978. Sowing rates of dill grown for essential oil production. Rasteniev Dni Nauki 14 (5): 57-63. Horticulture Abstract 48 (6): 5865.

32. Isabgol

G. S. Randhawa and B. S. Gill

Isabgol (*Plantago ovata* Fork.) belongs to the family Plantaginaceae. Isabgol is an annual herb about 30-50 cm tall. It is a stemless softy hairy plant. The plant bears sessile, filamentous, linear lanceolate to lanceolate leaves, entire or distantly toothed (Fig. 32.1). The flowers are sessile, small, bisexual, tetramerous, anemophilous and protogynous and as such favouring out crossing. The inflorescence is a spike, made up of many florets. The spikes are 1.2 to 4 cm long and about 0.5 cm broad, cylindrical to ovate in shape and bears 45 to 69 flowers. The seed of this plant are boat shaped about 3 mm long and 1.5 mm broad sunk on one side translucent and pinkish grey. The concave side of seed is covered with thin white membrane, produced by fusing of outer layer of ovule together with the inner epidermis, forming seed coat. The seed epidermis is made of polyhedral cells whose walls are thickened by a secondary deposit which is the source of mucilage. The coating of the seed provides the husk on mechanical milling. Isabgol husk is thus a membranous covering of the seed, white or light pink in colour, translucent, colourless. It absorbs moisture and forms a tasteless mucilaginous substance which constitutes the drug.

The Isabgol seed has a cooling and soothing action. It is a laxative and is used in stimulating discharge of urine. It is prescribed for catarrah, chronic dysentry, intestinal fluxes, diarrhoea, blenorrhoea functional, derangements of digestive organs and febrile conditions. Unrheumatic and gouty affections, glandular swellings and in irritable surfaces of skin or mucous membrane, the crushed seeds moistened with water from a good soothing and softening poultice.

The Isabgol seed husk in used in dysentery, diarrhoea, piles and menstrual disorders. It is considered a harmless laxative without any side effects on stomach and intestine. It is used with lime juice, tea, milk and sugar as a soothing drink in the hot season, ice-creams and sauces. It also finds a place in dyeing, calico printing and setting lotions.

The efficiency of Isabgol in medicine is due to large quantity of mucilage present in the husk. Isabgol action is purely mechanical. The husk swells into a jelly like mass with cold water and mechanically stimulates the intestinal peristalsis, so there is no harmful side effects. The mucilage has a remarkable property as a thickner.

1. CLASSIFICATION

The genus *Plantago* comprises of about 200 species, of which 10 species occur in India. Out of these 10 Species, only two species, namely; *Plantago ovata* Fork (Blond psyllium) and *Plantago psyllium* (French psyllium) are economically important. But *Plantago ovata* is the main source of Isabgol due to its quality and yield of husk.

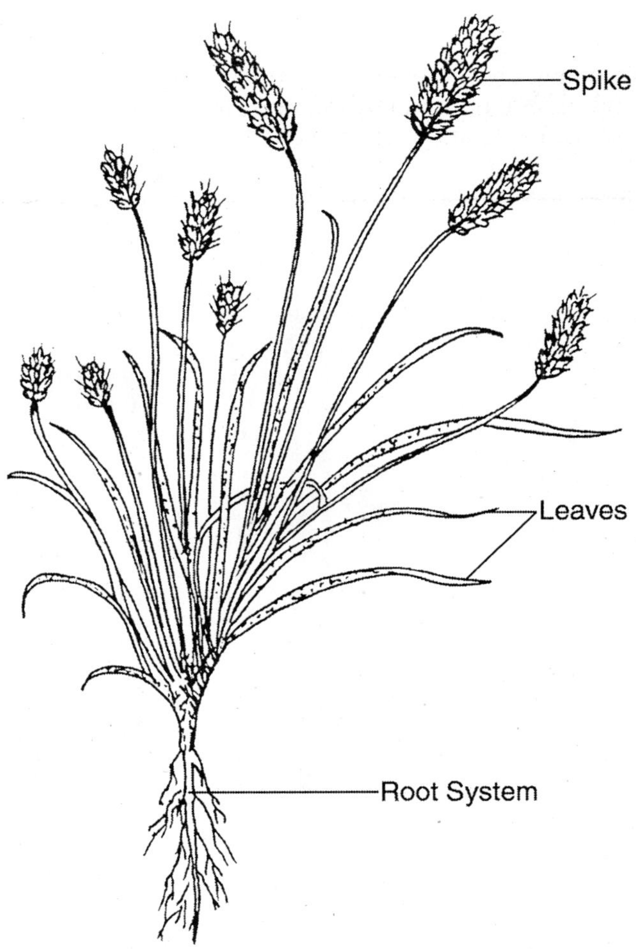

FIGURE 32.1: Isabgol plant.

2. ORIGIN

Isabgol is a plant of west Asian region. It was known to Persians and Arabs long ago and was introduced into India by Muslims. The name of Isabgol is derived from the Persian words *Isap* and *Ghol* meaning a horse ear, referring to the characteristic shape of seed. The French psyllium grown in France, Spain, Italy, Belgium and Brazil looks like Blond psyllium grown in India but is considered inferior in quality.

3. DISTRIBUTION

In India *Plantago ovata* is grown mainly in Gujarat, Rajasthan and parts of Madhya Pradesh and Haryana. India holds a monopoly in seed production and is a major exporter of seed and husk and earns about 20 crores of rupees from its export.

4. SOIL AND CLIMATIC REQUIREMENTS

Isabgol is grown in the *rabi* season under a cool and dry climate. Crop requires only low quantities of plant nutrients. It comes up best under irrigation in well drained sandy loam to loam soils. It is essential that there should not be any shower at the time of the crop maturity as even a light shower at this stage will cause shedding of seeds and destroys the husk which directly affects quality and yield. Excess rainfall or prolonged cloudiness with high relative humidity results in increased incidence of diseases and pests.

After the harvest of *kharif* crops the field is ploughed crosswise two to four times to remove the stubbles of the previous crop. A pre-sowing irrigation, if needed, is given to help preparatory tillage. After removing the stubbles, the land is harrowed three to four times. The last ploughing is followed by planking for making the land level and firm to create a good seed bed. Since the seed of Isabgol is very small it is essential that seed bed should have fine tilth.

5. MANURE AND FERTILISERS

Isabgol makes a moderate demand for nutrients. Usually 20 kg each of nitrogen and phosphorus is applied per ha (Table 32.1). The whole of phosphorus and half of nitrogen is applied at sowing and the remaining half of nitrogen is applied about 40-45 days after sowing (Randhawa at el., 1982).

TABLE 32.1: Effect of different levels of nitrogen and phosphorus on seed yield (kg/ha)

Nitrogen levels (kg/ha)	Seed yield (kg/ha)			
	1978-79	1979-80	1980-81	Mean
N_0	783	636	676	698
N_{20}	912	854	725	830
N_{40}	963	897	725	830
N_{60}	983	969	717	890
N_{80}	996	914	691	867
C.D. at 5%	116	131	Ns	-

The data (Table 32.2) indicates that Psyllium has merger nitrogen requirements and thus responded only to 20 kg N/ha. While at Sumerpur (Rajasthan) Singh and Choudhry (1994) reported 60 kg N + 30 kg P_2O_5 for transitional plain of Luni basin.

6. VARIETIES

6.1. Gujrat Isabgol No. 1 (GI-1)

Plants are 40 to 45 cm tall, having profuse tillering with 3-4 cm long compact spike. It does not respond much to nitrogen application. Takes about 110 days to mature. High seed yield (10 q/ha). It has synchronous flowering.

6.2. Gujrat Isabgol No.2 (GI-2)

Plants are about 40 cm tall, having profuse tillering and matures in about 110 days. Moderately resistant to downy mildew. Yields about 10 q/ha seed.

6.3. Haryana Isabgol No. 5 (HI-5)

This variety is 35-40 cm tall, profuse tillering with long and compact spikes. Matures in 135-140 days. Moderately resistance to downy mildew. Seed possess about 30% husk. Average yield is 10 q/ha seed. Recommended for commercial cultivation in areas with moderately fertile and restricted irrigation facilities.

7. SOWING

Though Isabgol seed remains viable for several years, but for uniform and good germination, fresh seed should be used. Eight to ten kg clean and bold seed will be sufficient to sow one hectare. Since seed is very small, thus the un-husked seed is mixed with sand before sowing for its uniform distribution.

Isabgol is sown in lines 30 cm apart. However, it can also be sown by broadcasting and then the seed is covered thinly by raking. The best time of sowing is from second fortnight of October to first week of November (Table 32.2.) The crop sown late is likely to be affected by hot weather at maturity which may cause the premature ripening leading to low yields. Care should be taken to sow the seed shallow as seed sown deep fails to germinate. The seed should be sown in fairly moist seed bed for good germination.

TABLE 32.2: Seed yield (kg/ha) under different sowing dates

Sowing date	Seed yield (kg/ha)	
	1974-75	1975-76
21 Oct.	-	1130
5 Nov.	1000	850
20 Nov.	880	540
5 Dec.	800	330
20 Dec.	350	260
5 Jan.	270	-

8. IRRIGATION

Isabgol can be grown under limited irrigation conditions. Give a pre-sowing irrigation to get proper stand of the crop. In case germination is poor and if available, a light irrigation given 7-10 days after sowing improves the plant stands. About 4-5 irrigation, including pre-sowing irrigation are needed. Crop yield is not much affected if 1-2 irrigation are withheld. Total water requirement of the crop is 350 mm.

9. WEED CONTROL

Isabgol is cultivated on light textured soils. It is a slow germination crop, therefore, weeds germinate first. Weeds grow luxuriantly and cause heavy losses in yield. The weed flora consists of *Chenopodium album, Chenopodium murale, Melilotus alba, Melilotus indica, Cynodon dactylon* and *Asphodelus teunifolius.* Weeds could be controlled by two hand hoeing at 15 and 30 days after sowing. Herbicide isoproturon 0.5 kg/ha as pre-emergence may be applied to check weeds. Singh (1993) reported seed yield of 3.11 q/ha in weekly check and 11.61 q/ha when treated with isoproturon 0.5 kg as pre-emergence and one hand weeding.

10. HARVESTING

The plant grows to a height of about 50 cm and each plant gives out 25 to 100 tillers depending upon the fertility of soil and weather conditions. The crop becomes ready for harvest in 140 days to 150 days. The spikes changes colour to brown and lower leaves turns yellow and then becomes dry while the upper one remains yellow. The crop can be tested for maturity by pressing a spike between two fingers in the matured plant. The mature seeds will come out. The crop should be harvested when about 80% of spikes mature. To avoid losses due to seed shedding, the crop should be harvested close to the ground in early morning hours. Collect the harvested crop immediately on the the threshing floor and is threshed manually. The straw can be used as fodder.

11. YIELD

Ten to fifteen kg seeds/ha, husk is 30% of seed weight and remaining consists of 65% *goli*, 3% *lali* and 2% *khakho*.

12. PLANT PROTECTION

12.1. Insect Pests

Aphid, thrips and mites sometimes appear as a pest and adversely affect the crop. Spray 1.25 L malathion 50 EC/ha and repeat after 15 days, if necessary.

12.2. Diseases

Powdery mildew sometimes attacks the crop. Spray the crop with karthane 40 EC @ 200/ha. Two or three sprays be given at 10 days interval.

13. PROCESSING

The seeds of Isabgol has been used for long as such and seldom in the form of separated husk. The large scale production of husk started only after 1930 when Indian Psyllium established itself as a permanent crop in Gujarat and became an exportable item. In recent years, the entire export of Isabgol is in the form of husk and even domestic consumption of seed is being replaced by husk.

The husk represents the epidermis of the seed coat. The seed epidermis is brittle and is loosely attached to the inner layers of seed coat and is therefore easily separable by milling and grinding. The process of manufacture has gradually evolved from hand pounding of seeds in a pestle and mortar to grinding in hand operated grinding stones (*chakkis*) and finally to most modern power operated automatic mills. The following processes are followed to manufacture the husk.

13.1. Cleaning

The produce contains sand, pebbles, gritty particles, fragments of capsules, fruit wall and some other extraneous materials. The fine sand particles often adhere to the groove on the under side of the seed. Cleaning involves sieving and polishing.

13.2. Sieving

This is done by a set of electric driven rocking sieves which grade the seed into three sizes and also removes the capsular fragments. The seed first pass over a 20 mesh sieve through which smaller seed (Grade III) and sand particles pass down. The remainder of the seeds pass a set of two sieves mounted one above the other. The upper sieve of this set has slits like perforations 45 mm x 1.5 m each. This allows major quantity of seed (Grade II) to fall on to the lower sieve. The large seeds (Grade I), pebbles and capsules are collected at the farther end of upper sieve. Through the lower any smaller seeds of Grade III or sand particles, if any present falls down while the portion of mesh is collected. This process of sieving yields 4 per cent Grade-I, 80 per cent Grade-II, 15 per cent Grade-III and 1 per cent comprising of sand and other matters.

Only Grade II seeds are used for manufacture of husk. After the cleaning and sieving process, The Seed still contains fine sand particles in their grooves. These seed undergo a mechanical rubbing between the stone and rubber polisher and thus any sand particles sticking to seeds are removed.

13.3. Grinding

A very carefully controlled process of grinding is applied to seeds. The seeds are subjected to a series of six to eight grindings. The space between the grinding stones being slightly reduced before each subsequent grinding. The seeds are initially fed by conveyer belts to hoppers of first grinder. The mixture of ground and unground seed is collected in receptacle box which has connection with a blower which sucks the lighter husk along with a few seeds. The lighter husk is thrown at a distance while any seed that may be present falls close to the blower. The major portion of partly ground and unground seeds of receptacle box is transferred to next grinder. This cycle is repeated six to eight times. The yield of husk is 28-30% of unhusked seed. The yield of husk at various grindings is approximately as under:

No. of grinding	% of husk yield
Ist	40
2nd	25
3rd	15
4th	10
5th	5
6th	5

The husk from 2nd and 3rd grinding is considered to be the best. The husk of first grinding has more foreign matters than produce of subsequent grindings. The husk collected from each grinding is sieved through 70 mesh sieve to separate *Khakho* which contains small husk powder and sand. From 5th and 6th grinding, *Lali* (embryo and pieces of seeds) is separated which is priced slightly more than husked grain (*Gola*). The *Gola, Lali* and *Khakho* are utilized as cattle feed.

14. QUALITY

The size, colour, presence of red scrapings, husk powder and dust are taken in mind while judging quality of husk.

14.1. Swelling Factor

Take 1 g Isabgol seed in 20 c.c. or 25 c.c. capacity graduated cylinder with stopper, add water to make volume to 20 ml. Agitate intermittently for first half hour. Read the final volume of seed after six hours and swelling factor is expressed as cc/g.

15. REFERENCE

Anonymous (1987). Handbook of Agriculture. Indian Council of Agricultural Research, New Delhi.

Gopani, D. (1969). Isabgol cultivation in India. Farm bulletin No. 68, pp. 1-18.Directorate of Extension, Ministry of food, Agriculture, Community Development and Cooperation, New Delhi.

Legha, P.K. and Sharma, G.D. 1993. Recent advances in Medicinal, Aromatic and under-utilized Plants. Tech. Bull. (MA&UUP), Department of Plant Breeding CCS HAU, Hissar.

Randhawa, G.S. Sahota, T.S., Bains, D.S. and Mahajan, V.P. 1978. The effect of sowing dates, seed rate and nitrogen fertiliser on the growth and yield of Isabgol (*Plantago ovata* F.). J. agric. Sci. Camb. 90: 341-343.

Randhawa, G.S., Mahey, R.K., Saini, S.S. and Sidhu, B.S. (1985). Studies on irrigation requirements of psyllium. Indian J.Agron. 30(2): 187-191.

Randhawa, G.S., Sidhu, B.S., Mahey, R.K. and Saini, S.S. (1982), Effect of nitrogen and phosphorus on the growth and yield of psyllium (Plantago ovata F.) J.Res. Punjab Agric. Univ. 19 (2): 121-122.

Singh, I. Bhati, D.S. and Sing G. (1993) Weed management in blood psyllium (*Plantago ovata*) Indian J. Agrn. 38(2) 348-349.

Singh, J and Chohan, G. S. (1994) Effect of mixtalol and different conbiriation: of nitrogen and phosphorous on growth and yield of blood psyllium (*Plantago ovata*). Ind. J. Agron. 39 (4): 710-711.

NARCOTIC CROP

33. Tobacco

Aditya Kumar Singh

Tobacco was first introduced in India by the Portuguese in the 17th century. Today, tobacco cultivation has become an industry which has made great strides and is playing a major role in the Indian economy. So much so that India has now become the fourth largest producer of tobacco in the world after USA and China. Of the total production, 33% is Flue Cured Virginia (FCV) tobacco and 67% non-virginia. The FCV quality of tobacco is used in the manufacture of cigarettes and non-virginia tobacco in the production of *beedies*, cigars, *zarda*, chewing tobacco, hookah and snuff.

Of the total production, 39% consists of *beedi* tobacco, 23% cigarette tobacco, 15% chewing tobacco, 3% cheroot tobacco and 2% snuff tobacco. At present, nearly 60% of the FCV tobacco produced is being utilised in the domestic cigarette industry and the balance is exported. World wide, 85% of the tobacco cultivated goes into the production of cigarettes but in India, only 20% of tobacco is used in cigarette making. *Beedi* is the most popular form in which tobacco is consumed in India.

Although, India is the fourth largest producer of unmanufactured tobacco in the world, accounting for 7% of the total world production, next to China and USA, it ranks ninth in the world tobacco export market. The main exporting countries are the USA, Brazil, Zimbabwe, Turkey, Greece and Italy. India earns about Rs. 7000 million central excise revenue during 1996-97.

1. ORIGIN

The information available on the origin and history of tobacco suffers generally from ambiguity and contradictions. Most of the scientists believe that tobacco was originated in America. Christopher Columbus brought some dried leaf and seed to Spain in 1492. Then it spread in Europe. However, some other scientist believe that the plant was first introduced into Europe in the year 1560 by a Spanish physician, who brought it from Mexico as medicine. Jean Nicot, the French ambassador to Portugal came to know of tobacco in Lisbon and introduced it to French Court, for giving honour to him, the botanical name of the plant was given as *Nicotine* and the word nicotine have been derived from his name. Tobacco is said to have been introduced into India in the beginning of 17th century by Portuguese.

2. DISTRIBUTION

In the world now-a-days China is having maximum area, production and average yields followed by USA, Brazil, India and Russia (Table 33.1). China is having more than 50% area and production in the world.

TABLE 33.1: Area production and yield of tobacco leaves in selected countries of the world (1992)

Country	Area lakh ha	Output lakh mt	Productivity kg/ha
China	30.65	101.00	3295
USA	3.14	7.64	2430
Brazil	3.44	5.83	1694
India	4.26	5.79	1358
Russia	2.29	4.60	2005

If we look into the distribution in our country we found that Andhra Pradesh is having nearly 50% of total area and production, followed by Gujarat and Karnataka (Table 33.2).

TABLE 33.2: State wise area, production and yield of tobacco (1995-96)

State	Area thousand ha	Production '000 t	Productivity kg/ha
Andhra Pradesh	134.3	171.9	1280
Gujarat	121.8	196.0	1609
Karnataka	58.3	48.9	839
Uttar Pradesh	18.5	87.0	4703
All India	392.9	556.6	1417

2.1. Tobacco Growing Areas of India

Guntur area: It is famous for cigarette and Natu types. The area consists of Guntur, Krishna, East-West Godavari and Telangana districts of A.P.

1. **Charottar area of Gujarat:** It is famous for the cultivation of Bidi tobacco and is limited to Kaira district.
2. **Nipani area:** This is bidi tobacco growing area which consists of Belgaon, Satara, Kolhapur, Songale etc districts.
3. **North Bihar and Bengal area:** Here mainly *hookah* tobacco is grown, but in limited areas chewing and snuff types are also grown. This area consists of Muzaffarpur, Darbhanga, Poornia, Cooch Bihar, Jalpaiguri, Berhampur and Dinajpur districts.
4. **U.P. and Punjab area:** The area is famous for hookah and chewing type of tobacco consisting of Farrukhabad, Mainpuri, Sitapur, Badau, Muradabad, Jullundhar and Firozpur districts.
5. **South Tamil Nadu area:** It includes Coimbatore and other southern regions of Tamil Nadu where chewing, cheroot, snuff and wrapper types of tobacco are grown.

3. CLASSIFICATION

The genus *Nicotiana* is one of the five large genera of Solanaceae and is represented by about 60 recognized species which are grouped in three sub-groups.

1. Subgenus - *rustica*
2. Subgenus - *tabacum*
3. Subgenus - *petunioides*

Out of the 60 species, only two species, i.e. *Nicotinana tabacum* and *Nicotiana rustica* are cultivated extensively. India grows both the species but by far the largest area is under *N. tabacum* which is grown all over the country. Since *N. rustica* requires cooler climate, its cultivation is confined mainly to the northern and north-eastern areas of the country, i.e. Punjab, U.P., West Bengal, Bihar and Assam.

The *N. tabacum* varieties known as *desi* types have tall plants with broad leaves and have usually pink flowers. The *N. rustica* varieties known as *Vilayati* and *Calcuttia* are characterised by short plants with round puckered leaf and yellow flowers. Specific varieties in *N. tabacum* have been developed for cigarette, cigar and cheroot, *bidi*, hookah and snuff tobacco. The varieties developed in *N. rustica* are used for only chewing hookah and snuff tobacco.

4. GROWTH STAGES

Tobacco is having four growth stages viz., (i) Seedling establishment (soon after transplanting), (ii) Early growth stage, (iii) Grand growth stage and (iv) Ripening (maturity) of leaves.

5. CLIMATE

The tobacco crop is grown in *kharif* season in A.P. and Karnataka mostly under rainfed conditions. While in U.P. and Gujarat it is grown in both the *kharif* and *rabi* season.

The crop needs around 28° C as an optimum temperature for germination but it may germinate at a temperature ranging between 15° to 35° C. The crop can not withstand frost. Crop starts wilting at a temperature above 35° C.

Tobacco needs 50-100 cm rainfall, which should be well distributed throughout the growing period. Areas having a rainfall, more than 125 cm are not fit for tobacco production. Whereas a rainfall of less than 75 cm results in production of feathery leaves having very high nicotin content. The leaves remain smaller and possess very poor burning quality.

A light showers in the early stage of crop after transplanting are boon. Heavy and unexpected rains late in the season when the crop is about to mature result in poor quality out turn, both in respect of bright grade and chemical composition of leaf. There is every likely hood that crop once again turns green and its quality is adversely affected.

6. SOIL AND SOIL PREPARATION

The type, grade and quality of tobacco produced are considerably influenced by the soil characteristics, particularly, the texture of surface and sub-soils. Light soils tend to produce a large and thin leaf, light in weight and colour, mild in strength and weak in aroma, whereas leaf produced on heavy soils is usually thick and heavy, dark coloured, strong and aromatic.

Due to interaction of the variety with the environmental complex, production of some of the main tobacco types by India has been concentrated in certain zones.

6.1. Flue-cured Tobacco Soils

For production of first grade flue-cured tobacco it is necessary to have a soil having free draining and which can be kept well-aerated throughout the growing period. The fertility status of the soil should not be high. In fact, nitrogen starvation condition should prevail at the time of maturation of leaf. In our country it is grown in black clay soils, sandy loams and red soils. The area under black clay soil is to, an extent of about 60 per cent of the total area under flue-cured tobacco. Efforts are being made to extend the area under the sandy loams and red soils as the tobacco produced from these light soils is found to be of exportable quality. It is mostly grown in A.P., Karnataka and some parts of Gujarat.

6.2. Natu-tobacco Soils

This local type of tobacco is grown in Andhra-Pradesh. Natu leaf which is used for manufacture of cigarettes has to be thin in texture with a pleasing aroma and moderate strength, whereas that used for cheroot, chewing and snuff can be stronger. Soils growing Natu tobacco range from heavy black clay soils in Guntur district to sandy and sandy loam soils in West Godavari and Srikakulam districts.

6.3. Cigar Filler Tobacco Soils

Cigar filler tobacco is a heavy type of tobacco with good taste, aroma and burning quality. It is mostly grown on sandy to loamy, well-drained red and brown soils of Coimbatore and Madurai districts of Tamil Nadu.

6.4. Cigar Wrapper Tobacco Soils

It is grown in Dinhata of Cooch Bihar district of West Bengal and to some extent in Banana and Coffee grades of the lower Palui Hills in Madurai district where surface soil texture is light silt loam and sub soil texture is heavy silt loam.

6.5. Bidi Tobacco Soils

Cultivation of bidi tobacco is confined mainly to two zones, viz., Charotar tract of Gujarat where it grows on sandy to sandy loam soils. Soils of Nipani (Karnataka) are silt loams to clays.

6.6. Hookah Tobacco Soils

It is grown mainly in U.P., Bihar and West Bengal. These soils are moderately alkaline silt loams and highly calcareous in nature.

6.7. Chewing Tobacco Soils

Hookah tobacco can also be used as chewing type and vice-versa. So the soils growing chewing tobacco in North India are the same as these used for hookah type. Besides, some alluvial sandy loam soils of Gujarat also produce chewing tobacco.

7. CROPPING SYSTEM

Continuous monocrop of tobacco develops *Orobanchee* infestation in the field hence suitable crop rotation should have to be followed. Common crop rotation adapted in different states is given below:

TABLE 33.3: Crop rotations followed in different part of India

State	Crop rotation
Tamil Nadu	Tobacco-Chilli, Onion-Tobacco, Tobacco-Sorghum/Pearlmillet, Tobacco-Arhar
Gujarat	Tobacco-Bajra, Cotton-Tobacco
West Bengal	Rice-Tobacco
Bihar	Maize-Tobacco-Maize
Uttar Pradesh	Maize-Tobacco, Maize-Potato-Tobacco (Summer crop)

8. MANURE AND FERTILISERS

On the basis of different experimental findings it is observed that the organic manure/compost/and oil cakes should be applied at the time of first ploughing so that they are mixed properly in the field before planting the seedlings.

Type of Tobacco	Org. Manure (q/ha)	Nutrients in kg/ha			Time of application
		N	P	K	
Flue cured virginia (in heavy soils of A.P.)	Green manuring with sun hemp/5-7.5 t/ha FYM	22	50	50	Org. manure 6 weeks and fertiliser 1 week before planting
FCV in light soils of A.P.	Green manuring 10-12.5 t/ha FYM	30	80	80	Org. manure 6 weeks, P. 1/2 N, 1/2 K one week before planting followed by top dressing of 1/2 N and 1/2 K 3-4 weeks after transplanting.
Natu tobacco of A.P.	15 t FYM/ha/7.5 FYM + 4 t G.N. cake	150	50	50	Org. manure 6 weeks, P. 1/2 N, 1/2 K one week before planting followed by top dressing of 1/2 N and 1/2 K 3-4 weeks after transplanting.
Bidi of Karnataka (Nipani area)	Press mud 10 t/ha	120	60	40	Org. manure 6 weeks, P. 1/2 N, 1/2 K one week before planting followed by top dressing of 1/2 N and 1/2 K 3-4 weeks after transplanting.
Bidi of Gujarat	Sun hemp/FYM @ 12.5 t/ha or 45 kg N through poultry manure	135	-	-	Basal 2 weeks before planting

Type of Tobacco	Org. Manure (q/ha)	Nutrients in kg/ha			Time of application
		N	P	K	
Cigar cheroot tobacco in T.N.	2-3 q of G.N. cake	50	-	-	Cake, 1/2 N 2 weeks before planting and 1/2 N as top dressing 6-7 weeks after transplanting
Chewing tobacco in Bihar	Green manure/ mustard cake @ 6-8 q + 600 q FYM	60	-	-	Cake, 1/2 N 2 weeks before planting and 1/2 N as top dressing 6-7 weeks after transplanting
Hookah chewing tobacco in W.B.	10 t FYM	150	150	-	Cake, 1/2 N 2 weeks before planting and 1/2 N as top dressing 6-7 weeks after transplanting
Wrapper tobacco in W.B.	2 t FYM	125	115	120	Cake, 1/2 N 2 weeks before planting and 1/2 N as top dressing 6-7 weeks after transplanting

The manure should be supplemented with the following quantities of fertilisers for various types as per recommendations of CTRI, Rajahmundry.

9. VARIETIES

TABLE 33.4: Tobacco varieties released in India

Variety	Year of release	Developed at	Cured leaf yield (kg/ha)	Area of adoption and salient features
FCV tobacco				
CTRI special (MR)	1980	CTRI, Rajamundri	1200	For TBS & SLS resistant to TMV
Godavari special	1982	CTRI, Rajamundri	1525	For TBS & SLS, resistant to TMV
Swarna	1984	Res. Stn; CTRI, Hunsur	1450	For KLS, resist to powdery mildew
Mc Nair 12	1986	CTRI, Rajamundri	1880	For NLS, tolerant to black shank
Jayasri (MR)	1986	CTRI, Rajamundri	1503	For TBS and SLS, resist to TMV
Hema	1987	Res. stn; CTRI, Guntur	1560	For TBS
Bhavya	1988	Res. Stn; CTRI, Hunsur	2000	For endemic black shank areas of KLS, resist to black shank and tolerate to root knot nematode.
Gauthami	1992	CTRI, Rajamundri	2000	For TBS and SLS
CM 12 (K4)	1993	CTRI, Rajamundri	2000	For NLS, tolerant to black shank
Virginia tobacco 1158	1993	CTRI, Rajamundri	2000	For TBS, resist to TMV

Variety	Year of release	Developed at	Cured leaf yield (kg/ha)	Area of adoption and salient features
Bidi tobacco				
NPN 190	1979	ARS, Nipani	1964	For Karnataka & Maharashtra
PL 5	1984	ARS, Nipani	2000	For Karnataka & Maharashtra
GT 5	1986	BTRS, Anand	3301	For Gujarat, tolerant to root knot and high nicotine
GT 7	1993	BTRS, Anand	2535	For rainfed areas in Gujarat, drought tolerant
GTH 1	1995	BTRS, Anand	3644	For Gujarat, tolerant to root-crust, high yielding
Chewing tobacco				
Thangam	1980	Res. stn; CTRI, Vedasandur	3226	For smoke-cured areas of T.N.
Bhagya Lakshmi	1980	Res. stn; CTRI, Vedasandur	3532	For sun cured areas of T.N.
Maragadham	1981	Res. stn; CTRI, Vedasandur	3013	For smoke-cured areas of T.N.
Pratha	1981	Res. stn; CTRI, Pusa	2200	For Bihar
G T 6	1986	BTRS, Anand	2712	For lal and kala chopadiu tobacco areas of middle Gujarat.
Pusa tobacco-76	1990	Res. stn; CTRI, Pusa	2600	For Bihar
Meenakshi	1992	Res. stn; CTRI, Vedasandur	4000	For sun cured areas of T.N.
Vaishali special	1993	Res. stn; CTRI, Pusa	2778	For Bihar
Rustica tobacco				
D D 437	1977	Res. stn; CTRI, Dinhata	1865	For Motihari areas of W.B.
Sonar Motihari	1977	Res. stn; CTRI, Dinhata	1690	For Motihari areas of W.B.
G C 1	1981	BTRS, Anand	2693	For rustica areas of Gujarat.
G C T-2	1994	BTRS, Anand	3512	For rustica areas of middle Gujarat.
Natu tobacco				
Prabhat	1977	Res. Stn; CTRI, Guntur	1500	Natu areas of A.P. suitable for cigarette Natu, resist to TMV
Vishwanath	1986	Res. stn; CTRI, Guntur	2429	Suitable for cigarette Natu areas of A.P.
Natu special	1992	Res. stn; CTRI, Guntur	1600	Suitable for cigarette Natu areas of A.P.
Cheroot tobacco				
Bhawani special	1980	Res. stn; CTRI, Vedasandur	2837	For Bhawani area of Coimbatore district of T.N.
Lanka special	1981	CTRI, Rajamundri	2780	For river island of East Godavary district of A.P., strong, pungent and aromatic

Variety	Year of release	Developed at	Cured leaf yield (kg/ha)	Area of adoption and salient features
Sendarapatty special	1986	Res. stn; CTRI, Vedasandur	2100	For Salem area of T.N.
Cigar wrapper tobacco				
S 5	1977	Res. stn; CTRI, Dinhata	1476	For cigar wrapper tobacco areas of northern W.B.
Cigar filler tobacco				
Krishna	1986	Res. stn; CTRI, Vedasandur	2250	For cigar filler area of T.N.
Burley tobacco				
Banket A-1	1994	BTRC, CTRI, Jeddangi	1800	For light soil agency areas of East Godavari, Vishakhapatnam, Vijaynagram districts of A.P., resistant to TMV

TBS = Traditional black soil of Andhra Pradesh, NLS = Northern light soils of Andhra Pradesh, SLS = Southern light soils of Andhra Pradesh, KLS = Karnataka light soils.

10. SOWING

Since the seed of tobacco is very small and initial development of seedling is very slow, the tobacco is not sown directly in the field. Instead its seedlings are first raised in nursery and then they are transplanted in the field when they attain a particular size.

10.1. Nursery Raising

Site of Nursery: Generally, tobacco nurseries are grown on sandy or sandy loam soils and located on the farm itself. The cigarette tobacco growing areas of Andhra Pradesh are an exception in that the crop is grown on a heavy black soil and the nurseries are generally raised on sandy to gravelly loams. The nursery site should have a good internal as well as surface drainage and should be situated at an elevated place. It is desirable to change the nursery site every year as it could minimise incidence of pests and diseases and also eliminate contamination by other varieties.

Systematic layout of nursery on raised beds with intervening channels helps in quick drainage of rain water. The beds are 1.0 m to 1.25 m wide to facilitate hand weeding and watering with rose cans. The beds can be of any convenient length along with the slope but generally not more than 10 m. The area of nursery should be around 1/100th part. For transplanting seedling in one hectare, 8 x 1.25 m wide, 10-12 rows would be sufficient.

For raising good seedlings organic and inorganic fertilisers should be used judiciously. For this purpose 80-100 kg ground nut cake along with 8-10 tons/ha FYM should be spread over raised bed. Then fertilisers are given according to different types of tobacco.

Cigarette tobacco	80 kg N, 60 kg P_2O_5 and 30 kg K_2O
Bidi tobacco	160 kg N, 80 kg P_2O_5 and 30 kg K_2O
Hookah tobacco	100-150 kg N, 80 kg P_2O_5 and 40 kg K_2O

10.2. Seed Rate

Generally, seed rate depends on the climate of an area. The seed of tobacco are very small. In one gram, there are 10,000 grains are present. Therefore, 3.5 kg/ha seed rate is generally sufficient, but in hot climates 4 to 5 kg seed is required.

10.3. Sowing Time

Sowing time for different types of tobacco is as follows

Types of tobacco	State	Time
FCV tobacco in TBS & SLS	Andhra Pradesh	Mid Oct.-mid Nov.
	Light soils of A.P.	Mid Sept.-mid Oct.
	Karnataka light soils	May-June
Bidi tobacco	Gujarat	Second fortnight of Aug.-first fortnight of Sept.
Karnataka	First fortnight of Aug.	
Burley tobacco	A.P.	Mid July - mid Aug.
Lanka tobacco	A.P.	Late Oct. - mid Nov.
Natu tobacco	A.P.	Irrigated - last week of Oct.
A.P	Rainfed mid Oct - early Nov.	
Cigar and cheroot tobacco	T.N.	Mid Oct. to mid Nov.
Hookah tobacco	W.B.	Mid Oct. to early Nov.
Hookah tobacco	U.P.	June – July, Sept.
Chewing tobacco	Bihar	Mid Sept. - mid Oct.
Chewing tobacco	T.N.	Late Oct.

10.4. Method of Sowing in Nursery

Usually the treated or water soaked seeds are sown by broadcasting on the surface of the beds and later they are mixed in the soil upto a depth of 1.5 to 2.0 cm with the help of rake. Before sowing, the seeds are mixed with ash or fine sand in 1: 15 or 1: 20 ratio for their proper or uniform distribution on the beds.

10.5. Care of the Seedlings in the Nursery

1. After sowing, sufficient moisture should be maintained in the nursery bed. Since mostly sowing is done in between May and September. For maintaining moisture beds should be covered by mulch. Once germination is initiated, the mulch should be removed.
2. The seedlings must be thinned out at 10 to 15 days after sowing in order to protect them from damping off. Resetting of 3 to 4 weeks old seedlings at a distance of 5 × 5 cm in new beds may be done for hardening of the seedlings.
3. Weeds must be taken out of the beds soon after their emergence.
4. A spray of diathane Z-78 at the rate of 11 to 12 Litres of 0.03 per cent solution for every 40 sq m area prevents spread of fungal diseases.
5. For control of pests, a spray of carbaryl 50 w.p. (sevin 50 g in 22 Litre of water) is useful.
6. Seedlings are transplanted when age is 7 to 9 weeks for *tabacum* and 5 to 6 weeks for *rustica*.

7. Irrigation of the beds should be stopped about 10 to 15 days before transplanting so that they may become drought resistant and may resist the set back.

10.6. Transplanting

Seedlings are transplanted at spacing given in Table 33.4.

TABLE 33.5: Spacing of different types of tobacco

Type	Spacing in (cm)	
	Row to row	Plant to plant
Flue cured virginia and natu of black soils of A.P.	80	80
FCV of light soils of A.P.	100	60
Cigar and cheroot in Tamil Nadu	75	50
Chewing in T.N.	75	75
Bidi in Gujarat	75	75
Bidi in Karnataka	105	45
Chewing in Bihar	90	60
N. rustica in W. Bengal	60	45
N. tabacum in W.B.	90	90
Hookah tobacco in U.P. *N. rustica*	45	45
N. tabacum	50	45

11. IRRIGATION

Tobacco is grown over a wide range of soil and climatic conditions and its irrigation requirements vary. Being a mesophyte, about 80 to 85% of the plant weight is water but it can tolerate drought more than excess of moisture as it has a fairly deep and vigorous root system. An early drought is also is also preferred for better establishment and it enables the plant to face any moisture stress during the growth of the plant. But once the stress is released it is advisable that the plants get good supply of moisture at regular intervals.

The total water requirements of tobacco under Rajamundry conditions as calculated by the climatological data is about 50 cm. A major part of this is supplied by the stored moisture in the soil and remaining quantity of 10 to 15 cm is to be met with partly by rain and partly by irrigation. The field should be irrigated when the soil moisture falls about 20 per cent field capacity because higher soil moisture reduces the leaf quality.

The flue cured Virginia tobacco grown on black cotton soils is very rarely irrigated whereas the one grown on light soils is given 5 to 7 irrigations. The cigar, cheroot and chewing tobacco in T. N. and chewing tobacco in Bihar are normally irrigated and about 15 to 18 irrigations are given. The *bidi* tobacco in Karnataka, Maharashtra and the hookah and chewing tobacco of West Bengal are grown rainfed but under drought one or two irrigations are desirable for production of better crop. The hookah and chewing tobacco in U.P. are irrigated and, generally 12 to 15 irrigations are given.

12. WEED CONTROL

About 20 days after transplanting, when the plants get established their young root lets they need a very fine and pulverised sub-surface soil for their growth and further expansion. Weeds can be kept under check by summer cultivation and three or four intercultures given during the period of crop growth.

The tobacco crop is very often infested by *Orobanchee*, a total root parasite. *Orabanchee* is found to be attached with the tobacco roots and they grow up through the soil round the tobacco plants. It is advisable to collect them or destroy them as soon as they are seen in the field. For chemical control of *Orobanchee* a pre-planting application of EPTC at the rate of 6.0 kg/ha (one month before planting) followed by post planting application of 2,4,5-T; 2,4-D and 0.1 per cent allyl alcohol have been recommended.

13. INTER-CULTURE OPERATIONS

13.1. Topping and De-suckering

The operation of topping consists of removal of the terminal buds just before or after the emergence of the flower head. Following topping the axillary buds become active and put forth shoots known as suckers. Removal of these is called de-suckering. The objective of topping and de-suckering is to divert the nutrients of the plant to leaves, instead of flowers and seeds. It results in the gain in the size and body of the leaf, thus increasing the yield and quality of the tobacco. Topping is done by two ways:

1. **Topping at top:** In this only terminal buds are removed.
2. **Lower topping:** In this, along with terminal buds 2-3 upper leaves are also removed.

For cigarette tobacco only topping at top is done, as light coloured and less pungent leaves are required. While for getting fatty and oily leaves lower topping is done. In cigarette tobacco 8 leaves, in *bidi* tobacco 10-12 leaves, in cigar, cheroot and chewing tobacco 12-16 leaves per plant are left and remaining leaves are topped.

De-suckering is a must if we are to derive maximum benefit from topping. Two or three de-suckering are generally done as the suckers reach a convenient size for handling. The manual de-suckering is a time consuming, and more expensive process, therefore use of chemicals may also be practiced.

13.2. Use of Chemicals for De-suckering

Application of coconut oil to the top six axils soon after topping suppresses the emergence of suckers in flue cured Virginia tobacco. In case of cheroot tobacco of T. N. application of 2 per cent naphthalene, acetic acid in triethenolemin to the topped portion suppressed all the suckers. Use of 2 per cent of maleic hydrazide or indol butyric acid are also recommended for an effective de-suckering.

14. PLANT PROTECTION

14.1. Plant Diseases

Tobacco is affected by several important fungal, nematode and viral diseases. Among them, damping off (*Pythium aphanidermatum*), frog-eye leaf spot (*Cercospora nicotianae*) and nematodes are major diseases in nursery, while frog-eye leaf spot (*Alternaria alternata*), root knot, mosaic, black shank and leaf curl (*Ruga tabaci*) are serious diseases of field crop. The root parasite orobanche reduces yield in all types of tobacco. Copper and copper oxychloride fungicides and dithiocarbamates are screened as bio efficacy and recommended for nursery and field for the control of damping off, leaf blight (*Phytophthora parasitica* sp. *nicotianae*), black shank, anthracnose (*Collectotrichum tabacum*) and frog-eye spot diseases.

Further, systematic fungicides recommended for control of damping off is metalaxyl (ridomil MZ 72 wp @ 0.2%, for frog eye spot and anthracnose is carbendazim (bavistin 50%) @ 0.03% and for powdery mildew is bayleton @ 0.02%.

Broom rape (*Orobanchee cernua*) is a complete root parasite on tobacco which causes losses by 24-52%. Though good amount of research was conducted with chemicals, mineral oils and vegetable oils and vegetable oils, the practices found effective and economical are:

1. Deep ploughing the field in summer
2. Burying the *Orobanchee* seed to deeper depths and thereby reducing the emergence of the parasite.
3. Growing trap crop like black gram, green gram, gingelly and red gram.
4. Periodical hand weeding with spear before flower setting offers good control of *Orobanchee*.
5. Root knot nematodes (*Meloidogyne incognita* and *Meloidogyne javanica*)
6. They are mostly prevalent in light soil nurseries and planted crop.
7. In Karnataka, soil formation with methyl bromide or durofume @ 15-18 mL/m^2 2 weeks before sowing is effective in controlling weeds, root knot nematode and soil insects up to 14 months.
8. In Gujarat and Andhra Pradesh root knot disease of tobacco as well as weeds can be managed effectively and economically by trapping the nursery area with 25 mm LDPE clear plastic film for 15 days during summer.
9. For field crop, deep cultivation in summer, sunhemp green-manure, growing tolerant variety, if available, and application of carbofuran or phenamiphos @ 1 kg/ha before planting are to be followed to overcome the problem.

14.2. Insect Control

1. **Tobacco caterpillar:** *Spodoptera litura* Fab.(Noctuidac: Lepidoptera) is by far the most important pest of tobacco in most areas. The caterpillars feed on the foliage gregariously in the early stages. Maximum damage is done to the nursery plants. When the plants are grown up, the caterpillars feed large portions of the leaves.

2. **Hairy caterpillars:** Species of *Amsacta* (Arctiidae: Lepidoptera) are serious defoliators of tobacco plants, especially in Gujarat. The other hairy caterpillar, *Diacrisia obliqua* W. (Arctiidae: Lepidoptera), is also known to cause considerable damage to the foliage.
3. **Grass hoppers:** Many grass hoppers have been reported to feed on tender leaves of tobacco at all stages. The more important being *Attractomorphis crenulata* F. and *Chrotogonus robertsoni* B.
4. **The white fly:** *Bemisia tabaci* Genn. (Aleyrodiae: Hemiptera) occurs widely and is an important pest. The nymph undergoes 3 instars, the total life cycle takes 16 to 40 days. White flies cause direct damage by feeding on the leaves in large numbers. Otherwise too, the adults transmit the virus causing leaf curl disease in tobacco. Due to the injuries, the leaves become curled, twisted, ruffled, puckered and thickened with abnormally prominent veins. The attacked plants are stunted.
5. **Aphids:** The green peach aphid, *Myzus persicae* Sulz. (Aphidadae: Hemitera), often causes considerable loss to tobacco. The aphid infests tobacco plants in large numbers and resultantly, the plants become sickly, the leaves curled up and faded; the attacked leaves are unfit for curing. The aphid also transmits the viruses causing the different diseases in tobacco. Species of *Macrosiphum* are also found infesting tobacco.
6. **Other sap sucking bugs:** The mirid, *Gallobelicus crassicornis* Dist. (Mirdae: Hemiptera), feeds on the leaves in large numbers; the pentatomid, Stibaraopus tabulatus schiodte (Pentatomidae: Hemiptera) is a medium sized, reddish brown bug feeding on the roots in southern states. As a result, the plants are stunted and gradual fading of the plants occur.
7. **Shoot borer:** The larvae of the moth, *Gnorimoschema heliopa* lower (Gelechiidae: Lepidoptera) is an important pest of tobacco in Andhra Pradesh and Maharashtra. The caterpillar mines the leaf and bores through the midrib and leaf stalk reaching the stem, which induces gall formation. Due to the galls growth is stunted and unusual branching of the attacked seedlings at the top is observed.
8. **The cut worm:** In the Gangetic plains and occasionally in Andhra Pradesh, the cut worm, *Agrotis ypsilon* Rott. (Noctuidae: Lepidoptera) is a serious pest. The young seedlings are cut off at the ground level during the night hours by these caterpillars.

The sap sucking insect pests can be managed effectively to a great extent in the nurseries before transplanting. Use of phosphamidon @ 0.04 per cent or monocrotophos @ 0.03-0.04 per cent is recommended. Synthetic pyrethroids, if used, should be alternated with the conventional insecticides. For the foliage feeders endosulfan (0.07%) or quinalphos (0.1%) or synthetic pyrethroid (like permethrin, fenvalerate, cypermethrin, decanthrin) can be effectively used @ 0.003 to 0.005 per cent. However, the use of systemic insecticides against sucking pests will also take care of these foliage feeding beetles. In tobacco phosphatic insecticides are preferred. Cut worms can be effectively managed by mechanical drilling after trapping them in trenches dug around the field. Use of contact insecticides like endosulfan, quinalphos and malathion are effective against the cut worms.

15. HARVESTING

Tobacco is grown for the production of quality leaves which largely depend upon the stage of the leaf at which they are harvested. The sign of maturity and method of harvesting differs from type to type of tobacco.

Cigarette tobacco is harvested by priming i.e. two or three leaves which have completed their growth and are turning from green to yellowish green or slightly yellow in colour are detached from the plant. Generally, it takes about 70 to 75 days from planting for the crop to be ready for the first priming. After the commencement of first priming, subsequent primings are taken in about 6 to 7 days intervals unless affected by rains. The entire harvesting is completed in 6 to 8 or 10 primings depending on the leaf number (which may vary from 18 to 26 leaves) and climatic conditions.

Cigar, cheroot, chewing hookah and *bidi* tobacco are harvested by the stalk cut method. In this method plants are cut close to the ground with a sickle and generally left in the field overnight for wilting. Subsequent handling of plants varies according to the method of curing adopted for different types of tobacco.

16. CURING

Curing is a process by which the harvested leaves are made ready for marketing. In more technical terms it may be stated that curing is a process of drying, decomposition of chlorophyll until the green colour disappears from the harvested leaves, hydrolysis of starch into sugars and respiration or fermentation of sugar thereafter. The main object of curing is to dry the green leaf in such a way that final product had required colour, texture and aroma. Flue curing, air curing, sun curing, fire curing and pit curing are the usual method adopted for different types of tobacco.

16.1. Air Curing

In air curing the source of energy is the atmospheric temperature carried through air. There are two stages of air curing: yellowing of leaves (the first stage) and development of brown colour (the second stage). As such this method has very little value for curing a high grade tobacco as the leaves give up their moisture slowly and turn finally to a brown colour. Air curing is commonly adopted in West Bengal, Bihar, U.P., M.P., Punjab, Karnataka, Andhra-Pradesh and Tamil Nadu.

16.2. Sun Curing

Of all the methods, the cheapest method is the sun curing where the source of energy is solar heat. In Tamil Nadu cigar and chewing tobacco plants are harvested by cutting the stalk and after an initial period of wilting in the field, are strung on bamboo poles and sun cured for about 15 to 20 days. Natu tobacco is harvested by cutting each leaf with a piece of stalk attached to it, strung on a jute thread with a needle and sun cured on scaffolds for 45 to 60 days. In bidi tobacco the plants are cut and left upside down to dry on the spot. In hookah and chewing tobaccos of Bihar, the whole plants are cut and left in the field for 4 to 16 days for drying. Later they are heaped in a curing yard for 2 days. They are then turned upside down and dried for another 4 days. Some what similar procedure, with slight modifications is followed in Punjab and Uttar Pradesh for hookah and chewing tobacco.

16.3. Fire Curing

Fire curing or smoke curing is done for Jaffna (A type of chewing) tobacco. Harvesting is done by priming, usually with a piece of stem, wilted for a few hours in the field, tied into bundles and hung on lathis in a hut. They are then smoke cured for 12 hours by burning coconut husk, than stacked for 3 days and again smoked.

16.4. Flue Curing

Flue curing is an improved method of curing. The curing consists of drying green leaves under artificial atmospheric conditions by adopting a process which does not allow the green leaf to come in direct contact with smoke or flames of the fuel and which permits the regulation of temperature and humidity.

The main feature of flue curing is drying of leaf under controlled conditions, where starch gets converted into sugar and the green tobacco becomes bright, aromatic and fine textured. This method is adopted in almost entire Virginia cigarette tobacco growing areas like Guntur, Godavari, Mysore and other areas at Hyderabad, Madras, U. P., M. P., and Orissa.

17. YIELD

Yield in tobacco refers to the weight of dry and marketable leaves of various grades pooled together. The yield of leaves differs from type to type as mentioned below.

Type of tobacco	Yield of leaves in kg/ha (cured leaves)
Cigarette	750-900
Bidi	650-900
Chewing type	800-1600
Hookah (Deshi)	750-850
Hookah (Villayati	1200-1600

18. QUALITY

There are three major factors affecting tobacco leaf quality:

18.1. Genetic Factors

Curing behaviour, flavour, aroma, body, texture, colour retention, burning qualities etc. expected to be genetically controlled to a greater or lesser extent. Position of leaves on stalk influences the leaf quality as the bottom leaves are relatively thin, light in weight and body, light in colour, mild and have good burning qualities, passing up the stalk, the leaves become progressively thicker, heavier in weight and body, more rough, more gummy, stronger, darker in colour, more aromatic, but of poorer burning qualities. The leaves approaching to the seed head can never reach to full maturity. As regards the chemical composition, in general, passing from the lowest to the top most leaves of fully matured plant there is a progressive increase in content of total nitrogen, protein nicotine and ammonia, a decrease in total ash, calcium, magnesium and pH value. While the content of soluble carbohydrates is lowest in the upper leaves and highest in the middle leaves of the plant. Thus it is observed from the same plant that on an average bottom 5-6, middle 2-4 and top 1-2 leaves formed the three grades or qualities viz., superior, medium and inferior respectively which is roughly in the proportion of 6:3:1.

18.2. Ecological Factors

Soil: Tobacco soil must be well drained, properly aerated with low inherent fertility. Soils with light sandy surface and friable sandy loam sub-soil surface, low content of organic matter, acidic reaction (desirable pH 5.0-6.0) and good amount of potash, phosphoric acid and iron are preferred for quality leaf production. Heavy soils produce tobacco with higher nitrogen than the sandy loam or sandy soils. Tobacco grown on saline soils absorbs moisture quickly before and after drying, changes colour rapidly during storage and has poor burning qualities which are attributed to large quantities of chlorine (more than 0.01 per cent in soil and more than 1.0 per cent in leaf) which plant takes up from the soil. Tobacco grown on paddy lands and tank beds becomes bleached.

Climate: Meteorological factors have considerable effect on the quality of leaf. High temperature and low relative humidity has harmful effect on leaf burn but the leaves are thick and aromatic, while at low temperature due to slower rate of metabolic processes, the leaf does not ripen fully and does not develop the desirable qualities. This desirable range of temperature for quality production of tobacco is 70-90° F. Rainfall desirably ranging from 50 to 125 cm during growing season of tobacco results in leaves with light weight, thin texture, light colour, less gummy, light body, reduced nicotine content and aroma but of better combustibility. While drought results in production of leaves having all the reverse qualities. Wind causes dryness of plants and may bring a temporary atmospheric and soil drought which may result in a poor quality leaves.

18.3. Cultural Factors

Crop rotation: The soils under continuous tobacco cultivation are depleted of humus because tobacco does not add much of organic matter to soil like most of the crops. Such soils produce poor quality tobacco leaves and have possibility of an increased incidence of diseases and *Orbanchee* (flowering parasite) infestation. Therefore, tobacco in rotation with legumes always gives higher and better quality leaves.

Manuring: Quality of tobacco, particularly of smoking depends on the balance of the nutrients in the leaf.

Nitrogen has the profound effects on larger leaf area, increased moisture content, increased organic acid, enzymic activity and nicotine content but leaves have a poor carbohydrate content in them, light green leaves of good quality are produced from plant partially starved of nitrogen. **Phosphorus** deficient plants produce narrower and smaller leaves of dark green colour, the rate of growth is slowed down and maturity is delayed, the leaves are dull and devoid of lusture on curing. **Potash** in the smoking tobaccos, is responsible for improving burning qualities like completeness of combustion and fire holding capacity that is also influenced by amount of potash in leaf. The physical qualities of the leaf, e.g. elasticity, smoothness and low prominence of veins and the smoking qualities are attributed to the high potash in the leaf. **Chlorine** in excess amount injures the growth of the plant and reduces the quality by producing thick and brittle leaves. It has an unfavorable effect upon the burning quality and stability of colour.

Time of planting: Early planting of all the tobacco types other than hookah in late October to early November is better than late planting during late November and early December. While in case of hookah tobacco the best planting time is 2nd fortnight of February than earlier or latter dates because the yield and quality both are found to be the best.

Plant spacing: Closer spacing gives higher yield of leaves but the closer spacing tends to reduce the size, body, thickness and weight per unit area of leaves. While in case of slightly wider spacing the plants produce larger, thicker and heavier leaves which are again not good quality leaves. From yield and quality point of view a spacing of 90 cm x 50 cm or 120 cm x 60 cm is found to be optimum.

Irrigation: There is general belief that heavy irrigation spoils the quality of leaf. As high moisture content of soils result in low content of nitrogen, nicotine, calcium and magnesium in tobacco leaf producing poor burning quality though yield may be high. While light irrigation at critical stages of the crop growth gives a sufficient boost without spoiling the quality of leaf.

Time of topping: The best quality leaves are found where the plants are topped at early flowering stage and the quality deteriorated as topping was delayed. This also decreases the sugar and nicotine contents in cured leaves.

19. CONSTRAINTS OF PRODUCTION

The values of national productivity average show a gradual increase from 1950-51 to 1990-91 for all tobaccos (730 to 1361 kg/ha), FCV tobacco (694 to 894 kg/ha) and non-Virginia tobacco (738 to 1559 kg/ha). Though good progress was achieved with the release of superior varieties and development of improved agronomic and plant protection practices at the research farms, farmers yields are stagnating. Analysis of constraints which limit yield in FCV tobacco are frequent drought spells, delayed monsoon leading to late planting and incidence of leaf curl disease in traditional black and southern light soils of Andhra Pradesh. Lack of variety resistant to root knot nematode and black shank diseases in northern light soils and Karnataka light soils areas. *Bidi* tobacco is also affected by drought stress and biotic stress of leaf curl as well as root knot disease in Gujarat, while aphid infestation, black shank disease and delayed and scanty rainfall in Karnataka. Chewing tobacco production is reduced by late plantings due to delayed monsoon incidence of tobacco mosaic virus and outbreak of leaf eating caterpillar in Tamil Nadu incidence of viral diseases in Bihar, leaf curl and root knot disease in Gujarat, brown spot disease is West Bengal and lack of high yielding varieties in Orissa.

To increase the productivity the following priority areas have to be vigorously pursued.

1. Exploitation of hybrids with CMS line.

2. Search for superior germplasm.
3. Integrated nutrient management with organic manures, inorganic fertilisers and bio-fertilisers.
4. Utilization of *nif* genes for biological nitrogen fixation.
5. Developing efficient, eco-friendly and less labour intensive sucker management practices to contain the suckers after topping the plant.
6. Evolving effective cultural and chemical means for management of *Orobanchee*.
7. Developing integrated disease and pest management practices which are farmer friendly and eco-friendly.

20. REFERENCES

Agricultural Situation in India 54(5): 1997.

CTRI, Rajahmundry. 1995. Tobacco varieties released by CTRI. Central Tobacco Research Institute, Rajahmundry.

CTRI, Rajahmundry. 1996. Annual Report 1995-96. Central Tobacco Research Institute, Rajahmundry.

DTD. 1997. Status paper on Tobacco. Directorate of Tobacco Development, Ministry of Agriculture, Government of India, Chennai.

N.C. Gopalachari. 1984. Tobacco, ICAR Pub.

R. Lakshminarayan. 1998. Tobacco (*Nicotiana* species) research in India. Indian Journal of Agricultural Sciences 68(8, special issue): 473-83.

Singhal, V. 1996. Indian Agriculture. Indian Economic Data Research Centre, New Delhi.

FODDER CROPS

34. Napier

Virendra Singh

Hybrid napier has been developed by crossing napier grass with bajra (*Pearl millet*). It grows throughout the year in tropical regions. In sub tropic too, it grown vigorously during warm season but remains dormant during cold months. It is a perennial grass. It contains 8.1% crude protein, 34.6% nitrogen free extract, 1.8% ether extract, 34.0% crude fibre, 10.0% ash and 0.37% phosphorus. It also contains 44.1% total digestible nutrients (Rai and Singh 1982). Napier grass forage which is also known as elephant grass is of poor quality. It has also lot of hairs on leaf blades so its acceptability by the animals has been very poor and hence it was crossed with bajra to improve the forage quality. It gives higher forage yield than many other grasses (Table 34.1).

TABLE 34.1: Comparative performance of hybrid napier vis-a-vis other grasses in hill at 900 m altitude

Grasses	Green forage t/ha
Hybrid napier	60.1
Nandi grass	51.9
Guinea grass	48.4
Kikyu grass	42.9
Rhode grass	38.6

1. ORIGIN AND DISTRIBUTION

It is perennial tropical grass and originated from southern part of Africa. Its name was given after col. Napier who has realized its importance in 1909. It was introduced in India in 1912. It can be grown all over the world which are not prone to water logging, frost and snowfall.

Napier grass (*Pennisatum purpureum* Schum.) belongs to graminaceae family and genus *Pennisetum*. It is tall perennial grass and grows upto 2 to 3 meter height. It produces 15-20 tillers. Stem thickness is about 2.5 cm. The leaves are green, 60-90 cm long and 2.5 cm wide. No seed is produced and hence propagated vegetatively.

2. CLIMATE AND SOIL

It suits well to warm and moist climate. However, it is not suitable for water logged condition. It can be grown upto the altitude of 1000 meter from mean sea level beyond that growth is affected.

3. SOIL AND SOIL PREPARATION

Hybrid napier needs sandy loam to loam soils which are perfectly drained. It needs neutral soil but can be grown up to the pH of 5.5. One ploughing followed by 3-4 harrowing are sufficient for this crop.

4. CROPPING SYSTEM

4.1. Intercropping

Hybrid napier being a perennial crop grows for 5-6 year with economic yield. During winter season particularly in north India, berseem (*Trifolium alexandrium*) or lucerne (*Medicago sativa*) can be grown as inter-crop. In summer cowpea suits well as inter-crop. This system increase both yield and quality of forage besides improving soil fertility. The comparative productivity of such system with that of other crop rotations of annual forage crops are mentioned in Table-34.2.

TABLE 34.2: Production potential of important forage crop rotation

Rotation	Forage yield t ha-1	
	Dry	Green
Hybrid napier + Berseem-Cowpea	52	200
Sorghum + Cowpea-Berseem + Oat-maize + Cowpea	49	186
Maize-Laha forage-Oat-maize + Cowpea	50	180
Sorghum-Berseem-Maize + Cowpea	47	173

5. MANURE AND FERTILISERS

If farm yard manure is available, 5-10 tonnes of this manure should be applied prior to planting and mixed uniformly. At planting 50 kg N and 50 kg P_2O_5 ha^{-1} should be applied as basal dose. Thereafter, 30 kg N/ha in sole crop of hybrid napier per cutting should be applied. Under inter cropping system with berseem during winter and cowpea during summer, 15 kg N/ha per cutting gives same productivity as that of 30 kg N/ha per cutting applied in sole cropping besides providing additional forage from berseem and cowpea inter crops (Table 34.3.).

TABLE 34.3: Effect of N on forage yield of hybrid napier and total forage production

Treatment	Green forage yield t/ha	
	Hybrid Napier	Total
Sole hybrid napier	92.6	92.6
Hybrid napier with legumes at 15 kg N/ha/cutting	117.9	188.6
Hybrid napier + legumes @ 30 kg N/ha/cutting	119.1	184.4

6. VARIETIES

The first napier × bajra (NB) hybrid released was Pusa Giant which has higher yield potential with better quality forage than that of napier grass, but Pusa Giant has been found susceptible to cold climate and hence has been given poor performance in north India (sub tropics). The hybrids developed thereafter such as NB-21 performed better than Pusa Giant. The former has lower mortality than the latter. The details of NB hybrid with their suitability and yield potential (Chaube, 1990) are mentioned below:

Hybrids	Suitability	Yield potential ($t\ ha^{-1}$)
Pusa Giant	Whole India, mainly tropics	100-130
NB-21	Whole India	120-150
Swethiha-1 (IGFRI-3)	North India including hills, more suitable for inter cropping	110-120
IGFRI-6	North India including hills, more suitable for inter cropping	120-130
IGFRI-7	North India including hills, more suitable for inter cropping	130-160
Yashwant (IGFRI-10)	Whole India	130-140

7. PLANTING

7.1. Planting Time

In irrigated conditions, planting should be done during summer season from February end to June in north India so that its vigorous growth can suppress growing weeds during rainy season. Under rainfed conditions, planting needs to be done on rainy days preferably in the month of June to first fortnight of July.

7.2. Planting Material

Two bud setts of stem or rooted slips (tillers) are used for planting in case of hybrid napier. Around 20,000 such planting materials are required to plant one hectare area.

7.3. Planting Method

In well prepared field, in case of rooted slips, the root should be placed in soil and shoots above the ground whereas in case of setts one node should be in soil and other buds above the grounds. The setts should be inserted in ground at 45° angle. Planting should be done at spacing of 1 m between rows and 0.5 m between plants. After planting field should be irrigated.

8. IRRIGATION

Two to three irrigations are desired after planting till its establishment. During summer season, irrigation at the interval of 15-20 days has been found optimum. During rainy season drainage is equally important.

9. WEED CONTROL

Except within two months after planting, no weeding is generally required. After each cutting cultivation between rows either by cultivator or deshi plough may be beneficial.

10. PLANT PROTECTION

No insect-pest and disease problem has been observed in hybrid napier grass.

TABLE 34.4: Effect of hybrid, cutting height and N-Level on forage production of hybrid napier

Treatments	Forage yield t/ha	
	Dry matter	Green
Hybrids	-	-
Pusa Giant	104	22
NB – 21	157	37
NB – 5	154	32
Cutting height (cm)		
5	120	25
10	138	28
15	132	26
CD (P= 0.05)	12	3
N - Levels (kg/ha)		
0	102	20
50	123	28
100	130	30
150	139	30
CD (P= 0.05)	12	3

11. HARVESTING

Under irrigated condition cuttings at 5-6 weeks interval with 10 cm stubble height has been optimum (Table 34.5) when plant height of the grass attains about 90 cm from ground. The cutting interval may be increased to 7-8 weeks under rainfed.

TABLE 34.5: Effect of cutting interval and stubble height on forage yield of hybrid napier

Treatments	Forage yield t/ha	
	Green	Dry
Cutting interval (weeks)		
3	56.6	14.5
4	62.2	16.0
5	68.0	18.4
6	70.8	20.2
CD	4.9	1.1
Stubble height (cm)		
5	61.1	16.4

Treatments	Forage yield t/ha	
	Green	Dry
10	64.6	17.6
15	67.4	17.8
C.D.	4.2	0.9

12. YIELD

The yield potential of hybrid napier is generally 150-200 t/ha green forage. However, under tropical conditions with better management it might be more than 250 t/ha green forage.

13. REFERENCES

Chaube, R.N. (1990) Improved varieties of cultivated forage crops. Summer Institute on Forage Production Technology, IGFRI, Jhansi, India.

Rai, S.D. and Singh Virendra (1982) Napier grass in Bharat Ki Pramukh Phaslein (Principal Forage Crops) Edited Dr. K.C. Sharma, G.B. Pant University of Agric. & Tech., Pantnagar, India.

35. Anjan Grass

A. S. Gill, R. L. Arya and A. Singh

Anjan grass (*Cenchrus ciliaris*) is also called *Pennisetum ciliare* (L) Linke, Buffel, tussock forming perennial (20 to over 122 cm high) and sometimes rhizomatous. Anjan grass mostly used as a cut fodder but it is an excellent for grazing in hot dry areas of the tropics and subtropics. It persists well, especially the spreading type, under rather close grazing and becomes less palatable if allowed to mature. It is cultivated for permanent pastures and is among various eastern hay species native to India (Kolukkatai). In Australia, *Cenchrus* pastures are utilized for off-season grazing. Since the grass retains its nutritive value even when ripe, It makes very good hay and also silage.

It is variable in habit, including spreading types for pastures, which will withstand fairly close grazing and erect types for hay. It is an excellent grazing perennial suited to pasture and range-lands. It is not very aggressive and is used for rehabilitating sand dunes and recuperating highly eroded areas. Its high soil building capacity is due to its clustered root system in the upper 8 to 10 cm layer of soil. The grass is very safe and does not have any harmful affect on the animals. It is relished by all type of livestock even at the maturity stage.

1. ORIGIN

Anjan grass is a native of Africa.

2. DISTRIBUTION

Anjan grass is cultivated in the tropics and subtropics specially Africa, India and Indonesia. It was introduced and naturalized in northern Australia, and is adopted to tropical and subtropical summer-rainfall areas with a long dry season. It is also grown in Puerto Rico and Other regions of the West Indies especially on sandy soils. In India *Cenchrus ciliaris* is grown over in 20 lakh ha area and average productivity is about 15-40 t/ha green forages. Anjan grass is also cultivated in the arid and semi-arid parts of north-west India. It is one of the principal grasses of the Haryana and Tharparkar tracts. *Cenchrus ciliaris* was accidentally introduced many years back into the semi-arid tracts of Australia, probably through seed contained in camel suddles. It has now spread extensively in that country and forms the staple grass for the flourishing sheep and cattle-rearing industry.

3. CLASSIFICATION

Anjan grasses are grouped according to rhizome (under ground stem) development and the height to which they grown (Yates, 1975).

1. **Tall cultivars with rhizomes**: These cultivars reach a height of 5 feet (1.5 m) Under favourable conditions and are suited to an area having annual rainfall of 15-35 inches (380-890 mm).
2. **Medium height cultivars**: Anjan grass grown to a height of 3 feet (0.9 m) and are leafier, more prostrate, have a greater number of tillers or stems and do not develop rhizomes. It is also adapted to similar rainfall areas as the tall varieties.
3. **Short cultivars:** Under this category the cultivar is a tussocky, short plant growing to a height of 15-30 inches (400-750 mm) with dense, fine leafed tillers. It is very early flowering and because of earlier seeding it is suited to lower annual rainfall conditions of 12-15 inches (305-380 mm), though it has become widespread in the north-west of western Australia.

4. CLIMATE

Anjan grass is suited well to a rainfall of 12-35 inches (300-900 mm). Some cultivars are more drought resistant than others. It survives extreme and prolonged drought but grows vigorously when favourable conditions set in. It thrives even at high temperature upto 45°C to 48°C, the optimum temperature of photosynthesis being about 35° C. The grass remains green even in the summer season with high temperature (48° C) but the stem becomes wiry.

5. SOIL AND SOIL PREPARATION

Light textured soil of high phosphorus are preferred but anjan grass is adopted to self mulching soils and hard, heavy textured shrub and forest soils. In India the grass can grow in different type of soils but grow well on alluvial laterite, red, medium black and well drained soils. It comes up best on calcareous red soils and also does well on forest lands. Good growth and high tonnage yield of anjan grass is obtained when sown on sandy to sandy loam soils. The optimum soil reaction is 7 to 8, but grows on soil with a pH as low as 5.5. Ramia and Fernandez (1974) showed buffel grass to be very sensitive to soil containing high level of aluminum. The buffel grasses are less tolerant to salt. W.B. Loela cultivar of anjan grass is more tolerant to salt than other cultivars, maintaining relatively high yields upto 80 m e.g, sodium chloride per litre (Graham and Humphreys, 1970).

Field operations with a mouldboard plough followed by 2-3 harrowing or disking is sufficient. Seed should be sown on a well prepared seed bed at the onset of the rainy season with appropriate legume mixture and fertiliser doses. Because of their fluffy nature seeds tend to clog in seeding machines, but the flow can be improved by mixing the seed with saw dust or cracked grain of cereals or sorghum.

6. CROPPING SYSTEM

Generally pure grass production is low and any attempt to increase the overall production will be considered an important steps in forage production technology. Intercropping of grass with suitable legume forages could be one way of achieving the target as grass-legume mixture yields more forage with higher percentage of crude protein than grass alone (Table 35.1).

Basavaraju and Rao (1996) reported that *Cenchrus ciliaris* and *Panicum maximum* were grown in a mixture with *Macroptilium atropurpureum* or *Stylosanthes hamata*. The *Cenchrus ciliaris* and *M. atropurpureum* mixture produced the highest mean green fodder (14.7 t/ha) and drymatter (4.89 t/ ha) yield as compared to pure grass. Ley farming with grass and/or legumes in rotation with arable crops meet twin objectives of feed availability and maintaining land productivity. Under different ley farming systems, mixed stand of *Cenchrus ciliaris* + *Stylosanthes hamata* produced higher green forage and dry matter yield than their pure stand. Sorghum produced significantly higher grain and stover yield when grown after 4 years of stylo. The plots under *C. ciliaris* + *S. hamata* mixture also gave significantly higher grain and stover yield of sorghum over sorghum continuously grown for 4 years. The organic carbon content was higher in *Cenchrus ciliaris* + *Stylosanthes hamata*, available nitrogen in the soil (202.6 kg/ha) in stylosanthes alone grown continuously for 4 years (Niranjan et al, 1995).

TABLE 35.1: Grass-legume mixture

Treatments	Forage yield (q/ha)	
	Green	Dry
Cenchrus ciliaris	132.4	58.2
C. ciliaris + Field bean	197.2	83.7
C. ciliaris + Horsegram	198.3	83.4
C. ciliaris + Clusterbean	211.0	89.2
C. ciliaris + Cowpea	239.3	100.3
C. ciliaris + mucana	236.5	97.8

Association of grass with tree resulted overall increase in the production of grass component as compared to the pure stand of the grass (Table 35.2).

TABLE 35.2: Grass-tree intercropping: yield from grass component only

Intercropping	Forage yield (q/ha)	
	Green	Dry matter
Anjan + bird wood hybrid	182.6	53.5
Anjan + bird wood hybrid + Soobabul	231.2	76.4

7. MANURE AND FERTILISERS

The usual practice for increasing the forage yield of the grasses is through the application of nitrogenous fertilisers as it is well known that cereals do respond favourably to N application and also the fertility status of the soils in the country are generally poor in nitrogen content. It is desired to apply 10 t/ha farm yard manure at the time of land preparation. It requires 25-30 kg N and 30 kg P_2O_5/ha as to be applied at the time of sowing (Patel et al, 1998). Hazara (1995) reported that 40 kg N and 20 kg P_2O_5/ha was optimum for 2-3 cuts of *Cenchrus ciliaris* under normal conditions. Kumar and Mishra (1995) found that highest level of fertiliser (60 kg N + 60 kg P_2O_5 /ha produced significantly more seeds (61.88 kg/ ha) as compare to lowest level (20 kg N + 20 kg P_2O_5 /ha), which produced 54.65 kg/ha seed of anjan grass. Hazra and Tripathi (1995) found that increase in dry matter yield of *Cenchrus ciliaris* was maximum at highest N level i.e. 40 kg N/ ha. The responses to dry matter yield of grass and grass + legume mixture were higher at 20 kg N/ha and decreased further with higher nitrogen level (40 kg N/ ha). Increasing dose of nitrogen (90 kg/ha) and phosphorous (60 kg P_2O_5/ha) has increased the yield of grass (Singh, 1991).

Application of nitrogen, irrespective of the sources increased the green and dry fodder yield significantly. Basal application of N through sulphur coated urea (SCU) registered maximum green fodder and dry matter yield of anjan grass. Basal application of urea and sulphur coated urea registered 119 and 163 per cent higher green forage and 109 and 160 per cent higher dry matter yield, respectively over control (Gill and Patil, 1988).

It is not only NPK, but the use of micro nutrients also produced encouraging results in various countries. In Australia the use of Cu and Mo had brought spectacular results in the overall increase in the pasture production. Soil application of Zn, Mn and B conjunction with nitrogen fertiliser significantly increased the green fodder and dry matter yields over nitrogen alone and the increase in green fodder yield was of the order of 25.9, 18.1 and 25.9% and for dry matter yield it was 24.4, 19.3 and 22.4% respectively (Gill *et. al.*, 1980).

8. VARIETIES

Some of the important varieties of anjan grass are:

1. **Bundel Anjan (IGFRI-3108):** It is suitable for arid and semi-arid areas of the country. On an average, productivity is in a range of 250-300 q/ha. Outstanding performance in respect of drymatter production, good establishment, fast regenerative capacity, drought and frost hardiness and its nutritive value is also excellent.
2. **Marwar Anjan (CAZRI-75):** It is suitable for arid region of country. Average productivity is 150-275 q/ha.

Besides above some of the important varieties of *Cenchrus ciliaris* are Pusa giant Anjan, Pusa yellow Anjan, Co-1, CAZRI-413, Liana, Nueces, Hybrid *Cenchrus* (*C. ciliaris* × *C. setigerus*), Molopo, CAZRI S-358, CAZRI S-303 etc.

9. SOWING

Under rainfed conditions, sowing may be done with the onset of monsoon showers in July. For a transplanted crop, nursery can be raised in early June and after four weeks seedlings are ready for transplanting.

When sown by drilling, the furrows are maintained at 60 to 75 cm apart using seed rate of 2 to 3 kg/ha. In sandy soils, the planting should be carried out at wider spacing of 75 × 75 cm and the spacing may be reduced to 60 × 60 cm in loamy soil under good moisture conditions. Crop should not be allowed to be grazed by the animals for first two years. In north India, it can be sown during June-July. However, in South India suitable time is March-Sept. at 45 × 45 cm plant to plant and row to row spacing (Hazra, 1995). The highest seed yield of 69.01 kg/ha was obtained at a row spacing of 60 cm which was significantly higher to 75 and 40 cm row spacing (Kumar and Mishra, 1998). For nursery raising about 40 gram seed is required for 6 m x 1.2 m beds. For direct sowing condition, 4-6 kg/ha seed rate of Anjan grass is required.

9.1. Dormancy

Removal of glumes improved germination, but it is risky in arid areas. Anthrocyanins and phenolins in the seed coat inhibit germination. Freshly harvested seeds have poor germination and therefore ripening the seed for 12 months is desirable. Seed older than two years should be tested for germination before sowing. Scarification will also break the dormancy (Humphreys, 1978).

10. IRRIGATION

Anjan grass is a crop of dry areas and does not tolerate water logging condition. Light irrigations during summer months show better growth. Under drought conditions, some irrigations may be required in the monsoon season.

11. WEED CONTROL

For initial establishment, one or two weedings or interculture can be done. In subsequent years, one interculturing may be done in summer season to destroy seed of weeds and stimulate the growth of the grass. In establishing anjan grass pasture, any broad leaves weeds can be controlled with 2, 4-D @ 1 kg /ha, if the anjan seedling have developed atleast four leaves (Paull and Lee, 1978). Atrazine @ 4 kg/ha used as a post-emergence spray arrested survival of *C. ciliaris* but @ 1 kg /ha it was less damaging (Seatoni, 1978).

12. PLANT PROTECTION

Almost all important cultivated and range grasses namely *Dicanthium annulatum*, *Cenchrus ciliaris*, *Chloris gyana*, *Pannisetum pedicellatum*, *Brachiaria* sp., Pannisetum sp., *Panicum maximum*, *Chrysopogon fulvus*, *Heteropogon centortus*, *Iseilema laxum*, *Sehima nervosum* etc. are attacked by insects, bacteria, fungi, viruses and nematodes (Pandey et al, 1995).

12.1. Diseases

Rusts (*Puccinia* and *Uromyces* spp.): This is a serious disease of *Dicantheum annulatum* and *Sehima nervosum* but it can be seen in *Cenchrus ciliaris* too. Typical rust pustules of yellow and dark brown colour are formed on the leaf blade and sheath. Heavy incidence of rusts hamper the quality of grass tremendously and also affect the perenniality of the grass stand.

Leaf spots (*Phylachora*, *Curvularia*, *Pyriculcularia*, Soros [orium, *Colletotrichum* spp.) leaf spots caused by *Pyricularia* species is another destructive diseases of grasses particularly *Pennisetum pedicillatum*, *Cenchrus ciliaris* and *C. setigerus*. Typical eye shaped necrotic lesions with reddish margins and greyish centre appears on the leaf blade and sheath. Early infection affect the growth of the grass thereby reducing the yields. Besides reduction in the yield it also hampers the quality of grass. The fungus survives on plant residues in the field thus the intensity of the disease increase in the subsequent years.

12.2. Insects

The grasshoppers are the main destructive pests of the grasses. The other pests of the grasses are yellow hairy caterpillars, hairy caterpillar, red hairy caterpillar. army worm, termites, aphid and leaf hoppers.

12.3. Nematodes

Several species of nematodes specially ectoparasitic and migratory endoparasitic have been reported to cause damage resulting in to considerable herbage and yield losses in many range species. The presence of these nematodes even in low numbers are more important during establishment phase of the range-lands.

13. HARVESTING

First cut is generally taken after four months of sowing in the first year. In second year, three cuts can be taken and thereafter from third years onwards, six cuts could be harvested.

TABLE 35.3: Dry matter and protein percentage at different cutting intervals of anjan grass

Cutting intervals (in days)	Dry fodder (t/ha)	Protein per cent
10	3.3	11.7
20	7.1	9.5
30	10.1	8.1
60	20.1	6.3
Height of cutting (cm)		
5	8.4	
10	10.0	
15	12.0	

It is evident from the data (Table 35.3) that with the increase in height of grass cutting, the yield also increases but the protein content is adversely affected. It is evident that as the height of cutting the grass was increased the yield also increases but when grass was cut at 5 cm height the yield decreased.

14. YIELD

Irrigated crop gives about 400-500 q/ha/year green forage yield, while rainfed pasture gives about 100 to 150 q/ha/years green forage yield. The first cutting can be taken in 3-3½ months after sowing, but it is desirable to delay the first cut till seed formation to allow the grass to get established better. Subsequent cuttings can be taken at 50-60 days intervals. Under irrigated condition 12 cuttings can be possible in a year. The well established rainfed crop of *Cenchrus ciliaris* gave 90 to 112 q/ha of green fodder per annum in arid tracts with less than 30 cm of annual rainfall. In regions receiving 38 to 76 cm of rainfall, the yield would be 225 to 280 q/ha per annum.

15. QUALITY

Cenchrus ciliaris ranks itself high in nutritive quality among the indigenous grasses. The protein content is about 11% in the early stages, with a 'well-balanced' ration of calcium to phosphorus. Anjan grass is relished by all classes of livestock (Table 35.4).

TABLE 35.4: Chemical composition of Anjan grass

Digestibility Nutrients (%)	Young	Prime	Ripe
Crude protein	11.2	9.5	9.1
Ether extract	1.9	1.5	1.6
N-free extract	47.6	49.6	49.5
Crude fibre	27.2	34.4	29.2
Ash	18.1	13.3	15.9
Calcium	0.50	0.48	0.43
Phosphorus	0.37	0.27	0.28
Potassium	3.41	2.52	2.27

16. REFERENCES

Basavaraju, T.V. and Rao, M.G.R. 1996. Performance of pasture grass - legume mixtures under dryland conditions of Central dry zone of Karnataka. Karnataka J. Agric. Sci. 9 (3): 536-537.

Gill, A.S. and Patil, B.D. 1988. Improvement of nitrogen fertiliser efficiency in grass production. Indian J. Range Mgmt. 9: 49-52.

Gill, A.S., Patil, B.D. and Yadava, C.L. 1980. Response of anjan grass pasture to micro nutrient elements. Indian J. Range Mgmt. 1 (2): 103-105.

Graham, T.W.G. and Humphreys, L.R. 1970. Salinity response of buffel grass. Aust. J. Exp. Agric. Anim. Husb. 10: 725-728.

Hazra, C.R. 1995, Advances in forage production technology. Technology Bulletin of IGFRI, Jhansi.

Hazra, C.R. and Tripathi, S.B. 1995. Effect of nitrogen fertilisation to anjan grass (*Cenchrus ciliaris*) with and without pasture legumes in silvipasture. Forage Res. 20 (4): 287-290.

Humphreys, L.R. 1978. Tropical pasture and legumes.

Kanodia, K.C., Rai, P, Shankarnarayanan, K.A. and Ravi, Kumar 1981. Response of defoliation management cum manuring on quantity and quality of the dhablu grass. Indian J. Agron. 26 (4): 251-256.

Kanodia, K.C., Rai, P. and Dwivedi, G.K. 1983. Effect of harvesting schedules on the above ground and below ground biomass of *Dicanthium annulatum* (Forsec) steff. Forage Res.9 (1): 13-19.

Kumar, D. and Mishra, L.P. 1998. Studies on fertiliser and row spacing in anjan grass (*Cenchrus ciliaris* L.). Paper presented at National Convention and Symposium on Sustainable Biomass Production and Rural Development held at IARI, New Delhi on March 17-18, 1988 (In abstr p 24).

Niranjan, K.P., Arya, R.L. Yadava, R.B. and Gangwar, K.S. 1995. Performance of different ley farming systems under rainfed conditions. Range Mgmt. & Agroforestry 16 (1): 93-99.

Pandey, K.C., Hasan, N., Bhaskar, R.B. and Hazra, C.R. 1995. Pest and diseases of major fodder crops. IGFRI Technical Bulletin p 20-21.

Patel, J.R., Trivedi, G.C., Patel. P.C., Sadhu, A.C., Parmar, H.P. Patel, C.C. and Gangani, M.K. 1998. Research in production technology of forage crops in Gujarat. Gayatri Computerized Printing Press, Anand pp 152-153.

Paull, C.J. and Lee, G.R. 1978. Buffel grass in Queensland. Queensl. Agric. J. 104: 57-75.

Ramia, M. and Fernandez, J.E. 1974. Pasture development in the humid tropics: its ecology and economy. Proc. Int. Meeting on use of ecological guidelines for development in the American humid tropics, Caracas Paper No. 10.

Seatoni, W.J. 1978. Atrazine tolerance in fine tropical pasture grasses. Trop. Grassland 12: 35-41.

Shankarnarayanan, K.A., Davadghao, P.M. Kumar, Ravi, and Rai, P. 1977. Effect of defoliation management and manuring on drymatter yields and quality in Sehima, *Cenchrus ciliaris* and *Cenchrus setigerus*. Annals of Arid Zone 16 (4): 441-454.

Shankarnarayanan, K.A., Davadghao, P.M., Upadhyaya, V.S., Kumar, Ravi, and Rai, P. 1976. Effect of defoliation mangement and manuring on the drymatter yield, quality in *Iseilema laxum* Hak. Forage Res. 2: 55-62.

Singh, P. 1991. *Bharat ke charagah avum chara phaslen*. Edited by Dr. P. Singh, pp 98.

Wilson, J.R. and Mannetje, L. 1978. Senescence, digestibility and carbohydrate content of buffel grass and green panic leaves in swards. Aust. J. Agric. res. 29: 503-516.

Yates, P.B. 1975. Better pasture of the tropics. Arthur Yates and Co. pty. Ltd. pp 17-18.

36. Cluster Bean

N. D. Yadava

Clusterbean is a main *kharif* legume crop which recently classified in arid legume group. it is commonly known as a Guar (Rajasthan, Punjab, Haryana), Ghurthi (UP) which is grown for vegetable, green fodder, green manure and as a grain crop. It is a main source of gum having 28-32 per cent gum. It is drought hardy crop grows well in rainfed condition of Rajasthan specially in arid parts of the state and also fitted well in the rainfall pattern of the region. India has become a leading exporter of guar gum especially to USA and earns about Rs.170 million annually.

Clusterbean is grown mainly for grain and vegetable purposes. Its seed contains 28-32% of Gum therefore it is main raw material for gum producing industries. Generally major supply of gum for paper and cotton industries, is manufactured by guar seeds. The green pods are used for vegetable purposes. Its green and dried pods both are used for vegetables. It has high caloric and nutritive value. The crop is also grown for fodder purposes. Its fodder is very much rich in nutrients. The guar seeds are also used for preparing the animal feed. Its pods and grains are also used as a medicine for diabetic patients. The toasted guar is used for poultry meal and feed for pigs. Different concentrate used for animals are also prepared by guar seeds. The guar straw is also used as nutritive fodder for animals.

1. ORIGIN

Clusterbean is grown in India from a very ancient times as a fodder and vegetable crop. Some wild plants were found growing in India so that it is believed that it has been originated from India. Vavilove (1951) has reported its origin in India. But on the basis of wild species, its center of origin in tropical Africa (Gillete, 1958). Cheivellier, (1939) reported that cultivated guar is originated from the region of Sindh. In USA, clusterbean was introduced from India.

2. DISTRIBUTION

India is a largest guar exporting country in the world. In India the Rajasthan state has maximum area as well as production which contribute about 65 % of total guar production of the country. The other guar producing states are Haryana, Punjab, Gujarat and Madhya Pradesh but have few pockets where guar is cultivated for grain purposes. In Rajasthan during 1995 total area under clusterbean was 17.7 lakh ha with the total production of 2.77 lakh tonnes. The area and productivity per ha decreased during 1989-90 in Rajasthan as well as in India. The area production of clusterbean in different states has been summarised as under.

TABLE 36.1: Area, production and productivity (Kg/ha) of clusterbean in different states of India (1989-90)

States	Area (000 ha)	Production (000 ha)	Productivity (kg/ha)
Rajasthan	1966.0	438.3	223
Haryana	182.0	131.2	720
Gujarat	80.1	46.8	584
Punjab	31.4	29.6	943
Uttar Pradesh	2.7	2.0	741
All India	2262.2	648.3	287

3. CLIMATE

Clusterbean crop has wide climatic adaptability from humid to arid climate. Generally the grain crop is grown during *kharif* season mainly as a rainfed crop. The crop performed well under rainfed condition even where the rainfall ranges between 250-500 mm under sandy soils. The normal growth and development of plants requires 27-35°C. It can tolerate shade and frost up to a greater extent. It is photosensitive crop and flowering occurs when it is sown in *kharif* season.

4. SOIL AND SOIL PREPARATION

Generally clusterbean is grown on sandy loam to loam soils, but well drained loamy sand soils having a pH range of 7.5 - 8.5 is well suited for its cultivation. The soils of acidic nature or having high level of salinity or alkalinity are not suited for its cultivation. The soils of low lying areas having water logging or excessive wetness for a longer period are not suitable for cultivation. The dune soils in arid zone of western Rajasthan have sand depositions with poor fertility status, even though the crop grows well with poor rainfall also.

For soil preparation, the sandy and sandy loam soils need not to be ploughed more but clusterbean being a deep rooted crop having tap root system the soils must be tilled 1-2 times by disc harrow followed by cultivator for making good tilth in soil. The land which was not under cultivation and was left as a fallow for a long period for more than 3 years, one ploughing by disc plough followed by one to two harrowing is sufficient for better seed germination and plant growth. It reduces the weed infestation up to a greater extent. Before sowing the weeds and unwanted vegetations and crop stubbles, must be eradicated from the field.

5. CROPPING SYSTEMS

In arid rainfed areas clusterbean is mostly grown as a mixed crop with pearlmillet, mothbean, Sesame. In the areas where irrigation facilities are available and land is sown during *rabi* and summer season the crop is grown in rotation with wheat.

In rainfed condition the most prominent rotations are Pearlmillet + Guar-fallow, guar-fallow/mustard, Sesame + Guar -fallow, Guar -taramira, Guar-Mustard, Grass + Guar- fallow etc.

Intercropping with pearlmillet have showed no adverse effect on pearlmillet but also gave the additional seed yield under rainfed condition (Singh *et. al.*, 1985). The pearlmillet grain yield was increased 22% by intercropping. Chauhan and Dungarwal (1982) reported that intercropping of one row of clusterbean between the rows of rainfed maize increased the total grain/seed yield, maize grain protein content and net profit, compared with maize yields in pure stand.

Dhillon *et. al.* (1980) found that intercropping of clusterbean in rows of ground nut 30 cm apart had no adverse effect on groundnut compared with its yield in pure stand and gave in additional seed yield of 195-222 kg/ha. Singh (1980) reported that intercropping of clusterbean with grass i.e. *Cenchrus ciliaris* and castor yielded more per unit area without affecting the yield of main crop. Parallel cropping of Castor + clusterbean in (1:2) system gave highest yield of both the crops (Yadava and Gupta, 1992). Cluserbean parallel cropped with castor produced highest drymatter (6.4 q/ha) as well as grain yield (2.03 q/ha) than other legumes under rainfed condition (Yadava, 1992).

Yadava and Gupta (1987) found that strip cropping of clusterbean with sesame gave higher yield and net profit over their sole crops. Singh and Joshi (1980) found that strip cropping of clusterbean with sesame gave higher yield and net profit over their sole crops.

6. MANURES AND FERTILISERS

Clusterbean is a leguminous crop so that it does not requires to apply heavy doses of nitrogen but it needs proper addition of phosphatic and potassium fertilization for better development of root nodules. The addition of well decomposed compost or Farm Yard Manure (FYM) at the rate of 20-25 t/ha must be mixed within the soil at 15-20 days before sowing at the time of last ploughing during soil preparation.

TABLE 36.2: Effect of nitrogen, inoculation and P and seeding rate on yield and water use efficiency in clusterbean (Adapted from Singh and Singh, 1989).

Treatments	Grain yield (Q/ha)	Stover yield (q/ha)	WUE (kg/g/m/sec)	Gum content (%)
Nitrogen and inoculation				
Control	14.2	30.1	7.15	25.75
Inoculation	18.9	38.35	8.85	29.1
20 kgN/ha	15.4	34.65	7.59	26.95
20 kgN/ha +inoculation	19.0	40.85	9.33	33.30
C.D. 5%	0.99	4.24	---	1.30
Seed rate (kg/ha)				
10	14.1	33.15	7.02	28.95
20	16.8	36.3	8.31	28.10
30	19.1	39.75	9.35	27.33

Treatments	Grain yield (Q/ha)	Stover yield (q/ha)	WUE (kg/g/m/sec)	Gum content (%)
C.D. 5%	1.18	2.89	---	0.90
Phosphorus (kg/ha)				
00	13.9	34.90	6.89	24.5
30	16.9	36.40	8.33	28.4
60	19.2	37.90	9.44	31.7
C.D. 5%	1.00	2.89	--	1.01

At the initial stages when *Rhizobium* bacteria, present in the root nodules are not well developed the application of starter dose of 15-20 N kg/ha is necessary for proper growth of the plant. All the fertilisers must be given on the basis of soil test. Phosphorus plays a major role in metabolic activity especially in protein synthesis. In general under normal fertility status of soil 100 kg Diammonium phosphate (DAP) per ha or 15 kg N and 40-50 kg P_2O_5 per ha has been recommended for getting the optimum and economical yield of crop. All the fertiliser doses must be given at the time of sowing as a basal dose in the furrows (Below or side) for their better utilization by crop plants. The broad casting method of fertiliser application should always be avoided.

The clusterbean crop grown on sandy loam soil during 1988 and 1989 showed that the crop yields were higher with single super phosphate than Diammonium phosphate over all the phosphorus doses (10, 20, 40 and 60 kg P_2O_5 /ha). Grain yield was higher where phosphorus was drilled under the seed in furrows.

6.1. Micronutrients

In clusterbean crop the effect of micronutrients were only seen in the soils which are deficient in a particular micronutrient or nutrients or due to some acidic or salinity problems. Generally Zinc and sulphur were found to be the most limiting micronutrients which has been identified in clusterbean crop. Gupta and Potalia (1988) reported that application of 5-10 ppm Zn increased average drymatter yields from 2.93 to 11.1 g per pot in a pot experiment, $ZnSO_4$ was the most effective followed by ZnO, Zn frits, $Zn_3(PO_4)^2$ and Zn NH_4 PO_4. A field experiment at Jodhpur on effect of S revealed that the application of 60 kg S/ha resulted increased dry matter yield, plant height seed and straw, yield (Shivran, et al, 1996).

7. VARIETIES

Clusterbean crop is grown for different purposes viz. vegetable fodder and grain. Therefore, on the basis of purposes the varieties are different having different characteristics in terms of their growth, development and yield. Mainly on the nature of growth clusterbean plants has been divided into two groups of varieties i.e., branched and erect type. The crop has positive correlation with vegetative growth and grain yield because pod bearing clusters are born linearly on branches of plants. The some varieties and their characteristics has been given as under.

Durgapura safed: This is branched variety, grown for grain and fodder both. It matures in 127 days and seed yield is 15.5 q/ha. It is recommended for cultivation of whole Rajasthan.

FS-277: It is a high yielding variety developed through selection from land race of Sojat and released in 1980. It has pubescent leaves with serrated margins. It is a single stem variety seeds are bold and light grey in colour and contains 28.4 % gum and 43.1 % protein. It matures in 95-105 days with an average yield of 15-16 q/ha. it is recommended for grain purposes for cultivation of whole Rajasthan.

RGC 936: It is an early maturing, photoinsensitive variety with white flowers developed from a cross of EC 248 × RGC 401 and released in 1991. Plants are branched and bushy. Seeds are medium bold and light pink in colour. It has synchronous maturity (85-90 days) with grain yield of 8-11 q/ha and is recommended for drought prone areas of Rajasthan, Gujarat and Haryana.

RGC 197: The variety was developed through selection of local races of Nagour and released in 1990. Seeds are round medium bold and grey in colour containing 33% gum. It is tolerant to drought and suitable for intercropping. It matures in 100-120 days with grain yield of 10-12 q/ha and recommended for whole country.

Durgajay: A high yielding, branched and disease tolerant variety developed through selection from local land races of Nagaur region and released in 1980. It has attractive grey seeds with lustrous seed coats and contains 30% gum and 43% protein. It matures in 110-120 days with grain yield of 8-10 q/ha and recommended for all guar growing areas of country.

Durgabahar: It is a vegetable guar variety developed through a cross between Pusa Navbahar x RGC 401 and released in 1985. It is photoinsensitive and hence suitable for both *kharif* as well as *Zaid* seasons. Plants are glabrous and un-branched with white flowers. Clusters of long fleshy pods are borne at nodes of the stem. Its green pod yield is 70-75 q/ha and is recommended for whole of the country.

RGC471: The variety was developed through selection from land races of Barmer and released in 1991. Seeds are light grey and kidney shaped. It is moderately resistant to bacterial blight and powdery mildew. It has synchronous maturity and gives grain yield of 10-14 q/ha. It is recommended for whole Rajasthan.

HG75: The variety was developed in 1982. it is branched variety with 7 branches/plant, high yielding (9 q/ha) with wide adaptability suits to high fertility soils. It flowers in 50 days with a plant height of 103 cm and test weight is 31 g. It is recommended for all guar growing regions of the country.

Naveen: It is high yielding variety developed in 1986. It is early maturing, fit well with wheat rotation in north India, plant has 5-6 branches/plant with 42-45 pod/plant. The seeds contains 36.8% gum, test weight of 28 g with an yield of 11.71 q/ha.

Maru Guar: The variety was developed in 1982, branched (6-7 branches/ plant) high yielder with an yield of 10.08 q/ha and suits to dry regions of western Rajasthan.

8. SOWING

Clusterbean is a *kharif* season crop which is generally grown as a rainfed crop at the onset of monsoon specially in arid regions of western Rajasthan. So that in general the onset of monsoon decides the sowing time of clusterbean. The optimum sowing time of crop in last week of July and 1st week of August. The crop sown after 15 August produce low yield. The summer crop is also grown for fodder or vegetable purposes, which is sown in the month of March. The crop sown early or very late from the recommended sowing time reduced the seed and fodder yield specially in rainfed condition. Very early sowing result in the profuse growth and less number of pods per plant where as in late sowing plants do not attains the proper growth and development ultimately reduces the crop yields. The study conducted at Hissar revealed that sowing on 15th July was found the optimum date with highest seed yields (Singh *et. al.*, 1979). In another study during 1980-81 it was found that clusterbean sown on 10 July gave higher seed, protein and gum yield than the crop sown on 25 June,25 July or 10 August (Jain *et. al.*, 1987).

The method of sowing affects the seed germination, plant population per unit area and crop yield. The most common practice of broadcasting should always be avoided because it gives irregular plant population, poor germination due to improper seed distribution and placement in soil and poor resource utilization. Therefore, line sowing by using seed drill or seed cum fertiliser drill or by *deshi* plough using *kera* or *pora* attachments should be done. The erect growing varieties should be sown at a row to row distance of 30 cm whereas spreading type varieties at a row distance of 45 cm.

Optimum plant population in the field is a key factor for obtaining optimum production and better utilization of natural and applied resources. The seed rate directly affect the plant population per unit area. In clusterbean seed rate depends upon the type and nature of variety, seed size and purpose (grain/ fodder) for which the crop is going to be grown. The recommended seed rates for vegetable type is 10-11, kg/ ha, for grain 15-20 kg and 30-40 kg/ha for fodder crop. In late sown condition the seed rate can be increased slightly by 3-4 kg/ha apart from the recommended doses.

8.1. Seed and Seed Treatment

Seed quality plays an important role in productivity of any crop. Selection of recommended variety for particular areas, healthy and bold seeds with better germination percentage should only be selected for sowing. To check the seed cum soil born diseases during germination and initial growth stage, the seed treatment by organo-murcurial compounds like thiram or captan or cereson at the rate of 2.5 g /kg seed or bavistine at the rate of 1.0 g / kg seed must be done before sowing. Clusterbean being a leguminous crop therefore, to increase the nodulation in plant roots, seed treatment by *Rhizobium* culture should be done. It helps in the development of root nodules as well as in fixation of atmospheric nitrogen (Table 36.2). For treatment of seed with *Rhizobium* culture, 100 g Gur or Juggary is boiled in one liter of water, solution is cooled down up to room temperature. After cooling one packet of *Rhizobium* culture is added and mixed in this solution. Mix the clusterbean seed and solution, allow it to dry in the shade for half an hour. Do not dry it in exposed sunlight.

The *Rhizobium* inoculation increased the nodulation drymatter production and nitrogenase activity under arid condition on sandy loam soils during *kharif* season (Venkateswarlu *et. al.*, 1982; Singh and Singh, 1989).

9. IRRIGATION

Clusterbean is a drought hardy crop and grown extensively under rainfed conditions. The moisture deficit at 25-30 days after sowing affects the crop growth and branching, considerably. At the most critical stages of crop growth sufficient moisture in soil is essential for better yield of crop. The most critical stages in clusterbean are as under.

1. 20-25 days after sowing (Seedling stage).
2. Branching initiation stage.
3. Grand growth pariod.
4. Pod formation stage.
5. Grain maturation stage.

Among all above critical growth stages the most critical stages are 20-25 days after sowing and grand growth period. The crop grown for fodder purposes requires 3-4 irrigations for better growth and foliage development of plants. Singh and Prasad (1977) reported that in different cultivars of guar at different growth stages the seedling stage was most vulnerable to drought and the cultivars yielded of 1.2-1.77 t seed /ha and consumed 397-404 mm water with a moisture use efficiency of 3.03- 4.38 kg seed/mm water.

Alexander *et. al.* (1988) in their experimentation on effect of irrigation timings (1-4 irrigations based on growth stages) on water use growth and yield of guar on sandy loam soils reported that three post establishment irrigations, one before flowering, one during initial pod development and one at mid-pod filling stage were found optimum. Guar crop has been found moderately tolerant to salinity and saline water irrigation. Lahiri *et. al.* (1987) reported that final plant growth and seed yield significantly reduced at 6 mmhos/cm and above. The crop is highly sensitive to water stagnation at every stage of crop growth. Therefore, water stagnation should always be avoided. Kumar, (1986) reported that the border irrigation in guar in semi arid zone 8 m wide border with 80% cutoff ratio would be required and found to be best on sandy loam soil.

10. WEED CONTROL

Weeds compete for light, space, moisture and nutrient with crop plants resulting poor yield of crops. Weeds have heavy competition at critical growth stages of crop. It was recorded that weed competition in guar crops decreased the seed yield by 70-90%. The most critical period of crop weed competition was 20-50 days after sowing (Sonani et al, 1985).

In rainfed condition where moisture is a limiting factor it becomes utmost important to timely control of weeds. After sowing up to 25-30 days the field should be free from weeds. The proper control of weeds before flowering, at grand growth period of crop increases the crop yield. The best method of weed control is to uprooted by hand weeding using *kassi/khurpi* as and when the weed plants appears in fields. Two hand weeding one at 20 days and IInd at 45 DAS is sufficient to control the weeds. It also helps in moisture conservation creating a layer of dust mulching on ground surface.

Chemical methods of weed control is also available in which basaline 1 kg ai/ha in 800-1000 lit water should be sprayed before sowing and must be mixed thoroughly in soil. Nitrofen proved effective in controlling the weeds in clusterbean by giving an equal yield to that obtained from hand weeded treatments.

11. PLANT PROTECTION

11.1. Diseases

Bacterial Blight - *Xanthomonas cyamopsidis*: The occurrence of disease shows circular water-soaked small spots on leaf, Later on these spots coalesced to form a bigger spots surrounded by chloratic area of old leaves. Finally defoliation and black spots on stem occurs. Seed treatment with hot water at 56°C for 10 minutes. Always grow disease resistant variety like HG75,182 etc. Balanced dose of N and P is necessary to control the incidence of diseases.

Alternaria leaf spot - *Alternaria brassicae:* Small dark brown colour spot appears on leaf surfaces and increased humidity spread becomes more serious. In severe condition concentric rings of dark brown conidiophores are developed. On the critical incidence percentage (>30%) spraying of 0.2% dithane Z-78. Treatment of seed with hot water at 50 C for 39 minutes before spraying of dithane M-45(0.2%) at 10 days intervals from onset of disease was found effective to control the diseases.

Powdry mildew - *Leveillula taurica:* The white powdery small spots appears on leaves stem and fruits. It reduces the photosynthesis activity of leaves the patches later on became pale and brown in severe infestation the whole plant dried up. Spraying with 0.1% karathane or dusting of sulphur @ 15 kg/ha must be done to control the disease.

Anthracnose - *Colletotrichum capsici F cyamopsicola:* During rainy season black spot appears on peteol, stem and leaves. The growth of plant is checked. Seed treatment with agrosan GN or ceresan @ 0.5 g/kg seed.

Root rot - *Pellicularia filamentosa:* On the incidence initially reddish brown discolouration starts on the stem. Later on in severe infection the tissue become darkened and may black sclerotia are formed on the plant roots.

Choose the best crop rotation to avoid soil born primary inoculam. The diseases can be controlled by spraying of 0.2% dithane 2-78.

11.2. Insect- Pest

Catterpillars: *Diacrisia obliqua* Walker: The bihar hairy catterpillers feed on the foliage and under severe attack the whole plant is destroyed and become dead. It can be controlled by spraying 1.25 litres of endosulfan 35 EC in 1000 L of water. Carbaryl 4% dusting has been also found effective to control catterpillar

Aphids: *Alphis crassivora* Koch.: The whole plant is covered by nymphs and adults. These aphids suck the plant sap and devitalize the plants. Plant flowering and seed development is severely affected which ultimately reduce the yield.

Two to three spraying of 0.1 per cent melathion 50 EC helps in controlling the aphids.

12. HARVESTING

Judgment of crop maturity is prime most important factor before harvesting the crop. The failure in maturity Judgment directly affect the yield. In clusterbean when the plants starts turning yellow and generally all the leaves fell down it is assumed that the crop is ready for harvest. The crop must be harvested before complete drying to avoid loss due to pod falling at the time of harvesting. The days to maturity is dependent on many factors and it can not be exactly decided to harvest the crops in a frame time period. Generally medium duration Guar varieties matures within 100-125 days. The crop physiological maturity occurs after 10 days of leaf fall.

The best method for crop harvesting is to cut by sickle or in western Rajasthan the plant is uprooted and left in field for sun drying for 8-10 days. The dried crop is threshed by using tractor or Power thresher. The harvested seed must be dried in sunlight to maintain the seed moisture between 8-10%. The seed must be cleaned properly and it should be free from crop stables stones and broken immature seeds. Thereafter, the seed must be graded and stored in dry and cool place.

13. YIELD

A good crop of clusterbean yields about 20-25 q/ha seed yield 25-30 q/ha of straw. The clusterbean crop grown for fodder yields 250-300 q/ha of green fodder.

14. QUALITY

TABLE 36.3: Chemical composition of fodder and grain of clusterbean

Fodder		Grain	
Constituents	**Content (%)**	**Constituents**	**Content (%)**
IVDMD	60-65	Protein	30- 35.0
Crude protein	25.7	Gum	26.8-32.2
Fat	03.5	Oil	06.1- 7.7
N Free extract	42.6	Menerals	02.99-3.75
Ash	05.7		
Fibre	12.5		

TABLE 36.4: Chemical composition of guar seed and straw

Guar Seed		Guar Straw	
Constituents	**Contents (%)**	**Constituents**	**Contents (%)**
Ash	6.1 - 6.5	Crude protein	8.8
Crude Fiber	7.0 - 8.5	Fiber	38.7
Crude Protein	40.6 - 46.5	Fat	1.8
True protein	38.4 - 41.6	N free extract	45.9

Guar Seed		Guar Straw	
Constituents	**Contents (%)**	**Constituents**	**Contents (%)**
Fat	6.7 - 7.0	Ash	4.9
Carbohydrates	24.0-30.7	Neutral dig. fiber	65.7
Calcium	0.15 - 0.20	Acid	49.3
Phosphorus	0.6- 0.73	Hemicellulose	16.4
Methionine/100 g	0.58 - 0.90	Cellulose	40.3
Caloric value	337.0-351.8	Total dig. nutrient	56.2

15. PRODUCTION CONSTRAINTS

The clusterbean is a major rainfed crop of arid zone. It is generally grown on neglected low fertile soils with poor care and management conditions. The main constraints are as under:

1. Unavailability of high yielding, drought hardy, short duration variety.
2. Poor care and management.
3. Use of low inputs viz., fertiliser and irrigation.
4. Lack of knowledge of advanced production technology.

16. REFERENCES

Alexander, W.L., Bucks, D.A., Backhaus, R.A. 1988. Irrigation and water management for guar seed production. Agron. J. 80: 447-453

Ali, M. 1983. Response of clusterbean to phosphate application and rhizobial inoculation under dryland conditions. Legumes Res. 6: 65-68.

Balyan, J.S. and Seth, J. 1985. Effect of planting gcomertar and N fertilisation on the intercropping of pearlmillet and clusterbean. Indian.J. Agic. Sci. 30: 522-523.

Chauhan, G.S., Dungarwal, H.S. 1982. Production potential of maize with different companion crops and their method of planting. Ind. J. Agril. Res. 16: 193-98.

Dhillon, G.S. Kler, D.S. and Walia, A.S. 1980. Studies on groundnut-clusterbean association. J. Res. HAU 17: 34-37.

Gupta S.P., Gupta, V. K., Ramkala and Kala, R. 1993. Response of S to some *kharif* crops grown in course textured soils of Haryana. Ann. Arid Zone 32: 255-256.

Gupta, N.K. and Potalia, B.S. 1988. Relative efficiency of Zinc carriers on clusterbean and their residual effect on succeeding wheat crop. Agril. Sci. digest India 8: 31-33.

Jain V, Yadav B. D., Sharma B.D and Taneja K.D 1987. Effect of date of sowing row spacing and varieties on yield and quality of clusterbean. Indian. J.Agron. 32: 378-82

Kumar, D. 1986. Border irrigation for guar in semi arid zone of Rajasthan. Ann. Arid zone. 25: 55-60.

Lahiri, A.N., Garg, B.K. Vyas, S.P. and Mali, P.C. 1987. Response of clusterbean to soil salinity. Ann. Arid. Zone 26: 33-42.

Shivran, A. C., Khangarot, S. S. Shivran, P. L and Gora, D. R. 1996. Response of clusterbean varieties to sulphur and phosphorus. Indian. J. Agron. 41: 340-342.

Singh, R. P. and Joshi, N. L 1980. Intercropping of pearlmillet in arid areas. Ind. J. Agric. Sci. 50: 338-341.

Singh, B. P. 1985. Effect of intercroping with pearlmillet on productivity, profitability and water use on arid soils. Ind. J. Agron. 30: 408-13.

Singh, G. S., Sharma, R. K. and Goverdhan, Singh.1982. Studies on companion cropping of clusterbean with sesame in presence of basal dose of N and P. In: Proc.Ind. Soc. Des. Tech.and Univ. Centre of Desert studies 7:29-31.

Singh, K., S. Kumar, Taneja, K. D. Kumar, S (1979).Effect of sowing dates on seed yield of different varieties of guar H.A.U.J. Res. 9: 312-16.

Singh, K.C., Prasad, M.V.R. 1977. Soil moisture depletion and moisture use efficiency of some guar varieties. Forage Res. 3: 67-69.

Sonani, V.V., Patel, R. R., and Patel, J. A. 1985. Critical period for crop weed competition in guar.Paper presented in Annual conference of Indian Society of Weed Science,1985.

Venkateshwarlu, B; Rao, A.V. and Lahiri, A. N. 1982. Symbiotic performance of Rhizobial Plastrainson guar grown in desertic soil. Egyptian J. Microbiology. 18: 1-2.

Yadav, N.D. and Gupta, I.C. 1987. Sesame and beans in strips. Intensive Agric.25:14.

Yadav, N.D. and Gupta, I.C. 1992. Crop production in hot desert of India. Advances in arid zone. 1: 142-43.

Yadava, N. D. 1992. Performance of legume under different cropping system in rainfed condition of arid western Rajasthan.(In) abstract Nat. Symp on Res. Management for sustain crop production.Feb 25-28, pp: 120-121.

37. Lucerne

S. K. Patil

Lucerne or Alfalfa is cultivated in all agricultural zones throughout India. Because of very quick growth for repeated, cuttings, high yields, good fodder qualities and agronomic importance. Alfalfa is popularly known as the "Queen" of fodder crops. The green mass of alfalfa is characterized by a high protein content (18-20 %.) well balanced with respect to amino acids. It is enriched with respect to vitally important vitamins (A, B, D, E, K,) and various micro elements essential for the normal growth and development of all species of cattle and birds. Besides, good fodder, alfalfa also performs other important economic and biological functions. It is a good predecessor for many agricultural crops, cleans the soil pathogens. It serves as a desalinatar crop in the case of secondary salinization of the soil and is a good green manure and nectar producing crop. It also reduces the devastating effect of water and wind erosion by binding the soil. The well developed root system of alfalfa in the second to third years of growth produces a large amount of root mass (80 to 120 q/ha) and stubbles in the arable layer of soil, which is equivalent to the application of 40 to 60 tonnes of farm yard manure. Cultivation of alfalfa in crop rotation, not only improves soil fertility but also increase the yield of subsequent crops. An excess of green fodder may cause bloat in animals.

1. ORIGIN

It is a native of Mediterranean region. Lucerne or alfalfa has been regarded as one of the world's valuable fodder legume which is cultivated under extremly diverse environments. In the ladakh region of India. Several variants of *M. sativa* and *M. falcata* are indigionous to the area where they have been in cultivation since ages. The crop is cultivated on a modest scale throughout India.

2. DISTRIBUTION

The lucerne crop is grown all the over world in general and in countries like India, Australia France, Japan, U.S.A., Czechoslovakia, Russia, Bulgaria, Italy and Hungary in particular. In recent years interest in alfalfa has considerably increased in every part throughout the India and the area under this crop is expanding very rapidly in the state of Maharashtra, Gujarat, Andhra Pradesh, Karnataka, Tamil Nadu and Haryana because of perennial varieties. Interest is much rising in the other states also specially Madhya Pradesh, Rajasthan, Punjab and Himachal Pradesh.

3. CLASSIFICATION

The cultivated lucerne is of tetraploid nature. The richness and diversity of alfalfa is illustrated by the fact that cultivated forms cover the entire range between extreme type *Medicago sativa*, *M. falcata*, *M. glutinosa* complex from which they are originally derived. The two closely related species M. Sativa and *M falcata* extensively been used over the past seventy years to breed superior cultivars in USA, Canada and Russia etc. Several genetically plastic ecotypes of M. Sativa and *M. falcata* of ladakh region in India, evolved through natural introgression are of tremendous value in the development of cold and drought hardy types and also disease and pest resistant varieties. Several wild forms of *M. sativa* and *M. falcata* from Mangolia, China and some parts of USSR have been used in the breeding of pasture and meadow types.

4. GROWTH STAGES

Whole plant of lucerne is used as fodder. Almost a complete range from self sterility to self fertility have been observed in lucerne. Tysdal (1977) have reported more than 80% out crossing. About 15 percent could be classed as highly self sterile clones, Male sterility in lucerne is controlled by interaction of sterile cytoplasm with one or two recessive nuclear genes (Gumanive 1976) Self incompatibility was observed some times due to low pollen fertility. The degree of out crossing and combining ability of individual plants differ widely depending upon the intra plant genetic variation, climatic conditions and availability of pollination. The population of plants with high combining ability is usually small (Tysdal 1942). The first cut will be available on 50/55 days after sowing.

5. CLIMATE

Though for vigorous growth of crop, it requires cold climate but can thrives well in arid, semi arid and dry climatic condition. It thrives well in medium to low rainfall area. Being a perennial crop it requires assured water supply during the year.

6. SOIL AND SOIL PREPARATION

Lucerne requires well drained medium to heavy soil containing calcium, phosphorus and potash. The pH of soil should be 7 to 8. It also thrives moderately in saline, water logged and acidic soils are not suitable for its growth. Field should be ploughed twice crosswise. Before harrowing the field, FYM @ 7-8 tonnes/ha should be spread and then harrowing should be done to bring soil in fine tilth.

7. CROPPING SYSTEM

Lucerne is a perennial crop. The most common rotation are maize-lucerne, pearl millet-lucerne, sorghum-lucerne, soybean-lucerne, green gram-lucerne and pearl millet (Fodder)-lucerne.

8. MANURE AND FERTILISERS

8.1. Manure

Apply 20 to 25 tons/ha before sowing along with 20-120-50 kg NPK/ha at sowing and for perennial crop 100 kg P_2O_5/ha every year. This crop gives highest response for application of 20 kg N,150 kg P_2O_5 and 80 kg. K_2O./ha at the time of sowing and thereafter, 100 kg DAP/ha every four months. The fertilisers should be applied by broadcast method. The nutrients should be applied through straight fertilisers namely urea, single super phosphate and muriate of potash. This crop yields 15% more green forage by application of sulphur 40 kg/ha. Spray of 2% boron increases 20% seed yield.

9. VARIETIES

Recommended Variety	Region of adoption	Year of release notification	Characteristic/ Yield (q/ha)
Chetak (S-244)	U.P. Punjab, Hariyana Gujarath, M.P. Delhi	1973	800 to 1000 GF plant height 70-90 cm, Bold Seed, Yellow in Colur, quick regeneration, tolerent to aphids, Free from jaor pests and disease, flowering after 180-190 days.
Anand-2	Rajasthan, Gujrath	1975	800 to 1000 (GF) plant, vigorous growing broad light green leaves.
Co.1	Tamilnadu and karnataka	1980	600 to 800 (GF) plant height 60-80 cm.
LL. Composite-3	Whole of India	1982	790 (GF) 188.00 (DM) Fast growth, resistant to downy mildew frost, and lodging, plant height 75 cm, broad leaves seed bold, first cut 65 days. Subsequent cut 40 days.
Anand-4	Gujarat, Maharashtra, Rajasthan, Uttar Pradesh, Haryana and Madhya Pradesh	1984	900-1000 (GF) 4.0 (S) Tall plants, vigorous growing, Bold Seed, more light green leaves.
LL. Composite-5	Punjab, Hariyana	1985	720 (GF) 150 (DM) 3.5 (S) Tall, erect, broad dark green leaves, thick stem, seed bold, highly resistant to downey mildew.
Sirsa 9 (Type-9)	Whole of India	1993	700 to 800 (GF) Plant vigorous, kidney shaped yellow seeds.
RL-88	All India (All zones)	1993	1000 to 1200 (GF), 287 q/ha (DF), 24 q/ha (CF), 5 q/ (S) Maturity 110-120 days (Seed to Seed) 60 to 75 cm plant height, broad leaves, quick growth, more vigorous, light to dark blue flower colour, more persistent, first cut 55/60 days subsequent cut 25 days.
LH-84	Haryana Punjab	-	1000 to 1100 (GF), 260 (DM), 20 (CP), More vigorus, Tall plants, persistent, bold seed.
Poona-1	Maharashtra	-	800 to 900 (GF), 60 to 70 cm plant, green leaves, thick stem, yellow seeds.

Recommended Variety	Region of adoption	Year of release notification	Characteristic/ Yield (q/ha)
S.627	Whole of India	-	900 to 950 (GF), Medium plant, bold seed, persistent

10. SOWING

Second fortnight of October is a best time of its sowing. However, it could be sown from end of September to first fortnight of November. Seed should be treated with *Rhizobium* culture before sowing.

10.1. Method of sowing

Crop should be sown by the broadcasting the seeds. Seeds are mixed with sand and broadcasted on well prepared seed bed and covered with 1-2 cm soil by wooden rack. For broadcasting 20-25 kg/ha seeds are required. Line sowing in rows spaced 20 cm a part by seed drill or *pora* method needs 15 kg seeds per hectare. In late sowing conditions 30-35 kg seeds are required. Seeds should not be sown more than 2-3 cm deep.

11. IRRIGATION

First irrigation should be light and with slow stream of water by which the sown seed will not move from one place to other. Second irrigation is given 3 to 4 days after first irrigation. Thereafter the irrigation interval should be 7 to 8 days in light-soil and in medium black soil it should be 10 to 12 days.

Lucerne crops needs to irrigate at 50% available soil moisture i.e. the crop should be irrigated 12 to 15 days interval according to soil. The irrigation requirement of lucerne crop is 600 delta of water mm with water use efficiency of 14 kg DM/ha/ mm. Lucerne crops is less sensitive to water deficiency and can be grown, with acceptable yields without meeting the full water requirements., Irrigation scheduling should be based on minimising water deficit during the most sensitive growth period. Irrigation in lucern may be scheduling at 50 mm CPE.

Irrigation in narrow furrows and in ditches 150 cm apart not only produced higher green and dry matter yield of lucerne but also resulted in the highest water use efficiency (10.69 kg DM/ha/mm) in medium black soil. Besides the irrigation ditches also served the purpose of drainage channels during rainy months providing congenial rhizospheric environment for root growth, stand, maintenance and crop persistence. The common method of irrigation and fertiliser combination for lucerne has been reported to be 50% available soil moisture 120 kg P_2O_5 and 40 kg K_2O/ha. with proper drainage and weed free environment proved beneficial for maintaining the persistency in production of lucerne in the medium black soils where its cultivation has been limited to annual crop only.

12. WEED CONTROL

One weeding at 20-25 days after sowing serve the purpose. Besides the common weeds amar bel (*Cuscuta reflexa*) is a parasitic weed of lucerne. It may be controlled by the removal of affected lucerne plant. Adoption of crop rotation checks its infestation. A problem is in patches herbicides paraquat one per cent spray as spot application, this damages both crop and weed. If pproblem is severe apply 2 kg/ha dichlobenil or chloropropham granular formulation. This can also be controlled by drenching 1% potassium permanganate.

13. PLANT PROTECTION

Insect pests are probably the main factor limiting fodder yield of lucerne in India. Large number of insect pest attack all parts of plants at all the growth stages

Losses due to pest attack in green fodder yield are 32.7, 39.1 and 41.5% in the first, Second and third cut respectively.

13.1. Insect Pests

13.1.1. Cultural

1. Early sowing in last week of October. to first week of November, saves the crop from severe injury of lucerne weevil. First cutting of early sown crop is done by the end of December to first week of January. The eggs and first instar larvae are carried in the field with first cut.
2. The crop must be irrigated just after each harvest so that grubs left in field are drowned away.
3. The fertiliser combination of 30 kg N_+ 50 kg P_2O_5 + 50 kg K_2O/ha offers resistance.
4. Intercropping with oats acts as a barrier for the movement of pests.
5. Least susceptible varieties viz; IGFRI-244 Anand-2 and RL-88 should be grown.

13.1.2. Chemical

1. Granular formation of insecticide malatation 5 G. or endosulfan 4 G. applied as a side dressing along the crop rows @ 5 grams/m row is effective.
2. If soil application of insecticide is not done and leaf hoppers incidence is severe, foliar application of phosphamidon, malathion or dichlorvos (0.04%) or carbaryl (0.1%) or endosulfan (0.07%) 20 days before first cut is useful.
3. The peak period of incidence of all three major pests (leaf hopper, lucerne weevil and aphids) is at the second cut stage of crop. The pests are controlled successfully by soil application of endosulfan (0.07%) after 15 days of first cut.
4. Foliar spray of endosulfan (0.07%) or dimethoate (0.04%) controls thrips. Endosulfan is safer for pollinators.

13.1.3. Residues

A waiting period of 5 days frees lucerne fodder from residues of insecticide endosulfan (0.07%).

13.1.4. Biological

Biological insecticide, *Bacillus thuringiensis* Berl. applied @ 0.84 kg/ha as a foliar spray after 15 days of first cutting is effective in controlling the pests.

13.1.5. Integrated

Insect pests of lucerne may be successfully managed by adapting improved cultural practices i.e. by sowing least susceptible variety IGFRI-244 during last week of October to first week of November with the fertiliser combination of 30 kg N + 80 kg P_2O_5 + 50 kg k_2O/ha. Crop Should be irrigated just after each harvest without any recourse to pesticide. However, the pest damage can be further reduced by the integration of these practices with insecticide endosulfan 0.07% applied after 15 day of first cutting. The adoption of these practices gives additional fodder production.

13.2. Diseases

Rust caused by *Uromyces striatus* produces radish brown pustules on the leaves and stem late in the season. The diseases can be controlled by spraying Dithane M-45 @ 0.2 percent at 15 days interval. Common leaf spot (*Pseudopeziza medicaquinis*) appears as circular spots on the leaves. The mature spot bear dark brown to black raised disc in the centre. Severe infection causes defoliation.

14. HARVESTING

First cutting is done at 50 to 55 days after sowing. Thereafter subsequent cuttings will be taken at interval of 20-25 days. The cuts should be taken at 5 to 6 cm. height from ground level. After first cutting the tillering is more, which results in increased yield, First year crop gives 10 to 11 cuts.

15. YIELD

The green forage yield is 100 to 120 Metric tonnes /ha. Seed yield is 4-5 q/ha.

16. QUALITY

The quality of green forage will affected by insects attack and heavy rainfall. The quality may be improved by following proper plant protection measures and selecting proper land. (The crop should be grown in well drained soil). The quality of green forage is reach and delicious. Animals relishes it nicely. Feeding, the green forage of lucerne reduces the expenditure on concentrates. The forage (fodder) of lucerne contains following ingredients.

Crude protein	20.24%	Crude fibre	30.13%
Fats.	2.32%	Carbohydrates	36.62%
Minerals	10.99	Calcium	1.24%
Magnesium	0.25%	Sodium	0.53%
Potassium	3.64%		

17. SEED PRODUCTION

It is beneficial to take a seed production from perennial lucerne crop. After continuous cutting for two years, leave the crop for seed purpose in the month of February. Then give the stress of watering for a period of a month and then apply irrigation water. Before irrigating the crop apply 100 kg P_2O_5/ha. At the time of 50 percent flowering of a crop spray 2% DAP and second spray should be given 15 days after first spray. This will result in 20% increase in seed yield.

Further the higher need of sugar (20.5%) was observed for the development of potential flower which was not met by its level in the collar region, because of poor translocation. The translocation mechanism responsible for higher seed yield was improved by the spray of borax 4.5 kg/ha. in three splits starting from flower initiation stage which resulted into higher seed yield.

2% foliar spray of P_2O_5 increased the seed production. Increased number of cuts from 2 to 3 increased seed yield, but further increases in cuts (3 to 4) reduces seed yield It is a cross pollinated crop. Seed setting is improved with activity of honey bees. Crop is harvested in May.

18. POST HARVESTED TECHNOLOGY

The excess green forage can be preserved after drying in shed. Also good silage be prepared and preserved for 3 to 4 months which will be useful at the time of fodder shortage.

19. CONSTRAINTS OF PRODUCTION

The major drawback in the wider scale lucerne cultivation in India is its poor seed production and unstable forage yields. Improvement in yield potential and adaptability of lucerne would depend much on the achievement made through indirect methods such as breeding for high seed production, stress tolerance, disease and pest resistance etc. For the vast tracts of the temperate Himalayas, there is a need for the development of cold and drought hardy lucerne with high degree of persistence for pasture and meadows. As the first step it would be to introduce genotypes from iso-climatic regions and cross it with locally adopted types to improve its genetic base adaptive fitness over wider areas and overall crop potentialities. The third step would be to develop hybrid populations from *M. Sativa*, *M. falcala*, *M. Corulea*, *M. glutinosa* and *M. barali* etc.

38. Berseem

S. K. Sharma

Berseem is a popular succulent, palatable, nutritious and valuable leguminous winter fodder crop. Because of its wider adaptability, diverse quality and high yielding characteristics, it is called as *King of fodder crops* for irrigated regions of north India. It has 20 per cent crude protein, 25-92 per cent fibre, 14.18 per cent ash, 1.98 per cent calcium and 0.28 per cent phosphorus on dry matter basis. Average digestibility of fodder is about 70-75 per cent. From first to third cutting, leaves are tender and stems are succulent and contain 25-28 per cent protein. As the air temperature increases number and area of leaves are reduced and stem become fibrous. Thus at this stage fodder becomes fibrous, less digestible and contain low amount of protein. Amount of dry matter ranges from 12-14 per cent from first to third cutting and 18-20 per cent at later stages.

Inclusion of berseem in crop rotation causes a positive change in physical, chemical and biological properties of soil. Residues of vigrous root system and nodules loose the compact soil, decrease bulk-density and flocculate the dispersed soil particles. Roots penetrate to a depth of 120-150 cm. in the soil though more than 80 per cent of root mass remains in upper layer of soil. Roots have 2.73 per cent N and stubbles about 1.8 per cent. One hectare crop produces 24 quintals of roots of 14 quintals of dry stubbles. Besides, there is continuous leaf fall and decay of rootlets. Thus, it improves physical condition of the soil. From soil fertility point of view, it adds 0.38 to 0.46 per cent organic carbon, 45 kg nitrogen and 15-16 kg phosphorus to the soil. Saline soils can be reclaimed by a paddy - berseem rotation.

Berseem as a green manure crop occupies an important place in the arable crop rotations of the country. It has been estimated that it adds nearly 224 kg of nitrogen to the soil and succeeding crops are often benefited to the extent of 45 kg of nitrogen per hectare.

1. ORIGIN

Asia minor is considered as the origin place of berseem. From here it spread to Seria and Philistines and then finally to Egypt. It was introduced into India from Egypt probably in 1904 by sir Fletcher for growing at Mirpur Khas farm of Sind province which is presently in Pakistan. From sind province, it was introduced in northern parts of India and to peninsular parts later on.

2. DISTRIBUTION

Berseem is a well known fodder crop of the Mediterranean and subtropical regions of the world. On large scale, it is cultivated in Egypt, India, Israel, Persia, Italy, Syria, South Africa, Cyprus, South America, Australia and Pakistan.

In India, it is comparatively a new introduction. It constitutes a prominent fodder legume in irrigated areas of northern and western parts of country. In peninsular India it is not as popular as lucern. Presently berseem is grown as a fateful *rabi* fodder crop in Punjab, Haryana, Delhi, Rajasthan, Uttar Pradesh, Gujarat and some parts of Bihar, Maharashtra and Andhra Pradesh.

3. CLASSIFICATION

Berseem belongs to the family leguminosae and genus *Trifolium*. It is a large genus containing nearly 290 temperate and subtropical species. Important species of this genus are - Berseem or Egyptian clover (*Trifolium alexandrium*), Shaftal or Persian clover (*T. resupinatum*), white clover (*T. repens*), red clover (*T. pratense*), crimson clover (*T. incarnatum*), Alsike clover (*T. hybridum*) and subterraneum clover (*T. subterraneum*) etc. Among all these berseem is the chief forage crops.

4. CLIMATE

Berseem prefers Mediterranean climate. It is successfully grown in subtropical areas where dry and cool weather prevails. Warm and humid climate as well as regions with very low temperature are not suitable for growth and development of Berseem. At the time of sowing, average daily temperature should range between 12.5 to 15.5° C. Best germination occurs at temperature of 25-30° C. Long duration of cool temperature (15-20° C) promotes growth and yield of crops. Flowering occurs at 30-35° C temperature.

Excessive low and high temperature retard the crop growth. Temperature below 5° C kills the plants. While temperature above 40° C is detrimental to crop growth and yield. It has been noted that severe frost destroys the above ground shoots and hence in Himalayan regions, lucerne, red clover and white clover have been recommended in place of berseem. Berseem requires an annual rainfall of 150 to 250 cm. for optimum yields.

5. SOIL AND SOIL PREPARATION

Well drained clay loam soils are best suited for berseem. Optimum pH of soil should be 6.5 to 7.5. Except germination stage, crop can tolerate fair amount of salts in the soil. Acid and water logged soils should be avoided. Berseem requires fine tilth and for this purpose, ploughings are done with desi plough and is followed by harrowing and planking. Finally small sized beds (1/10 to 1/20 of an acre or 400 to 800 sq.m. size) are prepared. In light soils, puddling of soil is done to prevent excessive loss of water and nutrients.

6. CROPPING SYSTEM

Berseem is grown in India with different grain and forage crops. Some important rotations are as follows:

Rice – Berseem, Rice –Berseem – Greengram, Maize – Berseem – Cowpea, Maize+cowpea - Berseem+Mustard/oats - Maize+cowpea, Napier + Berseem + Cowpea, Cotton-Berseem, Jawar + Cowpea - Berseem + Mustard-Maize + Cowpea, Hybrid Napier+Berseem-Cowpea+Maize.

7. MANURES AND FERTILISERS

Berseem is a legume crop and hence nitrogen is applied at rate of 20 kg/ha as a starter dose either in the form of ammonium sulphate or ammonium phosphate. For getting optimum yields of berseem, 24-25 t FYM, 60-80 kg P_2O_5 and 30-40 K_2O per hectare have been recommended.

Full dose of nitrogen and potash and half-dose of phosphorus should be applied as basal. Remaining half-dose of phosphorus should be applied as foliar spray @ 3 per cent water solution in the form of single super phosphate or triple super phosphophate at six to eight weeks after sowing. Experimental studies at IGFRI, Jhansi, recommended foliar spray of single super phosphate before flowering resulting in 50% increase in grain yield. For foliar spray, 40 kg P_2O_5 is dissolved in 800 liters of water. Mix and stir the solution for six days until it is sufficiently dissolved. After removing undissolved water, maintain volume of solution upto 100 L only and spray it 15 days after cutting and before flowering. Molybdenum and boron affect growth of berseem considerably. They have good effect on leafiness, carbohydrate formation, nodulation and phosphorus response. Hence 2 kg Mo and 5-6 kg borax per hectare are applied. Good yield responses have been reported at 0.50 ppm Mo and 0.5 ppm Mo + 2 ppm CO. Soil application of 2-3 kg $ZnSO_4$/ha or 1 kg $ZnSO_4$/ha through foliar spray helps in getting higher yields of green fodder.

8. VARIETIES

On the basis of stem branching habit, three types of varieties are recognised for berseem which are given in following Table 37.1.

Varietal type	Stem branching habit	Number of cuts taken
Mescavi type	After cutting short side branches develop at plant base	5-6
Fahl type	No side branching after cutting but branches develop freely from upper part of stem	2-3
Saidi type	Intermediate between above two types. It develops basal shorts for a shorter period and upper part of stem branches	One only

Most of the varieties grown in India belong to Mescavi type. Fahl types are also grown in northern India but saidi type are not suitable for cultivation in India.

Based on nature of breeding and chromosome number varieties have been further grouped in two groups viz.; tetraploid and diploid. In brief, distinguishing characteristics of these groups of varieties are given as under:

Character	Diploid variety	Tetraploid variety
Plant characters	Leaves normal, lighter and stems and petiole less thicker than tetraploid type.	Leaves thicker, darker, broader and multifoliate. Thicker stem and petiole.
Regeneration capacity	Low	High
Temperature tolerance	Heat tolerant and cold susceptible Good for frost free areas	Cold tolerant and heat susceptible
Growth rate	Growth in later stages is poor	Growth in Initial stages is poor
Seed size	Smaller	Bigger
Seed rate	25-35 kg/ha	35-45 kg/ha
Green forage yield	800-1000 q/ha	1000-1200 q/ha (20% higher than diploid varieties)

Important well known diploid and tetraploid varieties may be enlisted as given below:

1. **Diploid varieties:** Mescavi, BL-1, BL-2, BL-10, IGFRI-S-99-1, Khadrabi, UPB-101, UPB-103, UPB-104, UPB-105, IGFRI-S-54.
2. **Tetraploid varieties:** Pusa Giant, T-526, T-529, T-560, T-561, T-674, T-678, T-730, etc. In old varieties of berseem, Mescavi and khadrabi are very popular. BL-1 variety is very good for growing in Punjab and adjoining areas. JB-1 and C 394 are suitable for certain parts of M.P. and U.P.
3. **Pusa Giant Berseem:** It is a tetraploid variety developed from diploid variety C-10 at IARI, New Delhi. It gives almost 20 to 30% more forage yield than diploid varieties. Its dry matter content is slightly lesser in first cut but in subsequent cuttings this effect is not observed. It is a winter hardy crop so can easily withstand frost. Seed yield of this variety is almost 40-50 per cent higher than diploid varieties. It is very rich in calcium. Average yield is about 1400 to 1600 q/ha.

9. SOWING

9.1. Time of Sowing

Berseem requires 25-30° C temperature for higher germination. Therefore, sowing time should be adjusted according to the agro-climatic regions that neither it should correlate with higher temperature and mansoon period nor it should be delayed to face low temperature. In north India, first fortnight of October is optimum sowing time while in Bengal and eastern region, it is sown from beginning of November to the first week of December.

9.2. Method of Sowing

Berseem can be sown by following methods:

1. **Broadcasting in standing water:** This is the most popular, easiest and quickest method. After preparation of well leveled small size beds, water is filled up to 5-6 cm depth. Soil is leveled by a planker. Treated seeds are broadcast uniformly in a crosswise direction. When water dry up in beds, a

thin soil layer is deposited on the seeds. It protects the seeds against birds and maintains good soil moisture which results in higher germination.

2. **Dry sowing method:** This method is similar to the method used for *rabi* crops. For proper germination, field should be well pulverised and should have optimum moisture content. Dry sowing can be done by two methods
   a) **Broad casting method:** By broadcasting the seeds on soil, planking is done. Seeds should not be sown more than 1-1.5 cm depth in soil.
   b) **Line sowing method:** Seeds are sown at 15 cm. row to row spacing by seeddrill at not more than 1.5 cm depth.
   c) **Berseem sowing in standing paddy crop:** This method is suitable for late transplanted paddy fields. 12-14 hours water soaked berseem seeds are broadcast in standing paddy crop in last week of September. When paddy crop is harvested, berseem crop grows to 15-20 cm height. This way the cost of land preparation is saved and timely sowing becomes possible. But this method requires 10% higher seed rate than recommended rate.

9.3. Seed Treatment

In berseem, two types of seed treatment is done which is mentioned as under.

1. ***Rhizobium* culture treatment:** The seeds should be inoculated with *Rhizobium trifolii* bacteria before sowing.
2. **Seed treatment for elimination of kasani seeds (*Cichorium intybus*):** Berseem seeds get mixed with chicory (kasani) seeds (crop associated weed). For their elimination, soak the berseem seeds in 15-20 per cent salt solution. Kasani seeds being lighter float on the surface. Remove floating seeds of karani and treated berseem seeds should be washed twice or thrice in clean water. Dry the seeds in shade. Now seeds are ready for sowing.

9.4. Seed Rate and Spacing

For a diploid variety, 25 kg/ha and for a tetraploid variety, 35-40 kg seed per hectare have been recommended. But for Pusa Giant berseem which is bold seeded tetraploid variety, 50 kg seed per hectare is required. Experimental studies showed that Mescavi (diploid) varieties grow more vigorously during the spring season and Pusa Giant (tetraploid) in the winter season. Tetraploid varieties grow more vigorously during winter season but their growth become slow during later stages (that is in Feb-March). During this period (spring season), diploid or mescavi varieties yield vigorously. Therefore, it is always better to mix the seeds of diploid and tetraploid together to get higher fodder yield from first to last cutting. In general, a mixture of diploid and tetraploid seeds in the ratio of 2 : 1 out of a total seed rate of 35-40 kg/ha has been proved superior to their sole cropping due to different regeneration capacity and economic characters. Experiments conducted at NDRI, Karnal and other research institutes have led to the conclusion that mixture of mescavi (diploid) and Pusa Giant (tetraploid) in the proportion of 18.75 kg: 6.25 kg, 25 kg: 12.5 kg or 37.5 kg: 12.5 kg produced the higher yields. Broadcasting is the most common method. Even then, for line sowing method, 15-20 cm. row spacing is recommended.

9.5. Seed Rate for Crop Mixtures

Several crops like Japanese rape, Chinese cabbage, barley, oat and senji are good companion crops for berseem. For mescavi varieties sown in last September to early October, with recommended seed rate (25 kg/ha), an additional amount of 2 to 3 kg of Japanese rape or Chinese cabbage is mixed. But this crop mixture is not suitable for delayed sowing due to poor yield. For delayed sowing during last of October to middle of November, 50 kg oat seed are sown across the broad cast or 20 cm. apart rows of berseem @ 30 kg seeds/ha. Oat gives higher yields in first cut as berseem establishes very slowly in beginning. Berseem seeds can be broadcasted in the standing crops of maize, cotton, paddy and hybrid napier after middle of September to early November.

10. IRRIGATION

Total water requirement of berseem is about 80-110 cm. About 15 irrigations (1.35 ha/meter water) are required for obtaining higher yields. One pre-sowing irrigation of 100-150 mm. help in quick germination and better establishment of crop. After this, first two irrigations should be light (5 cm) and in quick succession of one week. Subsequent irrigations are applied, at 10-15 days interval depending upon season (About 7-8 cm. water in each irrigation). Interval varies as 15-20 days in winter, 10-12 days in spring and 7-10 days from middle of April to mid May. In multicut berseem, timing of irrigation is very important for better utilization of water. Irrigation should be given a week before cutting to ensure quick and better growth. Border strip and check basin methods are suitable methods. Total ET of berseem, varies from 850-950 mm depending upon season and place. In berseem, 90% moisture is extracted from 0 to 90 cm. soil layer. But as the soil moisture tension increases depth of soil moisture extraction increases. Based on IW/CPE a ratio of 0.9 to 1.2 is recommended for scheduling irrigation in berseem. As a general rule, optimum forage production is possible only when available soil moisture level remains above 50 per cent at all times.

11. WEED CONTROL

Chicory (*Cichorium intybus*), dodder (*Cuscuta* spp.), Pyazi (*Asphodelus* spp.), Bathua (*Chenopodium* spp.) and Hiran Khuri (*Convolvulus* spp.) are important weeds of berseem. Out of these chicory is the major and trouble some weed of this crop. It is so notorious that it even overturns the entire berseem field. Leaves and roots of chicory cause a bad effect on animal health and milk yield because it contains high silica, crude fibre and less protein content. Thus, for control of this weed, attention should be given from beginning and following steps can be followed:

1. Eliminate chicory seeds from berseem seeds by dipping all required seeds into 15-20 per cent salt solution. As the chicory plants are observed in the field, remove them regularly by uprooting.
2. In berseem field, following herbicides can be used for effective weed control.

Name of herbidice	Nature of weed control	Dose (kg/h)	Time and stage of application
EPTC	Annual weeds	1.5	Pre-plant incorporation
Butachlorr	Broadleaf weeds including chicory	1-2	Pre-emergence
MCPB	Broadleaf weeds	0.5	Post emergence at 3-4 trifoliate leaf stage of crop
Fluazifop-p	Problematic grasses	0.12-0.25	Post-emergence at 3-4 trifoliate leaf stage of crop
Dinoseb	Chicory	3.5-4.0	Post-emergence at 2 weeks after crop emergence

12. PLANT PROTECTION

12.1. Insect-Pest

Surface caterpillars (*Agrotis segetum*) and cottonworm (*Prodenia littoratis*) infest heavily early sown berseem. For their control, following measures should be adopted.

1. Delay the sowing of berseem about a fortnight in areas where these pests show a regular occurrence.
2. Apply liberal irrigation after harvesting of previous crop.
3. Remove associated host plant host plant.
4. In case of severe infestation, re-sowing of crop is better alternative.

13. DISEASE

Following diseases are very important for berseem crop.

Name of disease	Stage of attack	Symptoms	Control
Damping off or stem rot of berseem (*Rhizoctonia solani*)	At all stages of growth	At seedling stage, yellowish patches develop on hypocotyl which turn to dark brown at later stages. Finally plants die.	Follow good drainage Adopt clean cultivation practices
Root and brown rots (*Rhizoctonia solani* and *Fusarium spp.*)	At later stage of crop growth	---	No effective control, ploughing down of crop is recommended

14. POLLINATION

Honeybee is the most important and effective pollinator in berseem crop. So, for getting higher yields rearing of 3 to 12 hives of honeybee per hectare have been recommended.

15. HARVESTING

Berseem is a multicut fodder crop. Total number, timing of first and subsequent cutting and height of cut from soil surface affects the total green forage yield, dry matter and protein content of berseem. Therefore, harvesting of berseem may be discussed in heads given below:

1. **Timing of first cutting:** It depends upon the full establishment of all plants, soil fertility status and temperature during crop growth. In general, first cut of berseemn is taken 45 - 50 days after sowing when plants are 30 cm. tall. Before 40-45 days, all plants do not develop fully and hence in subsequent cuttings weak plants die and gaps are observed in the field. In winter season due to low temperature first cut is delayed by one week because of slow growth of plants.
2. **Timing of subsequent cuttings:** Experimental studies concluded that after initial cutting, active period of berseem growth ranges between 25 to 30 days interval is most suitable 50 cuttings at 25-35 days from February to middle of May.
3. **Number of cuttings:** If the crop is grown for green forage, then 6-8 cuttings are taken. Greater frequency of cuttings affects adversely the growth and yield of successive cuttings. In late cuttings, plants become fibrous and hard. They have low feeding value. For seed crop, 2 to 3 cuttings are sufficient.
4. **Height of cut from soil surface:** It should be harvested 6-10 cm. above the ground level. It protects the crown buds present in basal portion of stem which help in faster development of young basal branches. Further, remaining few leaves on left portion of stem help in photosynthesis and hence quicker growth of plants.
5. **Cutting management for a seed crop:** För a better seed crop of berseem, cutting management is very essential and following points should be kept in mind.
   a) Crop should be left for seed production after only 2 or 3 cuttings in the last week of February or first week of March should be taken for a Seed crop.
   b) Timing of last cut should be so adjusted that temperature as well as number of pollinating insects may be optimum. At high temperature flower will wither before pollination and the number of pollinating insect is also reduced. This results in poor seed yield.
   c) Seed crop should be harvested when seeds balls look coloured and leaves start drying up.

16. YIELD

16.1. Forage Yield

Average forage yield of berseem is about 1000-1200 q/ha. Green forage yield in different cuttings is given below:

No. of cutting	Av. Forage yield (q/ha)
1st	175
2nd	225
3rd	225
4th	175

No. of cutting	Av. Forage yield (q/ha)
5th	125
6th	75
Total	1000

By mixing of Japanese rape, Chinese cabbage, Oat and senji with berseem seeds, yield can be increased upto 1200 q/ha. Since it will boost up the yield in first cuttings.

16.2. Forage and Seed Yield

About 600 to 800 quintals of green forage from three to four cuttings and 6-8 quintals of seed can be obtained from a healthy crop.

17. QUALITY

Quality of berseem is affected by following factors:

17.1. Climatic Conditions

Dry and cool climate increases protein content and reduces amount of fibre in fodder. High temperature during February results into fibrous stems of poor digestibility.

17.2. Soil pH

Berseem requires neutral pH soil. Soil should be fairly rich in phosphorus, potash and calcium. Good amount of phosphorus, potash and calcium increases dry matter content as well as the protein content. Saline soils affects adversely the succulency and feeding value of plants.

17.3. Type and Availability of Nutrients

Effect of phosphorus on quality of berseem is well established. Basal dose of 30-40 kg P_2O_5/ha followed by 3% triple superphosphate spray 6-8 weeks after sowing maintain optimum ratio of Ca:P, Mo:P and N:P in the plants which govern the succulencey and degree of toxicity and hence feeding value of fodder. Narrow Ca:P ratio when corresponded with optimum level of Mo and Co resulted into more leafiness, more carbohydrate formation, higher protein content. It also enhanced palatability of fodder.

17.4. Number and Interval of Cuttings

Qualitative value of berseem changes according to number and interval of cuttings in berseem. It has been noted that dry matter and fibre content increased while protein content of berseem fodder decreased from first cutting to fifth cutting. Dry matter increased from 9.0 per cent in the first cut to almost double (18%) in fifth cut, the protein content decreased from 21 to 16 per cent and crude fibre rose from 18 to 28 per cent. Among nutrients, phosphorus and calcium remain constant but potash showed a decrease from first to fifth cut. Increase in cutting interval increased dry matter but reduced the phosphorus and protein content in fodder.

17.5. Oestrogenic Compounds

Berseem contains oestrogenous substances. Their amount vary according to maturity of herbage and season of year. Oestrogenous substances regulate the sexual activities in animals and affects the heat or heat cycle in females of mammalian species. It has been noted that in young age and spring season berseem have more oestrogenous activity. If fed continuously in excess, may cause bloat, more specifically called the tympanitis. Further, grazing of berseem when it is wet with due may cause same disorder. To prevent bloat, spray the crop before grazing, with linseed or mustard emulsion. Phosphorus deficiency increases oestrogenic isoflavones in berseem.

18. POST HARVEST TECHNOLOGY

Berseem can be preserved as hay and silage. Leguminous crops like berseem is less suitable for silage making because it contains low amount of soluble carbohydrates (less than 10 per cent), high moisture content and high buffering capacity. For good silage making, crop should contain over 10 per cent water soluble carbohydrates, for example, maize, millets, oats and grasses. Hay making from berseem is comparatively easy for Indian farmers in view of climatic conditions. Time of harvesting is important for harvested at early bloom to full bloom stage for hay making. The crop is dried in the field to 25% moisture or less with a little loss of leaves, green colour and nutrients as possible. Loss of fodder and feeding value may be upto 10% in field drying. In hay making dry matter loss in field drying may be upto 10 to 25%.

To obtain good quality silage, berseem should be harvested at proper stage when moisture content of plants may be 55 to 75% and carbohydrate contents are high. Lactic acid fermentation occurs properly when green chopped plants are tightly packed in earthen pits or iron silos without entry of air, Glucose and fructose, the major carbohydrate sources for fermentation, are required at a minimum level of 6-7% on dry weight basis. At these levels, fermentation provides pH lower than 4.0. Molassed (3-4 kg) or beet pulp (6-7 kg) or grinded cereal grains (7-12 kg) can be added per 100 kg of green fodder as preservative for inhibiting butyric acid type of fermentation.

SUGAR CROPS

39. Sugarcane

B. Sundara

Sugarcane is the most important source of sugar in the world. About 60 per cent of the world's white sugar is manufactured from sugarcane. It is grown is India since time immemorial. Our Vedic literatures (1000 BC) make mentions of sugarcane. Alexander the Great (326 BC) and his soldiers found honey yielding "reed" in the Indian sub-continent. In fact, its west-ward spread was due to his invasion. It is presently grown in as many as 100 countries of the world. India is the leading nation both in area and production. A great deal of research has gone into evolving high yielding hybrid varieties and production technologies. In this chapter, improved production practices as applicable to our country are detailed.

Sugarcane is an important commercial crop of India supporting the second largest agro-based industry - the Sugar Industry - next only to the Textiles. It provides raw material for around 450 sugar mills in the country which utilise 45 per cent of the sugarcane produced. The balance sugarcane is utilised by the Gur and Khandsari industries - the two important village and cottage industries. About 35 million farmers are engaged in its cultivation and an equal number of agricultural labourers depend upon it for their livelihood. About half a million industrial workers are engaged in the sugar industry in the country.

At present sugarcane is viewed not only as a raw material for sugar, gur and khandsari productions, but also for a host of other industrial uses and is considered as a major agro-industrial crop of the future.

Sugar industry by-products (more aptly and increasingly referred to as "co-products") have many uses. Bagasse, the fibrous material left behind after crushing is used as fuel in the sugar mills. Excess bagasse is increasingly used for co-generation of power which is fed to electricity grids and thus revenue is earned. Bagasse is also used for paper and pulp manufacture. The molasses is the raw material for potable and industrial alcohol production. It is also used in several cattle feed preparations. The 'press mud' is a rich source of organic matter and nutrients and thus has great value in agricultural production. Sugarcane has been considered as an alternate and renewable source of energy. In Brazil, alcohol produced from juice is extensively used as motor fuel. About two per cent of the sugarcane produced in the country is used for chewing and juice (beverage) purposes.

Sugarcane is a major commercial crop in many countries of the world having profound influence on their economies. Juice in cane is 65 per cent. The composition of cane juice is given below:

Water	77-88 per cent
Sucrose	8-21 per cent
Reducing sugars	0.3-3%
Organic substances	0.5-1%
Ash	0.2-1%

1. ORIGIN

Daniels and Roach (1987) have reviewed in great detail the origin and evolution of sugarcane. Places of origin of different species of *Saccharum*, presently accepted by most workers, are indicated in Table 39.1.

TABLE 39.1: Origins of different species of sugarcane

Species	*Origin* place
Saccharum officinarum L.	New Guinea and East Indonesia (East of Wallace Line)
Saccharum barberi Jesw.	North India
Saccharum sinense Roxb.	China
Saccharum spontaneum L.	Colder regions of Sub-tropical India
Saccharum robustum Brandes Jews Ex. Grassl	New Guinea
Saccharum edule Hassk.	New Guinea

2. DISTRIBUTION

Sugarcane is a tropical plant, hence grows best in the tropical regions. However, its cultivation is spread out both in the tropical and the sub tropical regions of the world. India, Brazil, Cuba, China and Thailand are the top five sugarcane growing countries of the world (Table 39.2).

TABLE 39.2: Area, production and productivity of some important sugarcane growing countries (1995)

Country	Area (m.ha)	Production (m.t.)	Cane yield (t/ha)
Brazil	4.54	301.59	66.37
India	4.14	282.95	68.37
China	1.19	70.92	59.71
Cuba	1.05	36.00	34.29
Thailand	0.92	50.60	54.84
World	18.32	1147.99	62.67

TABLE 39.3: Area, production and productivity of some important sugarcane growing states of India

State	Area ('000/ha)	Production ('000 t)	Yield (t/ha)
Uttar Pradesh	1974	119830	60.7
Maharashtra	580	46656	80.4
Tamil Nadu	326	32944	101.0
Karnataka	313	24918	79.6
Andhra Pradesh	214	15180	71.0
India	4139	282895	68.4

In India sugarcane is grown in most of the states excepting Jammu and Kashmir and other hilly regions. Uttar Pradesh has more than 50 per cent of the area, followed by Maharashtra, Tamil Nadu, Karnataka and Andhra Pradesh (Table 39.3). The other important sugarcane growing states are Punjab, Haryana, Bihar, Gujarat, Madhya Pradesh and Orissa.

3. CLASSIFICATION

Linnaeus classified sugarcane as *Saccharum officinarum* in 1753.

The genus Saccharum has 6 species which are briefly described below. The key to the classification of the species of *Saccharum* (Jeswiet, 1925 modified by Brandes, 1958) is given by Daniels and Roach (1987).

3.1. *Saccharum officinarum* L.

It is known as *noble cane* since it is rich in sucrose. Relatively low in fibre, the stems are thick and long, and vigorous in growth. The leaves are broad and long. It has been the cane used for sugar production and the species usually referred to when we speak of sugarcane. However, none of the original officinarum is commercially grown at present. It has 2n = 80 chromosomes.

3.2. *Saccharum barberi* Jesw.

The species derives its name from the scientist Dr. C. A. Barber who described these original canes of sub tropical India. This is known as Indian species. These are sturdier and disease resistant, and have reasonably higher sugar and fibre contents. These were under large scale cultivation in the sub tropical India for *Gur* and *Khandsari* manufacture. Clones of this species are highly tolerant to high and low temperatures, problem soils and waterlogged conditions. Because of their poor yields, they are not commercially grown at present. They have 2n = 81-124 chromosomes.

3.3. *Saccharum spontaneum* L.

This is also known as "wild cane". Considerable variation is noticed in the morphology of the species. In general the cane is very sturdy and resistant to most diseases. The sugar content is very low and the species is not used for sugar production, but is extensively use in breeding to evolve hybrids resistant to diseases and abiotic stresses. The Spontaneums have a range of chromosome numbers of 2n = 40-128.

3.4. *Saccharum robustum*

This species was discovered in New Guinea. The stalks are long, thick and vigorous, rich in fibre and poor in sugar. This is also a wild species and not used for sugar production. Their chromosome numbers are 2n = 60, 180, 114-205.

3.5. *Saccharum edule* Hussk.

The characteristics are similar to *S. robustum* except that *S. edule* rarely produces normal inflorescences. It is therefore not used in breeding. It is grown as a vegetable crop in New Guinea, the edible part of the plant being the swollen and aborted inflorescences. The chromosome number is 2n = 60-80.

3.6. *Saccharum sinense* Roxb.

Indigenous to North East India, cane is long, thin, with broad leaves. It contains low to medium sucrose with long and zigzag internodes and prominent nodes. It is early maturing.

4. GROWTH STAGES

Sugarcane, a C_4 plant, is considered as one of the most efficient converters of solar energy. It has essentially four growth phases as indicated below though it is difficult to recognize distinct duration of each.

4.1. Germination Phase

The germination phase is from planting to the completion of germination of buds. Under controlled conditions, the germination occurs within a week to 10 days. But under field condition, the germination phase usually lasts for about a month and may even prolong upto 45 days. In sugarcane germination denotes activation and subsequent sprouting of the vegetative buds present in the stem cuttings (setts) used for planting. The germination is influenced by the internal factors like bud health, sett moisture and sett nutrient status; and the external factors like soil moisture, soil temperature and aeration. Of the external factors, temperature exerts the greatest influence. Optimum temperature for sprouting is about 28-30° C and the absolute lowest temperature is around 12°C. Soil moisture between 15 and 25% in the top 15 cm. is optimum. Warm, moist soil ensures rapid germination.

4.2. Tillering Phase

Tillering provides appropriate number of stalks required for a good yield. Tillering phase lasts for about 3 months from the germination stage. The general course of tillering in sugarcane resembles that of tufted grasses. Soon after planting, the buds of cutting start developing shoots. These are the "mother shoots" or "primaries". The little stem of these primaries consist of many short internodes, each of which carries a lateral bud. These buds give rise to the "secondary" shoots or tillers which in their turn produce tertiary shoots and so on. Tillering is a continuous process. But under field condition, it is suppressed due to closing in of the canopy. Maximum shoot population reaches around 90-120 days. By around 150-180 days, at least 50 per cent of the shoots die and stable population is established.

Tillering is a varietal character and it is influenced by light, temperature and soil moisture. Light is the most important external factor. Adequate light should reach the base of the sugarcane plant for better tillering. Around 30°C temperature is optimum. Cultivation practices such as spacing, manuring, irrigation and weed control influence tillering. Though 6-8 tillers are produced from a bud, ultimately only 3-4 remain to form canes under the usual growing conditions.

4.3. Grand Growth Phase

Once the cane formation starts at the end of the tillering phase, the stalks grow rapidly almost at the rate of 4-5 internodes per month under favourable conditions. During this phase leaf production and their growth is frequent and rapid. A well grown crop may reach a leaf area index (LAI) of 7-8. The cane elongation is facilitated by adequate moisture and nutrients, and, warm, sunny and humid climatic conditions. A temperature of around 30°C with 80 per cent humidity is most conducive for growth. During the grand growth phase, actual cane yield build up and development of sink for later sugar storage take place. The grand growth continues upto 270 days in a 12 month crop.

4.4. Maturation and Ripening

Maturation and ripening phase or sucrose synthesis and accumulation phase lasts for about 3 months, i.e., from 270 to 360 days period in a 12 month crop (Eksali). During this phase, vegetative growth is reduced and rapid accumulation of sugar takes place. As ripening advances, simple sugars (monosaccharides, viz., fructose, glucose) are converted into cane sugar (sucrose, a disaccharide). Cane ripening precedes from bottom to the top. Ample sunshine, clear skies, cool nights and warm days and dry weather are highly conducive for ripening. In the country, highly favourable conditions are found in certain parts of Maharashtra, northern Karnataka and south Gujarat.

5. CLIMATE

Sugarcane being a tropical plant, grows best in hot, sunny areas. However, it is extensively cultivated both in the tropical and sub tropical climates. Mangelsdorf characterized an ideal climate for production of sugar from sugarcane as follows:

1. A long, warm summer growing season with adequate rainfall
2. A fairly dry, sunny and cool but frost free ripening and harvesting season
3. Freedom from typhoons and hurricanes

Biswas (1998) has comprehensively reviewed the influence of climate on sugarcane. Temperature, light and moisture are the principal climatic factors that control cane growth.

Optimum cane growth is achieved in temperatures between 24° and 30°C. Fluctuations in temperature have great influence on enrichment of sucrose; so does the daily amount of sunshine.

Sugarcane is a sun-loving plant and as such day length has a strong positive correlation with growth, yield and sugar accumulation.

For obtaining high yields, a rainfall of 2000 to 2500 mm per annum, evenly distributed is ideal. But in India, it is grown in areas with rainfall ranging from 600 mm to 3000 mm. During the growth phase, the rainfall encourages rapid cane elongation. But, during the ripening phase it is undesirable as it leads to poor juice quality by encouraging vegetative growth. It also interferes with the harvesting and transport operations.

High atmospheric humidity is favourable for active growth. But under adequate irrigation it may not have much influence. During the ripening phase moderate values (45-65%) with limited irrigation are favourable.

In the country, highest yields are recorded in Tamil Nadu and best sugar recoveries in Maharashtra. These are related to climatic conditions as indicated by a brief analysis made by the author (Sundara, 1998).

6. SOILS AND SOIL PREPARATION

Well drained loamy soils, neutral in reaction (pH 6.5 to 7.5) are the most ideal for sugarcane cultivation. However it is grown in soil types ranging from heavy clays to extremely sandy ones and, under wide variations in the soil reaction. A cation exchange capacity of 15 meq/100 g soil is considered optimum. At lower CEC leaching losses of nutrients are higher. An organic matter levels of 2-4% are sufficient.

A soil slope of 1-3 has been considered ideal. Higher values are preferred in heavier soils. This permits wider choice of field layouts and most economical cultivation. The best rooting medium is provided by about 1 m of stable, well structured loam to clay loam soil. A bulk density of 1.4 g/cc, pore space of a minimum 50%, good drainage facility with a groundwater table depth of not less than 1.5 to 2 m and available water holding capacity of 15 cm/m are ideal soil physical conditions (Black Burn, 1984).

In India, alluvial, black, red and laterite soils are the major soil, types on which sugarcane is grown. Alluvial soils occur mostly in Punjab, Uttar Pradesh, Bihar, Gujarat, West Bengal and to a considerable extent in the southern states also. The texture of these soils vary from sandy to clayey. The soil reaction is neutral to alkaline and it is moderately fertile.

Sugarcane is cultivated in black soils mostly in Deccan canal tract areas of Maharashtra and parts of Karnataka, Andhra Pradesh, Madhya Pradesh and Gujarat. These soils are alkaline in reaction and contain free calcium carbonate ranging from 1.0 to 20%. These are shallow to fairly deep, low to medium in fertility and moderately clayey to clayey (40 to 70 per cent clay). These soils have high water holding capacity and poor to moderate drainage capacity because of low infiltration and permeability rates. Compaction or hard pan formation, poor drainage and aeration, and salinity/alkalinity are the chief problems. Lime induced chlorosis is also a common problem. It requires improvement of structure through organic matter additions, occasional deep ploughing, and provision of drainage. Such soils are also seen in Tamil Nadu sugarcane growing tracts.

Sugarcane is mostly grown in red soil in parts of Karnataka, Tamil Nadu, Andhra Pradesh, West Bengal and Bihar. These soils are normally shallow to medium deep with low fertility status, have excellent drainage and poor organic matter and are near neutral to acidic in reaction. These are sandy loam to loam in texture and low in lime, magnesium, phosphate and nitrogenous contents. These respond well to manuring and nitrogen. Because of poor water retention capacity, the soils require frequent irrigation. In calcareous red soils, lime induced chlorosis occurs.

In Kerala, parts of Karnataka, Goa, Maharashtra, Andhra Pradesh, Orissa, Assam sugarcane in also grown on laterite and lateritic soils. These are fairly deep, highly leached, well drained, low in pH (5 to 6.5) and of low fertility status. Liming is important to get better yields.

Sugarcane crop remains in the field for 2 or 3 years due to the practice of ratoon cropping. Therefore, a thorough soil preparation to a depth of 45 to 60 cm is essential as about 75 per cent of the root system is present at this top layer. A good soil preparation should involve the following essential steps:

- management of preceding crop residues;
- tillage (ploughing, harrowing, sub-soiling, etc.);
- levelling;
- incorporation of organic manures and
- field layout with irrigation and drainage channels.

A 'rotovator' is a very useful implement to incorporate preceding crop residue and also for in situ incorporation of green manure crop. Special care is required to prepare the land after a rice crop, which is a common rational crop in sugarcane growing areas. Tractor drawn implements for ploughing and other soil preparatory cultivation give best results than bullock drawn implements. A proper field levelling is essential to ensure uniform crop stand and a good irrigation layout. Levelling can be done by tractor drawn leveller. In the sub tropics, soil is prepared by repeated ploughing using wooden ploughs and then planked. This type of soil preparation helps in the conservation of soil moisture.

7. CROPPING SYSTEM

7.1. Crop Rotations

In India, monocropping of sugarcane is not practice. Sugarcane plant and ratoon crops are preceded or succeeded by some rotational crops. The choice of rotational crops depend upon (1) agro-climatic conditions/regions; (2) available resources particularly water, labour and monetary; (3) farmer's need for food, fodder, etc., and (4) availability of market for cash crops and surplus food crops grown.

In the sub tropics sugarcane is grown in rotation with *rabi* crops such as wheat, potato, gram, mustard and peas; and also, rice, maize, sorghum and cotton. In the eastern part of the country, it is rotated with rice and jute. In Maharashtra, the common rotational crops are wheat, gram/sorghum and cotton. In many areas of tropical states, particularly in the canal irrigated tracts, and in the coastal areas rice is the most common rotational crop. During summer, summer pulses such as blackgram, greengram, are grown in rotation. In lift irrigated garden lands, potatoes, chillies, sweet potato, onion, turmeric are rotated. Near large cities and towns quick growing vegetables like radish, coriander, cucumber etc., are grown as rotational crops.

7.2. Intercropping

Sugarcane is grown in India in row widths ranging from 60 to 135 cm depending upon the region, soil type, soil fertility and variety. Canopy closure takes around 3-4 months. During this period, much of the radiant energy goes waste. The space between the cane rows remains unused giving way for weed growth thus depleting moisture and nutrients. Hence during this period intercropping of short duration crops using appropriate varieties is possible. Wheat and potato are the two most successful intercrops in the subtropics in the autumn planted sugarcane. Other winter crops are also grown as intercrops. In the tropical states, blackgram, greengram, soybean, onion, etc., can be grown as intercrops.

In managing intercrops, important consideration are: (1) use of short duration crops and crop varieties, which are somewhat tolerant to shade and do not produce much foliage, (2) maintaining adequate spacing between sugarcane and intercrop rows so that there is no shading of sugarcane seedlings and (3) application of fertilisers and copious irrigation to sugarcane soon after the harvest of the intercrop.

8. MANURES AND FERTILISERS

A tonne of sugarcane on an average removes 1.0 kg N, 0.6 kg P_2O_5 and 2.25 kg K_2O. Thus a 100 tonne crop per hectare removes 100 kg N, 60 kg P_2O_5 and 225 kg K_2O from the soil. Besides, being grown largely with irrigation, there is considerable loss of nutrients by leaching. In sugarcane both cane yield and juice quality are profoundly influenced by nutrition. Therefore, balanced and timely application of manures and fertilisers is important.

8.1. Role of Major Nutrient Elements

Nitrogen: Nitrogen is required for vegetative growth (tillering, foliage formation, stalk growth) and root growth and thus for yield of cane. Deficiency of N causes paleness of foliage, early leaf senescence, thinner and shorter stalks, and longer but thinner roots. Excess N prolongs vegetative growth, delays maturity and ripening, increases reducing sugar content in juice, and thus lowers juice purity. Higher soluble N in juice resulting from excess N application affects clarification process. Due to increased succulence crop becomes susceptible to pest and disease attack.

Phosphorus: Phosphorus is necessary for protein formation, cell division, metabolic activity and phyotosynthesis. Tillering, root growth and thus crop growth, depend very much on phosphorus supply. It interacts with nitrogen and thus influences ripening. Deficiency of phosphorus leads to reduced tillering and delay in canopy closure. Stalk elongation is affected and there is less of secondary and tertiary stalks, leaves grow close and leaf colour appear green violet. Presence of about 300-400 ppm phosphorus in juice is necessary for proper clarification while processing.

Potassium: Sugarcane is known as a 'potassium devourer' due to heavy uptake (600-800 kg/ha) under unlimited supply. Potassium is required for carbon assimilation, photosynthesis, translocation of carbohydrates and for various enzymatic activities within the plant. For sucrose synthesis and its translocation to the storage tissues K is highly important. Potassium gives resistance to sugarcane against pest and disease attack and lodging. It helps during moisture stress by maintaining cell turgidity. Dieback of the leaf tips is the first symptom of potassium deficiency, followed by similar dying downward along the leaf edges. In field cane, K-deficiency symptoms include a fan-shaped leafy top with the leaves standing out very stiffly. Narrow leaves and thinner stalks are other symptoms.

Organic manures: Either FYM, or pressmud or compost should be applied at a minimum rate of 10-15 tonnes per hectare before planting. Restricted application in the furrows before planting would help better utilization besides effecting economy and also, helping in the better utilization of basal P fertiliser.

Pressmud an important by product of sugar industry. It is a rich source of organic matter and nutrients (Table 39.4). It should be cured (that is, left heaped in some place for 1-2 months for microbial degradation) before application. By using *Trichderma viridae* it could be composted into a fine manure (SBI, 1992). Also, sugarcane trash can be composted and used as source of organic manure. Pressmud and distillery effluents can be mixed and a manure popularly known as 'Bio earth' can be prepared. Various composting techniques are furnished by Sundara (1998).

TABLE 39.4: Average nutrient content of sulphitation pressmud

Nutrient	Content
Organic carbon	20-24%
Nitrogen	1.26%
Phosphorus	3.83%
Potassium	1.46%
Calcium	11.00%
Magnesium	1.65%
Sulphur	0.23%
Copper	69 ppm
Manganese	898 ppm
Iorn (DTPA)	24 ppm

8.2. Fertilisers

8.2.1. Response to Applied Nutrients

A large body of literature available on the response of sugarcane to applied nutrients indicate more or less universal response to N and scattered and localized response to P and K (Zende, 1990; Sundara, 1985 and 1995). In the sub tropics sugarcane response to N levels was upto 150 to 200 kg/ha while in the tropics it is in the range of 250-300 kg/ha though there were varieties showing response upto 350 kg N/ha. Adsali crops in Maharashtra show response upto 400 kg N/ha. A study in Coimbatore with 25 sugarcane varieties showed that most varieties responded to 40-80 kg P_2O_5. The response to applied potassium was in the range of 125-150 kg/ha.

Dosages of nutrients presently recommended in important sugarcane growing states are indicated in Table 39.5. However it is suggested to decide the dosages based on field-wise soil tests and follow 'Field-based" approach to manuring to economize the fertiliser nutrients.

TABLE 39.5: **TABLE 39.5:** Recommended fertiliser nutrient dosages in different states

State	Dosage/ha		
	N	P_2O_5	K_2O
Andhra Pradesh	112-400	-	-
Assam	130	60	60
Bihar	90-170	70-85	30-130
Gujarat	250	125	125
Haryana	150	50	-
Karnataka	187-250	75-125	75-185
Kerala	165	80	80
Madhya Pradesh	300	80	45
Maharashtra	170-400	110-170	170
Orissa	220	100	60
Punjab	150-225	-	-
Rajasthan	150-170	-	-
Tamil Nadu	225-275	62	112
Uttar Pradesh	150-180	-	-
West Bengal	160	60	60

8.3. Time and Method of Application

Among the three major nutrients, phosphorus is best applied as basal before planting in the furrow bottom and slightly mixed with the soil so that it is available near the root zone. This is necessary as phosphorus is subject to fixation and is highly immobile. The N and K fertilisers are best applied as top dressing in two splits, at around 45 and 90 days coinciding with the beginning of tillering and grand growth phases, respectively in the tropical belt. However, in the sub tropics N and K may be applied at planting and then, upon the receipt of South-west monsoon rains in June. The N and K fertilisers are band applied and covered by earthing up operations either manually or by a ridger.

8.4. Source

Among the N fertilisers, urea is most commonly used though ammonium sulphate is still preferred by jaggery farmers. Super phosphate is the usual source of phosphorus. But now-a-days DAP is increasingly used. A study at Coimbatore (Sundara and Natarajan, 1996) had indicated no difference between super phosphate and DAP on cane yield or juice quality. In acid soils rock phosphate along with organic manures or phosphobacteria may be used. Muriate of potash is the usual source of potassium.

It has been well established that only 25-40% of the applied nutrients are utilized by the crop and the rest is lost by various means such as leaching, volatilization, fixation etc. To reduce such losses and to improve the fertiliser use efficiency, techniques like split application, pocket manuring, neem cake blending of urea, etc., can be followed.

8.5. Pocket Manuring

In pocket manuring fertilisers are placed in small holes (2.5-3 cm diameter, 8-10 cm deep) made about 8-10 cm away from the sugarcane clump and covered with the soil. To make the holes sharpened wooden sticks or preferably crow bars may be used. Pocket manuring reduces nutrient loss and helps in its controlled availability.

8.6. Neem Cake Blending

Urea can be blended with neem cake in 4:1 (4 parts urea and 1 part neem cake) and dried overnight in shade and then applied. Urea can also be blended with coal tar. 'Nemin' a neem based commercial product is also available for blending with urea.

8.7. Biofertilisers

Extensive studies of Coimbatore (SBI Ann. Reps.) and other places in the tropical India have indicated that *Azospirillum brasilense* is highly useful and can save 50-75 kg N/ha. *Azospirillum* at 5-10 kg/ha is applied in two equal splits on 30 and 60 days of planting. Recent studies at Coimbatore have shown the usefulness of phosphobacteria (*Bacillus megaterium* var. *Phosphaticum*) to improve P availability to sugarcane (Sundara and Natarajan, 1996).

8.8. Micro and Secondary Nutrient Problem

Among the micronutrients, deficiencies of iron and zinc are more common in sugarcane.

Iron: Presence of iron in the active ferrous form is important than its total level. About 15 to 20 kg of iron is removed for producing 100 tonnes of cane. Iron deficiency causes iron chlorosis' (interveinal chlorosis). The problem is severe in calcareous soils. To correct iron chlorosis foliar application of 0.25 to 0.5% ferrous sulphate along with time at weekly intervals in suggested. Soil application of 25 kg ferrous sulphate may also followed.

Zinc: The zinc deficiency is widespread particularly in areas where sugarcane is rotated with rice. Zinc requirement is around 1 to 1.5 kg/ha. The stalks of zinc deficient plants exhibit distinct pipe like structure in the centre with spongy and light brown tissues around the pipe. Lateral bud sprouting, short leaf blades, light yellow stripes along the vein, and chlorosis of young tillers are the visible symptoms. Correction is by repeated foliar spray of zinc sulphate (0.25-0.5%) with time at weekly intervals until the symptom vanishes. 25 kg $ZnSO_4$ soil application may also be done.

Sulphur: Among the secondary nutrients, sulphur deficiency is observed particularly due to use of high analysis sulphur free fertilisers. Response to sulphur application of 50-100 kg/ha has been noted at Lucknow.

9. VARIETIES

Choosing appropriate variety for the region, soil type and planting season is important to realize full yield and quality potential of varieties and thus to maximise sugar output. Improved sugarcane varieties are evolved by Sugarcane

Breeding Institute, Coimbatore in collaboration with the State sugarcane research stations. A National Hybridization Garden facility is available at Coimbatore for this purpose. Through the All India Coordinated Research Project on Sugarcane, identified varieties for different agro-climatic regions.

There are four groups of sugarcane varieties

1. ***Early maturing varieties:*** These varieties attain a minimum of 16% sucrose in juice with 85% purity in 10 months. Ex. CoC 671, CoJ 64, Co 7805.
2. ***Midlate varieties:*** These varieties attain a minimum of 18% sucrose with 85% purity in 12 months. Ex. Co 6304, Co 1148, Co 1158, Co 86032.
3. ***Late maturing varieties:*** The late varieties attain atleast 16% sucrose with 85% purity in 14 months. Ex. Co 62175, Co 62198.
4. ***Short duration varieties:*** These varieties attain a minimum of 16% sucrose with 85% purity in 8 months. These however are not popular because of their low cane yield levels. Ex. Co 8338, Co 8341.

The duration of different groups of varieties are under normal planting dates. Though several new varieties are being introduced, still many of the old varieties are occupying large areas. According to a recent survey the following are the top 10 varieties under cultivation in the country: CoC 671, Co 62175, Co 740, Co 7805, Co 6304 and Co 86032 in the tropical states and Co 1148, CoJ 64, CCS 767 and BO 1 in the sub tropics. Salient features of these varieties are described below:

9.1. Co 740 (P 3247 x P 4745)

Early/mid late; thick yellowish green cane; heavy yielding, profuse tillering; known as "second wonder cane" (Co 419 - the "First wonder cane"). Prominent variety of Maharashtra and northern Karnataka grown as an *adsali* cane. Resistant to smut. Resistant to drought and tolerant to flood. Prone to lodging.

9.2. Co 1148 (P 4883 x Co 301)

Mid late/late, high yielding, with around 17% juice sucrose. Does not deteriorate during late periods of crushing, good ratooner, suitable for late planting. Resistant to red rot. Most popular in the sub tropical sugarcane belt.

9.3. Co 62175 (Co 951 x Co 419)

Late maturing, thick cane, high yielding with moderate quality. Grown extensively in Karnataka, and also in parts of Tamil Nadu, Orissa. It is a good ratooner, good for jaggery, well adapted to water logged/excess moisture conditions.

9.4. Co 6304 (Co 419 x Co 605)

Mid late, thick, vigorous and erect growing high yielding variety. Performs better under garden land conditions. Major mid late variety of Tamil Nadu. Non-lodging; good ratooner. Tolerant to smut and pine apple diseases.

9.5. Co 7805 (Co 740 x Co 6806)

Early, medium thick to thick green cane. Found better than Co 6907 in Andhra Pradesh. Has good field stand. High yielding and high quality variety. Resistant to smut and susceptible to red rot.

9.6. CoC 671 (Q 63 x Co 775)

Early maturing, thick green/pinkish green, very attractive canes. High yielding and very high quality. Very vigorous and quick growing. Most popular early variety of tropical India. Presently being phased out in Tamil Nadu due to red rot; and also in Gujarat due to red rot and wilt. Profuse flowering. Canes lodge. Susceptible to red rot and resistant to smut. Has good field keeping quality.

9.7. CoJ 64 (Co 975 x Co 617)

Early, rich cane. Very popular early variety of subtropical India, particularly in Punjab, parts of UP and Haryana. Good germination and profuse tillering. Good ratooner. Moderately resistant to red rot.

9.8. CCS 767 (Co 419 x Co 313)

Mid late, medium thick, green to purplish cane. Found better than Co 1148 and Co 1150 in the sub tropics. Has tendency to lodge. Moderately resistant to red rot and smut. Resistant to drought.

9.9. BO 91 (BO 55 x BO 43)

Mid late; major variety of Bihar with very good yield. Good ratooner. Withstands floods in North Bihar, resistant to drought, alkalinity and salinity. Highly resistant to red rot and smut.

9.10. Co 86032 (Co 62198 x CoC 671)

Mid late, attractive cane with high yield and high sucrose in juice. Resistant to smut, tolerant to salinity spreading fast in the tropical belt.

10. PLANTING

10.1. Planting Seasons

In the sub tropical region, there are two planting seasons, viz., spring and autumn. The spring planting is done after the harvest of the winter crops like wheat, potato, gram, etc. About 85 per cent of the plantings in the sub tropics is during spring. The spring planted crop experiences severe moisture stress and high atmospheric temperature during the formative phase. The growing season is restricted to 4-5 months only. This is why lower yields are obtained.

The autumn planting is done during September-October soon after the cessation of the south west monsoon rains. Due to longer growth period, favourable soil moisture regime and moderate weather during the early growth phase, generally higher yields of cane and sugar are obtained. But autumn planting is not popular as it interferes with the important winter crops.

In the tropical region, sugarcane is planted from December to April, peak plantings being done during February-March. There are some pockets were 'special season' planting during June to September is also in vogue. In certain parts of Maharashtra, Karnataka and Andhra Pradesh, *adsali* cropping is followed where plantings are done from July to September and harvested after 15 to 18 months.

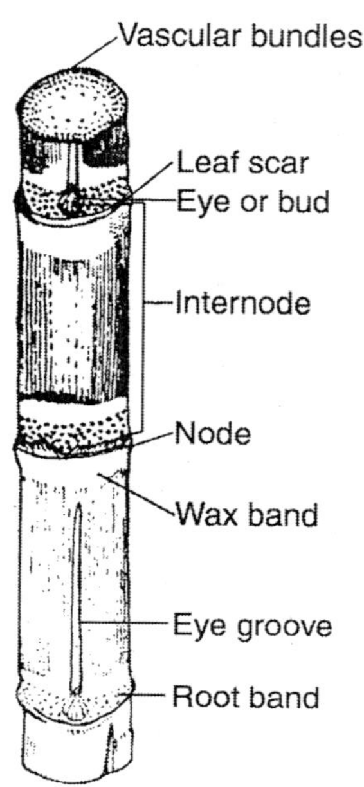

FIGURE 39.1: Sugarcane sett.

10.2. Planting Methods

Essentially there are two following systems of planting:

Ridges and furrows system: In this method ridges and furrows are formed using a tractor-drawn or bullock drawn ridger. The most common spacing is 90 cm between the furrows. Closer spacing (60-75 cm) is advocated for early varieties, shy tillering varieties and under adverse growing conditions (moisture stress, water logging, salinity/alkalinity, late planting, etc.). Wider row spacing of 100-120 cm is recommended for high yielding and high tillering varieties, under fertile soils. The depth of furrow should be around 25 cm. A furrow length of 10-15 m is ideal when guided irrigation is followed.

This system facilitates easy irrigation and good support to the crop. In this system, setts are placed in the furrow bottom either end-to-end or slightly overlapping manner depending upon the length of sett. The planting could be 'dry' (setts are placed, covered with soil and then irrigated) in light soils, or it could be 'wet' (furrows irrigated and then setts are pressed down in the furrow bottom) as in the heavy clay soils. In the coastal Andhra Pradesh U-shaped trenches are formed instead of furrows for sugarcane planting. This is known as 'trench system'.

Flat system: The 'flat' system is mainly followed in the sub tropical belt. In this system, repeated ploughings and plankings are done to conserve soil moisture. In finely prepared field, shallows furrows are opened with a country plough and setts are dropped and covered by planking. No irrigation follows planting. For the entire crop only 5-6 irrigations are given. Setts are sown by end to end method.

Special systems: Deep trench system, modified trench system, contour system, polybag seedling transplanting are some of the special systems of sugarcane planting. Of these, polybag transplanting technique is becoming popular.

Polybag seedling transplanting technique: In this technique, seedlings are raised by planting single bud setts in polythene bags (8 x 10 cm) containing FYM or pressmud, soil and sand in 1:1:1 proportion. One to one and a half month old seedlings, after clipping of the leaves are then transplanted (with rhizosphere soil intact, but removing the polythene cover) in the prepared main field in the furrows by opening a small pit at intra-row spacings of 30-45 cm. This method saves seed material by two thirds. This technique saves time, and permits better preparation of the main field.

10.3. Seed Material

In India, most common are the three-eye bud stem cuttings or 'setts'. It is not desirable to plant full long cane because of Apical dominance. Top one-third to 'two-thirds' of the cane is a good seed material. But is best to raise exclusive seed crops or 'short crops' which can be harvested at 8 months from the basic heat treated material. Additional fertiliser application (pre-fertilising) at the rate of 50:25:25 kg N, P_2O_5 and K_2O/ha 6 weeks prior to harvest gives better quality seeds (Sundara, 1995).

Three-tier nursery programme involving heat treatment is an essential requirement for obtaining healthy seeds and to prolong the commercial life of varieties.

10.4. Seed Treatment

Treating the setts with 0.1 per cent of a fungicide (bavistin, agrosan etc.) prevents setts rot caused by soil borne fungi.

10.5. Seed Rate

A seed rate of 60,000 to 75,000 two eye bud setts or 40,000 - 50,000 three eye bud setts are required to plant one hectare of sugarcane under normal growing conditions. However under adverse growing conditions higher seed rates, preferably through closer spacing, are desirable.

11. IRRIGATION

Water requirement of sugarcane varies between 2000 to 3000 mm per annum per hectare. Considerable variation in the requirement exists between regions (Table 39 6).

TABLE 39.6: Water requirement of sugarcane in some important sugarcane growing states

State	Water requirement (mm)
Uttar Pradesh	1400
Bihar	1400
Punjab	1500

State	Water requirement (mm)
Andhra Pradesh	1700-1800
Karnataka	2000-2400
Maharashtra (adsali)	3200-3500
Suru	2400-3000

Approximate water requirement of sugarcane in different growth phases are given below.

- Germination (0-45 days) 300 mm
- Tillering (45-120 days) 550 mm
- Grand growth (120-270 days) 1000 mm
- Ripening (270-360 days) 650 mm

11.1. Irrigation Scheduling

During germination phase "light frequent" irrigations facilitate better germination. Therefore irrigations at 25% depletion of available soil moisture (ASM) i.e., at 75% ASM level is ideal. Later during active growth phase irrigations should be given at 40 to 50 % depletion of ASM. During ripening phase, slight moisture stress is desirable and thus irrigations may be given at extended intervals. In terms of ASM, irrigation at 60-75% depletion i.e., at 40-25% ASM range is desirable.

Based on the above, the following irrigation schedule is suggested.

Growth phase	Irrigation interval (days)		
	Coarse textured soil	Medium textured soil	Fine textured soil
Germination	5-6	6-7	8-10
Tillering	6-7	7-10	12-15
Grand growth	Weekly	10	12-15
Ripening	10	12-15	15-20

The quantity of water applied during the germination phase at each irrigation is around 1 to 1.5 acre inches (2.5 to 3.75 cm) and during other phases it would range between 2 to 2.5 acre inches (5 to 6.25 cm).

The above schedule is largely applicable to tropical sugarcane belt. In the sub tropics sugarcane is grown with limited irrigation. In the sub tropics also summer irrigations at monthly intervals would help better crop growth and yield.

11.2. Irrigation Methods

In the ridge and furrows system of planting furrow irrigation is the most common method. In the canal irrigated tracts, long furrows are joined at the alternate heads forming a kind of 'serpentine' arrangement in which water is let in one end water is not guided. Irrigation continues till water comes out at the other end of the serpentine system. In lift irrigated areas a kind of 'check basin-cum-furrow' method is followed in which a set of 5-6 furrows are irrigated at a time and the water is let into another similar set. In the sub tropical belt under flat bed no particular irrigation layout is followed.

11.3. Drip Irrigation

In Maharashtra 'surface drip' irrigation system is followed to a limited extent. Studies at Coimbatore have indicated 40% saving in the irrigation water by drip system. However sugarcane yield benefits were not consistent (SBI, 1994). Some of the problems encountered were clogging, rodent damage, breakage etc. Besides, the system is expensive. For economy, field layout should be in paired rows at 60-120 cm (cane rows at 60 cm with 120 cm gap and drip line running in between the rows placed at 60 cm).

12. WEED CONTROL

Weeds causes 12 to 72 per cent reduction in cane yield. The row spacing adopted, the initial slow growth of sugarcane coupled with irrigation and manurial practices encourage abundant weed growth. The initial 3 months period is the most critical period of weed competition. In India about 63 weed species belonging to 29 families were found to occur in sugarcane (Sankaran, 1989). Grasses compete more with sugarcane followed by sedges and broad leaved weeds. The major weed flora are: *Cyperus rotundus, Cynodon dactylon* (Pers), *Sorghum halepense* Pers, *Panicum* spp., *Dactyloctenium aegyptium* Pers, *Chenapodium album* L., *Convolvulus arvensis* L., *Amaranthus viridis* L., *Portulaca oleraceae* L. and *Trianthema portulacastrum* L. In some areas such as southern Karnataka, *Ipomea* spp. is a serious problem in grown up cane. *Parthenium* is yet another very widely found weed.

Hand weeding and hoeing is still the most common method of weed control in sugarcane. To control first flush, blind hoeing is desirable. During the early growth phase about 3-4 hand weedings starting from 20-25 days may be required to keep the land free of weeds.

Herbicidal control of weeds is becoming more popular in the tropical states and also to some extent elsewhere. Atrazine at 1.75 kg/ha applied as pre-emergence spray after 3-4 days of planting sugarcane gives effective weed control for about 30-45 days. Persistent broad leaved weeds can be controlled by additional post-emergence spray of 2, 4-D at 1 to 1.5 kg/ha. Metribuzine (Sencor) is another herbicide which has been found very effective in sugarcane fields.

In certain areas, *Ipomea* spp is a serious problem. To control this weed, atrazine or Sencor may be sprayed to the base of plant at second earthing up stage.

Grass weeds are difficult to control. However directed spray of glyphosate gives effective control. Crop rotation with paddy and mechanical destruction of weeds are other means of controlling these weeds.

13. INTERCULTURAL OPERATIONS

13.1. Earthing

It is specially required when crop is sown by flat method. Soil is put around sugarcane stalks about 15 cm high. Best time of earthing is in June-July. It saves the crop from high velocity winds and prevents lodging.

13.2. Wrapping and Propping

Wrapping is done in 5 months old crop. Propping is practiced in Mysore. Both practices save crop from lodging specially from cyclonic storms.

14. PLANT PROTECTION

Sugarcane is affected by numerous pests and diseases, some of which at times cause huge losses. In the sub tropics root borer, top borer, pyrilla are the important insect pests. In the tropical states early shoot borer and internode borers are very common. Among the diseases, red rot is the most dreaded one throughout the country followed by wilt. In the following sections, brief accounts of important pests and diseases and their management practices are furnished.

14.1. Pests and their Management

Early shoot borer (*Chilo infuscatellus* Snellen.) - It attacks before internode formation and affects the initial crop stand. Most common in the late planted crop. The "dead heart" formation which can be easily pulled out is the characteristic symptom.

Control: (1) Avoiding hot weather planting, (2) when planted late, Lindane or Chlorpyriphos (both 20 EC) at 1 kg a.i/ha may be applied on the setts using a rose can as a prophylactic measure and also repeated upon noticing the incidence later. (3) Soil application of Sevidol at 12.5 kg/ha around the base of the plant at 30th day of planting, and if needed on 60th day, (4) spray application of Granulosis virus 106-107 inculsion bodies per ml.

Top Borer (*Scirpophaga excerptalis* Walker) - A major pest in sub tropical India. It infects during all the stages of growth. Leaves of the affected plants show 'shot holes'. There is also "dead heart" formation, but it cannot be pulled out easily. In the grown up canes dead hearts lead to bud sprouting producing 'bunchy top' appearance.

Control: (1) Change of planting date, good drainage, collection and destruction of egg masses, (2) Application of carbofuran 1.0 kg/ha or phorate 3 kg/ha granules during the last week of June and (3) Biological control by *Trichoderma japonicum* in the sub tropics and *Isotima javanesis* in the tropics.

Internode borer (*Chilo sacchariphagus* indicus Kapur.): It is a major pest in the tropical states. It damages the crop soon after internode formation and its activity continues till harvest. Most fresh attacks are found in top five internodes. Larvae tunnel upwards. Dead heart formation is observed. No offensive small emanates when the dead hearts are pulled out. Lodging, high N dosage, water logging and presence of water shoots favour the pest build up.

Control: (1) Use of borer free setts for planting, (2) Detrashing at 5, 7 and 9 months, (3) Removal of water shoots at 8th and 9th month, (4) Avoiding high N dosage and good field drainage, (5) Biological control by *Trichoderma chilonis* - inundative release at 3.5 cc per ha per fortnight from 4th month until a month before harvest, (5) Monocrotophos 0.75 to 1.0% spray after detrashing helps in reducing the borer population.

Stalk borer (*Chilo auricilius* Dudgeon): Major pest of UP, Haryana, Punjab and Bihar. The larval feeding of leaf causes longitudinal orange yellow streaks extending from tip to base on both sides to the midribs towards the margin.

The third instar larvae bores into internodes and canes. In the grown up canes on removing the leaf sheaths, borer holes can be seen on the internodes. The damaged internodes may show reddening and emit rancid odour.

Control: (1) Use of healthy setts for planting, (2) Destruction of crop debris and stubbles to prevent carry over pests, (3) Good earthing up, (4) field drainage and (5) application of monocrotophos spray at 0.75 to 1.0 kg a.i/ha in July after detrashing.

Root borer (*Emmalocera depresella* Swinhoe): Root borer infests sugarcane in the sub tropical belt, northern Gujarat, Maharashtra, Karnataka and Andhra Pradesh. It generally infests the young shoots and causes "dead hearts" which cannot be pulled out easily unlike the dead hearts of ESB. When pulled, the plant gets uprooted and larva can be seen. Near the base of the shoot only one entry hole could be seen. The infestation peaks during the south west monsoon.

Control: (1) Removal of dead hearts and killing, (2) destruction of affected shoots, (3) collection of moths by light traps, (4) Growing tall growing arhar around the cane fields, (5) Deep stubble shaving at ratooning, (6) Destruction of stubbles after harvest, (7) Puddling and crop rotation, (8) Chlorpyriphos or Ekalux 1-2 kg a.i/ha during May and August may give some control.

Scale insect (*Melanaspis glomerata* Green): Scale insect occurs in Andhra Pradesh, UP, Bihar, Karnataka, Gujarat, Haryana and Punjab. The scale appears in 5-6 month old crop mostly coinciding with the onset of the monsoon. Scale usually establish on internodes covered with leaf sheath. Nodal region is more infested than the internodal regions under heavy build up, stem appears greyish black with scales. The infested crop has poor vigour, canes shrivel, growth is stunted and the internodal length is reduced very much. Leaves of the infested canes show tip drying and appear unhealthy pale green. Scale insect injury attracts fungal infestation.

Control: (1) Seed material should not be taken from the infested fields, (2) Ratooning should be avoided in the infested fields, (3) Detrashing, trash burning and (4) Application of Malathion or Endosulfan at 0.1% after detrashing.

Pyrilla (*Pyrilla purpusilla* Walker): Pyrilla is the most destructive foliage sucking pest of sugarcane. It has been occurring in epidemic scale in the sub tropical states in the cycles of 5-6 years. The adults and nymphs suck leaf sap from the under surface of the lower leaves and the midrib resulting in yellowish white spots. The hoppers exude a sweetish sticky fluid known as 'honey dew' which promotes quick and luxuriant growth of fungus *Capnodium* sp. as a result, the leaves get completely covered by the shooty mould.

Control: (1) Trash burning to destroy unhatched eggs, and nymphs, (2) detrashing, (3) during pre-monsoon dusting methyl parathion (2%) at 12.5 kg/ha using rotary duster, (3) Malathion or Endosulphan 0.1% spray after detrashing, (4) During epidemics aerial spray has been carried out successfully using insecticides such as methyl demeton (1125 ml/ha), malathion (500 ml/ha) and endosulfan (750 ml/ha).

White flies (*Aleurolobus barodensis* Mask.): The nymphs of white flies suck the sap from the under surface of the leaves which turn yellow and pinkish in severe cases and gradually dry up. Heavily infested leaves are covered by sooty moulds caused by the fungus *Capnodium* spp. Summer droughts and dry spells during monsoon greatly favour build up of white flies. Water logging and N starvation causes severe outbreak of this pest.

Control: (1) Discouraging ratoon in the low lying areas, adequate manuring, improving drainage. Monocrotophos (40 EC) 0.04% or Endosulphan (35 EC) or Malathion 0.1 to 0.20% spray after stripping.

14.2. Diseases

Red rot: The red rot is caused by the fungus *Collatotrichum falcatum*. The disease is widespread in the sub tropics and the coastal areas of the tropical states. The most characteristic symptoms of the disease are observed when the affected stalks are cut open longitudinally. The internal tissue shows reddish discoloured patches or lesions interspersed with white horizontal patches. Crop drying is the external symptom.

Red rot management: (1) Growing tolerant/resistant varieties, (2) use of seeds devoid of primary infection, (3) drainage; crop rotation with paddy, (4) internal quarantine, i.e., no seed material should be brought from a red rot endemic area, (5) destruction of all crop residues of the affected crop, (6) avoiding ratoons from affected crops and (7) checking flow of irrigation water from the infected fields.

Smut: The smut is caused by a fungus, Ustilago scitaminea sydow. Smut is prevalent in all the states. The characteristic symptom is the production of long whip-like structure from the terminal bud of the stalk which is black in colour covered by thin silvery membrane. The disease is more in ratoons. The primary spread is through infected setts and the secondary spread is through wind borne teliospore.

Control: Rouging disease affected clumps before the silvery membrane brakes; covered with a paper or cloth bag, whips are buried or burnt. (2) Growing smut resistant varieties, (3) No seed material should be taken from the affected field and (4) Hot water treatment of setts at 50□ C for 2 hours in combination with fungicide Bayleton (0.1%).

Wilt: The sugarcane wilt is caused by *Cephalosporium sacchari* Butler and/ or *Fusarium moniliformae* fungi. At present wilt is very serious in sub tropics and in Gujarat. The fungi gain entry through injuries. Wilt affected plant appears wilted and conspicuously stunted. The canes lose their normal colour. On splitting open the canes at the early stage of infection, diffused reddish brown patches are seen. Later canes become light, hallow and shrink.

Control: When wilt appears in the field it is physically impossible to check its further spread (1) use of wilt free seed material, (2) crop rotation with paddy, (3) growing resistant varieties, (4) avoiding ratoons and (5) burning trash and other residues.

Ratoon stunting disease (RSD): The RSD is caused by a bacteria, *Clavibacter xyli*. This has been the most important cause for sugarcane varietal decline (Sundara, 1996). Diseased stools are stunted in growth, thin stalked with short internodes and have pale yellowish foliage. When split open longitudinally, pin head like orange coloured dots of bacteria can be noticed on the nodal region. The disease is more pronounced in ratoons and under adverse growing conditions.

Control: (1) Use of disease free setts for planting. The cutting knives for seed cane preparation should be disinfedted with 1% lysol and (2) healthy seed programme involving heat therapy.

Grassy shoot disease: GSD is a mycoplasma disease and is found in most sugarcane growing areas of the country. Profuse tillering is a typical symptom with narrow chlorotic leaves giving a grass like appearance. Varying degree of chlorophyll loss is observed. Very few tillers form canes which are thin and produce while shoots from the side buds.

Control: (1) Use of healthy seeds, (2) roguing diseased clumps and (3) heat therapy (hot water/hot air/moist hot air/aerated steam).

Mosaic: This is a viral disease. The affected cane leaves show dark green patches alternated with light green patches. Mild strains of mosaic do not cause yield loss while strains result in considerable loss of yield. Use of healthy setts is only way to prevent intestatin.

15. HARVESTING

In general maturing is judged by (1) the gradual withering up of lower leaves (2) Mature cane, if cut across reflects slight sparkling in its flesh. (3) Metallic sound on striking with finger.

Sugarcane does not exhibit any specific and reliable visible symptom of maturity. Therefore harvesting should be decided based on crop age and more scientifically in conjunction with maturity test.

The early, mid late and late varieties reach peak maturity at 10-11, 12 and 14 months, respectively and could be harvested accordingly. Maturity test using hand refractometer is easy, quick and reliable. For this purpose, juice from the middle portion of several canes is drawn using a pouch piercer and the Brix of the composite sample is recorded through the hand refractometer. HR Brix of about 20 and above at the specified age indicates the readiness of the crop for harvest.

Small mill tests (SMT) can also be carried out, Sucrose and Brix are recorded and purity is worked out (Meade and Chen, 1978). A sucrose level of more than 16% with a purity level of above 85% indicates the ripeness of the cane.

Maturity can also be assessed by sheath moisture determinations. A sheath moisture of around 76% in the sheaths 3, 4, 5 and 6 crop top indicates the crop maturity (Clements, 1978). Sugar factories give cutting based on 'crop age' alone or in conjunction with maturity tests. Jaggery farmers harvest the crop based on age.

Inversion of sucrose may occur if crop is not harvested after ripening or not milled within a reasonable time span. Polaris or Sodium metasilicate spray at ripening, 6-8 week before harvest improve recovery of sugar.

Sugarcane is harvested manually in India using various types of knives. Cane should be cut to the base (as the basal portions are rich in sucrose) and should be stripped of trash and topped to eliminate the immature top portions. The Indian Institute of Sugarcane Research has developed a stripper which can be used for de-topping also. All extraneous matter should be eliminated (tops, trash, roots water shoots, etc.).

15.1. Mechanical Harvesting

The Indian Institute of Sugarcane has developed a tractor operated harvesting machine which can harvest 2.5 ha/day. To employ harvesting machines used in other countries, crop spacing has to be increased to 150 cm Studies are being carried out by this author to standardize wide row spacing technique for mechanical harvest.

16. YIELD

Theoretically sugarcane can give an yield of 470 tonnes of cane per hectare per year (Moore, 1987). Many of the award-winning farmers have recorded yields in the range of 250-300 t/ha. But the average yield of the country is only around 70 t/ha. The average yields are higher (80 t/ha) in the tropical region than the sub tropics (55 t/ha). However, with the available technology, average yields could be increased to about 100 t/ha in the tropical region and to 75 t/ha in the sub tropical region.

The ultimate yield is the sugar. The average sugar recovery in the country is around 9.8 per cent. There is also considerable scope to increase the sugar recovery levels.

At harvest, stalks account for 75 per cent of the above ground fresh weight and 25 per cent is the tops and trash.

17. QUALITY

Sugarcane quality is assessed based on the following criteria:

1. The canes should be of optimum age.
2. The juice should have more than 16% sucrose with more than 85% purity, and low level of non sugars.
3. Cane should have optimum fibre content (11-14%) and should not have any pith.
4. There should be only negligible amount of extraneous matter (trash, mud, roots, etc.).

The juice quality depends upon variety, crop management practices, and climatic conditions during ripening stage. Delayed manuring as well as excess manuring of nitrogen is harmful. During the ripening phase cool and dry weather with ample sunshine is desirable. Lodging, over aged cane and under aged canes gave poor quality juice and thus lower sugar recoveries.

18. POST HARVEST TECHNOLOGY

After harvesting, sugarcane should be crushed within 24 to 48 hours. Delay leads to cane deterioration and thus loss of sugar. Therefore organised harvest plan, efficient transport and milling operations are essential.

Most sugar factories in India are of 1250-2500 TCD (tonnes cane/day) capacity. In the southern states sugar factories register the cane area well in advance to ensure adequate cane supply directly to the factory. While such a practice does not exist in the north. There are cane purchase centres which purchase cane from the farmers and then supply to the mills. In the process delay in crushing is common.

The crushing season lasts for an average of 5 months (Nov. to April). However, there are mills which operate for longer periods.

Sugarcane farmers are paid on the basis of cane tonnage unlike some other countries where payment is based on sugar. Statutory minimum price linked to a recovery of 8.5% is fixed by the Government every year.

About 45 per cent of the cane produced goes to white sugar manufacture. An equal quantity is utilised by the Gur and Khandsari sugar sectors. Gur and Khandsari units still function more or less in the age old traditional fashion. There is need to improve their efficiency, particularly in terms of sugar out turn and gur/sugar quality.

19. CONSTRAINTS OF PRODUCTION

Moisture stress during the formative phase is an important limitation throughout the country. In East UP, Bihar and coastal areas floods and water logging are common. Red rot is still the major disease followed by wilt. Early shoot borer and internode borer in the tropical region, and top borer and root borer in the sub tropics are the important pests. Low ratoon productivity particularly in the sub tropics is also a major constraint. Labour is becoming a serious limitation particularly during the harvesting season, more so in the southern states.

20. RATOON MANAGEMENT

Raising one or two ratoons is a common practice in our country. Ratoons are economical. However, ratoon productivity is lower by around 25 per cent in our country due to inadequate management practices and more incidence of pests and diseases. To improve ratoon productivity, important management practices are:

1. Raising a good plant crop.
2. Harvesting the plant crop at appropriate age when weather conditions are favourable for ratoon sprouting and early growth.
3. Stubble shaving, off barring and digging the interspaces.
4. Gap filling (when the gaps of 60 cm in a row is present) using poly bag seedlings.
5. Early manuring: Full P + 1/3 N + 1/3 K at ratooning followed by remaining N and K in two equal splits at 30 days interval. About 25% additional N dosage may be given to ratoons.
6. Use of more organic manures or in situ incorporation of trash with the addition of decomposing microbial culture.
7. Effective and need-based pest management.

21. REFERENCES

Alexander, K.C. 1990. Diseases of sugarcane and their management. In: P.N. Rao (ed.) 'Recent Advances in Sugarcane', pp. 364-415, KCP Ltd., Vuyyuru, AP.

Biswas, B.C. 1988. Agro Climatology of the Sugarcane Crop. Technical Note No. 193. World Meteorological Organization.

Daniels, D. and Roach, B.T. 1987. Taxonomy and evolution, In: Heinz, D.J. (ed) 'Sugarcane Improvement through Breeding', Elsevier, Amsterdam.

Dillewign, Van C. 1952. Botany of Sugarcane, The Chronica Botanica Co., Waltham, Mass USA.

Hapase, D.G., Gunjal, B.B. and Deshmukh, A.S. 1990. Irrigation management for sugarcane. In: P.N. Rao (ed), 'Recent Advances in Sugarcane', pp.248-270. KCP Ltd., Vuyyuru, AP.

Humbert, R.P. 1968. The Growing of Sugarcane. Elsevier, Amsterdam.

Johari, D. and Singh, R.G. 1991. Weeds of sugarcane, losses due to them and their control. A review. Bharatiya Sugar, 16(3): 33-39.

Lakshmikantham, M. 1983. Technology in Sugarcane Growing 2nd edition. Oxford & IBH Publishing Co., New Delhi.

Moore, P.H. 1989. Physiological basis for varietal improvement in sugarcane. In: K.M. Naidu, T.V. Sreenivasan and M.N. Premachandran (eds). 'Sugarcane Varietal Improvement'. Proc. Int. Symp. on Sugarcane Varietal Improvement. pp.19-55. Sept. 3-7, 1997, SBI (ICAR), Coimbatore.

Rao, K.C. 1987. Ripening and quality. In: K. Mohan Naidu and S. Arul Raj (eds), 'Sugarcane Technologies', pp.37-41. Sugarcane Breeding Institute (ICAR), Coimbatore.

Sankaran, S. 1988. Weed management in sugarcane. Keynote address. Proc. 16th Sugarcane R & D Workers' meeting of Tamil Nadu, Pennalam, Aug.4-5, 1988, SBI, Coimbatore.

Sankaranarayanan, P., Natarajan, B.V. and Marimuthammal, S. 1986. Sugarcane varieties under cultivation in India. ICAR, New Delhi.

Sankaranarayanan, P. and Natrajan, B.V. 1987. Sugarcane varieties in India (1979-1986). SBI (ICAR), Coimbatore.

Sreenivasan, T.V., Bhagyalakshmi, K.V. and Sahi, B.K. 1995. Sugarcane varieties. In: K.C. Alexander and S. Arulraj, (eds) 'Sugarcane Production Manual', pp.7-16, SBI (ICAR), Coimbatore.

Srivastava, S.C., Johari, D.P. and Gill, P.S. 1988. Manual of Sugarcane Production in India. ICAR, New Delhi.

Sugarcane Breeding Institute (SBI) Annual reports (1980-1996), SBI (ICAR), Coimbatore.

SBI. 1992. Enrichment of pressmud by microorganisms. Extension Pub. No.39. SBI (ICAR), Coimbatore.

Sundara, B. 1985. Effect of levels of phosphorus and potassiuma and their late application on sugarcane. Indian J. Agron. 30(1): 124-125.

Sundara, B., 1994. Phosphorus efficiency of sugarcane varieties in a tropical alfisol. Fertiliser research, 39: 83-88.

Sundara, B., 1989. Improving sugarcane productivity under moisture constraints and through cropping systems. In: K. Mohan Naidu, T.V. Sreenivasan and M.N. Ramachandran (eds). "Sugarcane Varietal Improvement". pp.221-251. SBI (ICAR), Coimbatore.

Sundara, B. 1994. Pre-fertilizing sugarcane nursery for higher yield of quality seeds. SISSTA Sugar Journal, 20(1): 157.

Sundara, B. 1995. Causes for sugarcane varietal yield decline and techniques of rejuvenation. Proc. 57th Ann. Conv. Sugar Technologists' Association of India, New Delhi, pp.153-167.

Sundara, B. 1998. Sugarcane Cultivation. Vikas Publishing House Pvt. Ltd., New Delhi.

Yadava, R.L. 1991. Sugarcane Production Technology. Oxford & IBH Publishing Co. Ltd., New Delhi.

40. Sugar Beet

P. S. Rathore

Sugar beet is a biennial plant, producing succulent root in first year and flowers during second year. World's 40-45 per cent sugar is manufactured from sugar beet. German Andrews, Marggraf first reported that sugar beet contains sugar similar to sugarcane. In the beginning of its cultivation its sucrose content was only 7.5 per cent restricting its commercial utilization. Louis Vilmorin of France succeeded in raising its sucrose content from 7.5-17 per cent. Thereafter sugar beet was used for manufacturing sugar around 1840. The comparative merit of sugar beet over sugarcane is given below:

Comparative merits of sugar beet over sugarcane

Particular	Sugar beet	Sugarcane
Maturity	6-7 months	12-18 months
Sucrose %	15-20	10-12
Purity %	84	-
Recovery %	12-14	6-8
Plant	Temperate	Tropical

Besides sugar beet tops, beet pulp and beet molasses are used as cattle feed. Lactic acid and vitamin B is extracted from beet molasses. Beet tops can also be used as a green manure. Sugar beet is cooked as turnip or used as salad. It is also used in preparation of pickles, *chutneys* and mixed in juice of tomatoes or carrots.

1. ORIGIN

Mediterranean region is considered as a centre of origin of sugar beet. After European countries it was introduced to USA. First sugar factory in USA was installed in Northanpton.

2. DISTRIBUTION

Unlike sugarcane, it is a temperate plant mainly cultivated in between 25-60° latitude. Russia, USA, Poland, Italy, France and Germany are the major producer.

Area, production and productivity of sugar beet in world is given in Table 40.1. In temperate regions it is grown in summers while in tropical and subtropical regions having hot summer, it is cultivated in winters. In India, it was first grown in Punjab 1914 and in 1968 beet sugar was prepared at Sri Ganganager, Rajasthan. Now sugar beet is cultivated in Rajasthan, Maharashtra, Uttar Pradesh, Punjab, Himachal Pradesh and Kashmir Valley with total area of 38.15 lakh ha and production of 2712.29 lakh and av. productivity of 71095 kg/ha(1994-95).

TABLE 40.1: Area production of sugar beet (1997)

Countries	Area (000 ha)	Production (1000 Mt)	Productivity (kg/ha)
Yugoslavia SFR	1300	17500	13462
Russia Fed	935	13800	14759
USA	578	27101	46902
China	550	14850	27000
Germany	504	28487	56521
Total	7476	263112	35196

3. CLASSIFICATION

Sugar beet is plant of family chenopodiaceae and botanically known as *Beta vulgaris* L. Its specie *Beta maritima* is considered as a wild progenitor.

4. CLIMATE

Sugar beet is *rabi* season crop. For its successful cultivation North Indian conditions are very congenial, seed germinates better at soil temperature of 12-15° C, Adequate moisture and bright sunshine are required for good growth. It is a long day plant. Sugar accumulation is high when temperature is in between 20-22° C. Sugar accumulation is adversely affected if temperature exceeds 30° C. Under adverse weather conditions, inflorescence may develop in first year. Flowering in sugar beet is known as bolting.

5. CROPPING SYSTEM

The main crop sequences associated with sugar beet are: green gram-sugar beet, sorghum-sugar, beet, maize-sugar beet, soybean-sugar beet, pearl millet-sugar beet, and cowpea-sugerbeat. Intercropping with autumn planted sugarcane is practiced in some parts of country.

6. SOIL AND SOIL PREPARATION

Well drained sandy loam to clay loam soils are good for its cultivation. Soils should be fertile and rich in organic matter. Root growth specially shape is adversely affected on heavy clay soils. Sugar beet thrive best in pH 6.0-7.5 but it can be grown very well in saline alkaline soils heaving pH 9.5. Crop is specially tolerant to high sodium and boron. Acidic soils are not suitable for its cultivation. Crop is also sensitive to water logging.

One deep ploughing with soil turning plough followed by 2-3 cross ploughing or harrowing and planking are needed to make soil well pulverized for sugar beet. Broadcast 4% endosulfan or 1.5% quinlolphos at 20-25 kg/ha to save crop from soil borne insects.

7. MANURE AND FERTILISERS

Crop respond appreciably to organic fertilisers. Apply 5-7 tonnes well decomposed FYM 15-20 days prior to sowing. To obtain good yield it needs 100-120 kg N 60-80 kg P_2O_5 and 75-100 kg K_2O. If soil test values are available, phosphorus and potassium should be given as per test values. Like other crops 1/2 amount of nitrogenous and full dose of phosphatic and potassic fertilisers should be given at the time of sowing. Remaining quantity of nitrogenous fertiliser should be given after thinning. If soils are very light, remaining N may be applied in two splits i.e. at thinning and earthing. Excess amount of nitrogen have an adverse effect on sugar synthesis and extraction. Potassium and copper increase sucrose content. In boron deficient soils, application of 15 kg boron/ha gives higher yield. Deficiency of boron leads to internal black spot. In standing crop deficiency can be corrected by spraying 0.2% borax at 30, 45 and 60 DAS. This not only increases sucrose but purity also.

Sodium is an essential nutrient for sugar beet. Crop growth and yields are good on saline-alkaline soils. Maganese deficiency causes speckled yellow of beet. Deficiency of Mn could be corrected by spraying 1-5% $MnSO_4$ with lime.

8. VARIETIES

Important varieties of sugar beets are mentioned below:

- **Ramonskaya:** It is a Russian variety with average yield of 35-40 t/ha, sucrose 14-16% and sensitive to leaf spot disease.
- **Maribo Magnapoly:** It is a variety from Denmark with average yield 36-40 t/ha and sucrose of 14-18%.
- **Tribel:** Average yield is 40-45 t/ha with sucrose of 14-16%.
- **I.R.S.R.Composit:** Developed at Lucknow. Gives average yield of 35-40 t/ha with sucrose of 14-18%.
- **Other important varieties:** Bigtriplex, Bush E, Maribo Resistapoly, USH 6, USH 7, GWH 14,GWH 66, MSH 123.

9. SOWING

Crop is sown in mid October to 1st fort-night of November. Delay in sowing after 30th October reduces crop yield. Crop is generally sown when soil temperature is around 10-15° C. Crop is sown either in flat beds or in ridges. Ridge sowing is better (Dargan and Chiller, 1973) Table-40.2.

After field preparation, ridges are prepared at 45-50 cm distance and 10-12 cm high. Seeds are either sown by drill or by dibbling. Seeds should be sown not more than 2.5 cm deep.

Seed rate depend on variety. In case of monogerm varieties 5-6 kg seed and in multigerm (4-5 seedlings/glomerule) 8-10 kg seeds are required/ha.

TABLE 40.2: Effect of method of sowing on sugar beet yield

Treatment	% Plant survived	Yield of root q/ha
Flat sowing	37.6	128.0
Ridge sowing	71.6	194.0
Alternate ridge sowing	67.8	157.8
Transplanting on flat surface	58.2	133.8
Transplanting on ridges	84.3	137.1

9.1. Seed Treatment

Seeds of sugar beet are covered with thick, hard layer. To improve germination seeds are soaked in water for 4-5 hours and then kept in cloth bag put over to drain water for 2-3 hours. Thereafter, seeds are dried in shade. After water soaking seeds are treated with bavistin or thiram. @2 g/kg seeds to prevent seed borne diseases.

10. IRRIGATION

Sugar beet is a *rabi* season crop maturing in 6-7 months. Crop needs 8-10 irrigations depending on soil and climatic conditions. Crop is sensitive to moisture stress. Crop requires two irrigation in formative stage, two in development stage and four in root development stage. Crop should not face moisture scarcity from February till harvest as this is the most critical period for root development. Crop may be irrigated as per schedule given in Table 40.3.

Scientifically interval between two irrigations may be 30-60% depletion of available water in 90-120 cm root zone, (Stagman et al, 1980). Water requirement of sugar beet from sowing to harvesting is 30-40 acre inch.

TABLE 40.3: Irrigation Schedule in Sugar beet

Irrigation No.	Stage
I	25-30 DAS
II	3 weeks after first
III	3 weeks after second
IV	3 weeks after third
V	4 weeks after fourth
VI	2 weeks after fifth
VII	2 weeks after sixth
VIII	2 weeks after seventh
IX	2 weeks after eighth

11. INTERCULTURE

11.1. Thinning

This is of special significance in sugar beet, specially in multigerm varieties. About 4-5 seedlings emerge from one seed ball (glomerule). When crop is of 4-5 weeks or 3-4 leaves, plant to plant space be adjusted to 20 cm by removing weak plants. Gaps if any in field may be filled by putting over night water soaked seeds. Transplanted seedling develop malformed or *fangy* root with less weight.

11.2. Earthing

Roots should be covered with soil. Earthing should be done in December when crop is 45-60 days old.

12. WEED CONTROL

Crop should be kept weed free till it develop canopy. Weed develop much faster than sugar beet during initial phases of crop growth. Hence crop is severely affected by weeds. Crop can be raised successfully if kept free of weeds upto first 40-50 days. Mani *et. al.* (1968) reported 70% loss in yield due to weed infestation. The weeds which develop after critical period are smooth red by dense foliage of crop. Two hand weeding at 25 DAS and 40 DAS control most of weeds. Hoeing should not be more deep, as it may damage beet roots. The common weeds found in sugar beet are *Trianthema* sp., *Portulaca* sp., *Avena fatua, Cyperus rotundus, Chenopodium album* and *Anagallis arvense*. Weeds in sugar beet can also be controlled by using herbicides as per doses and modes given in Table 40.4.

TABLE 40.4: Herbicidal Weed Control in sugar beet

Herbicide	Dose (kg/ha)	Mode of application
Cycloate	3-4	Pre-plant
Trifluralin	0.5-1.0	Pre-plant
Pyrazon	1.5-2.0	Pre-emergence
Barban	0.75	Post-emergence 2 leaf stage of weed
Betonal	2.0	Post-emergence 25-30 DAS

13. PLANT PROTECTION

Important insect-pests and diseases of sugar beet are mentioned below:

13.1. Insect-Pests

Bihar Hairy Caterpillar: Larvae feed on leaf epidermis from February-April and make hole. In severe infestation complete left is damaged. Insect is controlled by dusting 4% endosulfan @20-25 kg/ha or by spraying 1.5 L endosulfan (35 EC) in 750-1000 L of water.

Army Worm: Larvae feed on leaf, insect eat the plant foliage leaving midrib. It is controlled by spraying phosphomidon 250 mL/ha or spraying 2% methyl parathion dust @10-15 kg/ha.

13.2. Diseases

Leaf spot: It is caused by *Cercospora beticola*. Symptoms minute circular spots all over leaf blade. In advance stage leaf becomes greyish with deep red margins. Spray bavistin at @0.50 kg/ha at 15-20 days interval.

Root knot: It is caused by *Meloidogyne javanica*. The nematode make knotting of lateral roots. Plants remain stunted and chlorotic.

Heart rot: This is caused by deficiency of boron, resulting into necrosis of leaf petiole, rosetting and browning of leaves. To overcome this apply 25-30 kg borax at the time of sowing along with other fertilisers.

14. HARVESTING

Crop sown in October, is ready for harvest in March-April. At maturity beet consists of crown, the neck and the root and leaves. Beet roots are cone shaped and closes into slender tap root. Crop should be irrigated 3-4 days before harvesting. To ease in harvesting, run plough in between crop rows. The roots are removed by using lifting fork and shaken to remove sticking soil. Beet roots are topped at base of lower leaf.

15. YIELD

Roots 30-50 t/ha, green fodder 5-10 t/ha.

16. PROCESSING

Harvested produce should be sent to factory within 24 hours. If it is detained, it must be stored by covering with foliage. Sugar beet is extracted for sugar only, by diffusion process. Beet pulp and residue are obtained after extraction. The composition of sugar beet is given in Table 40.5.

TABLE 40.5: Composition of sugar beet

Constituents	Composition (%)
Water	75
Sugar	17.5
Other solids	unsoluble(mark) 5% ; cellulose 24
	Hemicellulose 22
	Pectic Substances 48
	protein-2
	Saponic Substances 2
	Mineral matter-2
	Soluble 2.5%
	Protein 58; Other N.Substances 12%

17. SEED PRODUCTION

National Seed Corporation has established seed production centre in India for commercial seed production at Kashmir and Kalpa valleys. Seed production is not feasible in plains because of high temperature. For producing seed, healthy beet roots are selected. About 1/3 bottom root length is cut off. These beet roots are sown in August-September in well prepared and fertile field. Crop is given irrigation and plant protection measures as usual to root crop. Seed crop mature in about 12 months in June-July. Average seed yield is 800 kg/ha.

18. REFERENCES

Anonymous, (1995) *Rabi* ke pramukh phasle unnat krishi vidhya. Agric Information Bull. Agric. Deptt. Govt of Rajasthan.

Durgan, K.S and Chillar, R.K (1973) A technique for raising sugar beet and vegetables in saline alkanine soils. Indian farming 22(12) : 13-16.

Mani, V.S., Gautam and Chakravarthy, T.K (1968). Losses in crop yields due to weed growth. PANS (e) 14 : 142-158.

Stagman, E.C, Musick. J.T. and Stewart J.I (1980). Irrigation water management. In : Design and operation of farm irrigation systems. M.E. Jensen (ed.). ASAE pp 761-816.

Yadav, R.L, Singh, P, Prasad, R and Ahlawat, IPS (1998). Fifty years of Agronomic Research in India. Indian Society of Agronomy, New Delhi.

TUBER CROP

41. Potato

S. S. Lal and K. C. Sud

Potato produces more food per unit area and time and can supplement the food needs of our country in a substantial proportion. It is a versatile food which can be prepared in a great variety of dishes and can be eaten as a stable food, as a vegetable or as a snack food. Potato has wide adaptability and can be grown in a wide range of agro-climatic conditions. It is possible to grow potato in our country all the round the year in one part or other. Being a short duration crop having a wide flexibility in planting and harvesting time, it can be successfully grown in different multiple and intercropping systems.

Potato is a non-cereal food crop which fulfils all the criteria for a healthy food and offers a great potential for increasing global food production and contributing much to the national economy of the country. Potato crop though occupies 0.3 per cent of total cropped area in the country yet it contributes upto 1.9 per cent of the total value of the agriculture output. It produces more total dry matter and proteins per kg/ha/day when compared to wheat, rice and maize. The yield of energy and proteins from this crop is 678 (10 K Cal) and 176 kg, respectively.

1. ORIGIN

The potato belongs to high land tropics of Andean mountain in Bolivia and Peru in South America. It was selected as an article of food by Mayas, Incas and Red Indians. Later on it was taken to Europe in the sixteenth century by the Spaniards where it was adapted as a commercial food crop for temperate climate under long day growing conditions. It is believed that it have been introduced in India in early seventeenth century by the Portuguese. Although there have been some reports that potato might have been taken to India directly from South America across the Pecific Ocean. In the early period, the potato was primarily grown under long day conditions during summer months in the higher altitudes of the hills of northern and southern India from where it spread to the plains of India as an autumn, winter and spring crop.

2. DISTRIBUTION

East Europian countries and China accounted for more than half of the total area and potato production in the world (Fig. 41.1). The yield per unit area is highest in Netherlands (45.0 tonnes/ha) followed by United Kingtom (41.3 tonnes/ha) and United States (37.4 tonnes/ha). The average yield in India is 19.2 tonnes/ha which is although quite low when compared to developed countries but is still higher than the world average of 16.1 tonnes/ha. The per capita consumption is highest among Europians (above 300 kg/year) where it is used as a major food. In India, however, the per capita availability of potatoes is only 20 kg/year which is almost 1/3 of the world average.

In India, a consistent increase in area, production and productivity of potato have been observed during the last three decades. Today, the potato ranks 23rd in area (1.49 million ha) but fourth in production (28.68 million tonnes) after sugarcane, paddy and wheat. In India, it is grown almost in all the states and under diverse agro-climatic conditions. Nearly 82 per cent of potatoes are grown in the plains during short winter days from October to March, about 10 per cent in hills under long day conditions during summers (April- September) and rest 8 per cent in the plateau region in the south-eastern and peninsular India as a rainfed crop (July- October).

Uttar Pradesh accounted for over 35.3 per cent area and 41.1 per cent production closely followed by West Bengal with 22.5 per cent area and 32.5 per cent production (Fig. 41.2). Among the other states, Bihar, Punjab, Madhya Pradesh and Gujarat together constitute over 30.4 and 20.7 per cent area and production, respectively. Average productivity varies widely and differ from state to state. It is highest in Tamilnadu followed by West Bengal, Punjab, Gujarat and Uttar Pradesh.

3. CLASSIFICATION

Potato belongs to the family *Solanaceae* and the genus *Solanum*. The tuber bearing species have been identified as *Hyperbasarthrum* under section *Tuberarium* (now called *Petota*) of sub-genus *Potatoe.* The genus comprises more than 2000 species out of which 19 series and 235 species are under sub-section *Potatoe*. However, new species continue to be discovered with area of botanical explorations extended into unexplored areas.

Assuming that the basic number of the genus *Solanum* is x = 12, about 74 per cent of tuber bearing species are diploid (2n =24) while about 12 per cent are tetraploids (2n = 48), pentaploids (2n = 60) and hexaploids (2n = 72). Tetraploids includes *S. tuberosum ssp. tuberosum* which is cultivated in most parts of the world. Its ancestral form *ssp. andigena*, a typical Andean potato, is grown extensively at high elevations in most of Latin America. The other species cultivated are the diploids, S. *stenotomum*, S *phureja* and S. *ajanhuiri*.

4. GROWTH STAGES

After the termination of dormancy period, buds on the tuber get physiological activated and starts to emerge from the tuber's eyes and develop into small sprouts. Under assured irrigated conditions, potato plant takes about 10-15 days for emergence whereas, under rainfed conditions, the first emergence takes about 15-35 days depending upon the availability of *in situ* soil moisture and soil type. The tuberisation starts towards middle of the active vegetative growth and extends over a period of 50-70 days depending upon the cultivars.

In potato, three critical growth stages have been identified i.e. stolon formation, tuber initiation and tuber bulking. The first two stages are important from the nutritional point of view and serve as an index of final potato yields while the last stage i.e. tuber bulking is generally linked to the appearance of fungal and bacterial diseases. In hills, where the crop is generally taken under long day conditions during summer months, the moisture stress at stolon formation stages causes considerable reduction in potato yields. Similarly during late tuber bulking stages, appearance of late blight (*Phytophthora infestans*) affects foliage thereby resulting in loss of yields in case the variety is susceptible to late blight.

5. CLIMATE

Potato plant is sensitive to several climatic factors, such as temperature, rainfall, humidity and photoperiod which exert a considerable influence on its growth and tuber development. The optimum temperature for sprouting is 18-2° C. Good crop growth is observed when days are sunny and nights are cool. The highest tuberization is obtained when day and night temperatures are about 20° C and 14° C, respectively. Low temperature, high light intensity and short days are conducive for early initiation of tuberization and subsequent tuber development. In high hills of north and east Himalayas, the long day conditions during summers (April- October) are most favourable for production of disease free seed.

High temperature which prevails throughout north Indians plains during autumn season (September-October) adversely affects, plant emergence, leaf growth and tuber formation at early growth stages. Seed tubers produced at 16-22° C have been found to be superior to those produced at temperature above 27°C. The maximum tuberisation takes place at the mean temperature of 18 to 20° C and inhibited at temperatures above 29° C. In nut shell, frost free, sunny days and cool nights are ideal for potato crop.

6. SOIL AND SOIL PREPARATION

The soil type exert a great influence on growth, tuber size and shape, maturity of the crop, yield, cooking and keeping quality of the tubers. The potato grows well on well-drained sandy loam to clay loam soils (Ustochrept), and soils rich in humas (Haplastalf). Fine texture black soils (Rhodustalfs) are prone to cracking on drying and problems of cracking and greening are likely to occur and may expose tubers to the ravage of tuber moth. In general, potato can be grown well in sandy loam to silty clay loam soil having good drainage facilities.

Potato plant prefers soil in acidic to neutral range (pH 5.5 to 7.5). High or low pH hinders plant growth besides affecting nutrients availability to the plant. In problematic soils having pH more than 8.5, but electical conductivity less than 1 mmhos/cm^2, the crop can be grown after adding suitable ammendments viz., gypsum or pyrites. In laterite and highly acidic hill soils having low pH < 4.5, the problem of phosphorus fixation is likely to come up.

A fine textured soil with minimum clods is considered good for potato cultivation. Clods are kept minimum by not working the land when either too wet or too dry. The field is ploughed to a depth of 20 cm and planking is done immediately to avoid soil moisture loss.

In the hills, two ploughing with mouldboard plough before snowfall in Oct.-Nov. and two ploughings before potato planting give desirable tilth. Where as in the plains, after deep ploughing and burying the green manure crop, single disc harrowing and planking before potato plainting are sufficient. Hot weather cultivation consisting of ploughing the field during summer months (May-June) reduces the incidence of soil borne diseases and pests and also controls perennial weeds.

7. CROPPING SYSTEM

Potato being a short duration crop and highly responsive to fertilisers and irrigation has been recognised as one of the most suitable crops for inclusion in multiple, relay and inter cropping systems. Potato is grown in different seasons in India depending upon the climatic conditions. In hilly areas, it is grown during summers whereas in Indo-Ganegetic plains, a winter crop is taken. In Deccan palateau, potato is grown in *Kharif* as well as *Rabi* seasons. In Nilgiris hills, even three crop of potato can be raised in one year. Consequently, potato finds its place in a variety of cropping systems followed in different agro-climatic zones in the country (Table 41.2). Studies have shown that the need for application of additional nutrients to wheat or paddy is also considerably reduced after potato as the nutrient requirement is reduced by almost half in case of N and no P and K is needed.

The emphasis is also being laid on intercropping of potato in potato + wheat and potato+ sugarcane cropping system. Sugarcane is an important crop grown in northern plains and is planted either in autumn or spring season. Since the autumn planted crop competes with potato. Intercropping of two has been found to be highly remunerative. Similarly, wheat is another crop widely grown in *Rabi* season in northern plains and competes with potato. It can also be grown after early crop of potato, harvested by the end of November.

TABLE 41.1: Important potato based cropping systems in India

Region	Cropping Sequence
North-western plains (Punjab, Haryana and northern parts of Rajasthan)	Potato-wheat-paddy/maize, Potato-paddy/maize, Potato-sunflower-paddy
Western and central Indo-Gangetic plains (Western UP. MP and Rajasthan)	Potato-wheat-paddy, Potato-paddy
Eastern Gangetic plains (Eastern UP, Bihar, West Bengal and Orissa)	Potato-greengram-paddy, Potato-maize-paddy, Potato-onion-paddy, Potato-groundnut-paddy, Potato-jute-paddy
Plateau region (northern peninsular and mid plateau)	Potato-paddy, Potato-groundnut, Potato-potato
North-western hills	Potato-peas, Potato-wheat-maize
North-eastern hills	Potato-maize, Potato-cauliflower, Potato-radish
Nilgiri hills	Potato-potato, Potato-cabbage
Intercropping system	Potato+sugarcane, Potato+wheat, Potato+mustard

8. MANURES AND FERTILISERS

Potato being a short duration crop having sparse root system is highly responsive to applied nutrients. It produces highest dry matter per unit area and time as such requires adequate supply of plant nutrients for optimum growth. The nutrients dose depends upon the soil type, crop season, source, time and method of application. For judicious use of fertilisers, the nutrients doses have to be applied on the basis of soil tests particularly for N, P, K, S and micro nutrients. Studies have shown that the critical limits for P, K and to some extent micronutrients, are almost double than that of cereal crops thus the soils which contains sufficient available P and K are likely to be deficient in these nutrients for potato crop. Depending upon the soil test values and critical limit, the nutrient dose can be adjusted accordingly.

8.1. Manures

In the hills, application of farm yard manure at 10-15 tonnes /ha along with half of recommended N dose and P and K are to be applied in furrows at the time of planting. The remaining N dose is applied at the time of final earthing up. The application of farm yard manure improves the soil structure and also soil fertility. Fertiliser N need of seed crop is 25-30 per cent lower than that of ware crop as excess of N delay tuber initiation and development and also increases the yield of large size tubers. The P and K dose can be adjusted according to dose of farm yard manure applied. Studies conducted on various soil types have indicated that 30 tonnes of farm yard manure not only meets full P and K need of potato but also takes care of entire P and K need of succeeding crop.

In the plains, growing of geen manure crops like sannhemp (*Crotalaria junscea*) or dhaincha (*Sesbania aculeata*) during summers and subsequent burying *in situ* for decomposition with in soil or 20-25 tonnes of farm yard manure before potato planting improves the fertility and physical condition of soil and economise on P and K dose.

8.2. Response

Potato response to applied fertilisers and manures depends upon soil capacity to supply nutrients, soil type, cultivar and climatic factors (Table 41.3). Nitrogen, phosphorus and potassium have been designated as macronutrients as they are required in large quantity whereas calcium, magnesium and sulphur come under secondary nutrients and choice of micronutrients viz., zinc, copper, manganese, iron, molybdenum and boron depends upon the extent of deficiency and their critical level in the soil.

- **Nitrogen:** Nitrogen has been found to be a first limiting nutrient in potato production. In soils, otherwise rich in organic matter, may not release nitrogen at a rate required to maintain the rapid activity of the growth processes. The responses are high in alluvial soils followed by acidic, black and red soils. The optimum dose has been found to be 156, 135, 140 and 122 kg/ha, respectively with corresponding response of 13.7, 8.2, 7.3 and 8.1 tonnes /ha. Studies have shown that long duration varieties like Kufri Sindhuri, Kufri Badshah, Kufri Lalima and Kufri Jyoti are more responsive than short duration varieties viz. Kufri Chandramukhi and Kufri Bhhar.

- **Phosphorus:** The response to phosphorus mainly depends upon not only on soil type but also on the available phosphorus content of the soil. The response to phosphorus is high in acidic and laterite soils, medium in alluvial soil and low in black soils. The optimum dose has been found to be 124, 94, 78 and 64 kg P_2O_5 /ha in acidic hill, red laterite, alluvial and black soils with average response of 7.7, 4.5, 2.2 and 1.6 tonnes/ha. The soil test value for available P plays a greater role in determining the potato response to applied fertiliser. The critical P level in the acidic soils by Bray P test has been found to be 35 ppm whereas for alluvial soils, the critical P by Olsen's method has been fixed at 10 ppm. In alluvial soils, Kufri Jyoti shows higher reponse to P than Kufri Badshah, Kufri Bahar and Kufri Lalima.
- **Potassium:** Keeping in view the role played by potassium in maintaining the tone, vigour and efficiency of the plant, the potato requires high doses of potassium fertilisers. The potato response to applied K has been found to be high in alluvial hill and red soils while black soils accounted for lowest response. The optimum dose for these soils have been found to be 113, 103, 125 and 69 kg K_2O/ha with average response of 5.2, 4.5, 3.5 and 1.0 tonnes / ha. Like P, the potato response to K depends upon soil available K status. The alluvial soils containing 105 ppm K and less by ammonium acetate method are highly responsive whereas acidic hill soils having less than 120 ppm K respond well to applied K. Potato varieties differ greatly in their response to K indicating a strong interaction of the varieties to K response and yield potential particularly with regard to yield of large size tubers. In general, short duration and rapid bulking varieties producing greater number of large size tubers require high doses of K than varieties with medium and small size tubers. The contribution of N in modifying K response also plays an important part in quantifying the responsiveness of the variety to applied fertilisers.

Micro nutrients: The potato response to micronutrients depends upon the magnitude of their deficiency in the soil. The responses of potato to micronutrients are strikingly higher on micronutrient deficient soils than on rich soils. The response to micronutrients are high in alluvial soils followed by hill, black and red soils. Application of organic manures viz. farm yard manure and green manuring result in obtaining low response to micronutrients. On the other hand, use of high analysis fertilisers in areas of high crop intensity also modify the response to applied micronutrients

8.3. Nutrient Source

- **Nitrogen:** For potato, among various sources, ammonium sulphate is considered to be the best but because of its high cost, it is seldom used. The most popular N carriers used are urea and calcium ammonium nitrate. However, urea has been found to be 10-20 per cent less efficient than ammonium sulphate and calcium ammonium nitrate. This has been attributed to its harmful effect on young sprouts hence plant emergence particularly when applied at higher doses. Thus, care should be taken that N dose through urea should not exceed 50 Kg /ha at planting time as there are chances of injury to young sprouts which may delay crop emergence.

- **Phosphorus:** The most common P fertiliser used in potato cultivation is single super phosphate and has been found to be equally good in all type of soils. The other P carriers being currently in use are diammonium phosphate and rock phosphate. The later can be used in acid soils by mixing it with the single superphosphate in 3:1 ratio. Tracer studies with different P carriers have shown that nitro phosphate, pyro phosphates and poly phosphates are equally good for potato in acidic soils.

TABLE 41.2: Recommended doses of nutrients (kg/ha) for ware crop of potato

Location	State	Variety	N	P_2O_5	K_2O
Jalandhur	Punjab	K Jyoti	220	150	150
		K Badshah	220	200	180
		K Jawahar	240	50	50
Patna	Bihar	K Lalima	240	130	200
Modipuram	UP	K Jyoti	240	100	120
Kanpur	-	K Badshah	210	100	180
Pantnagar	-	K Badshah	180	120	200
Hissar	Haryana	K Badshah	210	140	50
Kalyani	West Bengal	K Badshah	170	170	125
Shimla	HP	K Jyoti	130	120	90
Shillong	Meghalaya	K Joyti	95	140	70
Jorhat	Assam	K Jyoti	150	140	50
Shalimar	J&K	K Jyoti	130	150	130
Odhagamandalam	Tamilnadu	K Jyoti	115	145	115
Kota	Rajasthan	K Bahar	120	100	100
Rajgurunagar	Maharashtra	K Jyoti	170	50	50
Deesa	Gujarat	K Badshah	240	200	200
Bangalore	Karnataka	K Jyoti	160	50	100
Chhindwara	MP	K Badshah	120	100	100

- **Potassium:** Among the two potassium source viz. potassium chloride and potassium sulphate, the later has been found to be best in term of its beneficial effect on tuber quality viz., dry matter, ascorbic acid and sugar content. But because of its high cost, the potassium chloride is most commonly used in potato crop an excess of chlorides deteriorates keeping quality of potato.
- **Sulphur and Micronutrients:** Among, the secondary nutrients, the role of sulphur is gaining importance in recent years because of its role in oil seed crops and tuber quality. The ammonium sulphate and gypsum can supplement the S needs of the crop. The application of micronutrients is advised only on the basis of soil tests and area specific problems. The most common deficiency of zinc is in the areas of high crop intensity and also in soils having high available P content. Application of 25 kg zinc sulphate is recommended in zinc deficient soil and in areas of high crop intensity. The deficiency of others micronutrients is often area specific.

8.4. Method and Time of Application

- **Nitrogen:** Split appplication of N, 1/2 at planting and rest at the time of earthing up has been recommended. Broadcasting of urea is equally effective when compared with furrow and band placement in alluvial soils. However, band placement of urea gives better results in term of tuber yield and N use efficiency. In hills, application of mixture of urea and calcium ammonium nitrate in 1:1 ratio at the time of planting and urea at earthing up is highly economical and checks adverse effect of urea. In case of N deficiency, the corrective measures can be taken by applying N dose on the basis of tissue tests at stolon formation stage at 40-45 days in hills and 25-30 days in plains.
- **Phosphorus:** In view of its low mobility, whole of P has to be applied at the time of planting in furrows. Mixing of 5 tonnes of farm yard manure per ha with P fertiliser can economise upto 25 % P dose. Another economical way is to soak seed tubers in a solution of 1.5% of single super phosphate and 0.5 % urea thus saving in 25 Kg of P/ha for the crop.
- **Potassium:** Like P, application of K at planting time in furrows is better than its application above or below seed tubers or broadcasting. In ware potato crop, split application i.e. half at planting and rest at earthing up along with N dose gives higher yield of large size tubers. Mixing of 5 tonnes of farm yard manure with K fertiliser also economises K dose by 25 per cent. Tissue tests for K at stolon formation stage can help in taking corrective measures in case of K deficiency for applying K dose based on tissue test at earthing up.
- **Micronutrients:** In case of micronutrients deficiency, soaking of seed tubers in 0.05% respective salt solutions having sulphate as accompaning anion is best method for meeting miconutrient requirement of potato. Zinc requirement can also be met by dipping seed tubers in 2% ZnO suspention and drying the tubers in shade before planting. Foliar application is another way in meeting micronutrient deficiency in the standing potato crop.

9. VARIETIES

The original introductions by European and Britishers were of Andigena types and were known by local names as Phulwa, Satha, Gola and Darjeeling Red Round. From early 1814-1939, several Tuberosum and Andigena varieties were introduced but because of degeneration due to virus infection, they were not successful. However, several introductions viz., Great Scot, Craig's Defiance, Magnum Bonum and Up-to-Date in hills and Phulwa, Satha and Gola in Plains remained popular for a long time.

TABLE 41.3: Characteristics of important potato varieties developed by CPRI, Shimla

Variety	Year of release	Tuber Characters	Maturity	Keeping Quality	Desirable resisteance	Adaption
Kufri Sindhuri	1967	Medium, round, red, eyes medium deep	Late	Good	Moderately resistant to early blight and tolerant to leaf roll	North Indian plains

Variety	Year of release	Tuber Characters	Maturity	Keeping Quality	Desirable resisteance	Adaption
Kufri Chandramukhi	1968	Large, oval, white, eyes fleet	Early	Good	Susceptible to late blight	North Indian plains and plateau region
Kufri Jyoti	1968	Large, oval, white, eyes fleet, cracking in tubers	Medium	Good	Resistant to late blight, immune to wart and moderately resistant to early blight	North Indian plains and hills
Kufri Lauvkar	1972	Large, round, white, eyes fleet with prominent eye brows	Medium	Good	Susceptible to early and late blight	Plateau region of peninsular India both for *Rabi* and *Kharif* seasons.
Kufri Badshah	1979	Large, oval, white, eyes fleet, tuber turns purple on exposure to light	Medium	Average	Resistant to late blight, early blight and PVX	North Indian plains and plateau regions
Kufri Bahar	1980	Large, oval-round, white, medium deep eyes	Medium	Poor	-	North Indian plains
Kufri Lalima	1982	Medium, oval, red, eyes medium deep	Medium	Good	Moderately resistant to early blight	North Indian plains
Kufri Jawahar	1996	Medium, round-oval, white, eyes shallow,	Early	Good	Moderately resistant to late blight	Northern and Central plains, plateau region
Kufri Sutlej	1996	Large, oval, white, eyes shallow	Medium	Good	Moderately resistant to late blight	Northern and Central plains
Kufri Ashoka	1996	Large, oval-long, white, eyes shallow	Early	Good	Susceptible to late blight	Central and Eastern plains

The major breakthrough in potato improvement programme came with the introduction of seed plot technique which made it possible to raise, select and multiply breeding material under disease-free conditions in the plains. Starting from 1958 to date, Central Potato Research Institute has released a total of 34 high yielding varieties for different agroclimatic regions in the country. The salient features of 10 most popular potato varieties are given in Table 4. Most recently, the Institute has released a late blight resistant variety Kufri Giriraj for north western hills as a replacement for Kufri Jyoti which has become susceptable to late blight. Kufri Chipsona 1 and Kufri Chipsona 2 having high dry matter and low sugar content have been released for processing industries.

10. PLANTING

Potato planting and harvesting time varies with agroclimatic conditions of the region and the purpose for which the crop is grown (Table 41.4). Field should be clear of stubbles and perennial weeds at the time of planting. The seed material of the variety should be pure so that plants develop and mature uniformly and tubers produced confirm shape, size and colour, besides keeping quality and cooking behaviour.

TABLE 41.4: Planting and harvesting schedules for potato under different agroclimatic regions

Region	Season	Planting time	Harvesting time
A. North western Hills			
Very high hills	Summer	May-June	Sept.-Oct.
High hills	Summer	Mid March-April	Sept.-Oct.
Mid hills	Spring	Feb.-March	July-August
B. North central high hills	Summer	Mid Feb.- March	August- Sept.
C. North eastern high hills			
Darjeeling area high hills	Spring	Mid. Dec.- Mid. Jan.	July-august
Shillong hills	Summer	Feb.-March	July-August
D. Southern hills	Autumn Spring	August-Sept. Jan.-Feb.	Dec.-Jan. May-June
E. North western plains			
Punjab	Early Autumn	Mid. Sept.	Mid. Nov.-Dec.
	Spring	Mid. Oct.	Feb.-March
Jammu, Punjab, Western UP, Haryana and Rajasthan	Winter	Oct.	Jan.-March
F. North central plains	Winter	Mid. Oct.	Feb.-March
G. North eastern Plains			
Bihar	Winter	Oct. end- 2nd week of Nov.	Jan.-Feb.
West Bengal	Winter	Early Nov.	Jan.- Feb.
Orissa	Winter	Early Nov.	Jan.-Feb.
H. Plateau region	Kharif	June-July	Sept.-Oct.
	Rabi	Oct.-Nov.	Feb.-March

10.1. Pre-sprouting

Seed tubers are pre-sprouted before planting which helps in developing multiple, thick, stout, green and healthy sprouts. This also reduces the emergence time and maintains uniform plant stand, beside early maturity. In the hills, the seed tubers are pre-sprouted by keeping them in small slotted wooden trays of 15-20 kg capacity or spread in a thin layer on semi-pucca floor in the shade. The tubers having no sprout or thin sprout should be rejected. While in the plains, potato seed tubers should be taken out from the cold storage 10-15 days before planting and kept in baskets, wooden or plastic trays in the shade having diffused light.

10.2. Method of Planting

More than 90 per cent of area under potato is planted manually after opening the shallow furrows, broadcasting or placement of fertiliser mixture, placing seed tubers followed by ridging with spade or bullock drawn or tractor drawn ridger. Another method is that first ridges are formed after application of fertilisers and insecticides and then tubers are dibbled with *Khurpa* (a narrow blade tool) 5-7 cm deep on the ridges manually. Care has to be taken that tubers do not come in direct contact with the fertilisers and should remain 4-5 cm above or side of fertiliser zone.

Seed potatoes are planted with a tractor drawn semi and automatic potato planters which combine few or all the operations viz., opening the furrows, application of fertilisers, placement of seed tubers, ridging and spray of insecticides in one operation.

FIGURE 41.1: Pre-sprouted tuber and flowering in potato

10.3. Seed Size, Spacing and Seed Rate

Seed rate depends on the size of tubers used in planting for ware as well as seed crop. Tubers of different sizes can be used effectively without increasing seed rate by space adjustment. In general, 30-45 g tubers are recommended for ware as well as seed production in the plains. About 25 to 35 q/ha seed rate is recommended depending on the size of seed tubers. The best intra row spacing has been found to be 15, 20, 30 and 40 cm for 20, 40, 60 and 80 g seed tubers when inter row spacing is kept constant at 60 cm. However, inter row spacing for seed crop for manual, bullock drawn and tractor drawn implements can be kept at 45, 50 and 60 cm, respectively. A plant population of one to 1.11 lakh per ha is maintained to derive the maximum benefit and stem density of at least 40-45 stem/ m^2 for optimum number and yield if the crop is to be taken for seed purpose.

11. IRRIGATION

Water is the key input in potato production and the problems of water management varies from region to region. The water requirement of crop varies from 350-550 mm depending upon the length of growing season, atmospheric conditions, soil type and crop variety.

In high and mid hills, the potato crop is mostly taken under rainfed condition exception being of Lahaul valley of Himachal Pradesh where potato crop is taken under assured irrigation. In low hills and plains, the crop is grown during winter season when seasonal rainfall is not sufficient to meet entire water requirement of crop. Hence, potato crop requires frequent but light irrigations.

In the hills, the rainfall although is more than evapotranspiration yet the crop suffers from moisture stress during early stages of its growth and excess water during bulking period. To overcome the stress, the use of pine needles mulch not only conserve soil moisture but also decreases soil temperatures by 2 to 3□ C. Moreover, owing to cumulative effect of pine needle mulch on hydrothermal regimes and reduction of soil and runoff losses, its application increases potato yield and checks the weed growth. Another technique in hills is water harvesting, in which the runoff water from adjoining sloppy fields/plots (kept fallow) is collected and used for potato crop.

Frequent and light irrigations to potato crop at low moisture tensions, except in riverbed cultivation (Gujarat) are essential for raising a successful crop in the plains. Pre-planting irrigation is beneficial for quick and uniform emergence. Second irrigation is generally given after 5-6 days in coarse textured soils and 8-10 days in fine textured soils. Irrigation just before earthing up should be little heavier and subsequent irrigation should be given upto half of the ridge height. Irrigation in light soils should be at the interval of 8-9 days and in heavy soils 12-15 days for maintenance of wet regimes. Otherwise irrigation should be given at 0.2-0.3 atmosphere tension at 15-20 cm soil depth or at 15-20 mm cumulative pan evaporation (CPE). The drip and sprinkler systems are modern tools which save water to a great extent but are still not within the reach of ordinary farmers.

12. WEED CONTROL

Weeds pose a serious problem in potato crop in the plains as well as in the hills. They compete with the main crop for moisture, nutrients, light and space during crop growth period and in addition make difficult the intercultural operations at earthing up and towards maturity and harvesting. They also serve as an alternate host of several insects/pests and diseases. In potato, 10 per cent yield losses have been attributed to the harmful effect of weeds. The common weeds competing with potato are given in Table 41.6. The critical period for crop-weed competition is 4-6 weeks after planting in the plains and 6-8 weeks in the hills. Weeding normally commence at 20-25 days after planting when the plants attain height of about 10-15 cm. Early control of weeds is also obligatory from the plant protection point of view as they serve as host for viruses and vector. After weeding and hoeing, earthing up is done immediately to conserve soil moisture and proper ridge formation. Use of herbicides for weed control is of great significance in potato. It checks early weed growth, reduces mechanical damage to the plants, quick and less laborious farm operations. It also minimises the spread of mechanically transmitted viruses like virus X and S by omitting manual weed control during crop growth period. The herbicides used for weed management in potato along with their time of application and type of weed flora check are given in Table 41.6.

TABLE 41.5: Major weed flora of potato crop in plains and hills

Scientific name	Common name	Local name	Scientific name	Common name	Local name
Amaranthus viridis L.	Pigweed	Jangali chaulai	*Amaranthus viridis* L.	Pig weed	Jangali chaulai
Anagallis arvensis L.	Pimpernel	Krishnaneel	*Bidens pilosa* L.	Begger's sticks	Dipmal
Asphodelus tenuiflotus Cavan	Wild onion	Piazi	*Chenopodium album* L.	Lamb's quaters	Bathua
Chenopodium album L.	Lambs quaters	Bathuai	Chenopodium murale L.	Lamb's quaters	Bathua
Convolvulus arvensis L. Scop.	Field bindweed	Bathua	*Commelina benghalensis* L.	Tropical spider wort	Kanakaua
Cynodon dactylon L. Pers.	Bermuda grass	Dub	*Cynodon dactylon* L. Pers	Bermuda grass	Dub
Cyperus rotundus L.	Purple nutsedge	Motha	*Rumex* species	-	Jangali palak
Melilotus sp.	Sweet clover	Senji	*Echinochloa crusgalli* L. Beauv	Water grass	Savank
Oxalis corniculata L	Indian sorrel	Khati-buti	*Melilotus indica* L. All	Yellow sweet Clover	Pili Senji
Poa annua L	Blue grass	Buin	*Oxalis corniculata* L.	Wood sorrel	Khati-buti
Setaria glauca L.	Foxtails	Banra, Banari	Pennisetum clandestiinum	Kikuya grass	Kikuya grass
Trianthema monogyna L.	Carpet weed	Patharchatta	*Setaria glauca* L. Beauv	Foxtails	-
Vicia sativa L.	Common vetch	Ruari, Ankari	*Spergula arvensis* L	Corn spurry	Bundhania/ Matkan

TABLE 41.6: Herbicides for weed management in potato

Chemical name	Name of herbicide	Dose (kg a.i./ha)	Time of application	Type of weed flora controlled
Fluchloralin	Basalin	0.70-1.00	Pre-planting*	Annual grasses and broad leaf weeds
Plendimethalin	Stomp	0.50	Pre-planting*	Annual grasses and broad leaf weeds
Alachlor	Lasso	1.00-1.50	Pre-emergence	Annual weeds, grasses and nutsedge
Atrazine	Aatrex & Gesaprin	0.50	Pre-emergence	Annual grasses and broad leaf weeds
Isoproturon	Arelon, Tolkram & Graminon	0.75-1.00	Pre-emergence	Broad leaf weeds
Linuron	Lorox & Afalon	0.50-0.75	Pre-emergence	Annual grasses and broad leaf weeds
Methabenz-thiazuron	Tribunil	1.00	Pre-emergence	Annual grasses and broad leaf weeds
Metribuzin	Sencor	0.75-1.00	Pre-emergence	Annual grasses and broad leaf weeds
Oxyflurofan	Goal	0.10-0.20	Pre-emergence	Annual grasses and broad leaf weeds
Simazine	Hexazin & Princep	0.50	Pre-emergence	Annual grasses and broad leaf weeds
2, 4-D	Weedox, Weedar-96	0.50	Pre-emergence	Broad leaf weeds
Parquat	Gramaxone	0.40-0.60	Post-emergence**	Annual grasses and broad leaf weeds
Propanil	Stam F-34	1.00	Post-emergence**	Annual grasses

* Spray on soil before planting and incorporate in soil to reduce volatalization
** At 5 per cent plant emergence

13. PLANT PROTECTION

The potato is prone to attack by various diseases and insects/ pests. Integrated pest management package has been developed for the control of foliage and tuber borne diseases. Among the fungal diseases, late blight is most serious and is a regular feature in hills and to some extent in northern plains of India. Aphids, leaf hoppers, cutworms and defoliating caterpillars have been found to be potential pests affecting the potato crop directly or indirectly. In case of tuber borne diseases, bacterial wilt, soft rot, common scab, dry rot etc. are the major diseases affecting potatoes. Many viral diseases also affect the potato crop and reduce the yield to a great extent.

13.1. Late Blight

Late bilght (*Phytophthora infestans*) is a most serious disease affecting potato crop in the northern, eastern and southern hills. In eastern and northern plains, it occurs in mild to severe form depending upon the season. The disease affects the foliage, stem and tubers. Moisture and temperature are the main factors limiting the development and spread of late blight. In hills, where the crop is grown under rain fed conditions, the decease appears within a week when the relative humidity is above 80% for two consecutive days associated with temperatures of 10-20 °C. In plains, where the crop is grown under irrigated conditions in autumn / spring, the disease appears (i) when the temperature dose not go below 18°C and maximum not above 24°C, (ii) the sky remains cloudy for at least two days, and (iii) there is some rain or drizzle.

As soon as the conditions for the disease prevail, spraying the crop with either mancozeb (Dithane M-45) @ 2.0 Kg/ha or copper oxychloride @ 3.5 Kg/ha is to be given as a prophylatic measure. In hills, incorporation of sticker is a must along with fungicide. After a week, combined spray of systemic fungicide (metalaxyl or oxydixyl) and mancozeb can be applied followed by repeated sprays after 10-15 days interval in plains and weekly interval in hills.

13.2. Aphids

Aphids transmit viral diseases and their control is essential in seed crop. Phorate 10 G @ 10 kg/ha at planting time controls jassids, leaf hoppers and white flies, otherwise they spread vector transmitted viruses. An additional quantity of phorate 10 G @ 7.5 kg/ha can be applied at the time of earthing up to prevent the secondary infection of aphid vector particularly if the crop is grown for seed purposes. Two sprays of dimethoate 30 EC or dimeton 25 EC @ 1.0 L/ha at an interval of 12-15 days depending upon the appearance of leaf hoppers, white flies etc. may be given.

13.3. Viral Diseases

Among several constraints in potato production, viral diseases constitute a major factor as these bring about progressive reduction in yield. The commonest among viral disease symptoms are mosaics which include large number of distinct viral diseases invoked by different viruses such as super-mild, transient, typical green or yellow, mild/severe mosaics or mottle. Potato being a vegetatively propagated crop, harbours over 30 viruses and pathogens. Most potato viruses get transmitted in a rapid succession through contact of diseased with healthy plants while others spread within and outside fields by the insect-vectors. A virus infected potato plant do not recover and passes the infection on to daughter tubers in successive generations.

Among latent mosaic, potato virus Y (PVY) is both mechanically and non persistentily aphid transmitted, cause heavy yield depressions. The common symptoms are leaf drop streak or acropetal necrosis and can be readily detected by ELISA. Potato virus A (PVA) causes mild mosaic. when compared to PVY. PVA produces interveinal and veinal chlorotic mottle and symptoms are more pronounced in cloudy weather. The diseased leaves are shiny exhibiting slight rugosity with wavy margins. Potato virus X (PVX) spread naturally through rubbing contact of plants and tubers; seed cutting knives, farm implements and sprouts contact also. The symptoms ranges from latency to stunting, mild mottling, interveinal mosaic and necrosis of top foliage and tubers. Potato virus S (PVS) is probably the most difficult to eradicate. The virus mainly spread through diseased tubers, rubbing contact of diseased and healthy plants. The symptoms are bronzing with fine necrotic spots on the upper leaf surface. Sometime, slight deepening of veins, rugosity and stunting may occur. Resistant cultivars, early roguing, use of certified seed and avoidance of vector aphids are certain measures which control the spread of the viruses.

Potato Leaf role virus (PLRV) is the most commonly occurring virus in Indian plains, except in seed producing areas of Punjab and Haryana. The virus is transmitted through infected tubers and aphid vector, mainly *Myzus persicae* and *Aphis gossypii* in circulative manner. Symptoms comprise rolling of top younger leaves. The rolled leaves are thick, leatherly and brittle. The diseased plant are severely stunted with interveinal chlorosis. The PLRV can easily be managed by using certified seed and adopting Seed Plot Technique.

13.4. Bacterial Soft Rot

Bacterial soft rot is one of the major disease affecting potatoes in field, post harvest handling, transit and also storage. Over 90 per cent of potato in the country is grown under short day conditions in winter months. The produce from the crop, harvested in February, has to be stored under warm and hot temperatures prevailing during warm spring, summer and monsoon seasons. The disease is mainly caused by *Bacillus polymyxa* and *Pseudomonas marginalis* in subtropics and by *Erwinia carotovora* in temperate hills. These bacteria are carried latently in potato tuber tissue under favourable conditions i.e. high humidity, anaerobosis, tissue injury etc.

The skin of immature tubers peels off at harvest and such tubers have high chance of developing soft rot. The tubers transported in poorly ventilated wagons during a warm weather also rot heavily. Similarly immature late blight exposed tubers are likely to have high incidence of soft rot. Under normal package of practices, the average soft rot incidence is 4, 10 and 4-8 per cent in north-western plains, eastern and peninsular India and hills, respectively.

Several antibiotics, fungicides and chemicals have been found to be effective in checking the soft rot. Dip treatment of seed tubers in 3.0 per cent solutuion of Streptocycline or Agrimycin-100; Methoxyethyl mercuric chloride at the rate of 0.5 per cent solution for 10-30 minutes is quite effective. Stable bleaching powder (0.10%) and boric acid (3.0%) should be used for treating produce ment for human consumption. The boric acid treatment has also been found to be effective for other tuber diseases like common scab, black scurf and dry rot. Crop rotation and application of high N dose (200-300 kg N/ha) also influences the soft rot.

13.5. Bacterial Wilt

Bacterial wilt or brown rot in potato caused by *Pseudomonas solanacearum* is most destructive disease and is wide spread in tropics, subtropics and warm temperate regions of the world. In India, the disease is endemic on wet coast, Deccan and central Karnatka, western Maharashtra and Madhya Pradesh, plains of Assam, West Bengal and Orissa, Bihar plateau and upto a height of 2200 m in north wetern hills. The disease may damage the the crop by (i) premature wilting of standing crop and (ii) rotting of tubers in field or in stores. The infected seed potatoes or infested soil are the two most effective inoculum source for short and long distance dispersal of the disease. From infected plants, secondary spread can take place by root contact, irrigation water and farm implements. Knife used for cutting seed tubers and river water further aggravates the seriousness of this disease.

A strong relation has been observed between soil moisture and severity of bacterial wilt as soil moisture influences the disease by (i) increasing survival of the bacterium in soil (ii) increases infection (iii) increases exit from the host and its spread to the soil and (iv) increases the magnitude of disease. Temperature and soil moisture in combination have a synergistic effect on the disease development. Similarly, soil type and physico-chemical properties also influence the survival of pathogen. Soils with high clay content favours long survival of pathogen whereas disease incidence is high in acidic soils compared to alkaline soils.

Agronomical-practices and stable bleaching powder can reduce the disease incidence by 70-100 per cent. Crop sequence having i.e. crops like wheat, barley, maize and keeping fallow in between along with the use of disease free seed are other measures which are effective in controlling the bacterial wilt and brown rot.

13.6. Cutworms

These cause damage to the crop both in hills and plains. To check these, the ridges should be drenched with the solution of chloropyrifos 20 EC @ 2.5 l/ha available in the form of Coroban. The crop may be sprayed with endosulfan 35 EC @ 1.5 l/ha to control leaf eating caterpillars or defoliators. Spraying of insecticides and fungicides can be combined to save time as well as cost on spraying. Parathion dust (2%) @ 2.5 kg/ha or thiodon 35 EC @ 1.5 l/ha sprays are given in case of attack by *Epilachna* beetles

13.7. Potato Tuber Moth

Potato tuber moth (PTM) is the most destructive insect pest of potato both in fields and in the country stores. It has been recognised as the important pest in the plateau region and in pockets in the hills of northern and eastern Himalayas where cold storage facilities are lacking and harvested potatoes are stored in country stores. This is attributed to moth fly which lays eggs in the stores. When the larvae tunnel into stored potatoes, they not only are rendered tubers unfit for seed or human consumption but also led to secondary attack by fungi and bacteria. Temperatures in range of 20 – 30°C are conducive for the development of its immature stages. Dusting the fields with toxaphene (5%) at 40 kg/ha and chlordane, malathion or parathion at 2 kg/ha gave effective control of pests. The dried leaves of Lantana and Eucalyptus are most effective in reducing tuber damage in the country stores in hills. Potatoes covered with sand, saw dust, ash together with incorporation of parathion, isobenzen (telodrin) or malathion provide good protection against PTM.

14. HARVESTING

The time of harvesting of the potato crop is decided on the basis of prevailing market prices, purpose for which the crop is grown and the succeeding crop which is to be taken up after potato.

In the plains, an early varieties, like Kufri Chandramukhi matures in 80-90 days, medium varieties viz., Kufri Jyoti and Kufri Bahar in 90-100 days and late varieties i.e. Kufri Sindhuri and Kufri Badshah in 100-120 days. Sometimes, for early marketing, Kufri Chandramukhi is also harvested at 65-70 days of planting to fetch high price in the market and to some extent sowing of the succeeding crop i.e. wheat at the optimum time. The ware crop meant for storage should be harvested at full maturity. The skin of the tuber can be hardened by withholding the irrigation at least 10 days before harvesting. Haulms could also be cut to enforce maturity particularly in case of seed crop. Utmost care is taken to see that the tuber skin is hard enough to withstand impact of harvesting tools and machinery. Therefore, crop is to be harvested at 15-20 days of haulm cutting and at proper soil moisture. The harvesting in plains must be completed before March to avoid charcoal rot and damage by high temperature.

Potato harvesting manually is most time and labour demanding operation and requires about 110 man-days /ha. It may be done by tractor drawn one or two row potato digger or by the bullock drawn one row digger with the help of spade or *Khurpi*. With tractor drawn single row potato elevator digger the man power is reduced to 50 man-days/ha and covers 1.0 ha per day and exposes over 85 per cent tubers to the surface. On the other hand, with animal drawn single row potato digger, a whole day is required to cover one ha and about 3.0 per cent potatoes are cut during digging operation.

In the hills and regions where crop is taken under long day conditions and is rainfed, the harvesting is completed after the offset of rains. The produce is mainly ment for seed purposes. In the hills, the harvesting is done either manually with small tools or with bullocks mainly because of small terrain and holdings. The harvest is dried in open.

15. YIELD

In potato, yield is a function of number of stems, number of effective stolons, number and weight of tubers/plant and varies with variety and agro-environmental conditions under which crop is grown. These parameters vary with cultivars. The yield of seed crop ranges from 20 to 25 tonnes/ha depending upon the duration of crop and time of haulms cutting. For ware crop, an average yield of 25-30 tonnes/ha is reasonable and easily obtained with the use of healthy seed of improved varity at good management practices. In areas of Lahaul valley in Himachal Pradesh and river bed cultivation in Gujarat, the potato yields up to 40-50 tonnes per ha could be achieved easily with proper crop management.

16. QUALITY

Composition of Potato given in Table 41.7.

TABLE 41.7: Nutritive value of potato (per 100 g of edible portion)

Particular	Content
Edible portion (%)	85.00
Moisture (g)	74.70
Protein (g)	1.60

Particular	Content
Fat (g)	0.10
Minerals (g)	`0.60
Fibre (g)	0.40
Carbohydrates (g)	22.40
Energy (Kcal)	97.00
Calcium (mg)	10.00
Phosphorus (mg)	40.00
Iron (mg)	0.70
Carotene (ug)	24.00
Thiamine (mg)	0.10
Riboflavin (mg)	0.01
Niacin (mg)	1.20
Vitamin C (mg)	17.00

17. POST HARVEST TECHNOLOGY

Grading: After harvesting, cut and diseased tubers are sorted out and heaped for skin curing. In the plains, the heaps upto 1 m height and 3-4 m wide at the base are covered with straw or mat to protect tubers from sun and rains. In the hills, harvested tubers are heaped for skin curing in house itself due to risk of rains. Great care is exercised to avoid bruising of tubers while harvesting and transporting.

For better returns in the market, the produce meant for seed or ware purpose is properly graded in four grades, viz. small below 25 g (<25 mm), medium 25-60 g (25-35 mm), large 60-100 g (35-50 mm) and extra large above 100 g (above 50 mm). In general, grading is done manually by hand picking which is very slow. Simple hand and power operated mechanical graders are also used widely particularly in plains to handle harvested potatoes mainly meant for seed purposes.

Storage: Purpose of storage of seed potatoes is to maintain the tubers in right physiological stage at the time of planting by preventing excessive storage losses. The storage methods vary in different parts of the country. In the plains, the potatoes meant for seed purposes are stored under refrigerated storage where temperature is maintained below 2-5° C and humidity about 90 per cent. Proper ventilation is also desirable. The effect of storage temperature has a marked effect on the subsequent growth and yield. If the temperature are high then sprouting development is fast and the tubers due to excessive vigour of sprouts are rendered physiologically old thus affecting the viability of seed tubers. If the potatoes are stored under optimum environmental conditions, the tubers are at right physiological stage at planting time.

In the hills, potato are stored under uncontrolled conditions, the tubers are subjected to chilling in the store due to low temperatures prevailing in the months of November-January. It not only results in gaps in the fields but vegetative growth is also slow. The physical condition of the seed tuber, as judged by the firmness of the tuber and the extent of sprouting, is ideal if the seed material is stored in semi-improvized stores with insulation in roofing and walls.

Processing: Potatoes are processed into many types of products: (i) fried products, such as chips and french fries: (ii) dehydrated products, such as dice and granules and (iii) canned. For such types of products, the raw material requirment is different. While the Bureau of Indian Standard has laid down specific standard for the products like starch and canned potatoes; the processing industries has not formulated any standards on its own. In the Indian subcontinent, many potato based fast foods viz., *massala dosa*, *tikkies* and *samosas* and *potato bonda* are readily available in many parts of the country. Dehydrated and fried potato based snacks like chips, french fries etc. are also finding place in the market in recent years. However, processing for such products is being done by individuals or small scale entrepreneurs with the result the rates are high and beyond the reach of a common man. There is a great demand for some processed products such as french fries, chips, potato starch and also potato flakes in urban areas of India.

For processing potatoes into dehydrated dice and pieces, it is desirable, that varieties have long dormancy and have oval tubers with fleet eyes. The dry matter content should be high so as to obtain higher yield of dehydrated products. The sugars content must be low and tuber should be free from after cooking darkening. For the production of dehydrated dice or pieces, firm to mealy texture is essential while for granules, waxy texture is preferred. The CPRI has developed two varieties namely Kufri Chipsona 1 and Kufri Chipsona 2 for processing purpose especially for chips making.

18. SEED PLOT TECHNIQUE

The seed plot technique developed by the Institute has brought about significant increase in area and production in the country. With this technique, seed potatoes can be successfully grown in plains by adjustment of planting and harvesting time along with the use of plant protection measures. Disease free sprouted seed tubers are planted in first week of October in western Indo-Gangetic plains and a week to fortnight later in central and eastern Indo- Gangetic plains so that enough time is available for growth and bulking (80-90 days) before critical aphid levels reached in the respective regions. The seed tubers are planted at closer spacing in moderately rich soil with the recommended levels of nutrients so as to obtain maximum proportion of seed size tubers.

In order to prevent the spread of viral and mycoplasmal diseases and mixture of different varieties, the crop is inspected twice i.e. (i) at 25-35 days after planting and (ii) before haulms cutting. The plants showing mosaic infection and off type are rogued out. The care is taken that healthy plants are not touched while roguing and the rogued plants are destroyed.

Apart from the normal plant protection measures against insect/pests and diseases, the seed crop is specially protected against aphids and leaf hoppers to prevent it from viral diseases. Soil application of granular phorate 10G @ 10 kg/ha followed by spraying the crop with dimethorate (30EC) or methyl demeton (25EC) at 1.0 L/ha by mid December is undertaken to control aphids and leaf hoppers.

Irrigation is restricted before 10-15 days of haulms killing which is done by cutting/pulling or destroying the foliage by use of paraquat 0.5 kg/ha before aphid population reaches the critical limit (20 aphids/100 compound leaves). The re-growth if any should be removed immediately. The crop is harvested 10-15 days after haulm killing when the tuber's skin become firm. Freshly harvested tubers are cured for 10-15 days, graded and treated against surface borne diseases before keeping in cold stores.

19. TRUE POTATO SEED

The non-availability of good quality seed tubers, the high seed cost, the problem of transport, storage and virus infiltration in seed tubers leading to degenration of seed stocks have restricted the potato cultivation to a greater extent. True potato seed (TPS) technology has a great potential in supplementing the availability of healthy propagules for potato production. For greater adoption of TPS technology among the farmers, intensive efforts are required for making TPS production economically viable. The male parent should preferably produce abundant pollen of high fertility while female parent should be male sterile and give high berries/seed set so that large quantities of TPS seed can be produced. This will give rise to a crop with superior seedling vigour beside its capacity to withstand transplanting shock and give a high yielding homogenous population.

Vector transmission of virus (es) in TPS crop is high without severe symptoms and consequent fast degeneration. Another major problem lies in selection of suitable populations which can produce sizeable yields of uniform size tubers in short duration as the population currently available have long duration thus are likely to be exposed to high aphid period. The joint work conducted by CPRI, AICPIP and International Potato Center (CIP) on TPS has led to the identification of two high yielding TPS population, viz. C-3 and HPS I/113. Studies have shown that the transplanting seedlings in field and use of seedling tubers as seed are successful approaches for the introduction of commercial potato production in potential areas i.e. Karnataka, Maharashtra, Madhya Pradesh, Orissa and states of North-eastern regions where good quality seed tubers are either not available or too expensive because of high cost of tranportation.

20. CONSTRAINTS OF PRODUCTION

Potato being a semi-perishable commodity can not be stored under ordinary conditions particularly in plains and tropics where high temperatures and dry weather prevail after harvest. The shrinkage due to loss of water and microbial attack are the major constraints in potato storage.

Good disease free seed, adequate inputs like fertilisers and water are three important factors for obtaining high yield potentials. Seed alone constitutes 40 per cent of total cost of cultivation. At present, hardly 30 per cent farmers have assess to the good quality seed of improved varieties, if replacement takes place after three years. Fertiliser is the second most important input in potato production and hence, optimal and judicious application of nutrients along with use of organic manures are need of the day. Losses due to severe late blight incidence due to unfavourable weather conditions is also one of the constraints in certain years. The other major constraints in potato production are recurrent price fluctuations and price crashes, high marketing, storage and transport costs and non availability of adequate storage and competitive marketing system.

21. REFERENCES

Advances in Horticulture- Vol. 7 (1993) Eds. K.L. Chadha and J.S. Grewal. MPH, New Delhi.

Agro-techniques for Seed Potato Production (1992) Eds. V.S. Khushwa & J.S. Grewal, Tech. Bull. 37. CPRI, Shimla.

Bacterial Soft Rot of Potato in India (1990) Eds. A.K. Somani & G.S. Shekhawat. Tech. Bull. 21. CPRI, Shimla.

Nitrogen Management in Potato (1991) Eds. R.C. Sharma *et. al.* Tech. Bull.32., CPRI, Shimla.

Phosphorus and Potassium Nutrition of Potato (1991) Eds. J.S. Grewal *et. al.* Tech. Bull. 31. CPRI, Shimla.

Potato Bacterial Wilt in India (1992) Eds. G.S. Shekhawat & S.K. Chakraborti. Tech. Bull. 38. CPRI, Shimla.

Potato in India (1992) Tech. Bull. 1, CPRI, Shimla.

Potato in Sub-tropics (1976) By. Puskarnath. Orient Longman, New Delhi.

Potato Late Blight (1991) Eds. S.K. Bhattacharya *et. al.* Tech. Bull. 27. CPRI, Shimla

Recent Technology in Potato Improvement & Production (1977) Ed. B.B. Nagaich, CPRI, Shimla.

True Potato Seed for Potato Production in India (1990) Ed. J.S. Grewal, Tech. Bull. 24. CPRI, Shimla.

Viral Diseases of Potatoes (1971) Ed. B.B. Nagaich, ICAR Tech. Bull. 28

Water Management in Potato (1990) Eds. J.S. Grewal, M.C. Sood & R.C. Sharma, Tech. Bull. 28. CPRI, Shimla.

Weed Management in Potato (1991) Eds. S.S. Lal & J.S. Grewal. Tech. Bull. 33. CPRI, Shimla.

APPENDIX

42. Cultivated Field Crops

Scientific name	Common name	English name
Cereals & millets.		
Oryza sativa L.	Dhan, chawal	Paddy/Rice
Triticum aestivum L.	Gehu	Wheat
Zea mays L.	Maize, Makka	Corn
Sorghum bicolor L. Moench	Jowar	Sorghum
Pennisetum glaucum L.	Bajra	Pearlmillet
Hordeum vulgare L.	Jao	Barley
Eleusine coracana Gaertn	Mandua/Ragi	Finger millet
Panicum miliaceum L.	Cheena	-
Paspalum scrobiculatum L.	Kodo	-
Setaria italica L.) Beauv.	Kakun	Italian millet
Triticale officinale	Triticale	Tritrcale
Euchlaena mexicana schrad	Teosinte	Teosinte
Oilseeds		
Brassica juncea (L.) Czern & Loss	Rai/Raya/aha	Mustard
Brassica campestris var yellow sarson	Turnip rape/yellow sorghum	Indian colza
Brassica compestris var brown sarson	Turnip rape/Brown sorghum	Indian colza
B. Campestric var yellow toria	Indian rape/yellow toria	Rape
B. Campestris var black toria	Indian rape/black toria	Rape
B. juneca var rugosa	Pahari Rai	Rugosa
B. nigra	Black mustard	Black mustard
B. teunifotil	chitti sorghum	mustard
Linum usitatissimum L.	Alsi	Linseed
Carthamus tinctorius L.	Kusum, Kurdi	Safflower
Helianthus annuus L.	Suraj mukhi	Sunflower
Arachis hypogaea L.	Mungphali/groundnut	peanut
Sesamum indicum L.	Til	Sesame
Ricinus communis L.	Irndi	Castor
Gnizotia abyssinica (L.F.)	Cass. niger	Niger
Eruca sativa. Mill.	Taramira	Pocket salad
Brassica alba boiss	White mustard	White mustard

Scientific name	Common name	English name
Pulses		
Cicer arietinum L.	Chana/Bengal gram	Chick pea. (Deshi)
Cicer kabulium L.	Kabuli chana	Chick pea.
Lens esculenta Moench	Masur	Lentil.
Pisum satium L.	Matar	Field pea.
Cajanus cajan (L.) Millsp.	Arhar	Pigeon pea.
Vigna radiata (L.) Wilczek	Mung/moong	Green gram
Vigna mungo (L.) Hepper	Urd/mash	Black gram
Vigna unguiculata (L.) Walp.	Chaula/Lobia	Cow pea
Phaseolus aconitifolius Jaiq.	Moth	Dew bean
Glycine max (L.) Merril	Soyabean	Soyabean
Dolichos biflorus (Lam) Verde	Kulthi	Horse gram
Lathyrus sativa L.	Khesare	Chickling hetch
Phaseolus calcaratus	Rice bean	Rice bean
Sesbania aculeata Poir	Dhaineha	Dhaineha.
Fibre crops		
Gossypium arboreum L.	Deshi cotton	Cotton
G. herbaceum L.	Deshi cotton	Cotton
G. hirsutum L.	American cotton	Cotton
G. barbadense L.	American cotton	Sea island cotton
Corchorus capsularis L.	White jute	Jute
C. Olitorius L.	Tossa jute	Jute
Crotolaria juncea L.	San	Sann Hemp
Hibiscus cannabinus L.	Mesta or kenat	Hemp
Boehrevia nivea (L.) Gaud	Ramie	Ramie
Spices & condiments		
Cyminum cyminum L.	Zeera	Cumin
Coriandrum sativum L.	Dhahiya	Coriander
Syzygium aromaticum (L.) Merr.	Lung	Clove
Cinnamomum zeylaricum	Cinnamom	Cinnamom
Triganella foenum graecum L.	Methi	Fenugreek
Foeniculum vulgare Mill.	Sonf	Fennel
Tuber crop/Root		
Solanum tuberosum L.	Aalu	Potato
Manihot esculenta Crantz	Tapico/Cassawa	
Narcotics		
Nicotiana tabacum	Tambaku	Tobacco
N. rustica	Tambaku	Tobacco
Paparer somniferum	Affim	Opium poppy
Medicinal & aromatic		
Plantago ovata Forsk	Isabgol	-
Anethum sowa Kurg	Sowa	Dill
Anethum graveolens L.	Ajwain	Aniseed

Scientific name	Common name	English name
Sugar		
Saccharum officinarum L.	Ganna/santha	Sugarcane
Beta vulgaris L.	Chukander	Sugarbeet
Forage		
Medicago sativa L.	Rijka	Lucerne
Trifolium alexandrinum L.	Berseem	Egyptian clover
Avena sativa	Jai	Oats
Cyamopsis tetragonoloba L.	Guar	Cluster bean
Pennisetum purpureum Schum	Napier grass	Napier
Cenchrus ciliaris L.	-	White anjan
C. setigeris	-	Black anjan
Lesiurus sindicus		

43. Important Herbicides

Herbicide/s		Manufacturing firms
2, 4-D formulations and Diuron	–	Agromore P. ltd., Mysore Road, Bangalore-26.
Diquat, Dalpan and grammoxine	–	Allcali & chemical corporation of India, 19 walchand, Hirachand Marg-1, Mumbai.
Fluchloralin (Basalin), Dinoseb and pyrazon	–	BASF India P. ltd. May & baker house, PB. no. 19108, Mumbai-25.
Metribuzin and menthabenz thiazuron	–	Bayer India ltd. 82, Veer Nariman Road, Mumbai-20.
Pendimethalin (Stomp)	–	Cynamid India ltd. Nylock house, PB. no. 9109, Mumbai-25.
Dalapan, Diuron, 2, 4, 5-T, & 2, 4-D	–	Dow chemical international, Kasturba Gandhi Marg, New Delhi-1.
Atrazine, chloroxuron, Isoproturon, Terbutryn simazine and propazine	–	Hindustan CIBA - Geigh ltd., Khetan Bhawan, 198, J. Tata Road, Mumbai-20.
Paraquat and Diqhat	–	Imperical chemical industries, 18 Strand Road, Calcutta-1.
Oxyfluorofen & Propanil	–	Indofil chemical ltd. E. Moses Road, PB. no. 6280, Mumbai-1.
Propachlor, Butachlor and Alachlor	–	Monsanto chemicals P. ltd. Wake field house, PB. no. 344 A, Mumbai-1.
Oxadiazon, Isoproturon & Terbacil	–	Phone - 804 lenu. Agric. clinic, May & Baker house, Worli, Mumbai-25.
Atrazine and simazine	–	Rallis India ltd. Ravelin Street. Mumbai-1.
Dicamba	–	Velsicol corporation. C-32, Panchsheel Enclave, New Delhi - 17.
Isoproturon, linuron and anilofos	–	Hoechst pharmaceuticals ltd. 193, PB. no. - 11123, Mumbai - 21.

44. Availability of Biofiterlizer in India

1. Bose Research Institute, Calcutta, West Bengal.
2. Chandra Shekhar Azad University of Agric. & Technology, Kanpur (U.P.).
3. G.B. Pant Univ. Agric. & Technology, Pantnagar (U.P.).
4. IARI, New Delhi.
5. J.N.K. Vishwa Vidhyalaya, Jabalpur (M.P.).
6. C.C.S. H.A.U., Hissar (Haryana).
7. Punjab Agric. Univ. Ludhiana, Punjab.
8. Tamil Nadu Agric. Univ., Coimbatore (T.N.).
9. Andhra Pradesh Agric. Univ. Rajendra Nagar, Hyderabad.
10. Rajendra Agric. Univ. Dholi (Bihar).
11. Univ. of Agric. Sciences, Hebbal, Bangalore.
12. Vimpex Agencies, Bangalore.
13. Rajasthan Bio-fertilisers Industries, Bhopal Ganj, Bhilwara (Raj.).
14. Microbac India, Shyam Nagar, 24 Pargans, West Bengal.
15. A.R.S., Durgapur (Raj.).

45. Common Weeds

Botanical name	Common name	English name
Gramineae (Grass family)		
Agropyron repens	-	Quack grass
Andropogon annulatus	Padhar	-
Eremopogon faveolatus	Dab	-
Dactyloctenium aegypticum	Makra	Crow's foot grass
Cenchrus catharticus	Berboot	-
Eleusine crusgali	Kodon	Barnyard
Pbragmites kakra	Narkul	-
Saccharum munja	Moonj	-
Saccharum spontaneum	Kans	Tiger grass
Setaria glauca	Bandranbandri	Yellow fox-tail
Sorghum halepense	Barru	Johnsongrass
Phalaris minor	Gulidanda	Little canary grass
Avena fatua	Jangli Jai	Wild oat
Cynodon dactylon	Doob	Bermunda grass
Poa annum	-	Annual speargrass
Malvaceae (Mallow family)		
Abutilon indicum	Kanghi or Kharinti	-
Malvastrum coromandelianum	Bankapas	-
Cyperaceae (Sedge family)		
Cyperus rotundus	Motha	Nutsedge
Malva neglecta	-	Round leaved mallow
Amaianthaceae (Pigweed family)		
Achyranthus aspera	Chirchita or latjira	Pricklychaff flower
Amaranthus spinosus	Spiny amaranthus	Pigweed
Amaranthus viridis	Janglicholai	Pigweed
Amaranthus blitum	Cholai	Pigweed
Celosia argentia	Safed murga or chilmili	Quailgrass or white cock's comb
Digera arvensis	Lahsua	-
Gomphrina globasa	-	-
Primulaceae (Primrose family		
Angallis arvensis	Krishna neel	Pimpernal

Botanical name	Common name	English name
Neguminosae (Legume family)		
Alhagi camelorum	Jawasa	Camel thorn
Cassia obtusifolia	Panwar	-
Cassia occidentalis	Kasaundi	-
Lathyrus sativus	Khesari	-
Medicago denticulata	Jangli rijhaka	Wild alfalfa
Melilotus alba	Senji sefed	Indian clover
Melilotus indica	Pilisenji	Indian clover
Tephrosia purpure	Dhamasa	-
Clitoria ternatea	Titli matar	Conch flower creeper
Psoralea corylifolia	Bawchee	-
Compositae (Sunflower family)		
Blumea lacera	-	Kukrondha
Carthamus oxycantha	Pohli	Wild safflower
Cichorium intibus	Kasni	Chicory
Cirsium arvense	-	Canada thistle
Eclypta erecta	Kala bhangra	
Launea asplenifolia	Jangli gobhi	-
Pluchea lanceolata	Bayansuri/chagus	Arrow wood
Tridex procumens	Pradeshi Bhangra	
Vernonia cisaria	Phulni	Ash coloured flaebane
Xanthium strumarium	Bichu	Cocklebut or Burweed
Acanthospermum hispidum	-	Starbur
Wedelia calendulosa	Bui	
Artimisia scupoonia	Bannu	
Chrysanthemum leucanthemum	-	Poor land flower
Helianthus sp.	-	Wild sunflower
Paraveraceae (Poppy family)		
Argemone mexicana	Satyanashi	Prickly poppy
Fumaria parviflora	Gajari or pitpapra	
Lillaceae (Lily family)		
Asphodelus tenuifolius	Piazi	
Allium vineale	Jangli lahson	Wild garlic
Allium calendulosa	Jangli onion	Wild onion
Nyctaginaceae		
Boerhavia diffusa	Biskhopra	Hogweed
Aselepiadaceae (Milk weed family)		
Calitropis procera	Aak/madar	
Acathaceae		
Barloria dillenii	Kala Bansa	
Justicia diffusa	Darhua	
Ocimum canum	Bantulsi	Hoary Basil
Urticaceae (Nittle family)		
Canabis sativa	Bhang	Hemp.

Botanical name	Common name	English name
Chenopodiceae (Choose foot family)		
Chenopodium album	Bathur dow lembsquarters	
Chenopodium murale	Khartua	Nettle leaf goose foot
Tillaceae		
Corchorus acuntaglis	Jangli Jute	
Cappanilaceae (Caper family)		
Clone viscosa	Nulhul	Cleome
Convolvuiaceae		
Convolvulus arvensis	Hira khuri	Bind weed
Cuscuta reflexa	Amarbel	Dodder
Ipomea purpurea	-	Morning. glory
Ipomea corcea	Vilayati oak	
Conmelinaceae		
Commelina benghalensis	Kankana	Dayflower
Nuphohbiaceae (Spungea family)		
Crotoa sparciflora	Ban mircha	
Euphorbia hirta	Bari budhi	Pill pod spurge
Euphorbia macrophylla	Chouil	
Phyllanthus niruri	Hazardana	
Cucurbitaceae (Gourd family)		
Cucumis mello	Jangli Kharbooz	
Sonanaceae (Night shade family)		
Datura stramonium	Datura	Minson weed
Nicotiana plumbagnifolia	Jangli tambakoo	
Physalis minima	Chirpotal	
Solanum nigrum	Makoy	Black night shade
Solanum xanthocarpum	Kanteli	
Pontederiaceae		
Echinops echinatus	Kanteli	
Eichornia crassipes	Jalkumbhi	Water hyacinth
Umbeiliferae (Parsely family)		
Hydrocotyl asiatica	Brahmibooti	Assiatic penny wort
Daucas carota	Jangli gazer	Wild carrot
Pastinaca sativa		Wild parsnip
Verbenaceae		
Lantana camera	Phullakri	Lantana
Labiatae (Mint family)		
Leucas aspera	Gumma	Leucas
Cactaceae (Cactus family)		
Opuntia dillenii	Nasphati	Prickly pear

Botanical name	Common name	English name
Orobanchaceae (Broom rape family)		
Orobanche aegyptiaca		Broom rape
Portulacaceae (Pursalane family)		
Portulaca olaracea	Bada lunia	Purslane
Portulaca quadrifida	Lunia	Purslane
Cruciferae (Mustard family)		
Sinapis alba	Sita sarson	
Brassica Kaber	Jangli Mustard	Wild mustard
Scrophulariaceae (Figwort family)		
Striga lutea	Rukhari	Witch weed
Striga asiatica	Rukhari	Witch week
Aizoaceae (Carpet weed family)		
Trienthema monogyna	Patharchata	Carpetweed
Trienthems portulacustrum	Satha	Carpetweed
Zyzyphus rotundifoia	Jharberi	Wild ber
Zygophyliaceae (Caltrop family)		
Tribullus terrestris	Gokhru	Puncturevine
Boraginaceal (Borage family)		
Heliotropium ellipticum	Kameda	
Heliotropium ovalifolium	Kameda	
Amsinekia intermedia	-	Tarweed
Naivelaiceae		
Gynandropsis gynedra	Red flowered	Carvelle seed
Gynandropsis pentaphyla	White flowered	Hulhul
Caryophyalaceae (Pink family)		
Spergula arvensis	Satganthia	Corn spurry or sand weed
Fanunculaceae		
Hanunculus acutus	Jangli palak	
Anacardiaceae (Cashew family)		
Rhus vernix	-	Poison sumac
Hydrochantaceae		
Hydrilla verticulatla	Hydrilla	Hydrilla
Leguminoceae		
Crotolaria burbia	Lageria	
Leptodenia pyrotechnica	Kheep